T0329304

EPIGENETIC GENE EXPRESSION AND REGULATION

Translational Epigenetics Series

EPIGENETIC GENE EXPRESSION AND REGULATION

Edited by

SUMING HUANG
Department of Biochemistry and Molecular Biology
University of Florida, College of Medicine
Gainesville, FL, USA

MICHAEL D. LITT
Center for Medical Education
Indiana University-Ball State University
Muncie, IN, USA

C. ANN BLAKEY
Department of Biology
Ball State University
Muncie, IN, USA

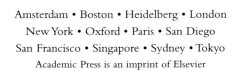

Amsterdam • Boston • Heidelberg • London
New York • Oxford • Paris • San Diego
San Francisco • Singapore • Sydney • Tokyo
Academic Press is an imprint of Elsevier

Academic Press is an imprint of Elsevier
125 London Wall, London EC2Y 5AS, UK
525 B Street, Suite 1800, San Diego, CA 92101-4495, USA
225 Wyman Street, Waltham, MA 02451, USA
The Boulevard, Langford Lane, Kidlington, Oxford OX5 1GB, UK

Notices
Knowledge and best practice in this field are constantly changing. As new research and experience
broaden our understanding, changes in research methods, professional practices, or medical treatment may
become necessary.

Practitioners and researchers must always rely on their own experience and knowledge in evaluating and
using any information, methods, compounds, or experiments described herein. In using such information
or methods they should be mindful of their own safety and the safety of others, including parties for
whom they have a professional responsibility.

To the fullest extent of the law, neither the Publisher nor the authors, contributors, or editors, assume any
liability for any injury and/or damage to persons or property as a matter of products liability, negligence
or otherwise, or from any use or operation of any methods, products, instructions, or ideas contained in
the material herein.

ISBN: 978-0-12-799958-6

British Library Cataloguing-in-Publication Data
A catalogue record for this book is available from the British Library

Library of Congress Cataloging-in-Publication Data
A catalog record for this book is available from the Library of Congress

For information on all Academic Press publications
visit our website at http://store.elsevier.com/

Typeset by TNQ Books and Journals
www.tnq.co.in

Printed and bound in the United States of America

CONTENTS

CONTRIBUTORS

Blake Atwood
Stem Cell Institute, Department of Biochemistry and Molecular Genetics, University of Alabama at Birmingham, Birmingham, AL, USA

Aissa Benyoucef
The Sprott Center for Stem Cell Research, Regenerative Medicine Program, Ottawa Hospital Research Institute, Ottawa, ON, Canada; Department of Cellular and Molecular Medicine, University of Ottawa, Ottawa, ON, Canada

C. Ann Blakey
Department of Biology, Ball State University, Muncie, IN, USA

Marjorie Brand
The Sprott Center for Stem Cell Research, Regenerative Medicine Program, Ottawa Hospital Research Institute, Ottawa, ON, Canada; Department of Cellular and Molecular Medicine, University of Ottawa, Ottawa, ON, Canada

Jason O. Brant
Department of Biochemistry and Molecular Biology, Center for Epigenetics, Genetics Institute, University of Florida College of Medicine, Gainesville, FL, USA

Jörg Bungert
Department of Biochemistry and Molecular Biology, Center for Epigenetics, Genetics Institute, Powell Gene Therapy Center, College of Medicine, University of Florida, Gainesville, Florida, USA

Yun Chen
Department of Pathology, University of Oklahoma Health Sciences Center, Oklahoma City, OK, USA

Changwang Deng
Department of Biochemistry and Molecular Biology, University of Florida College of Medicine, Gainesville, FL, USA

Nora Engel
Fels Institute for Cancer Research, Temple University School of Medicine, Philadelphia, PA, USA

Alex Xiucheng Fan
Department of Biochemistry and Molecular Biology, Center for Epigenetics, Genetics Institute, Powell Gene Therapy Center, College of Medicine, University of Florida, Gainesville, Florida, USA

Ekaterina Gavrilova
Department of Biochemistry and Molecular Biology, Center for Epigenetics, Genetics Institute, Powell Gene Therapy Center, College of Medicine, University of Florida, Gainesville, Florida, USA

Shrestha Ghosh
Department of Biochemistry, Li Ka Shing Faculty of Medicine, The University of Hong Kong, Hong Kong, China

Keith E. Giles
Stem Cell Institute, Department of Biochemistry and Molecular Genetics, University of Alabama at Birmingham, Birmingham, AL, USA

Suming Huang
Department of Biochemistry and Molecular Biology, University of Florida College of Medicine, Gainesville, FL, USA

Mir A. Hossain
Department of Biochemistry and Molecular Biology, Center for Epigenetics, Genetics Institute, Powell Gene Therapy Center, College of Medicine, University of Florida, Gainesville, Florida, USA

Gangqing Hu
Systems Biology Center, National Heart, Lung, and Blood Institute, NIH, Bethesda, MD, USA

Xin Hu
Edmond H. Fischer Signal Transduction Laboratory, School of Life Sciences, Jilin University, Changchun, China

Wei Jian
Department of Anatomy and Cell Biology, College of Medicine, University of Florida, Gainesville, Florida, USA

Alta Johnson
Department of Anatomy and Cell Biology, College of Medicine, University of Florida, Gainesville, Florida, USA

Priscilla Nga Ieng Lau
Leukemia and Stem Cell Biology Group, Department of Haematological Medicine, King's College London, London, UK

Michael D. Litt
Center for Medical Education, Indiana University-Ball State University, Muncie, IN, USA

Xuehui Li
Department of Anatomy and Cell Biology, College of Medicine, University of Florida, Gainesville, Florida, USA

Yangqiu Li
Institute of Hematology, Jinan University Medical College, Guangzhou, China

Yuanyuan Li
Department of Medicine, Division of Hematology and Oncology, University of Alabama at Birmingham, Birmingham, AL, USA; Comprehensive Cancer Center, University of Alabama at Birmingham, Birmingham, AL, USA; Nutrition Obesity Research Center, University of Alabama at Birmingham, Birmingham, AL, USA

Xiumei Lin
Department of Hematology, Guangzhou First People's Hospital, Guangzhou, China

Jianrong Lu
Department of Biochemistry and Molecular Biology, College of Medicine, University of Florida, Gainesville, FL, USA

Huacheng Luo
Department of Biochemistry and Molecular Biology, College of Medicine, University of Florida, Gainesville, FL, USA

Christine M. McBride
Department of Biology, The University of Alabama at Birmingham, Birmingham, AL, USA

Benjamin B. Mills
Department of Biology, The University of Alabama at Birmingham, Birmingham, AL, USA

Bhavita Patel
Department of Medicine, University of Florida College of Medicine, Gainesville, FL, USA

Yi Qiu
Department of Anatomy and Cell Biology, College of Medicine, University of Florida, Gainesville, Florida, USA

Félix Recillas-Targa
Instituto de Fisiología Celular, Departamento de Genética Molecular, Universidad Nacional Autónoma de México, Ciudad de México, México

Nicole C. Riddle
Department of Biology, The University of Alabama at Birmingham, Birmingham, AL, USA

Chi Wai Eric So
Leukemia and Stem Cell Biology Group, Department of Haematological Medicine, King's College London, London, UK

Jared Stees
Department of Biochemistry and Molecular Biology, Center for Epigenetics, Genetics Institute, Powell Gene Therapy Center, College of Medicine, University of Florida, Gainesville, Florida, USA

Ming Tang
Department of Biochemistry and Molecular Biology, College of Medicine, University of Florida, Gainesville, FL, USA

Jessica L. Woolnough
Stem Cell Institute, Department of Biochemistry and Molecular Genetics, University of Alabama at Birmingham, Birmingham, AL, USA

Bowen Yan
Department of Anatomy and Cell Biology, College of Medicine, University of Florida, Gainesville, Florida, USA

Thomas P. Yang
Department of Biochemistry and Molecular Biology, Center for Epigenetics, Genetics Institute, University of Florida College of Medicine, Gainesville, FL, USA

Yurong Yang
Department of Anatomy and Cell Biology, College of Medicine, University of Florida, Gainesville, Florida, USA

Keji Zhao
Systems Biology Center, National Heart, Lung, and Blood Institute, NIH, Bethesda, MD, USA

Zhizhuang Joe Zhao
Edmond H. Fischer Signal Transduction Laboratory, School of Life Sciences, Jilin University, Changchun, China; Department of Pathology, University of Oklahoma Health Sciences Center, Oklahoma City, OK, USA

Lei Zhou
Department of Molecular Genetics and Microbiology, UF Health Cancer Center, College of Medicine, University of Florida, Gainesville, Florida, USA

Zhongjun Zhou
Department of Biochemistry, Li Ka Shing Faculty of Medicine, The University of Hong Kong, Hong Kong, China; Shenzhen Institute of Research and Innovation, The University of Hong Kong, Shenzhen, China

PREFACE

The definition of epigenetics has changed a number of times over the past decades, reflecting our increasingly detailed understanding of regulatory mechanisms in eukaryotes. The chapters of this book effectively point the way to a contemporary definition. Epigenetics, as defined in recent years, concerns itself with the transmission through cell division, and in some cases through the germline, of phenotypic information that is not contained in the DNA sequence itself. The search for biochemical mechanisms that could implement an epigenetic program has, reasonably, focused on reactions that lead to covalent modifications of DNA, and on the identity and biochemistry of proteins and nucleic acids that might bind to DNA tightly enough to remain attached through cell division. That search initially focused on DNA methylation, and more recently on the nucleosome, the octamer histone complex that binds tightly to DNA and packages virtually all of the genome.

DNA methylation provides potentially the most straightforward mechanism for epigenetic transmission of information. As first pointed out over 40 years ago, cytosine methylation at CpG sites can be propagated during replication by an enzyme that recognizes the hemimethylated site and methylates the CpG on the newly synthesized strand. Subsequent studies, notably at imprinted loci, have shown that DNA methylation can affect, directly or indirectly, the binding of transcription factors. The DNA methylation model, in fact, provided the impetus for the present definition of epigenetics, because it appeared to represent the first plausible pathway for transmission through DNA replication of information not encoded in DNA sequence. Since then, however, discovery of the role of chromatin structure in regulation of gene expression, and, during the last twenty years, of the role of noncoding RNAs, has greatly expanded the scope of the epigenetic enterprise.

A large number of articles in this book examine the now extensive literature on the relationship between chromatin structure and gene expression. At the level of individual nucleosomes, the obvious first question is whether nucleosome stability and placement play an important regulatory role. One major research direction has addressed the identity, biochemical consequences, and biological function of histone covalent modifications. The identified distinct modifications include acetylation, mono-, di-, or trimethylation, phosphorylation, ubiquitylation, sumoylation, ADP ribosylation, and more. Each modification targets specific residues on the individual histones of the nucleosome, but the effects, it is now clear, cannot be described in terms of a single code. As with other regulatory mechanisms in eukaryotes, organisms appear to take advantage of the combinatorial versatility afforded by the wide variety of modifications. As described in several chapters, these modifications are an integral part of mechanisms that control initiation of transcription, transcriptional pausing, transcript elongation, termination, and RNA splicing. Other modifications are involved in DNA replication and DNA damage repair. Some marks are predominantly associated with transcriptionally active or inactive chromatin. Certain combinations of marks tend to appear at promoters, others at enhancers. In pluripotent stem cells, some promoters carry both activating and silencing modifications, in principle making the cells ready for rapid changes in expression, depending on the developmental path they take. Histone modifications also play a critical role in transcriptional elongation. When RNA polymerase II transcribes a gene, it carries with it histone modifying enzymes, signaling the formation of a chromatin structure that discourages inappropriate initiation of transcription from within the gene coding region.

All of these reactions are carried out by an array of enzyme complexes specialized to add or remove the individual modifications at particular sites on the histones. Typically, these complexes carry, in addition to the subunit containing the active site, a variety of other protein cofactors that can confer further target specificity. Notable among these modifying complexes are the Polycomb and Trithorax groups, associated respectively with transcriptional silencing and activation. Other complexes are devoted to the task of removing modifications: enzymes and pathways have been identified not only for deacetylation, but for removal of methyl groups from histone lysine and arginine residues. Modifications are thus dynamically regulated in response to regulatory signals and are an essential part of the mechanism by which genetic information contained in DNA is selectively expressed.

How are these modifications used to transmit that information? Modified histones can, in some cases, recruit transcription factor complexes to promoters and enhancers. Other large families of protein complexes use ATP to move nucleosomes away from important DNA regulatory sites or to position them so that they cover these sites. Some histone modifications can alter the strength of interaction between histones and DNA. Certain modified histones can recruit enzymes involved in DNA methylation, and 5-methylcytosine can recruit histone modifying enzymes, thus coupling the two kinds of potentially epigenetic signals. In every case, the protein complexes involved can vary in details of subunit content and specificity, making it necessary to study each regulatory pathway as a separate problem. This has become especially clear as we have begun to understand the role of noncoding RNAs (ncRNAs) in regulatory processes and especially in epigenetic mechanisms involving chromatin. In particular, ncRNAs help to deliver Polycomb or Trithorax group complexes to specific sites, thus guiding the delivery of silencing or activating marks. Other ncRNAs function as part of regulatory protein complexes, and are essential for their activity.

Although much of this work has focused on epigenetic effects at the level of individual genes or gene families, recent studies have revealed the importance of large scale organization within the nucleus in the control of gene expression. It has long been known that enhancers can in some instances act over very great distances to activate specific promoters. It has not been clear how the enhancer chooses its target. Chromatin conformation capture technology (3C, 4C, 5C, and Hi-C) has revealed the presence of large scale organization of the genome into loop domains, with interactions within loops favored over those between loops. How such loops are established and maintained through cell division is a question that remains to be explored and may well be an important part of the epigenetic machinery.

One question raised by all these mechanisms is the extent to which they fit the present definition of epigenetics. There is, at least in principle, a way for DNA methylation patterns to be transmitted through cell division, as described above. Patterns of histone modifications could also be preserved: for example, the Polycomb complex, PRC2, which methylates lysine 27 on histone H3, can be recruited specifically to histones carrying that mark, and could then modify adjacent histones newly deposited at the replication fork. Similarly, one could propose that some part of the loop domain organization of the genome is maintained, although it is known that many features of this higher order structure are not preserved during mitosis. Demonstrating the actual use of such mechanisms in vivo has, however, been difficult, particularly for those mechanisms involving chromatin structure. To qualify according to the strict definition, chromatin structure at a gene should be preserved through cell division and should be determinative of the state of activity of that gene in the daughter cells. From one point of view, changes in histone modifications or nucleosome position could be viewed simply as a part of the mechanism of gene expression–a consequence, rather than a cause of the activity state. For example, an active gene could be maintained in that state because relevant transcription factors remain at high concentrations through mitosis. Afterward, these factors might be sufficient to reestablish activity, and appropriate histone modifications could be regenerated.

Because it is difficult to distinguish between cause and effect, there has been resistance to the idea that chromatin structure and histone modifications convey epigenetic information. Given their evident significance for cell function, this is perhaps not such an important objection. As has been suggested before, it may be more appropriate to revise the definition of epigenetics, which in earlier times referred to the developmental processes that led from a single fertilized egg to a complete organism. There can be no doubt about the role of histones, chromatin, and DNA methylation in those processes. At the same time, it should be clear that mechanisms do exist for inheritance of information that is not carried in the DNA sequence. Position effect variegation in *Drosophila* provided the first evidence for a change in phenotype that was connected with location of a gene within a chromosome, rather than with changes in the gene itself. A clear example exists even in *Saccharomyces*, where genes located near telomeres can be maintained in a silent state through many generations, then switch to an active state, which is likewise stable through many cell divisions. In neither case is there any change in the DNA sequence itself; changes in chromatin structure provide the necessary signals.

Although other examples of transmission through cell division exist, epigenetic inheritance through the germline has been more difficult to demonstrate. A clear example is provided by the agouti mouse, in which coat color, determined by DNA methylation patterns of a retrotransposon inserted near the *agouti* gene, is transmitted to offspring. Other examples exist in plants. There is also evidence for multigenerational epigenetic transmission of phenotypes in flies, mice, and humans. It should be noted that the fidelity of transmission is not as high as for genetic inheritance. However to the extent that these phenotypic changes help to stabilize certain mutations, they may contribute to more permanent changes marked directly in the genome. Because DNA methylation and probably chromatin chemistry can be influenced by environment (e.g., diet), epigenetically controlled phenotypes can reflect environmental signals. The old arguments about heredity vs. environment can now be seen in a new light. But whatever the contribution of these epigenetic signals to inherited phenotype, their importance transcends definitions. They are the manifestations of the array of biochemical pathways that modulate all DNA function in eukaryotes and make possible the enormous variety of cell behaviors required for the success of multicellular organisms.

Gary Felsenfeld
Laboratory of Molecular Biology
National Institute of Diabetes and Digestive and Kidney Diseases
National Institutes of Health, Bethesda, MD, USA

ACKNOWLEDGMENT

This work was supported by the intramural research program of the NIH, National Institute of Diabetes and Digestive and Kidney Diseases.

CHAPTER 1

Epigenetic gene expression—an introduction

C. Ann Blakey[1], Michael D. Litt[2]

[1]Department of Biology, Ball State University, Muncie, IN, USA; [2]Center for Medical Education, Indiana University-Ball State University, Muncie, IN, USA

Contents

1. INTRODUCTION

1.1 Epigenetics – a unifying concept for genetic mechanisms

One could argue that every aspect of a person's health is influenced by the expression of their genes. There are approximately 20,000 protein-coding genes, though a seemingly ever-changing number, controlling the fate and development of cells within the human organism. This cacophony of differential gene expression influences an organism from the point of initial development upon fertilization, onward through each stage of growth and maturation, and includes all interactions—both genetic and environmental—at each of these stages. Consequently, understanding the mechanisms that control and regulate gene expression is pivotal

Epigenetic Gene Expression and Regulation
http://dx.doi.org/10.1016/B978-0-12-799958-6.00001-9

to expanding our knowledge and tools to manipulate the health of the human organism. As a result, interpretation of genetic and environmental data, and how best to analyze and utilize that data, has become a key focus area in biomedical research. The implications for the development and application of treatments in clinical medicine are still in their infancy. However, current studies targeting diseases, nutritional-based issues, and environmental-associated damage have made remarkable advances through the use of the tools of genomics and bioinformatics, revealing a convergence into the field known as epigenetics.

At the birth of genetics, the underlying mechanisms of these processes may have eluded us; nonetheless, observational data analysis has led to significant discoveries tying the various fields of biological science together. In his 2014 review, *A Brief History of Epigenetics,* Felsenfeld [1] provides an excellent synthesis and perspective of epigenetics, from both a historical and theoretical view of genetics. Epigenetics was originally used to describe the processes involved in the development of a zygote to a mature organism. Through a clear retrospective description of our stepwise progression from brilliant insights and clouded uncertainties to our current level of understanding, Felsenfeld [1] weaves together the rich tapestry that is our current understanding of epigenetics. The depth of epigenetics' role becomes visible through the examination of the landmark discoveries and significant contributions of researchers across a range of organisms and questions. Past research discoveries allowed for an examination of connections and interactions from the gene to cellular to organismal levels. Molecular and genomic studies now allow for further refinement and redefining on all levels. A new perspective is therefore needed in order to consider the active role of genes, genomes, and gene/genome expression and interactions in the health of the organism and its future offspring. Thus, we now find ourselves looking upon a new epigenetic horizon with a far more complex landscape, encompassing and reaching beyond that originally envisioned by Waddington [2–6].

1.2 Ideas and questions

Many ideas explored in this text can be approached as questions, though more arise to the reader; most are yet to be fully resolved:

1. How does chromatin change and evolve, not just in the sense of evolution?
2. What are the combinatorial effects of these modifications on the chromatin?
3. How is methylation (DNA, histones, or other) coordinated to provide control across cell generations?
4. In what role(s) are RNAs critical to the models of epigenetic regulation?
5. What are critical protein complexes critical to epigenetic processes?
6. How are these critical epigenetic protein complexes structured and regulated?
7. How does an epigenetic system develop and is regulated in an organism?
8. How does an epigenetic system change/evolve within the organism, including modification and regulation within the vast array of tissues and activities?

9. How is epigenetics research changing biomedical approaches to disease?

10. How is the concept of transgenerational epigenetics, including environment and nutrition effects, changing our view of biology and genetics?

Initially, when viewing the human organism, we see the physical features that comprise the unique characters that identify one individual from another and differentiate the human organism from other species. As we delve deeper into the molecular nature of a single organism, we find that the genetic makeup (hereditary material) of the organism and each cell, with the exception of gametic cells, is *relatively* identical. However, it is the relative identical nature that keeps the organism from being one homogenous mass of cells. Thus, we arrive at a very old question: how does the organism take a single set of instructions and derive the vast array of different tissue types and organs and organ systems that comprise a typical functioning organism [7]?

Thus, when we examine a eukaryotic organism beginning with the gamete, all of the key processes in the organism's life cycle must be tightly regulated both temporally and spatially. These regulatory mechanisms must monitor and control all the critical processes within the organism of the current generation (beginning with sperm and egg) through to its ultimate senescence and death and include the production of the next generation, as well. These processes include all associated biological activities involved in fertilization, zygote formation, and growth and development of the embryo, and growth, development, and maturation of the adult organism. Therefore, the magnitude of the role of epigenetics, and subsequently transgenerational epigenetics, is just beginning to be understood [8].

2. EPIGENETICS

2.1 Regulating the process

To understand the role epigenetic mechanisms play in the life cycle of an organism, we must consider the wide range of different types and categories of molecular structures at work within a cell and organism, where the major categories are DNA, RNAs, and special classes of proteins. We must consider the mechanisms capable of modifying the initial set of information delivered by the sperm and egg in the formation of the zygote, to the functional sets found in each of the tissues, organs, and systems that comprise the organism at each stage of development. Epigenetics mechanisms provide not only a way but also a means by which these modifications can occur without the need for alteration of the original DNA sequence of the organism [1]. The resulting temporal and spatial expression of the genetic information allows for the appropriate growth and development of a healthy organism.

2.2 DNA methylation

Altering the organism without changing the primary sequence of the DNA can be accomplished through several mechanisms. One category of these mechanisms involves the chemical modification of deoxycytidine residues within specific cytosine-guanine-rich sequences, typically via methylation and/or demethylation, but others include hydromethylation and the derivatives of demethylation, formylation, and carboxylation [9]. Renewed and growing interest in the potential role of highly conserved modification to deoxyadenine residues, once relegated by many as mechanisms solely of prokaryote systems, has attained a stature as an important mechanism in spatiotemporal regulation of eukaryotic development [9–11]. These modifications to the DNA sequence have the potential to both alter the three-dimensional structure of the molecule of the DNA strand and inhibit or allow for accessibility to the DNA structure by other molecules. Accessibility, in turn, plays a role in gene expression and therefore, the ability of the cell to perform necessary functions at critical junctures in the organism's life cycle [12].

In an oversimplified example, the methylation of the fifth carbon in the cytosine base of a cytosine-guanine-rich promoter region can result in limited accessibility by the transcription machinery [12]. In the converse case, the demethylation of these 5-methylcytidine residues results in the previously inaccessible region becoming accessible to regulatory factors. This may subsequently promote active expression of the downstream gene versus repressing it. Thus, methylation has effectively silenced the gene. While this is a highly oversimplified example, it serves to emphasize the point that methylation and demethylation do not alter the sequence of the DNA yet still have the potential to dynamically change the accessibility of the structure and expression of gene sequences [13–15]. Elegant examples, with far greater descriptions of mechanisms associated with imprinting and embryonic development and disease progression, are presented in this volume chapters 3, 7, 8, 15, 17, and 18.

2.3 Chromatin
2.3.1 Histone modifications
When considering epigenetics modifications, methylation and demethylation of DNA are the first mechanisms that must be understood. Recognition of DNA methylation and demethylation, alone, does not provide a sufficiently complete explanation of the epigenetic phenomena at work; therefore, researchers must consider the regulation of the genome from the larger context of chromatin structure. Here, chromatin structure takes on additional levels of complexity that can begin to account for some of the observed regulator mechanisms at work in the genome. This complexity includes both double-helical DNA linker regions (upto ~90 bp in length), and regions composed of both double-helical DNA and histone proteins called a nucleosome. More precisely, the nucleosome is composed of approximately 147 bp of double-helix DNA coiled 1.65 times around a histone octamer protein core [16,17].

The histone octamer component of the nucleosome, as an independent structure relative to the double-stranded DNA molecule, is of critical importance in the epigenetic control of the genome [18–21]. This unique, eight-subunit structure is composed of two units each of the histone proteins H2A, H2B, H3, and H4 [16,18]. A fifth histone protein,

H1, is not part of the nucleosome core particle, but retains important functional roles within the structure and dynamics of chromatin [22–24]. Variant forms of the octamer core exist and are intimately associated with specific regions of the genome [25,26].

The main four posttranslational modifications of the nucleosome proteins, methylation, acetylation, phosphorylation, and ubiquitination, are most often discussed in textbooks. But there remain a host of other posttranslational modifications (Table 1) that can occur, which, depending upon the temporal and spatial location of the nucleosome, will have different effects in different cell types during the developmental program. The octamer can have far-reaching effects on the overall organization and structure of the DNA to which it is associated, including the regulation of access to the double-stranded DNA molecule such as by methylation/demethylation of the DNA or by transcriptional machinery of the nucleus. The specific arrangement and chemical modifications of each of these proteins can have a major impact on gene expression and thus, the well-being of the organism [25–36]. These additional forms of histone modifications, as well as new forms, require further exploration and research.

2.3.2 Chromatin organization and remodeling

The nucleosome and linker regions of DNA coil into different hierarchical levels of organization, depending upon the histone variants of the region and the modifications to the DNA and the histone components of the nucleosomes. Together, these components of chromatin function as integral components of the genome or more precisely, when contemplated relative to their coordinated roles in gene expression, the epigenome.

Table 1 Major histone modifications, not all-inclusive

Acetylation
ADP-ribosylation
Butyrylation
Citrullination
Crotonylation
Formylation
0-GlnNAc-glucosamine (0-GlcNAc)
Glutathionylation
Glycosylation
Hydroxylation (or 5-Hydroxylation)
Malonylation
Methylation
Phosphorylation
Proline isomerization
Proprionylation
Succinylation
Sumoylation (similar to ubiquitination)
Ubiquitination

Excellent publications on chromatin structure, organization, and remodeling have begun to appear with great regularity. Current literature of the biosciences has become dominated by investigations into possible connection of epigenetics to disease and cancers. These works have significantly impacted our ability to commence to understand the depth and range of mechanisms of histone and chromatin modifications that is possible [37–41] and therefore, their subsequent contributions to chromatin dynamics and remodeling within the genome [33,42–46].

Extensive discussions on histone modifications, chromatin dynamics and remodeling can be found in this volume in chapters 3, 4, 5, and 14.

2.4 RNA

As with any biological process, the correct timing of gene expression will play a pivotal role. Facilitating the expression of genes and the mechanisms that allow for expression are a host of molecules, involving a wide range of ribonucleic acids (RNA, see Table 2). The roles of these RNAs make it possible for the myriad of cascading processes necessary for homeostasis, as well as the adaptive nature of the organism to survive environmental stresses. RNAs are involved in the methylation of DNA [47], as well as having other structural, catalytic, and regulatory functions within chromatin modification machinery [48,49]. Detailed discussions of the various mechanisms associated with epigenetics and different RNAs can be found in this volume in chapters 3, 6, 10, 11, and 13.

Table 2 Some classes of RNAs involved in epigenetics gene expression, not all-inclusive

Coding RNA:	hnRNA	mRNA	
Noncoding RNA:			
ncRNA	Noncoding	macroRNA	macro
rRNA	Ribosomal	sRNA	small
tRNA	Transfer	snRNA	small nuclear
circRNA	Circular	snoRNA	small nucleolar
eRNA	Enhancer	siRNA	small interfering
ceRNA	Competing endogenous	piRNA	Piwi-interacting
asRNA	Antisense	mascRNA	MALAT1 small cytoplasmic
lncRNA	Long noncoding	iRNA	interfering
lincRNA	Long intergenic noncoding	siRNA	small interfering
alincRNA	Activating long noncoding	22G RNA	Class of small interfering (siRNA)
as-lincRNA	Antisense long noncoding	26G RNA	Class of small interfering (siRNA)
NATs	Natural antisense transcripts	microRNA or miR	micro

2.5 Clinical implications

Through the study of epigenetics, we now begin to see the connections to biological processes that had heretofore remained a puzzle to medical science. In many ways, we are at the infancy of where epigenetic research may take biomedical research as we begin to elucidate the full range and coordination of the epigenetic mechanisms at work in eukaryotic organisms. Our understanding of these processes continues to expand as we examine the disturbances observed in utero cells undergoing differentiation, to the establishment of imprinting patterns within the sperm genome contribution of a newly fertilized egg, or the modifications of histones during X-chromosome inactivation. These topics are discussed in greater detail in this volume in chapters 3, 7, 8, 11, and 13.

Historical records can provide important data in epigenetic discovery, as has been clearly shown by the example of Dutch Hunger Winter analyses of health data of subsequent generations within that population [50]. This famine occurred between September/October 1944 and May 1945 and ended with the end of World War II. A total of 18–22,000 deaths were associated with this famine, although a total of 4.5 million individuals were actually affected by a 400–800 calorie per day diet [51]. "Studies that draw upon multigenerational data sets allow for the analyses of extremes of environmental influences that cause resultant disturbances in the primordial germ cells of an organism. In the case of the Dutch Hunger Winter, researchers were able to follow epigenetic changes in several generations. For example, the women (F_0 generation) who carried female progeny (F_1 generation) in gestation during the months of the Dutch Hunger Winter, as well as follow those changes in the subsequent generation of in utero primordial germ cells (F_2 generation) [52]. The transgenerational epigenetics effects of the Dutch Hunger Winter are discussed in greater detail in this volume in chapter 15.

Evaluation of changes in patterning that differ from those who did not undergo an environmental or nutritional stress, and the persistence of pattern alterations, not just DNA methylation but also modified nucleosomal patterning, add greater depth to our knowledge and understanding of the impact of these long-term stresses. The ramifications of alterations in the genetic mechanisms at work have become an area of intense research in the biomedical sciences [53,54]. Companies collaborating with public–private partnerships have established one of the largest joint ventures of its kind, called the EpiGen Consortium (www.epigengrc.com/).

When gene expression does not follow the expected program or when a disease develops, understanding the epigenetic changes associated with these disease states may help further our ability to develop clinical approaches to the treatment of diseases. Explorations of epigenetic phenomena associated with various cancers and diseases (for examples, see Table 3) add emphasis to the relationship between the fine level of regulation of chromatin structure, accessibility, regulation of gene expression, and the transcriptional

Table 3 Some epigenetic-influenced diseases, syndromes, and cancers, not all-inclusive

Developmental with craniofacial features	Cancers
Silver–Russell Syndrome (SRS) Wolf–Hirschhorn Syndrome (WHS)	Breast cancer 1, early onset (BRCA1) Breast cancer 2, early onset (BRCA2) Clear Cell Carcinoma (CCC) Squamous Cell Carcinoma (SCC)
Developmental/progressive	
Duchenne Muscular Dystrophy (DMD) Hutchinson–Gilford Progeria (HGPS) Werner Syndrome (WRN)	**Disorders/diseases and cancers of the blood**
Complex metabolic/organ disorders/diseases	Hereditary Persistence of Fetal Hb (HPFH) Myelodysplasia Syndrome (MDS) B-cell Lymphoma (BCL) Diffuse Large B-cell Lymphoma (DLBCL) Myeloproliferative Neoplasia (MPN) Acute Myeloid Leukemia (AML) Chronic Lymphocytic Leukemia (CLL) Chronic Myeloid Monocytic Leukemia (CMML) Promyelocytic Leukemia (PML) Stem Cell Leukemia (SCL)
Chronic Kidney Disease (CKD) Cardiovascular Disease (CVD) Metabolic Syndrome (MetS)	
Complex neurodegenerative disease	
Alzheimer's Disease (AD) Creutzfeldt–Jakob Disease (CJD) Neural Tube Defects (NTDs)	
Developmental with tumor formation	
Beckwith–Wiedemann Syndrome (BWS)	

machinery. Detailed discussions and analyses of the role of epigenetics in cancers and disease can be found in this volume in chapters 9, 12, 14, 16, and 18.

3. FUTURE DIRECTIONS AND CHALLENGES

3.1 Analytical tools

The quantity and quality of data produced in epigenetics has increased exponentially since the 2010. The vast array of techniques and technologies that were initially developed to further enhance our ability to rapidly sequence the genome and its products has led to a need for greater bioinformatics and biocomputational capabilities to handle the large, nearly overwhelming volumes of resultant data (Tables 4–6). The difficulty arises when attempting to source the best tools for a specific application from the range of tools available, as well as the location or deposition of these tools for utilization in the case of software algorithms and packages.

As excellent as these resources have proven to the discovery of new data relationships, there remains an issue of complete interconnectedness across databases. Valuable time and labor is lost without continual improvements in workflows for automated data handling, processing, and database access. These latter challenges focus primarily on software and hardware communications. Thus, there will remain a continual need for more highly

Table 4 Molecular techniques utilized in epigenetic studies, not all-inclusive

3C	Chromosome conformation capture
Hi–C	Extension of 3C technology, for investigations of long–range interactions
4C	Circular chromosome conformation capture
5C	Chromosome conformation capture carbon copy
oxBs–Seq	Oxidative bisulfate sequencing, discerns between 5mC and 5hmC
Tab–seq	Tet-assisted bisulfite sequencing, discerns between 5mC and 5hmC
ChiA–PET	Chromatin interaction Analysis by Paired-End Tag Sequencing
ChIP–seq	Chromatin immunoprecipitation combined with massively parallel sequencing to identify interactions between DNA and RNA and proteins
nano–ChIP-seq	A Type of ChIP-seq using fewer cells
ChIRP	Chromatin isolation by RNA purification
CAGE–seq	Capped RNA sequencing
3P–Seq	Polyadenylated Ends RNA Sequencing
RNAe	RNA-induced Epigenetic Silencing
RNAi	RNA interference

Table 5 Biocomputational algorithms for genomic analyses, not all-inclusive

Name	Algorithm type/use	Website/location
Genomic		
Bowtie 2 [56]	Short-read backward alignment, uses a reference genome	
BWA [57]	Burrows–Wheeler Aligner, used with Bowtie to increase speed and accuracy of alignments	bowtie-bio.sourceforge.net/bowtie2/index.shtml
GRACOMICS [58]	Graph	bibs.snu.ac.kr/software/GRACOMICS/gracomics.php
Code		
CNCI [59]	Classification, based on composition	www.bioinfo.org/software/cnci/
CPAT [60]	Logistic regression using linguistic features	lilab.research.bcm.edu/cpat/
CPC [61]	Code classification based on features	cpc.cbi.pku.edu.cn/
Histones		
histoneHMM [62]	Bivariate Hidden Markov Model	histonehmm.molgen.mpg.de/
SICER [63]	Transcript inference, NGS bin approach	
Chromatin		
ChroModule [64]	Supervised machine learning, annotation of epigenetic state	
ChromHMM [65]	Hidden Markov model, unsupervised automated analysis of epigenetic states	compbio.mit.edu/ChromHMM
MACS [66]	ChIP data interpretation	liulab.dfci.harvard.edu/MACS
SeqMINER [67]	ChIP data interpretation	bips.u-strasbg.fr/seqminer/
Spectacle [68]	Spectacle-learning approach to analysis of chromatin states	github.com/jiminsong/Spectacle

Table 6 Biocomputational algorithms for transcriptomics analyses, not all-inclusive

Name	Algorithm type/use	Website/location
RNA		
IsoLasso [69]	Assembler based on LASSO [70]	alumni.cs.ucr.edu/~liw/isolasso.html
Oases [71]	Assembler	www.ebi.ac.uk/~zerbino/oases/
Polyester [72]	Sequence simulation	bioconductor.org/
Codons		
PhyloCSF [73]	Alignment/substitution frequencies	compbio.mit.edu/PhyloCSF
Transcripts		
Cufflinks v7 [74]	Assembler, with bipartite graph and reference base	www.broadinstitute.org/cancer/ software/genepattern/ modules/docs/Cufflinks/7
GSNAP [75]	Short-read alignment	research-pub.gene.com/gmap/
MapSplice [76]	Splice junction	www.netlab.uky.edu/p/bioinfo/ MapSplice
MITIE [77]	Reconstruction	bioweb.me/mitie
ORMAN [78]	Transcript product potential and mapping	orman.sf.net
Scripture [79]	Assembler	www.broadinstitute.org/ software/Scripture/
STAR [80]	Alignments to a reference	code.google.com/p/rna-star/
StringTie [81]	Assembler	ccb.jhu.edu/software/stringtie/
Tophat [82]	Alignments	tophat.cbcb.umd.edu
Traph [83]	Expression estimation	www.cs.helsinki.fi/en/gsa/traph/
Trinity [84]	Assembler	trinityrnaseq.sf.net
Velvet [85]	Graph	www.ebi.ac.uk/~zerbino/velvet

trained researchers and personnel with up-to-date knowledge and skills in both computer science and the biomedical sciences.

Identification of appropriate computational/biocomputational tools, even if freely available for use to researchers, often requires a continual surveillance of the bioinformatics literature in order to remain up-to-date on the newest, or to find the latest developments. These tasks have become even more relevant, both in their use within the biomedical/biological field of specialization and as biocomputational tools.

Many researchers have attempted to facilitate and share their own interconnectivity with data and data workflows with their colleagues through the development of Web sites and database structures which, in turn, cross-reference the multiple external databases, algorithms, and resources utilized in their own endeavors. But, these approaches become cumbersome and maintenance of computer links are labor intensive and withdraw resources from the primary funded areas of their research projects. Resources for research are a scarce commodity, and highly trained personnel is an ever-increasing expense.

So, the impetus to create more efficient workflow algorithms and data handling methodologies will continue. As with most areas of genomics, technology for data production far exceeds our ability to process it, and backlogs are created. The improvement of workflow will remain a constant battle as the deluge of data continues to mount. It is admirable that so many researchers share their tools and approaches beyond publications. However, the challenges of accessibility and integration for improved workflow without loss of researcher productivity still remain key issues that have yet to be resolved.

3.2 Databases

The major databases and portals such as NCBI, EMBL, GenBANK, ENCODE, and GENCODE have provided a significant venue for researchers to access large volumes of data and associated information beyond that which individual research projects can provide. Recognition of the need for continual work toward integration of cytogenetic, molecular, and cellular data for more rapid discovery and application of knowledge in the biomedical sciences remains clearly evident.

There are numerous individual databases available to the researchers interested in pursuing epigenetic studies or simply wishing to understand the interconnections of available genomic information (Table 7). Through portals, it is possible to find a vast array of information and bioinformatics tools. Online datamining resources have become much more accessible, such as YeastMine (yeastmine.yeastgenome.org), MouseMine (www.mousemine.org), FlyMine (www.flymine.org), or metabolicMine (www.metabolicmine.org), to name only a few, all of which allow for rapid access to depths of information both within and across species.

No one portal or database can provide all the resources currently required by the scientific community. In addition, changes in funding can dramatically alter or even curtail accessibility to key database resources for the entire research community. The fields of bioinformatics and biocomputational analyses continue to work diligently toward greater accuracy and efficiencies, but continual funding for maintenance, improvements, and new developments in database structure, automation, and accessibility will remain key issues.

3.3 Nomenclature

One of the difficulties in contemplating the enormous volume of research that is being produced in this rapidly expanding area is that of nomenclature. Cross-species nomenclature issues slow discovery when every alias of a particular gene or transcript needs to be searched using multiple search engines or mining algorithms in order to obtain the fullest degree of information on known or putative functions and interactions. While some researchers do provide tables with regard to their particular transcripts or genes of interest when they publish in order to facilitate the reader and future discovery, these are not typical. Some authors in various publications have provided tables of a limited number of homologs within their publication to improve cross-species access to important findings.

Table 7 Some key databases, from which additional data-mining tools and resources can be accessed

| General | Databases on medicine and molecular biology | www.meddb.info/index.php.en?cat=6 |
| COSMIC | Catalog of Somatic Mutations in Cancer | http://cancer.sanger.ac.uk/cosmic |

Genomic/transcriptomic/epigenetic database resources

| ENCODE | Encyclopedia of DNA Elements Consortium | genome.ucsc.edu/ENCODE/ |
| GENCODE | Encyclopædia of Genes and Gene Variants Consortium | www.gencodegenes.org/ |

Epigenetics databases

epigenie	Epigenetic Databases, Tools, and Resources	www.epigenie.com/epigenetic-tools-and-databases/
NCBI/Epigenomics	NCBI Epigenomics	www.ncbi.nlm.nih.gov/epigenomics
NCBI/Gene	NCBI Gene	www.ncbi.nlm.nih.gov/gene/
RefSeq	NCBI Reference Sequence Database	www.ncbi.nlm.nih.gov/refseq/

DNA methylation databases

DiseaseMeth	The Human Disease Methylation Database	202.97.205.78/diseasemeth/
MethDB	DNA Methylation and Environmental Epigenetic Effects	www.methdb.de/
MethHC	DNA Methylation and Gene Expression in Human Cancer	MethHC.mbc.nctu.edu.tw
MethylomeDB	Brain Methylome Database	www.neuroepigenomics.org/methylomedb/

Histone and histone modification databases

HHMD	Human Histone Modification Data	202.97.205.78/hhmd/index.jsp
HIstome	The *Histone Infobase*	www.actrec.gov.in/histome/
HistoneDB	Histone Database @ NHGRI	genome.nhgri.nih.gov/histones/
HistoneHits	Histone Systematic Mutation Database	54.235.254.95/histonehits/#home

Model organism databases

Ensembl	Genome Database for Vertebrates and other Eukaryotes	www.ensembl.org/index.html
FANTOM	Functional Annotation of the Mammalian Genome Consortium	fantom.gsc.riken.jp/
FlyBase	Database for *Drosophila* Genes and Genomes	www.flybase.org
MGI	Mouse Genome Informatics	www.informatics.jax.org/
PlantDGB	*Arabidopsis* Genome Database	www.plantgdb.org/AtGDB/
PomBase	Pombase, Resource for Fission Yeast	www.pombase.org
RGD	Rat Genome Database	rgd.mcw.edu/
SGD	*Saccharomyces* Genome Database	www.yeastgenome.org
TAIR	The *Arabidopsis* Information Resource	www.arabidopsis.org
WormBase	Nematode Information Resource	www.wormbase.org
Xenbase	*Xenopus* Genome Database	www.xenbase.org/entry/
ZFIN	Zebrafish Model Organism Database	zfin.org/

Still, there remains a significant issue with regards to the acronyms and cross-referencing of genes and gene fragments. An example of the problem can be illustrated with the Trithorax group protein (*trx*) from *Drosophila*, where the human homolog is *KMT2A* with the aliases *HTRX1, HRX, MLL,* and *MLL1* (Table 8). As research has progressed, there have been greater efforts to keep the same acronym across species. We find many of the homologous genes in humans and mouse, with the exception that the human acronym is all capital letters, and the mouse has only the first letter capitalized. In the case of the human gene aliases *TRX, TRX1,* and *TRDX*, these acronyms are for the cytoplasmic thioredoxin gene (*TXN*), which is also known as *TRX1* and *LMA1* in *Saccharomyces* and *TRX2* in *Drosophila*.

Therefore, clear identification of the organism(s) and the primary gene name will help reduce some of the sources of confusion in literature and discussions. Thus, greater clarity of the true breadth of data coverage in this field will become more apparent and accessible, and facilitating faster discovery and utilization of information through striving with care to avoid the use of secondary aliases and non-organism-specific acronyms, identifying the organism in question within discussions, and continuing to provide cross-references to key model organisms where possible.

4. SUMMARY

4.1 Core concepts of the text

The chapters in this text are arranged to provide the reader both an overview of current developments in this exciting and dynamic field of research and to function as a resource for graduate students and biomedical professionals.

In the first part of the text, the reader will be presented with an overview, as well as in-depth reviews of major components of epigenetics research. A review of the current histone modifications models is presented. Next, a discussion on chromatin dynamics and interactions provides additional depth. This is followed by a close examination of chromatin organization and the role of noncoding RNA in epigenetics. Subsequent chapters focus on the tools of epigenetic research and their application, thereby providing readers a solid background from which to venture further.

In the second part of the text, the reader will be introduced to epigenetic effects from the basis of our understanding of its heritability to its role in imprinted-associated gene expression. Next, chapters present discussions of epigenetic mechanisms in stem cells and tissue development, which ultimately lead the reader to examine the effects of epigenetics on the aging process. Afterward, discussions consider the possible epigenetic effects on future generations (transgenerational epigenetics), and those of the environment.

Finally, in the third part of the text, discussions consider the role of epigenetics on a specific area of biomedical research—cancer epigenetics. Beginning with apoptosis, the reader is presented mechanisms of regulation and roles of epigenetics in cancer. The

Table 8 A comparison of the gene names and aliases is given for the Trithorax Group Protein (TRX-G) for *Lysine (K) Methyltransferase 2A/B* across 4 species versus the cytoplasmic form of the Thioredoxin *(TXN)* gene. In each case, the primary gene name is provided as well as several aliases, though every gene alias is not included. As shown in red, the gene name *TRX*, *TRX1*, and *TRX2* are utilized across all species and within a species *(Drosophila)* for two distinctly different genes

Organism	Saccharomyces	Drosophila	Mus	Homo sapiens
Trithorax group Proteins (TrxG)				
Gene:	*SET1*	*Trx*	*Kmt2a*	*KMT2A*
Location:	**ChrVIII**	**chr3R**	**chr9A5.2**	**chr11q23**
Aliases:	*KMT2*	*trxI and trxII*		*TRX1*
	YTX1	*Trx-G*	*Mll*	*MLL*
	YHR119W	*NR0A5*	*Mll1*	*MLL1*
		DMTRXIll	*ALL-1*	*ALL-1*
		R-bx	*Cxxc7*	*CXXC7*
		Rg-bx	*HTRX1*	*HTRX1*
		tqrx	*Hrx*	*HRX*
		Tqrx-g		*TET1-MLL*
				MLL1A
Gene:	*SET1*	*trx*	*Kmt2b*	*KMT2B*
Location:	**ChrVIII**	**chr3R**	**chr7B1**	**chr19q13.1**
Aliases:			*Trx2*	*TRX2*
			Mll2	*MLL2*
			Mll4	*MLL4*
			Wbdp7	*WBP7*
				MLL1B
				HRX2
				HTRX2
Thioredoxin, cytoplasmic (TXN)				
Gene:	*TRX1*	*Trx-2*	*Txn1*	*TXN*
Location:	**chrXII**	**chr2L**	**chr4B3**	**chr9q31**
Aliases:	*Trx-2*[a]	*Trx*	*Trx1*	*TRX1*
	LMA1[b]	*DTrx-2*	*ADF*	*TRX*
	PBI2[c]	*DmTrx-2*		*TRDX*
		DmelTrx-2		
Gene:	*TRX3*	*dhd*	*Txn1*	*TXN*
Location:	**chrIII**	**chrX**	**chr4B3**	**chr9q31**
Aliases:	*YCR083W*	*Trx-1*	*Trx1*	*TRX1*
		DmTrx-1	*ADF*	*TRX*
		Dmeldhd		*TRDX*
		Dmdhd		

[a] *Trx2* (location: chrVII) in yeast has a paralog, *TRX1*.
[b] *LMA1* is used both as aliases in yeast for *TRX1, TRX2,* and *PBI2* and for the complex formed between *TRX1* and Pbi2p.
[c] *PBI2* (location: chrXIV) = Pbi2p, also known as *I2B* or *IB2*.

specific role of major protein complexes, such as the Polycomb group and Trithorax group proteins, is examined. These protein complexes, along with an array of noncoding RNAs, are considered throughout with regard to gene expression pathways associated with carcinogenesis and they provide expert insights into the development of new therapies.

4.2 Final thoughts

E.B. Wilson made the following observation in 1895 [55]:

"Now, chromatin is known to be closely similar to, if not identical with, a substance known as nuclein… which analysis shows to be a tolerably definite chemical composed of nucleic acid (a complex organic acid rich in phosphorus) and albumin. And thus we reach the remarkable conclusion that inheritance may, perhaps, be effected by the physical transmission of a particular chemical compound from parent to offspring."

With the advent and secure establishment of DNA as the hereditary information molecule, there still remained the processes—the heritable processes—that could not be explained solely based on DNA alone. Questions start to involve a paradigm shift from what molecule was heritable to what mechanisms are at work. We begin to view the genome and its structures in a different, very mechanistic, light as we delve into the temporal–spatial of what, how, and when these heritable processes play out their role in gene expression. So, we find ourselves presented with an inescapable view of a new perspective of genetics, where "the physical transmission of a particular chemical compound from parent to offspring [55]" is merely the first consideration. All other considerations require the incorporation of modifications to that original "particular chemical compound" that we know now to be DNA. Each generation contributes to or alters the collection of heritable modifications given to the next generation. It is the study of these heritable modifications, including their associated mechanisms, and their intimate roles in gene expression and regulation that has become the primary focus of what is now called epigenetics [1].

Thus, from concepts to functional roles in growth and development to interactions and mechanisms of disease, understanding of the role of epigenetics is necessary to fully grasp how a single fertilized egg is capable of giving rise to the level of complexity that is the adult organism, with all its varied cell and tissue types [1], as well as all its heritable potentialities for disease and survival for future generations. Mechanistically, we are just beginning to understand and find explanations for the ties between environmental effects and genome changes, which for quite some time, had remained just out of reach. Now, we have just begun to elucidate the role of epigenetics in transgenerational nutritional and environmental effects. Yet, this is only the beginning. Knowledge of how these mechanisms related to growth and development, as well as to medical diseases, has revised our approaches to these processes and potential new treatments. We have a constantly improving set of molecular and computational tools that allow us to attempt the explanation of both the structures and functions of many of the molecular components involved. So we find ourselves at the cusp of learning how many of these epigenetic mechanisms function, interact, change, and evolve. But, there is still more as yet unknown.

Epigenetics research continues to provide amazing new insights into the hereditary effects of the genome, no longer viewed as simply sets of genes. Rather, the genome is seen as an epigenome, a rich source of heritable genetic diversity with multiple potentialities. The ability to predict the outcomes of these interactions and to apply this knowledge is of pivotal importance to biomedical science. However, just as chromatin is a vibrant and dynamic structure, our models of the epigenome must be recognized and maintained flexibility to account for the vast differences observed in phenotypes, for changes within and to chromosomal structures, and for the adaptive nature in molecular interactions. Only by remaining open to the broad spectrum of possibilities can the true understanding of epigenetics be used to reap future benefits for human health.

ACKNOWLEDGMENTS

This work is supported by the Department of Biology and the College of Sciences and Humanities, Ball State University, as well as the Center for Medical Education, Indiana University-Ball State University, Muncie, IN, USA.

REFERENCES

[1] Felsenfeld G. A brief history of epigenetics. Cold Spring Harbor Perspect Biol 2014;6(1).
[2] Waddington CH. Organisers and genes. Cambridge: Cambridge University Press; 1940.
[3] Waddington CH. Canalization of development and the inheritance of acquired characters'. Nature 1942;150:563–4.
[4] Waddington CH. Principles of embryology. New York: MacMillan Co; 1956.
[5] Waddington CH. The strategy of the genes; a discussion of some aspects of theoretical biology. London: Allen & Unwin; 1957.
[6] Waddington CH. Canalization of development and genetic assimilation of acquired characters. Nature 1959;183(4676):1654–5.
[7] Wilson EB. The cell in development and inheritance. New York: MacMillan; 1896.
[8] Daxinger L, Whitelaw E. Transgenerational epigenetic inheritance: more questions than answers. Genome Res 2010;20(12):1623–8.
[9] Heyn H, Esteller M. An adenine code for DNA: a second life for N6-methyladenine. Cell 2015;161(4):710–3.
[10] Zhang G, Huang H, Liu D, et al. N6-methyladenine DNA modification in *Drosophila*. Cell 2015;161(4):893–906.
[11] Fu Y, Luo G-Z, Chen K, et al. N6-Methyldeoxyadenosine marks active transcription start sites in *Chlamydomonas*. Cell 2015;161(4):879–92.
[12] Suzuki MM, Bird A. DNA methylation landscapes: provocative insights from epigenomics. Nat Rev Genet 2008;9(6):465–76.
[13] Messerschmidt DM, Knowles BB, Solter D. DNA methylation dynamics during epigenetic reprogramming in the germline and preimplantation embryos. Genes Dev 2014;28(8):812–28.
[14] Gkountela S, Zhang KX, Shafiq TA, et al. DNA demethylation dynamics in the human prenatal germline. Cell 2015;161(6):1425–36.
[15] Wu SC, Zhang Y. Active DNA demethylation: many roads lead to Rome. Nat Rev Mol Cell Biol 2010;11(9):607–20.

[16] Luger K, Mader AW, Richmond RK, Sargent DF, Richmond TJ. Crystal structure of the nucleosome core particle at 2.8A resolution. Nature 1997;389(6648):251–60.

[17] Luger K. Structure and dynamic behavior of nucleosomes. Curr Opin Genet Dev 2003;13(2):127–35.

[18] Luger K, Hansen JC. Nucleosome and chromatin fiber dynamics. Curr Opin Struct Biol 2005;15(2):188–96.

[19] Chakravarthy S, Park YJ, Chodaparambil J, Edayathumangalam RS, Luger K. Structure and dynamic properties of nucleosome core particles. FEBS Lett 2005;579(4):895–8.

[20] Hamperl S, Brown CR, Garea AV, et al. Compositional and structural analysis of selected chromosomal domains from *Saccharomyces cerevisiae*. Nucleic Acids Res 2014;42(1):e2.

[21] Li G, Reinberg D. Chromatin higher-order structures and gene regulation. Curr Opin Genet Dev 2011;21(2):175–86.

[22] Woodcock CL, Skoultchi AI, Fan Y. Role of linker histone in chromatin structure and function: H1 stoichiometry and nucleosome repeat length. Chromosome Res 2006;14(1):17–25.

[23] Brush GS. Evidence that histone H1 is dispensable for proper meiotic recombination in budding yeast. BMC Res Notes 2015;8:275.

[24] Cheutin T, McNairn AJ, Jenuwein T, Gilbert DM, Singh PB, Misteli T. Maintenance of stable heterochromatin domains by dynamic HP1 binding. Science 2003;299(5607):721–5.

[25] Pusarla RH, Bhargava P. Histones in functional diversification. Core histone variants. FEBS J 2005;272(20):5149–68.

[26] Kamakaka RT, Biggins S. Histone variants: deviants? Genes Dev 2005;19(3):295–310.

[27] Henikoff S, Furuyama T, Ahmad K. Histone variants, nucleosome assembly and epigenetic inheritance. Trends Genet 2004;20(7):320–6.

[28] Campos EI, Reinberg D. Histones: annotating chromatin. Annu Rev Genet 2009;43:559–99.

[29] Marino-Ramirez L, Levine KM, Morales M, et al. The Histone Database: an integrated resource for histones and histone fold-containing proteins. Database (Oxford) 2011;2011:bar048.

[30] Simon M, North JA, Shimko JC, et al. Histone fold modifications control nucleosome unwrapping and disassembly. Proc Natl Acad Sci USA 2011;108(31):12711–6.

[31] Karch KR, Denizio JE, Black BE, Garcia BA. Identification and interrogation of combinatorial histone modifications. Front Genet 2013;4:264.

[32] Zentner GE, Henikoff S. Regulation of nucleosome dynamics by histone modifications. Nat Struct Mol Biol 2013;20(3):259–66.

[33] Fischle W. Molecular mechanisms of histone modification function. Biochim Biophys Acta 2014;1839(8):621–2.

[34] Huang H, Sabari BR, Garcia BA, Allis CD, Zhao Y. SnapShot: histone modifications. Cell 2014;159(2):458. e1.

[35] Kebede AF, Schneider R, Daujat S. Novel types and sites of histone modifications emerge as players in the transcriptional regulation contest. FEBS J 2015;282(9):1658–74.

[36] Sadakierska-Chudy A, Filip M. A comprehensive view of the epigenetic landscape. Part II: histone post-translational modification, nucleosome level, and chromatin regulation by ncRNAs. Neurotox Res 2015;27(2):172–97.

[37] Adkins NL, Niu H, Sung P, Peterson CL. Nucleosome dynamics regulates DNA processing. Nat Struct Mol Biol 2013;20(7):836–42.

[38] Allfrey VG, Faulkner R, Mirsky AE. Acetylation and methylation of histones and their possible role in the regulation of RNA synthesis. Proc Natl Acad Sci USA 1964;51:786–94.

[39] Anderson JD, Lowary PT, Widom J. Effects of histone acetylation on the equilibrium accessibility of nucleosomal DNA target sites. J Mol Biol 2001;307(4):977–85.

[40] Banerjee T, Chakravarti D. A peek into the complex realm of histone phosphorylation. Mol Cell Biol 2011;31(24):4858–73.

[41] Barth TK, Imhof A. Fast signals and slow marks: the dynamics of histone modifications. Trends Biochem Sci 2010;35(11):618–26.

[42] Langst G, Manelyte L. Chromatin Remodelers: from function to dysfunction. Genes (Basel) 2015;6(2):299–324.

[43] Bloom KS. Centromeric heterochromatin: the primordial segregation machine. Annu Rev Genet 2014;48:457–84.

[44] Brzeski J, Jerzmanowski A. Plant chromatin – epigenetics linked to ATP-dependent remodeling and architectural proteins. FEBS Lett 2004;567(1):15–9.

[45] Boskovic A, Torres-Padilla ME. How mammals pack their sperm: a variant matter. Genes Dev 2013;27(15):1635–9.

[46] Cheedipudi S, Genolet O, Dobreva G. Epigenetic inheritance of cell fates during embryonic development. Front Genet 2014;5:19.

[47] Matzke MA, Mosher RA. RNA-directed DNA methylation: an epigenetic pathway of increasing complexity. Nat Rev Genet 2014;15(6):394–408.

[48] Bierhoff H, Postepska-Igielska A, Grummt I. Noisy silence: non-coding RNA and heterochromatin formation at repetitive elements. Epigenetics 2014;9(1):53–61.

[49] Liebers R, Rassoulzadegan M, Lyko F. Epigenetic regulation by heritable RNA. PLoS Genet 2014;10(4). e1004296.

[50] Heijmans BT, Tobi EW, Stein AD, et al. Persistent epigenetic differences associated with prenatal exposure to famine in humans. Proc Natl Acad Sci USA 2008;105(44):17046–9.

[51] Lumey LH, Stein AD, Kahn HS, et al. Cohort profile: the Dutch Hunger Winter families study. Int J Epidemiol 2007;36(6):1196–204.

[52] de Rooij S, Roseboom T, Painter R. Famines in the last 100 years: implications for diabetes. Curr Diab Rep 2014;14(10):1–10.

[53] Dogra S, Sakwinska O, Soh SE, et al. Dynamics of infant gut microbiota are influenced by delivery mode and gestational duration and are associated with subsequent adiposity. MBio 2015;6(1).

[54] Godfrey KM, Sheppard A, Gluckman PD, et al. Epigenetic gene promoter methylation at birth is associated with child's later adiposity. Diabetes May 2011;60:1528–34.

[55] Wilson EB. An atlas of the fertilization and karyogenesis of the ovum. New York: MacMillan; 1895.

[56] Langmead B, Salzberg SL. Fast gapped-read alignment with Bowtie 2. Nat Methods 2012;9:357–9.

[57] Li H, Durbin R. Fast and accurate short read alignment with Burrows-Wheeler transform. Bioinformatics 2009;25(14):1754–60.

[58] Seo M, Yoon J, Park T. GRACOMICS: software for graphical comparison of multiple results with omics data. BMC Genomics 2015;16:256.

[59] Sun L, Luo H, Bu D, et al. Utilizing sequence intrinsic composition to classify protein-coding and long non-coding transcripts. Nucleic Acids Res 2013;41(17):e166.

[60] Wang L, Park HJ, Dasari S, Wang S, Kocher JP, Li W. CPAT: Coding-Potential Assessment Tool using an alignment-free logistic regression model. Nucleic Acids Res 2013;41(6):e74–81.

[61] Kong L, Zhang Y, Ye ZQ, et al. CPC: assess the protein-coding potential of transcripts using sequence features and support vector machine. Nucleic Acids Res 2007;35(Web Server issue):W345–9.

[62] Heinig M, Colome-Tatche M, Taudt A, et al. histoneHMM: differential analysis of histone modifications with broad genomic footprints. BMC Bioinformatics 2015;16(1):60.

[63] Zang C, Schones DE, Zeng C, Cui K, Zhao K, Peng W. A clustering approach for identification of enriched domains from histone modification ChIP-Seq data. Bioinformatics 2009;25(15):1952–8.

[64] Won KJ, Zhang X, Wang T, et al. Comparative annotation of functional regions in the human genome using epigenomic data. Nucleic Acids Res 2013;41(8):4423–32.

[65] Ernst J, Kellis M. ChromHMM: automating chromatin-state discovery and characterization. Nat Methods 2012;9(3):215–6.

[66] Feng J, Liu T, Qin B, Zhang Y, Liu XS. Identifying ChIP-seq enrichment using MACS. Nat Protoc 2012;7(9):1728–40.

[67] Ye T, Krebs AR, Choukrallah MA, et al. seqMINER: an integrated ChIP-seq data interpretation platform. Nucleic Acids Res 2011;39(6):e35.

[68] Song J, Chen KC. Spectacle: fast chromatin state annotation using spectral learning. Genome Biol 2015;16:33.

[69] <cmb.2011.0171.pdf>.

[70] Tibshirani R. Regression shrinkage and selection via the lasso. J R Stat Soc B 1996;58(1):267–88.

[71] Schulz MH, Zerbino DR, Vingron M, Birney E. Oases: robust de novo RNA-seq assembly across the dynamic range of expression levels. Bioinformatics 2012;28(8):1086–92.

[72] Frazee AC, Jaffee AE, Langmead B, Leek JT. Polyester: simulating RNA-seq datasets with differential transcript expression. Bioinformatics 2015;31(17):2778–84.

[73] Lin MF, Jungreis I, Kellis M. PhyloCSF: a comparative genomics method to distinguish protein coding and non-coding regions. Bioinformatics 2011;27(13):i275–82.

[74] Trapnell C, Williams BA, Pertea GM, et al. Transcript assembly and quantification by RNA-Seq reveals unannotated transcripts and isoform switching during cell differentiation. Nat Biotechnol 2010;28:511–5.

[75] Wu TD, Nacu S. Fast and SNP-tolerant detection of complex variants and splicing in short reads. Bioinformatics 2010;26(7):873–81.

[76] Wang K, Singh D, Zeng Z, et al. MapSplice: accurate mapping of RNA-seq reads for splice junction discovery. Nucleic Acids Res 2010;38(18):e178.

[77] Behr J, Kahles A, Zhong Y, Sreedharan VT, Drewe P, Ratsch G. MITIE: Simultaneous RNA-Seq-based transcript identification and quantification in multiple samples. Bioinformatics 2013;29(20):2529–38.

[78] Dao P, Numanagic I, Lin YY, et al. ORMAN: optimal resolution of ambiguous RNA-Seq multimap-pings in the presence of novel isoforms. Bioinformatics 2014;30(5):644–51.

[79] Guttman M, Garber M, Levin JZ, et al. Ab initio reconstruction of cell type-specific transcriptomes in mouse reveals the conserved multi-exonic structure of lincRNAs. Nat Biotechnol 2010;28(5):503–10.

[80] Dobin A, Davis CA, Schlesinger F, et al. STAR: ultrafast universal RNA-seq aligner. Bioinformatics 2013;29(1):15–21.

[81] Pertea M, Pertea GM, Antonescu CM, Chang TC, Mendell JT, Salzberg SL. StringTie enables improved reconstruction of a transcriptome from RNA-seq reads. Nat Biotechnol 2015;33(3):290–5.

[82] Trapnell C, Pachter L, Salzberg SL. TopHat: discovering splice junctions with RNA-Seq. Bioinformatics 2009;25(9):1105–11.

[83] Tomescu AI, Kuosmanen A, Rizzi R, Makinen V. A novel min-cost flow method for estimating tran-script expression with RNA-Seq. BMC Bioinformatics 2013;14(Suppl 5):S15.

[84] Grabherr MG, Haas BJ, Yassour M, et al. Full-length transcriptome assembly from RNA-Seq data without a reference genome. Nat Biotechnol 2011;29(7):644–52.

[85] Zerbino DR, Birney E. Velvet: algorithms for de novo short read assembly using de Bruijn graphs. Genome Res 2008;18(5):821–9.

CHAPTER 2

Histone modifications—models and mechanisms

C. Ann Blakey[1], Michael D. Litt[2]

[1]Department of Biology, Ball State University, Muncie, IN, USA; [2]Center for Medical Education, Indiana University–Ball State University, Muncie, IN, USA

Contents

1. INTRODUCTION

The term epigenetics roughly translates to "outside the field of genetics," and it was used by Waddington [1–5] to describe the interactions of both genes and environment in the development of the phenotype of an organism. Early clues to the heritable molecular mechanisms of the variegated cellular phenotypes of an organism were discovered within the chromatin of the cell. Currently, epigenetics might be best defined as "the study of mitotically and/or meiotically heritable changes in gene function that cannot be explained (solely) by changes in DNA sequence" [6]. Thus, it should be noted, the definition of the term has evolved with the depth of our knowledge of genetic mechanisms.

Epigenetic Gene Expression and Regulation
http://dx.doi.org/10.1016/B978-0-12-799958-6.00002-0

Regardless, the central question still remains, "How can a single fertilized egg give rise to a complex organism with cells of varied phenotypes?" [7]. It is this question that researchers in the field of epigenetics seek to address.

To gain a better understanding of the processes associated with chromatin dynamics, knowledge must first be gained regarding the specific factors and spatiotemporal events involved in active chromatin remodeling within regions of the overall structure. The *how* and *when* changes are fundamental to understanding the mechanisms that control alternations in the structure from a state refractory to gene transcription to one which is conducive to gene expression, and vice versa. One approach to these questions can be addressed by examining the primary hierarchical state, and therefore, the smallest structural unit of chromatin—the nucleosome—and the modification to the octamer of histone proteins at its structural core. This chapter focuses on histone modifications and their role in regulating gene expression via chromatin remodeling.

2. CHROMATIN

In eukaryotic cells, DNA exists within the nucleus in a highly compacted structure known as chromatin that includes both the nucleic acids and histone proteins. Chromatin structure can be roughly divided into three hierarchical states based on the degree of compaction: primary, secondary, and tertiary. The primary state was first described in human spermatozoa as *beads-on-a-string* under the electron microscope in 1979 [8] and 1980 [9]. The 11-nm diameter fiber that was observed consisted of repeating units of naked DNA [10] and nucleosomes [11] to form arrays of nucleosomes within the fiber [8,9,12]. The secondary level has been shown to involve the addition of linker histones, H1 or H5, and exhibit further condensation of the 11-nm fiber to form a 30-nm chromatin fiber through nucleosomal interactions [9,12–17]. However, the exact arrangement of the linker DNA and nucleosomes in the 11-nm fiber at the secondary level remains the subject of further investigation [9,12,13,15–19]. In the tertiary level of compaction, the chromatin is folded, resulting in inter- and intrachromosomal interactions of the 30- and 11-nm fibers [12,20–22]. The chromatin is a dynamic structure, changing its degrees of compaction and three-dimensional features to accommodate its regulatory effects, roles, or functions. It is within this chromatin scaffold that the fundamental nuclear processes of transcription, replication, and DNA repair occur [22–26].

The chromatin is a dynamic structure, changing its degrees of compaction and three-dimensional features to accommodate its regulatory effects, roles, or functions [27]. Two general states of chromatin have been studied extensively, heterochromatin and euchromatin. Heterochromatin is a highly compacted form of chromatin and typically lacks much, if any, degree of active gene transcription. It is distributed nonuniformly throughout the genome, with more extensive regions located near centromeres, telomeres, and gene desert regions. During meiosis and mitosis, DNA and histone form a typical conformation of chromatin, where the chromatin takes on a 100- to 300-fold level of compaction as compared to euchromatin [27]. Euchromatin is also distributed throughout the genome and is closely associated with active gene expression/regulation or gene rich regions [28]. Consequently, the regulation and expression of all genes are intricately linked to the fluid changes in chromatin structure. Thus, a central question of the biosciences is, "How are the changes in chromatin structure regulated?"

3. THE NUCLEOSOME

3.1 Nucleosome structure

The nucleosome, which has been crystallized [29], consists of roughly 147 bp of DNA wrapped approximately 1.65–1.7 times around a core protein octamer composed of two units of each of the histone proteins, H2A, H2B, H3, and H4 [24,29–32]. Histones have been highly conserved across all eukaryotic species with virtually no amino acid residue changes observed between the worm, the fly, the chicken, the mouse, and humans. That said, there are known variant forms of the histones which have been studied, for example yeast varies by only a few residues in H2A and H3, resulting in the H2A.Z and CenH3 forms, respectively [33–35]. This high degree of conservation reflects their overall importance in the regulation of cellular functions. A key regulatory feature of histone involves the levels and types of posttranslational modifications (PTMs). As early as 1960, Allfrey identified acetylated, phosphorylated, and methylated histones [36]. Histones were also among the first proteins shown to be ubiquitinated. Although Allfrey proposed that histone modifications played a role in RNA transcription, it was not until the 1990s that their association with gene regulation became clear [37–40]. As of 2015, a large number of different histone posttranslational modifications (HPTMs) have been identified.

In addition to acetylation, phosphorylation, methylation, and ubiquitination, the characterization of histone polypeptides and their residues indicates that they can be formylated, proprionylated, butyrylated, crotonylated, malonylated, succinylated, 5-hydroxylated, sumoylated, ADP-ribosylated, citrullinated, and glycosylated [41] (see Table 1). The roles each of these modifications plays in mediating chromatin functions and in regulating nuclear processes, such as transcription, DNA replication, and DNA repair, are still being deciphered [25,42].

3.2 Nucleosome structure and stability

The nucleosome poses several problems for DNA-based template processes. Transcription factors need to bind to DNA that is wrapped around the histone octamer, RNA polymerases need to move through the nucleosomes to transcribe RNA, and DNA polymerases need to negotiate the nucleosome during replication. Nonhistone factors are required to reposition the nucleosomes in order to clear regions of the DNA helix for the binding of regulatory factors. Several histone modifications have been shown to

Table 1 Major histone modifications, affected residue, and general description of mechanism(s) or property(ies). There are over 60 additional histone marks that have been identified to date, but their functional roles remain to be fully investigated

Modification	Basic aa and/or External			Neutral aa and/or Internal			Mechanism(s) and Property(ies)
	K	R		S	T	Y	
Citrullination		R					Chromatin structure; gene regulation
Methylation	K	R					Heterochromatin; gene regulation/ chromatin structure; gene silencing; DNA repair
Ubiquitination	K						Heterochromatin; gene regulation/ DNA repair
Sumoylation (similar to ubiquitination	K						Gene regulation
ADP-ribosylation	K						Heterochromatin; gene regulation; imparts a negative charge to modified residues; also can include glutamic acid residues
Formylation	K						Gene regulation
Proprionylation	K						Gene regulation; coenzyme A cofactor (butyryl-CoA); fine-tuned gene expression in response to metabolic needs, read by bromodomain-containing proteins
Butyrylation	K						Gene regulation; coenzyme A cofactor (propionyl-CoA); fine-tuned gene expression in response to metabolic needs, read by bromodomain-containing proteins
Crotonylation	K						Chromatin structure; gene regulation; highly conserved, linked to active transcription and gene action, linked to nucleosome disruption, enriched at gene promoters and predicted enhancers
Malonylation	K						Gene regulation
Acylation	K						Gene regulation; link between cellular metabolism and epigenetic mechanisms; neutralizes positive charge of lysine
Succinylation	K						Gene regulation
Acetylation	K			S★	T★	Y★	Gene regulation/chromatin structure; gene activation; low abundance★
0-GlcNAcylation	K			S★	T★	Y★	Chromatin structure; gene regulation; DNA repair; mitosis; sensor of nutrient status; occurs in cell cycle-dependent manner (mammals); mapped to gene promoters associated with metabolism, growth, development, stress, and aging★
Phosphorylation			H	S	T	Y	Gene regulation; DNA repair; Mitosis; imparts a negative charge to modified residues
Glycosylation				S	T		Gene regulation
Hydroxylation						Y	Gene regulation

Major table Refs [41,95,96].

alter the nucleosome structure and stability to assist in these processes. These modifications reside at critical junctures, such as the entry–exit point of the DNA on a nucleosome and the dyad axis of the nucleosome. Modifications can alter histone–DNA interaction, histone–histone interactions, and histone–chaperone interactions, and therefore, all have the capability of affecting nucleosome structure and stability.

Modifications of specific residues on the H3 histone, H3K56 and H3R42, located at the DNA entry–exit region of the nucleosome, can significantly alter the activity of the region. Hyperacetylation of H3K56 is associated with actively transcribed regions, whereas hypoacetylation is associated with transcriptionally silent regions [43]. Studies with the acetylation mimic H3K56Q have been shown to increase by sevenfold DNA breathing (the partial unwrapping of the DNA from the nucleosome core, allowing for binding factor access) [44], but did not affect in vitro DNA compaction [45]. These data are consistent with the role of acetylation in mediating the regulatory effects of H3K56 by enhancing the unwrapping of the DNA close to the DNA entry–exit site. Similarly, as seen with H3K56ac, the asymmetric dimethylation of H3R42 (H3R42me2) is also associated with active transcription [46].

4. HISTONES AND GLOBAL MECHANISMS
4.1 Histone modifications
4.1.1 *Histone acetylation*
The first histone modification discovered was histone acetylation [47]. Acetylation involves the transfer of an acetyl group from acetyl-CoA to histone lysine (K) ε-amino groups. Numerous studies immediately followed this discovery and suggested a relationship between actively transcribed genes and hyperacetylation, thus indicating a role for acetylation in facilitating gene transcription [36]. Several cellular processes, in addition to transcription, are now associated with the acetylation of histone lysine residues, including DNA replication and DNA repair [48-50]. As shown in Figure 1, the majority of modified lysines are located in the N-terminal tails of nucleosomal histones. It is important to keep in mind that histone tails are rapidly undergoing acetylation and deacetylation, with the average histone acetylation events having half-lives of only a few minutes [51]. This rapid modification turnover is mediated by histone acetyltransferases (HATs), which add acetyl groups to the histone tails, and by histone deacetylases (HDACs), which remove them. These HATs and HDACs often reside in large multisubunit complexes with multiple enzymatic activities. Each of these protein complexes has a coordinated function that both recognizes the specific region(s) of chromatin and performs the necessary modification/regulatory function.

The addition and removal of acetyl groups are very dynamic. Acetylation of the four core histones has a high turnover rate, with the average acetyl group having a

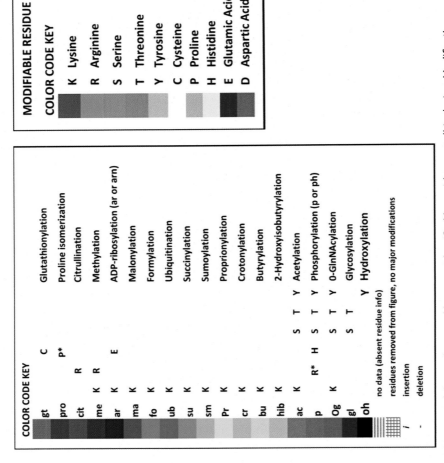

Figure 1 Major modifications of the nucleosomal histones identified by residue, not all-inclusive. Modifications are color-coded in descending order from primarily inactivating PTMs to those with more activating functional roles. The histones are presented in the following order: **H2A, H2B, H3, and H4** [34,41,93,96–106].

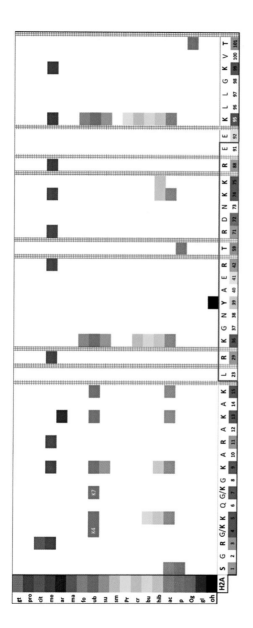

Histone H2A

Figure 1 Continued.

Histone H2B

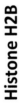
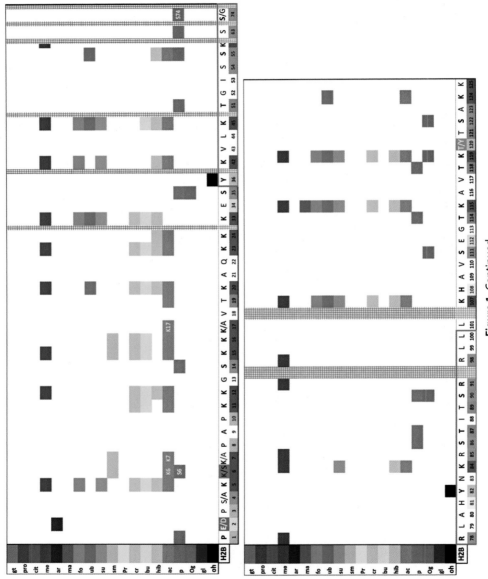

Figure 1 Continued.

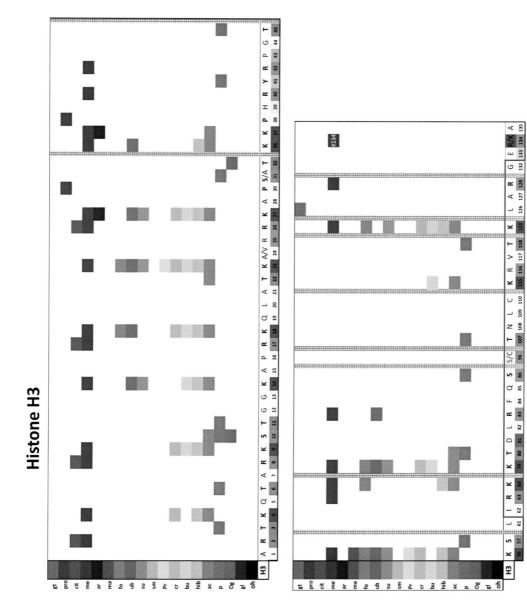

Figure 1 Continued.

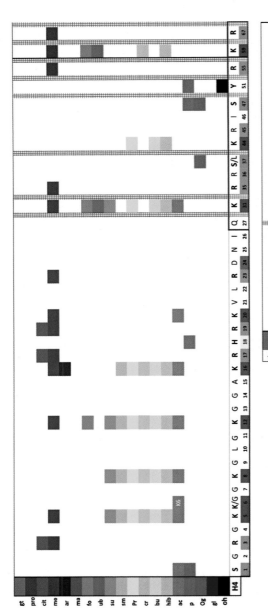

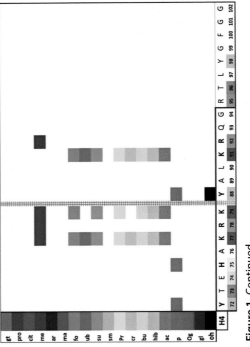

Histone H4

Figure 1 Continued.

half-life of less than 15 min [52]. This overall average turnover rate is actually a complex mixture of different half-lives, dependent on the histone and organism [51,53]. The bulk of histone acetylation has a very short half-life of only 2–3 min. Between 20 and 45% of histone acetylation has a significantly slower half-life of 30–40 min. These two distinct rates support two ideas by which histone modifications mediate their regulatory effectors on chromatin. For sites involving rapid exchange, it is not simply the presence of acetylation at actively transcribed genes that regulates transcription, but the rate of exchange of the modifications. This is supported by data linking histone acetylation and deacetylation and the process of nucleosome remodeling during active transcription [54]. For sites that turn over slowly, it is the basal level of histone acetylation that is important for recruiting factors and regulating transcription [51,55]. This complex dynamic exchange rate of histone acetylation is consistent with and complements the role of changes in nucleosome charge and recruitment of effector proteins in regulating gene transcription.

HATs can be separated into two major classes based on their catalytic domains: the GNAT (Gcn5 N-acetyltransferase) family and the MYST (Moz, Ybf2[Sas3], Sas2, and Tip60) family. The HATs p300/CBP represents third class of HATs whose catalytic domains show less conservation. The GNAT family targets histone H3, and the MYST family targets H4, while the p300/CBP complex targets both H3 and H4. HDACs can be separated into three distinct families based on their catalytic activity: Type I, Type II, and Type III. Type I and Type II show homology to the yeast enzymes Rpd3 and Hda1, respectively, and have similar deacetylation mechanisms that do not require a cofactor. Type III, the Sir2/sirtuin group, requires the NAD^+ cofactor to function. For deacetylation, Type I targets H3 and H2B substrates, Type II targets H3 and H4 substrates, and Type III targets H4K16ac and H3K56 [56].

In isolation, both the HATs and HDACs generally are unable to properly recognize and modify the chromatin substrate [57]. Several studies suggest that regulating transcription is dependent not only upon the extent of histone acetylation but also upon the acetylation of specific histone residues [58]. Therefore, understanding the state of histone modifications, as well as the role of HAT and HDAC complexes, is crucial for understanding the role of histone modifications chromatin in regulating cellular functions.

4.1.2 Histone methylation
Methylation of the histone residues lysine (K) and arginine (R) results in distinct modification states. Both residues can each present four distinct states: the unmodified state, as well as three distinct degrees of methylation. Lysine residues can be either mono-, di-, or trimethylated, whereas arginine residues can be either mono- or dimethylated symmetrically or asymmetrically. These modifications do not change the overall basic charge of the residues; however, they do alter the hydrophobic character and the overall size of the modified residues.

Different states of methylation are recognized by different effector proteins [59], and are associated with distinct regulatory functions. For example, genome-wide analysis of H3K36 methylation showed that mono- and dimethylated H3K36 occurred near the 5' region of gene bodies and that di- and trimethylated H3K36 occurred primarily at the 3' end of gene bodies [56,60]. Significantly of interest, several methyltransferases appear to act progressively at the ends of gene bodies, thus increasing the degree of methylation over time. This imparts a temporal role to the modifications, with trimethylation requiring more time than dimethylation, and dimethylation taking more time than monomethylation [61]. Consequently, the regulatory effects of the specific methylation state of a genic region may involve temporal controlling elements.

For arginine methylation, asymmetrical dimethylation of H3R2 (H3R2me2) has been shown to block the binding of certain effector protein complexes to H3K4me3, a modification associated with active transcription [62]. Conversely, symmetrical arginine dimethylation has been demonstrated to enhance the binding of certain effectors to H3K4me3 [63]. Therefore, not only is the type of PTMs important in regulating chromatin, but also is the distinct state of certain modifications involved in mediating the timing of those regulatory effects.

4.1.3 Histone phosphorylation

Similar to histone acetylation, histone phosphorylation is a dynamic process with rapid turnover. Isotopic pulse labeling studies indicate that histone phosphorylation has a half-life of roughly 30 min or 2–3 hours. In contrast to histone acetylation and phosphorylation, histone methylation has a relatively slow turnover rate. Instead of minutes and hours, histone methylation, depending on the target sight, has a half-life ranging from 0.297 to 4.809 days [51,64]. Of interest is the association of the target methylated residues with the longest half-lives, such as H3K27me3 (approximately 3 days), H4K20me3 (approximately 5 days), and transcriptionally silent chromatin. Marks associated with active gene transcription such as H3K4me1 (approximately 1 day) and H3K9me1 (~1/3 of a day) have among the shortest half-lives of histone methylation marks. These differences may reflect the establishment of different transcriptional states of the chromatin, with the slowest residue turnover sites of histone methylation establishing transcriptionally silent heterochromatin, and the fastest residue turnover sites poising chromatin for active transcription [64].

Mechanistically similar to acetylation, histone phosphorylation may directly regulate nucleosomes' dynamics by altering the charge of histones. Phosphorylation of serine, threonine, and tyrosine residues on histones imparts a negative charge. This can result in charge repulsion between the negatively charged DNA and the negatively charged residues. Early studies showed that the addition of unphosphorylated histone tails to nuclei protected the DNA form DNase I digestion more efficiently than the addition of phosphorylated histone tails [65], suggesting that charge repulsion on the histone tails made the nuclear DNA more accessible. Phosphorylation of H3 at T118, a site located in the DNA-histone interface near the dyad symmetry axis, reduced the DNA-histone binding free energy by 2 kcal/mole and increased nucleosome mobility by 28-fold. However, substitution of T118 with glutamic acid to impart a negative charge at the site did not result in changes in the DNA–histone binding free energy or nucleosome mobility [66], implying that not only the charge, but the type of moiety present was important mechanistically.

The transient nature of histone acetylation and phosphorylation argues against their role in directly maintaining stable gene expression. This differs from histone methylation with its long half-lives and high target specificity of a reader (discussed in Section 4.2); it is likely that this modification is directly involved in the establishment of district stable chromatin states that regulate gene expression.

4.1.4 Other large modifications

Where most histone modifications involve small chemical moieties such as acetyl, methyl, or phosphate, some modifications involve larger molecules. Both ubiquitination and sumoylation increase the size of histones by approximately 66%. Ubiquitination adds a 76-amino acid globular domain to the lysine residue, and sumoylation adds approximately 100 amino acids. The regulatory effects and changes in nucleosome dynamics of these modifications are diverse and complex.

Monoubiquitination of H2BK123 has been shown to increase nucleosome stability [67]. This increased stability might be mediated by the direct interaction of the DNA and the charged residues on the ubiquitin moiety [68]. For this modification, stabilization enforces gene repression by inhibiting H2A–H2B eviction of nucleosomes that are positioned over regulator regions, thereby interfering with transcriptional initiation [67]. The H2BK123 modification has also been shown to impair chromatin fiber compaction and increase chromatin accessibility [69]. This induction of an open accessible chromatin may explain H2B roles in the promotion of transcription elongation [70] and its importance in double-strand DNA repair [71].

In contrast to H2BK123ub1, monoubiquitination of H2AK119 (H2AK118ub1, in *Drosophila*) appears to inhibit transcriptional elongation via an indirect mechanism. H2A monoubiquitination by a PRC1-like complex generates a positive feedback loop by creating a binding site for the Jarid2-Aebp2-Polycomb Repressive Complex 2 (PRC2). This recruitment of the PRC2, in turn, promotes H3K27 trimethylation and results in a repressive chromatin state, as well as further PRC1 binding to monoubiquitinated H2A [72].

Histone sumoylation, which is usually associated with transcriptional repression, occurs on all four core histones. Like ubiquitination, it probably influences nucleosome dynamics both directly and indirectly. Modification of H4K12 by SUMO-3 may mediate transcriptional repression by interfering with chromatin compaction. The effects of this modification are more pronounced on long-range internucleosomal interactions by the destabilization of compacted dinucleosomes [73]. This suggests that sumoylation might directly interfere with long-range promoter/enhancer interactions. For indirect mechanisms, there are numerous proteins that interact with sumoylated histones, which are capable of regulating chromatin. However, these interactions still need to be further investigated to elucidate their mechanisms.

Like histone phosphorylation, ADP ribosylation imparts a negative charge on the modified residue. Chromatin fractions enriched for ADP ribosylation are more sensitive to nuclease digestion than chromatin fractions depleted of ADP ribosylation [74]. Sedimentation velocity studies indicated that poly (ADP-ribosylated) chromatin was less condensed than chromatin depleted of ADP ribosylation [75]. Furthermore, ADP-ribosylated histone tails on chromatin were more accessible to antibody binding [76]. This suggests that ADP ribosylation may mediate its regulatory effects by relaxing DNA-histone interactions, possibly via a charge repulsion mechanism. Clearly, several HPTMs mechanistically mediate their regulator effects on chromatin by manipulating the charge of histones.

4.2 Readers, writers, erasers

Histone modification readers are recognizing and interpreting their target sites in a complex chromatin landscape. Hence, several mechanisms are involved in the mediation of their regulatory effects. The reader can interact with both *cis* and *trans*. For *cis*, readers interact with PTMs on a single tail. For *trans*, readers interact with PTMs intranucleosomally on different tails or interact with PTMs internucleosomally. For both *cis* and *trans* readers, interactions with a target PTM can be influenced by adjacent PTMs.

Presently, a large number of readers have been discovered and characterized. There exists a multitude of readers for nearly every HPTM. Some readers, the most highly characterized, are capable of recognizing multiple modifications, such as histone acetylation, histone methylation, and histone phosphorylation. At least four distinct domains recognize histone acetylation. These include the bromodomain, the double bromodomain (DBD), the double PHD finger (DPF), and the double pleckstrin homology domain (Double PH). For histone lysine methylation, there are more than a dozen different domains currently characterized. Among these domains are ATRX-DNMT3-DNMT3L (ADD), ankyrin, bromo-adjacent homology (BAH), chromo barrel, chromodomain, double chromodomain (DCD), malignant brain tumor (MBT), plant homeodomain (PHD), Pro-Trp-Trp-Pro (PWWP), tandem Tudor domain (TTD), Tudor, WD40, and the zinc finger CW (zf-CW). The ADD, Tudor, and WD40 domains have also been shown to bind histone-methylated arginines. Histone phosphoserine or phosphothreonine sites are recognized by the tandem BRCT domain, the 14-3-3ζ domain, and the BIR domain [59]. The number and variety of domains capable of recognizing HPTM provides a complex signaling network of proteins that are able to regulate many DNA template processes. It is important to note that these domains of readers and writers frequently reside in multisubunit proteins complexes, containing several different readers, catalytical domains, and scaffolding domains. Together, these mechanisms allow the readers and writers to navigate a complex chromatin environment to implement regulatory responses.

There are several ways reader and writer domains can interact with HPTMs. Depending on the context, these domains can bind and distinguish the state of a PTM and interact with a specific single PTM, several or multiple specific single PTMs, a set or group of PTMs. For example, a high degree of specificity for K4me3 is observed with the PHD finger [77]. The bromodomain is fairly promiscuous, being capable of recognizing numerous acetylated lysine sites on H3, H4, H2A, and H2B [78]. The WD40 domain also recognizes a group of modifications, interacting with H3K27me3, H3K9me3, H4K20me3, and H1K26me3 [79]. As for multiple modified sites, the DBD of TAF1 shows a higher affinity for the double acetylated peptides H4K5acK8ac and H4K8acK16ac than the monoacetylated peptides [80]. In the case of the state of methylation, the TDRD3 Tudor domain recognizes asymmetrically methylated arginines [81] (H4R3me2a), whereas the WD40 domain of WDR5 prefers symmetrically methylated arginines [79]. Similar specificity for mono-, di-, and trimethylation of lysine is also observed [59].

This is only a small set of examples. However, from this, one can see that these domains are capable of interacting and distinguishing a vast array of different HPTMs. The result is a signaling network of HPTMs that act like the tumblers in a lock, the lock and key model, where complexes of proteins recognize specific sets of HPTMs to be targeted to unique sites on the DNA to mediate their regulatory effects.

4.3 Regulatory modules

Posttranslational modifications (PTMs) that do not affect the histone charge act less directly to regulate chromatin architecture. One model proposes that some of these modifications mediate their effects as regulatory modules. In this case, the interactions of PTM with an effector protein modify its properties. Several studies of histone methylation support this mechanism. Histone tails containing trimethylation of histone H3 at lysine 4 (H3K4me3) appear to trigger allosteric inhibition of the PRC2 HMTase activity, limiting its enzymatic processivity [82]. Methylation of histone H3 at lysine 36 (H3K36me) is necessary for the HDAC Rpd3S activity. Furthermore, the modification was thought to recruit Rpd3S. However, it was found that Rpd3S was in fact recruited by phosphorylation of the RNA polymerase II complex. This suggests that H3K36me functions by altering the catalytic activity of Rpd3S [83]. Other PTMs can also act and function as regulatory modules. Ubiquitinated H2A (H2Aub) specifically inhibits the enzymatic activity of H3K36-specific methyltransferases ASH1L, NSD1, NSD2, and HYPB [84]. Therefore, some histone modifications may not be involved in directly recruiting effector proteins; rather, their presence alters the catalytic properties of effector proteins, which are subsequently recruited by other factors or modifications. Consequently, regulator modules are another route among the molecular mechanisms by which PTMs are capable of modifying the regulatory properties of chromatin.

4.4 The models

Frequently, two basic conceptual models are considered when examining the mechanisms of chromatin regulation by HPTMs. These models can be divided into two categories: *cis*- and *trans*-acting effects. Mechanistically, the histone modifications discussed previously also appear to regulate chromatin both directly (Model 1) and indirectly (Model 2) see Figure 2.

4.4.1 Conceptual model 1 - HTPM chromatin regulation, cis effects

In this Model 1, HPTMs regulate chromatin via *cis* effects. Here, the modifications directly change the chromatin, possibly by altering the charge of the histone residues such that the DNA is more or less accessible to transcriptional machinery (such as with

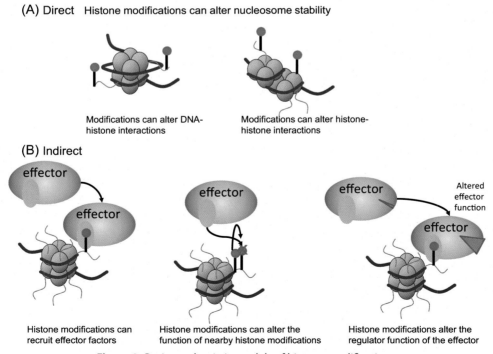

Figure 2 Basic mechanistic models of histone modifications.

the DNA breathing in acetylation). For example, hyperacetylated nucleosomes show an average increase of 1.4-fold in exposure of nucleosomal DNA target sites [85]. Or, the modifications can result in an altered state of the chromatin by mediating changes in the secondary and tertiary chromatin configuration (such as the positive feedback loop mechanism of PRC1/PRC2 in ubiquitination and methylation and of H2A and H3), which are conducive or inhibitory to gene expression.

One *cis* mechanism by which PTMs mediate their regulatory effects on chromatin involves changing the charge of histone tails. Several histone modifications are capable of altering the net charge of histones. Acetylation and acylation can neutralize the positive charge of modified lysines, whereas phosphorylation and ADP-ribosylation impart a negative charge to modified residues [86]. The charge neutralization model involves a *cis*-acting mechanism by which PTMs regulate chromatin by altering the charge of histone tails. Several studies have shown that alterations in the charge of histone tails change the DNA-binding properties of histones and nucleosomes. In vitro studies showed that acetylated peptides bind to DNA significantly more strongly than unacetylated H4 terminal tail peptides, and hyperacetylated nucleosome had a lower affinity for DNA than unacetylated nucleosomes [85,87].

These studies imply that charge neutralization of the histone tails makes the DNA more accessible for regulatory factors, such as transcription factors and transcription machinery. In vivo studies suggest that it is not the specific lysine residue modified, but the accumulation of charge that is associated with changes in gene expression. In the yeast site, directed mutagenesis of lysine residues at 5, 8, 12, and 16 on the H4 tail showed that only the absence of all four lysines generated a phenotype change. One modifiable lysine in any of the four positions did not generate a phenotype [88]. Microarray analysis of gene expression changes in yeast associated with different sites of H4 lysine modification further support the charge neutralization model. In this study, all 15 possible combinations of H4 lysine modification at sites 5, 8, 12, and 16 were examined. For sites 5, 8, and 12, changes in overall gene expression correlated not with the specific site of modification, but with the accumulation of charged sites. Lysine 16 showed site specificity, but only in less than 10% of the genes [89]. Mutational studies of modifications on H3 at lysines residues 4, 9, 14, 18, 23, and 27 showed redundant roles regulating subtelomeric gene repression in yeast [90]. Together, these studies suggest that the accumulated patterns of charge changes in histone tails are more influential in regulating chromatin than the specific site of modification on the tail [91].

4.4.2 Conceptual model 2 - HTPM chromatin regulation, trans effects

In this model, HPTMs regulate chromatin via *trans* effects. Model 2 demonstrates the inhibitory effects of HPTMs. Here, the HPTMs can demonstrate either an inhibitory or positive effect on gene expression. Thus, histone modifications repress the binding of accessory factors during inhibitory effects [92]. Alternatively, histone modifications can actively recruit the binding of accessory factor. For example, the methylation of H3 at lysine 9 (H3K9) recruits the binding HP1, resulting in the formation of heterochromatin [93].

Consistent with the *trans* model effects, histone modifications can act to recruit regulatory complexes to distinct loci. A variety of different protein domains are capable of recognizing specific PTMs. These effectors have been termed readers. The bromodomain of the HAT p/CAF (p300/CBP associated factors, also known as PCAF in some literature), which recognizes acetylated lysines, was the first histone reader discovered. Experiments on p/CAF showed that the bromodomain can contribute to localized acetylation by tethering itself to specific acetylated chromosomal sites [94]. This provided proof in the support of these *trans* effects as a critical mechanism for how histone HPTMs regulate chromatin.

It is important to consider each of these models with respect to our growing understanding of the role of chromatin in the regulation of nuclear processes and the ever growing number of histone modifications that fine-tune these processes.

5. SUMMARY

As illustrated above, the structure of chromatin can be greatly influenced by the presence or absence of different histone modifications. These modifications can, in turn, have a tremendous effect on the regulatory mechanisms at work on the chromatin, as well as on gene expression. The reading, writing, and erasing of histone modifications are highly dynamic processes with multiple protein complexes involved in definitively unique and/or more generalized interactions with kinetic variance being observed. The half-life of a specific histone modification will depend on both the site of the modification as well as the type of modification. This implies that the temporal fixation of specific gene expression patterns is influenced by both the type and target of posttranslational histone modifications.

Thus, we are brought to the conclusion that as we learn more regarding histone modifications and plethora of protein complexes associated with fine-scale interactions with the chromatin, we are just beginning to understand the depth and robustness of epigenetic control of gene expression.

ACKNOWLEDGMENTS

This work is supported by the Department of Biology and the College of Sciences and Humanities, Ball State University, as well as the Center for Medical Education, Indiana University-Ball State University, Muncie, IN, USA.

REFERENCES

[1] Waddington CH. Organisers and genes. Cambridge: Cambridge Univ Press; 1940.
[2] Waddington CH. Canalization of development and the inheritance of acquired characters. Nature 1942;150:563–4.

[3] Waddington CH. Principles of embryology. New York: MacMillan Co; 1956.

[4] Waddington CH. The strategy of the genes; a discussion of some aspects of theoretical biology. London: Allen & Unwin; 1957.

[5] Waddington CH. Canalization of development and genetic assimilation of acquired characters. Nature 1959;183(4676):1654–5.

[6] Russo VEA, Martienssen RA, Riggs AD. Epigenetic mechanisms of gene regulation. Plainview, NY: Cold Spring Harbor Laboratory Press; 1996.

[7] Felsenfeld G. A brief history of epigenetics. Cold Spring Harbor Perspect Biol 2014;6(1).

[8] Wagner TE, Yun JS. Fine structure of human sperm chromatin. Arch Androl 1979;2(4):291–4.

[9] Gusse M, Chevaillier P. Electron microscope evidence for the presence of globular structures in different sperm chromatins. J Cell Biol 1980;87:280–4.

[10] Oudet P, Gross-Bellard M, Chambon P. Electron microscopic and biochemical evidence that chromatin structure is a repeating unit. Cell 1975;4:281–300.

[11] Olins AL, Olins DE. Spheroid chromatin units (v bodies). Science 1974;183(4122):330–2.

[12] Woodcock CL, Dimitrov S. Higher-order structure of chromatin and chromosomes. Curr Opin Genet Dev 2001;11:130–5.

[13] Felsenfeld G, McGhee JD. Structure of the 30 nm chromatin fiber. Cell 1986;44:375–7.

[14] Robinson PJ, Rhodes D. Structure of the '30 nm' chromatin fibre: a key role for the linker histone. Curr Opin Struct Biol 2006;16(3):336–43.

[15] Tremethick DJ. Higher-order structures of chromatin: the elusive 30 nm fiber. Cell 2007;128(4): 651–4.

[16] Grigoryev SA, Arya G, Correll S, Woodcock CL, Schlick T. Evidence for heteromorphic chromatin fibers from analysis of nucleosome interactions. Proc Natl Acad Sci USA 2009;106(32):13317–22.

[17] Grigoryev SA, Woodcock CL. Chromatin organization - the 30 nm fiber. Exp Cell Res 2012;318(12): 1448–55.

[18] Robinson PJ, Fairall L, Huynh VA, Rhodes D. EM measurements define the dimensions of the "30-nm" chromatin fiber: evidence for a compact, interdigitated structure. Proc Natl Acad Sci USA 2006;103(17): 6506–11.

[19] Maeshima K, Hihara S, Eltsov M. Chromatin structure: does the 30-nm fibre exist in vivo? Curr Opin Cell Biol 2010;22(3):291–7.

[20] Rattner JB, Hamkalo BA. Higher order structure in metaphase chromosomes I. The 250A fiber. Chromosoma 1978;69:363–72.

[21] Rattner JB, Hamkalo BA. Higher order structure in metaphase chromosomes II. The relationship between the 250A fiber, superbeads and beads-on-a-string. Chromosoma 1978;69:373–9.

[22] Bian Q, Belmont AS. Revisiting higher-order and large-scale chromatin organization. Curr Opin Cell Biol 2012;24(3):359–66.

[23] Zlatanova J, Leuba SH. Chromatin fibers, one-at-a-time. J Mol Biol 2003;331(1):1–19.

[24] Luger K, Hansen JC. Nucleosome and chromatin fiber dynamics. Curr Opin Struct Biol 2005;15(2):188–96.

[25] Chakravarthy S, Park YJ, Chodaparambil J, Edayathumangalam RS, Luger K. Structure and dynamic properties of nucleosome core particles. FEBS Lett 2005;579(4):895–8.

[26] Woodcock CL. Chromatin architecture. Curr Opin Struct Biol 2006;16(2):213–20.

[27] Felsenfeld G, Groudine M. Controlling the double helix. Nature 2003;421(6921):448–53.

[28] Messner S, Hottiger MO. Histone ADP-ribosylation in DNA repair, replication and transcription. Trends Cell Biol 2011;21(9):534–42.

[29] Luger K, Mader AW, Richmond RK, Sargent DF, Richmond TJ. Crystal structure of the nucleosome core particle at 2.8A resolution. Nature 1997;389(6648):251–60.

[30] Weintraub H, Worcel A, Alberts B. A model for chromatin based upon two symmetrically paired half-nucleosomes. Cell 1976;9:409–17.

[31] Richmond TJ, Davey CA. The structure of DNA in the nucleosome core. Nature 2003;423(6936): 145–50.

[32] Luger K. Structure and dynamic behavior of nucleosomes. Curr Opin Genet Dev 2003;13(2):127–35.

[33] Kamakaka RT, Biggins S. Histone variants: deviants? Genes Dev 2005;19(3):295–310.

[34] Campos EI, Reinberg D. Histones: annotating chromatin. Annu Rev Genet 2009;43:559–99.

[35] Eriksson PR, Ganguli D, Nagarajavel V, Clark DJ. Regulation of histone gene expression in budding yeast. Genetics 2012;191(1):7–20.

[36] Allfrey VG, Faulkner R, Mirsky AE. Acetylation and methylation of histones and their possible role in the regulation of RNA synthesis. Proc Natl Acad Sci USA 1964;51:786–94.

[37] Arany Z, Sellers WR, Livingston DM, Eckner R. E1A-Associated p300 and CREB-Associated CBP belong to a conserved family of coactivators. Cell 1994;77:799–800.

[38] Mizzen CA, Yang X-J, Kokubo T, et al. The TAFII250 subunit of TFIID has histone acetyltransferase activity. Cell 1996;87(7):1261–70.

[39] Brownell JE, Zhou J, Ranalli T, et al. *Tetrahymena* histone acetyltransferase a: a homolog of yeast Gcn5p linking histone acetylation to gene activation. Cell 1996;84:843–51.

[40] Blanco JCG, Minucci S, Lu J, et al. The histone acetylase PCAF is a nuclear receptor coactivator. Genes Dev 1998;12:1638–51.

[41] Rothbart SB, Strahl BD. Interpreting the language of histone and DNA modifications. Biochim Biophys Acta 2014;1839(8):627–643.

[42] Adkins NL, Niu H, Sung P, Peterson CL. Nucleosome dynamics regulates DNA processing. Nat Struct Mol Biol 2013;20(7):836–42.

[43] Xu F, Zhang Q, Zhang K, Xie W, Grunstein M. Sir2 deacetylates histone H3 lysine 56 to regulate telomeric heterochromatin structure in yeast. Mol Cell 2007;27(6):890–900.

[44] Ngo TT, Ha T. Nucleosomes undergo slow spontaneous gaping. Nucleic Acids Res 2015;43(8):3964–71.

[45] Neumann H, Hancock SM, Buning R, et al. A method for genetically installing site-specific acetylation in recombinant histones defines the effects of H3 K56 acetylation. Mol Cell 2009;36(1):153–63.

[46] Casadioa F, Lub X, Pollockc SB, et al. H3R42me2a is a histone modification with positive transcriptional effects. Proc Natl Acad Sci USA 2013;110(37):14894–9.

[47] Phillips DM. The presence of acetyl groups of histones. Biochem J 1963;87:258–63.

[48] Unnikrishnan A, Gafken PR, Tsukiyama T. Dynamic changes in histone acetylation regulate origins of DNA replication. Nature structural & molecular biology 2010;17(4):430–7.

[49] Groth A, Rocha W, Verreault A, Almouzni G. Chromatin challenges during DNA replication and repair. Cell 2007;128(4):721–33.

[50] Wurtele H, Kaiser GS, Bacal J, et al. Histone H3 lysine 56 acetylation and the response to DNA replication fork damage. Mol Cell Biol 2012;32(1):154–72.

[51] Barth TK, Imhof A. Fast signals and slow marks: the dynamics of histone modifications. Trends Biochem Sci 2010;35(11):618–26.

[52] Jackson V, Shires A, Chalkley R, Granner DK. Studies on highly metabolically active acetylation and phosphorylation of histones. J Biol Chem 1975;250(13):4856–63.

[53] Waterborg JH. Dynamics of histone acetylation in vivo. A function for acetylation turnover? Biochem Cell Biol 2002;80(3):363–78.

[54] Reinke H, Gregory PD, Horz W. A transient histone hyperacetylation signal marks nucleosomes for remodeling at the *PH08* promoter in vivo. Mol Cell 2001;7:529–38.

[55] Clayton AL, Hazzalin CA, Mahadevan LC. Enhanced histone acetylation and transcription: a dynamic perspective. Mol Cell 2006;23(3):289–96.

[56] Pokholok DK, Harbison CT, Levine S, et al. Genome-wide map of nucleosome acetylation and methylation in yeast. Cell 2005;122(4):517–27.

[57] Luo Y, Jian W, Stavreva D, et al. Trans-regulation of histone deacetylase activities through acetylation. J Biol Chem 2009;284(50):34901–10.

[58] Agalioti T, Chen G, Thanos D. Deciphering the transcriptional histone acetylation code for a human gene. Cell 2002;111:381–92.

[59] Musselman CA, Lalonde ME, Cote J, Kutateladze TG. Perceiving the epigenetic landscape through histone readers. Nat Struct Mol Biol 2012;19(12):1218–27.

[60] Rao B, Shibata Y, Strahl BD, Lieb JD. Dimethylation of histone H3 at lysine 36 demarcates regulatory and nonregulatory chromatin genome-wide. Mol Cell Biol 2005;25(21):9447–59.

[61] Patnaik D, Chin HG, Esteve PO, Benner J, Jacobsen SE, Pradhan S. Substrate specificity and kinetic mechanism of mammalian G9a histone H3 methyltransferase. J Biol Chem 2004;279(51):53248–58.

[62] Iberg AN, Espejo A, Cheng D, et al. Arginine methylation of the histone H3 tail impedes effector binding. J Biol Chem 2008;283(6):3006–10.

[63] Yuan CC, Matthews AG, Jin Y, et al. Histone H3R2 symmetric dimethylation and histone H3K4 trimethylation are tightly correlated in eukaryotic genomes. Cell Rep 2012;1(2):83–90.

[64] Zee BM, Levin RS, Xu B, LeRoy G, Wingreen NS, Garcia BA. In vivo residue-specific histone methylation dynamics. J Biol Chem 2010;285(5):3341–50.

[65] Mirsky AE, Silverman B. Blocking by histones of accessibility to DNA in chromatin. Proc Natl Acad Sci USA 1972;69(8):2115–9.

[66] Banerjee T, Chakravarti D. A peek into the complex realm of histone phosphorylation. Mol Cell Biol 2011;31(24):4858–73.

[67] Chandrasekharan MB, Huang F, Sun ZW. Ubiquitination of histone H2B regulates chromatin dynamics by enhancing nucleosome stability. Proc Natl Acad Sci USA 2009;106(39):16686–91.

[68] Vijay-Kumar S, Bugg CE, Wilkinson KD, Vierstra RD, Hatfield PM, Cook WJ. Comparison of the three-dimensional structures of human, yeast, and oat ubiquitin. J Biol Chem 1987;262(13):6396–9.

[69] Fierz B, Chatterjee C, McGinty RK, Bar-Dagan M, Raleigh DP, Muir TW. Histone H2B ubiquitylation disrupts local and higher-order chromatin compaction. Nat Chem Biol 2011;7(2):113–9.

[70] Batta K, Zhang Z, Yen K, Goffman DB, Pugh BF. Genome-wide function of H2B ubiquitylation in promoter and genic regions. Genes Dev 2011;25(21):2254–65.

[71] Moyal L, Lerenthal Y, Gana-Weisz M, et al. Requirement of ATM-dependent monoubiquitylation of histone H2B for timely repair of DNA double-strand breaks. Mol Cell 2011;41(5):529–42.

[72] Kalb R, Latwiel S, Baymaz HI, et al. Histone H2A monoubiquitination promotes histone H3 methylation in polycomb repression. Nat Struct Mol Biol 2014;21(6):569–71.

[73] Dhall A, Wei S, Fierz B, Woodcock CL, Lee TH, Chatterjee C. Sumoylated human histone H4 prevents chromatin compaction by inhibiting long-range internucleosomal interactions. J Biol Chem 2014;289(49):33827–37.

[74] Hough CJ, Smulson ME. Association of poly(adenosine diphosphate ribosylated) nucleosomes with transcriptionally active and inactive regions of chromatin. Biochemistry 1984;23(21):5016–23.

[75] de Murcia G, Huletsky A, Lamarre D, et al. Modulation of chromatin superstructure induced by poly(ADP-ribose) synthesis and degradation. J Biol Chem 1986;261(15):7011–7.

[76] Huletsky A, de Murcia G, Muller S, et al. The effect of poly(ADP-ribosyl)ation on native and H1-depleted chromatin. A role of poly(ADP-ribosyl)ation on core nucleosome structure. J Biol Chem 1989;264(15):8878–86.

[77] Musselman CA, Kutateladze TG. Handpicking epigenetic marks with PHD fingers. Nucleic Acids Res 2011;39(21):9061–71.

[78] Filippakopoulos P, Picaud S, Mangos M, et al. Histone recognition and large-scale structural analysis of the human bromodomain family. Cell 2012;149(1):214–31.

[79] Margueron R, Justin N, Ohno K, et al. Role of the polycomb protein EED in the propagation of repressive histone marks. Nature 2009;461(7265):762–7.

[80] Jacobson RH, Ladurner AG, King DS, Tjian R. Structure and function of a human TAFII250 double bromodomain module. Science 2000;288(5470):1422–5.

[81] Yang Y, Lu Y, Espejo A, et al. TDRD3 is an effector molecule for arginine-methylated histone marks. Mol Cell 2010;40(6):1016–23.

[82] Schmitges FW, Prusty AB, Faty M, et al. Histone methylation by PRC2 is inhibited by active chromatin marks. Mol Cell 2011;42(3):330–41.

[83] Drouin S, Laramee L, Jacques PE, Forest A, Bergeron M, Robert F. DSIF and RNA polymerase II CTD phosphorylation coordinate the recruitment of Rpd3S to actively transcribed genes. PLoS Genet 2010;6(10):e1001173.

[84] Sun L, Luo H, Bu D, et al. Utilizing sequence intrinsic composition to classify protein-coding and long non-coding transcripts. Nucleic Acids Res 2013;41(17):e166.

[85] Anderson JD, Lowary PT, Widom J. Effects of histone acetylation on the equilibrium accessibility of nucleosomal DNA target sites. J Mol Biol 2001;307(4):977–85.

[86] Rousseaux S, Khochbin S. Histone acylation beyond acetylation: terra incognita in chromatin biology. Cell J 2015;17(1):1–6.

[87] Hong L, Schroth GP, Matthews HR, Yau P, Bradbury EM. Studies of the DNA binding properties of histone H4 amino terminus. Thermal denaturation studies reveal that acetylation markedly reduces the binding constant of the H4 "tail" to DNA. J Biol Chem 1993;268(1):305–14.

[88] Megee PC, Morgan BA, Smith MM. Histone H4 and the maintenance of genome integrity. Genes Dev 1995;9(14):1716–27.

[89] Dion MF, Altschuler SJ, Wu LF, Rando OJ. Genomic characterization reveals a simple histone H4 acetylation code. Proc Natl Acad Sci USA 2005;102(15):5501–6.

[90] Martin AM, Pouchnik DJ, Walker JL, Wyrick JJ. Redundant roles for histone H3 N-terminal lysine residues in subtelomeric gene repression in *Saccharomyces cerevisiae*. Genetics 2004;167(3):1123–32.

[91] Kasinathan S, Orsi GA, Zentner GE, Ahmad K, Henikoff S. High-resolution mapping of transcription factor binding sites on native chromatin. Nat Methods 2014;11(2):203–9.

[92] Fischle W, Tseng BS, Dormann HL, et al. Regulation of HP1-chromatin binding by histone H3 methylation and phosphorylation. Nature 2005;438(7071):1116–22.

[93] Cheutin T, McNairn AJ, Jenuwein T, Gilbert DM, Singh PB, Misteli T. Maintenance of stable heterochromatin domains by dynamic HP1 binding. Science 2003;299(5607):721–5.

[94] Dhalluin C, Carlson JE, Zeng L, He C, Aggarwal AK, Zhou MM. Structure and ligand of a histone acetyltransferase bromodomain. Nature 1999;399(6735):491–6.

[95] Tan M, Luo H, Lee S, et al. Identification of 67 histone marks and histone lysine crotonylation as a new type of histone modification. Cell 2011;146(6):1016–28.

[96] Lauberth SM, Nakayama T, Wu X, et al. H3K4me3 interactions with TAF3 regulate preinitiation complex assembly and selective gene activation. Cell 2013;152(5):1021–36.

[97] Kouzarides T. Chromatin modifications and their function. Cell 2007;128(4):693–705.

[98] Magnani L, Eeckhoute J, Lupien M. Pioneer factors: directing transcriptional regulators within the chromatin environment. Trends Genet 2011;27(11):465–74.

[99] Wang X, Bai L, Bryant GO, Ptashne M. Nucleosomes and the accessibility problem. Trends Genet 2011;27(12):487–92.

[100] Shindo Y, Nozaki T, Saito R, Tomita M. Computational analysis of associations between alternative splicing and histone modifications. FEBS Lett 2013;587(5):516–21.

[101] Harmston N, Baresic A, Lenhard B. The mystery of extreme non-coding conservation. Philos Trans R Soc Lond B Biol Sci 2013;368(1632):20130021.

[102] Arnaudo AM, Garcia BA. Proteomic characterization of novel histone post-translational modifications. Epigenet Chromatin 2013;6:24–30.

[103] Capell BC, Berger SL. Genome-wide epigenetics. J Invest Dermatol 2013;133(6):e9.

[104] Fischle W. Molecular mechanisms of histone modification function. Biochim Biophys Acta 2014;1839(8):621–2.

[105] Huang H, Sabari BR, Garcia BA, Allis CD, Zhao Y. SnapShot: histone modifications. Cell 2014;159(2):458–e1.

[106] Brehove M, Wang T, North J, et al. Histone core phosphorylation regulates DNA accessibility. J Biol Chem 2015; Papers in Press - Manuscript M115.661363 (Published July 13, 2015).

CHAPTER 3

Genomic imprinting in mammals— memories of generations past

Nora Engel

Fels Institute for Cancer Research, Temple University School of Medicine, Philadelphia, PA, USA

Contents

1. INTRODUCTION

Mammals are diploid organisms consisting of two parental genomes that are functionally distinct. The maternal and paternal genomes are different in their potential for gene expression because of sex-specific specializations that are imposed on the DNA during oogenesis and spermatogenesis. Additional differences may occur after fertilization, before the two genomes are united. These differences are determined by *epigenetic* mechanisms, i.e., reversible chemical and structural modifications that are applied to the DNA and can affect the transcriptional competence of genes, without altering the DNA sequence *per se*. Many of the epigenetic asymmetries between the maternal and paternal genomes are erased during early embryogenesis, restoring the ability of most genes to be expressed from both parental genomes. However, there is a group of genes that have an unusual behavior: they retain different expression potentials after fertilization, with the two parental copies displaying differential effects on the phenotype. These are the so-called imprinted genes.

Findings from several independent lines of investigation during the 1980s and 1990s converged to support the hypothesis that genes could be consistently inactive when

Epigenetic Gene Expression and Regulation
http://dx.doi.org/10.1016/B978-0-12-799958-6.00003-2

derived from one parent in mammals. The first clues came from nuclear transplantation studies in which embryos with two maternal or two paternal genomes were unable to develop [1,2]. Genes responsible for this phenomenon were initially mapped in mice with partial or whole chromosomal uniparental disomies [3]. Discovery of imprinted X inactivation in murine placentas confirmed that parental-specific genomic effects existed in mammals [4]. Parent-of-origin effects were also apparent in certain genetic diseases in which entire chromosomes or chromosomal regions originated from a single parent (reviewed in Ref. [5]). Finally, identification of the first imprinted genes gave definitive proof of maternal- and paternal-specific expression potential [6–9]. (For a comprehensive history of the discovery of the phenomenon of genomic imprinting see Ref. [10]). To date, approximately 120 imprinted genes have been identified in both humans and mice, most of them clustered into 16 domains (for current catalogs of imprinted genes, see *igc.otago.ac.nz; mousebook.org; and* geneimprint.com). Since the focus of imprinting studies has been on transcribed units, it is not known if sequences with other genomic functions (or with no function whatsoever) are imprinted.

2. THE LIFE CYCLE OF IMPRINTED REGIONS

Genomic imprinting is the process by which DNA sequences, including genes or regulatory regions, are labeled epigenetically as paternal or maternal, allowing them to be distinguished during development. By mapping the imprinting process onto the life cycle of the mammalian organism, several logical steps can be discerned (Figure 1). During gametogenesis, when male or female gametes are produced, the genome undergoes sex-specific epigenetic modifications that in turn direct the cells to differentiate into either oocytes or sperm. This stage is conceivably the best opportunity for imprints to be established independently on the parental genomes. After fertilization and before the two genomes begin to coexist within a single nucleus, each pronucleus undergoes distinct reprogramming processes. For example, in the sperm, protamines are replaced by histones, DNA demethylation occurs by an active process, and it is possible that access of transcription factors to each genome may be different (see Ref. [11] and references therein). In theory, this stage, in which the genomes are still in separate compartments, also allows for differential modifications, although there is no evidence for this as yet. Elegant experiments with mouse androgenetic and gynogenetic embryos showed that intergenomic transactions or counting mechanisms to inactivate different parental alleles after fertilization were unlikely [12]. Most of the epigenetic differences are erased in the ensuing cell divisions, equalizing the two parental genomes and conferring pluripotency on the embryo. Imprints, however, are retained in all or most cell types, depending on the individual gene [13], and continue to distinguish the chromosomes originally derived from the sperm and oocyte throughout development. However, imprints are erased in the primordial germ cells, the embryonic precursors to sperm and oocytes. Shortly after implantation, these cells are sequestered in the soma [14,15] and undergo the erasure of

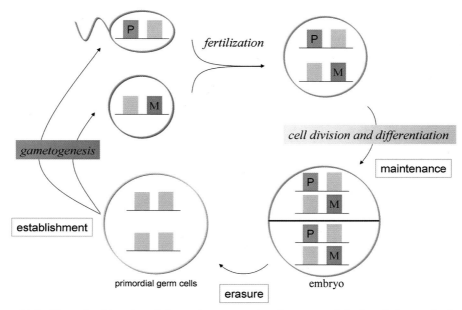

Figure 1 The life cycle of imprints. Schematic representing the stages of the imprinting cycle through fertilization, embryogenesis, and gametogenesis. The P and M represent the paternal and maternal imprints, respectively, within each genome. Processes of imprint maintenance, erasure, and establishment are indicated.

their inherited paternal and maternal imprints. Imprinting is subsequently reestablished based solely on the sex of the developing embryo [16–18], thereby setting up the next generation of heritable parental imprints from each sex. To summarize, mammals inherit two genomes, one maternally and another paternally imprinted, but they only transmit one imprinting state to their progeny, depending on whether they are male or female.

This cycle of events underlies the three main requirements for gene imprints: they must be (1) stable, (2) heritable through mitosis, and (3) reversible. It also underscores that the establishment of imprints in each gamete involves sex-specific mechanisms: in the sperm, specific DNA sequences must be recognized and marked somehow whereas in the oocyte, a different set of sequences is selected for labeling. Conventionally, the silenced allele is the one designated as imprinted. It is likely that identification of sequences to be imprinted in the gametes is a separate mechanism from the actual marking. The molecular identity of the mark is discussed below.

It is important to emphasize that the context in which imprinting occurs in each germ line is completely different: whereas oocytes are imprinted while in meiotic arrest and postnatally, male gametes are imprinted before meiosis, during a phase of prenatal mitotic arrest. Figure 2 depicts the timeline of murine gametogenesis as representative of this process in mammals.

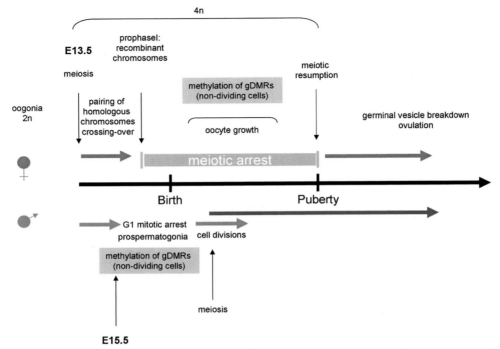

Figure 2 Timeline of gametogenesis for male and female mice, indicating acquisition of gamete-specific methylation (gDMRs) for each germ line. Processes occurring during this stage are depicted above the timeline for female gametogenesis and below the timeline for male gametogenesis. E13.5 and 15.5 indicate days postcoitum in murine embryogenesis. Note that meiosis initiates prenatally for females and postnatally for males. The methylation of gametic differentially methylated regions (gDMRs) occurs in nondividing cells for both germlines, but in different developmental contexts: for females, gDMRs are established postnatally during meiotic arrest; for males, establishment of gDMRs occurs in mitotically arrested cells.

3. MECHANISMS OF GENOMIC IMPRINTING

The epigenetics field is very much the legacy of the study of genomic imprinting, since investigation into the molecular mechanisms of monoallelic expression yielded a wealth of information on heritable regulatory mechanisms. In addition, and together with the study of X inactivation, genomic imprinting as a model for transcriptional regulation generated discoveries such as the role of noncoding macroRNAs in silencing, the domain-wide effects of *cis*-acting sequences, and the existence of mammalian insulators. Elucidation of the mechanisms of genomic imprinting may also have a direct bearing on random monoallelic expression and genotype-dependent expression differences.

Imprinting at genes gives rise to a difference in expression levels between the parental alleles, ranging from decreased RNA abundance to complete repression of one allele relative to the other. Since both the active and the inactive copies of imprinted genes

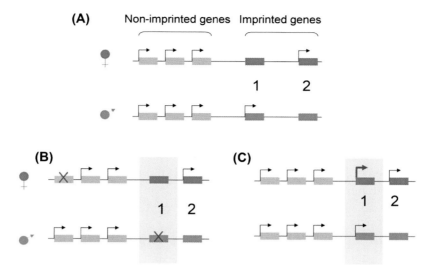

Figure 3 Imprinted genes and potential disease mechanisms. (A) Maternal and paternal genomes are represented, with nonimprinted genes in gray. Actively transcribed genes exhibit broken arrows to indicate RNA production. Gene 1 is a maternally silent gene and is only expressed from the paternal copy, whereas Gene 2 is paternally silent and maternally expressed. (B) Large X represents inactivating mutation. When mutation occurs on a nonimprinted gene (gray boxes), there is still an active copy on the other parental chromosome. If imprinted Gene 1 is mutated on the paternal copy, normally the sole expressed allele, there is complete loss of the gene product. (C) If the maternal copy of Gene 1 is activated (represented by red arrow) by a mutation or epimutation, such as loss of methylation, there is now double dosage of the gene product.

coexist in the same environment—i.e., within reach of appropriate transcription factors— the epigenetic differences between the alleles of imprinted genes or their regulatory sequences must account for the parental differences or biases in transcriptional status (Figure 3(A)). In fact, DNA methylation and histone modifications, typically characterized as "epigenetic," do distinguish the parental alleles, in the same way that they distinguish the status of a gene that switches from active to repressed (or vice versa) during development or in response to external cues [19,20].

But what is the primary mark of an imprinted gene and how does it differ from other developmentally dynamic marks? A common feature of imprinted genes, both in humans and mice, is the presence of DNA sequences that are consistently methylated on one parental allele and preserved throughout cell divisions [21]. In seven imprinted domains, these **d**ifferentially **m**ethylated **r**egions, or DMRs, have been shown to have regulatory functions by targeted mutations in the mouse, and are also critical for parental-specific monoallelic expression of the genes in the domain [22–29]. These DMRs have been designated **i**mprinting **c**ontrol **r**egions (ICRs). Methylation of *cis*-regulatory sequences can exclude binding of proteins to their recognition sites, with an outcome on gene expression dependent on the role of the sequence element in the activation or repression of transcription. Indeed, DNA methylation at ICRs associated with imprinted genes has the dual role of retaining a memory of parental origin and, if located in a region involved in transcriptional regulation, of affecting gene expression. Accordingly, the terms "imprint" and "DNA methylation mark" are sometimes used synonymously.

Methylation of discrete DNA elements is established in a gamete-specific manner by *de novo* DNA methyltransferases, *Dnmt3a* and *Dnmt3l*, which are both expressed in the germ line [30,31]. Deletion of these enzymes in mouse models disrupts establishment of imprints. Some gametic methylation marks are undoubtedly contingent on the phenotype of the specialized cells that the genomes inhabit, and most are not maintained after fertilization, at least in oocytes [32], whereas methylation imprints subsist throughout development. How are imprints targeted to specific sequences? This is one of the most intriguing questions in the field. Since the enzymes responsible for *de novo*, and maintenance methylation of imprinted regions during gametogenesis also exert genome-wide methylation in somatic tissues and have no sequence specificity, we are obviously missing a crucial piece of the puzzle. One possibility is that there is no sequence-specificity or defining feature of imprinted sequences and that methylation occurs by default during gametogenesis at loci that are accessible to the methyltransferases, naturally different in each sex. Another possibility is that there are one or more factors that recognize imprinted genes and function as labels recognized by methyltransferases, and that somehow collaborate to tighten the mark. A less satisfying but possible alternative is that each imprinted region may be ruled by a unique mechanism. Interestingly, differences in methylation acquisition between alleles have been documented, where the allele that was methylated previously was more readily remethylated than the unmethylated one, at least in sperm [33]. This suggests an unidentified distinguishing factor or conformation that facilitates DNA methylation and is inherited from the embryonic cells that give rise to the germ line. Thus, silencing of imprinted genes may not accurately model tissue-specific gene repression.

Studies at the *Rasgrf1* locus have yielded evidence that interactions between a noncoding RNA (ncRNA), produced by the ICR, and the PIWI machinery help recruit *Dnmt3a* and *Dnmt3l* and are required for methylation [34]. An ncRNA molecule is likely an ideal vehicle for monoallelic targeting of epigenetic enzymes to a single locus, because it has the sequence specificity and, if it only acts in *cis*, the locus specificity as well [35]. In this case, as in many others, there is still no mechanistic definition of "recruitment," a word commonly used to indicate the black box of how nonspecific epigenetic enzymes can target their specific DNA binding sites.

During preimplantation, there is a transition from the highly specialized genomes of the germ cells to totipotency, with loss of methylation and reconfiguration of histones in many regions. Only imprinted regions conserve their marks until passage through the germ line erases them. This brings up the question of how methylation at imprinted regions resists this massive genomic reorganization. The protection of some, but not all, methylation marks at DMRs during phases of reprogramming requires *Pgc/Stella* [36] and *Zfp57* [37], although there is still no molecular insight into exactly what "protection" is. Also, it remains to be seen if other proteins or histone modifications somehow protect the methylation at imprinted genes.

Somatic DNA methylation patterns are conserved because of the semireplicative mode of action of the enzymatic machinery involved (for detailed reviews on DNA methyltransferases and their biochemical mechanisms, see Refs [38,39]). Specifically, deletion of *Dnmt1* leads to deregulation of all known imprinted genes and generalized genomic hypomethylation [40]. Although, in theory, chromatin modifications could be imprinting marks, histone modifications have not been shown to be epigenetic in the strictest sense of the word, i.e., there is no known enzymatic machinery that heritably interprets and copies modified histones to newly synthesized DNA strands. Chromatin features of the active and inactive alleles of imprinted genes do follow the pattern expected of transcribed and silent sequences [41], but seem to be a consequence of the transcriptional status, as shown for the *H19* domain.

To summarize, there is considerable overlap between the life cycle of imprints and the dynamics of DNA methylation during development [42]. However, methylation at imprinted genes is not erased when genome-wide methylation loss occurs, but rather at a different developmental stage and only in germ cells—the cells that ensure the continuity of the life cycle. This suggests that DNA methylation *per se* is not enough to distinguish imprinted from nonimprinted regions.

4. TOP-DOWN DISSECTION OF IMPRINTED DOMAINS

It would be very enlightening to be able to put together elements piecemeal to generate a synthetic imprinted gene or region. Yet efforts to dissect existing imprinted domains have not afforded us a full understanding of how DNA sequences confer imprinting. Transgenes harboring intact and mutated versions of several imprinted regions have been the closest to finding minimal sequences required to establish imprinting. For example, a 16-kb transgene, harboring the mouse *H19* gene, required both upstream and downstream flanking sequences and recapitulated imprinting when present in low copy numbers [43]. However, deletion and mutation experiments at the endogenous locus have as yet failed to disrupt establishment of paternal methylation of the *H19* DMR, indicating that the puzzle has not been completely solved [22,44–46]. Interestingly, 10 copies of a transgene carrying a portion of the Rous sarcoma virus long terminal repeat, the *Igα* heavy chain gene, and the mouse *c-myc* gene recapitulated the imprinting cycle to some degree: 50% of the progeny of female transgenic mice had the same hypermethylation pattern as the mother [47]. Whether or not this faithfully mimics the imprinting process is still not clear.

Based on bioinformatic inspection and comparative studies of DMRs, it has been proposed that CpG islands, certain repetitive sequences, and transposons are common elements of imprinted regions [48–50]. Nevertheless there is currently no single feature of imprinted genes or their surrounding sequences that distinguish them from nonimprinted genes, and no single DNA sequence has imparted imprinting on a nonimprinted gene.

5. TRANSCRIPTIONAL INTERPRETATION OF THE IMPRINT

Although genetic processes other than transcription may be affected by parent-of-origin effects, most of the studies in the field have focused on the transcriptional outcome of imprinting. To date, all genetically characterized ICRs are regulatory sequences, with methylation acting as an on–off switch by impeding access of proteins to the recognition sites therein. The resulting effect on transcription depends on the nature of the regulatory sequence and, in some cases, is subject to stage- or tissue-specific events, i.e., the DNA methylation imprint may not be manifested as monoallelic expression until cell-type-specific factors are present. In addition, the degree of silencing of the methylated copy varies between different domains, different genes within a domain, different cell types, and even between different individuals.

To date, two types of DNA elements are involved in controlling transcriptional outcomes at imprinted loci: promoters and insulators. In the first case, a promoter is active on one unmethylated copy, but inactive on its methylated counterpart. All the examples involving this mechanism are maternally methylated and paternally unmethylated, and these constitute the majority of known imprinted genes. The two known outcomes to this scenario are (1) silencing of coding genes by ncRNAs and (2) interrupted translation of coding genes by antisense transcription.

In the first case, in the *Igf2r* and *Kcnq1* imprinted domains, an antisense ncRNA RNA is produced from the paternally unmethylated promoter [51,52] (Figure 4(A)). These are unusually large ncRNAs and are often called macroRNAs to distinguish them from the "long" ncRNAs (lncRNAs). Transcription of the paternally expressed 108-kb *Airn* and the 92-kb *Kcnq1ot1* results in *cis*-silencing of some of their neighboring protein-coding genes. On the maternal allele, methylation represses the activity of the promoter of the macroRNA, relieving the surrounding genes from inactivation. Control over several neighboring genes by an ncRNA has also been shown for the *Gnas* and *Snrpn* domains [53,54]. The second case, illustrated by the *H13/Mcts2* domain, these two genes overlap in their coding region, but there is differential methylation of the *Mcts2* promoter, such that transcription of *Mcts2* on the unmethylated allele interferes with the production of the full *H13* RNA, which translates into an inactive peptide [55]. The same mechanism has been observed at *Herc3/Nap1/5* [56].

How does transcription of ncRNAs silence the sense counterparts, and even flanking genes? X inactivation by the Xist ncRNA in females, although not antisense to a protein-coding gene, is one paradigm for repression, with spreading along the chromosome and recruitment of chromatin-condensing factors [57]. *Airn* and *Kcnq1ot1*

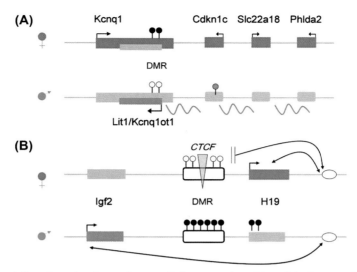

Figure 4 Transcriptional mechanisms of monoallelic expression at imprinted genes. Active genes are represented with an arrow indicating direction of transcription and inactive genes lack arrows. (A) A partial schematic of the *Kcnq1* domain, with the differentially methylated region, DMR, showing maternal methylation (filled lollipops) in an intron of the *Kcnq1* gene. On the paternal allele, the same region is unmethylated, and gives rise to a noncoding RNA (*Lit1/Kcnq1ot1*), represented as a wavy line, that acts in *cis* as a repressor of neighboring genes. Thus, *Cdkn1c*, *Slc22a18*, *Phlda2* are only expressed from the maternal copies. (B) Representation of the *H19/Igf2* imprinted region. In this case, the paternal chromosome carries a methylation mark upstream of the *H19* gene, in a sequence that on the unmethylated maternal allele can bind *CTCF* to establish a genomic insulator (gray inverted triangle). The insulator blocks *Igf2* from access to the enhancers downstream of the *H19* gene (ovals). Thus, *Igf2* is only active on the paternal copy.

do not neatly fit this model though, and there is evidence that the ncRNA molecule *per se* is not required for silencing to occur [58,59]. At least for *Airn*, one interpretation is that, since the RNA is transcribed in the antisense direction, there is transcriptional interference with the *Igf2r* promoter. Another possibility is that the promoter of the ncRNA, when active, is an insulator, impeding access of the *Igf2r* promoter to downstream enhancers required for full expression. We do not know how the promoters of the ncRNAs escape silencing themselves—perhaps these promoters are "stronger," or germline factors preestablished there counteract the effects of the RNA.

It has been difficult to account for tissue- or stage-specific imprinting, especially since reports analyzing imprinted gene expression typically limit their studies to a subset of tissues or stages, and there is little overlap between them. The *Kcnq1* domain has been a useful model to study stage-specific loss of imprinting. There are several similarities between the *Kcnq1* and *Igf2r* domains. The ncRNA emerges from an intron of *Kcnq1* in an antisense direction. The *Kcnq1ot1* molecule or its promoter region is required for repression of the upstream *Cdkn1c*, *Slc22a18*, and *Phlda2* genes [26]. However, there is a stage-specific pattern of monoallelic expression early in embryogenesis for the sense *Kcnq1* gene, with a transition to biallelic expression by midgestation [60]. Our studies suggest that enhancers can compete with ICRs for local gene-specific regulation, overcoming the domain-wide control of imprinted expression [61].

A large group of imprinted genes regulated by *Kcnq1ot1* are monoallelic only in placenta. Although important for their role in regulating growth of the embryo, the mechanisms may be distinct and the specific features of extraembryonic tissues (for example, the presence of polyploid cells and the intermingling of different cell types and origins) may be irrelevant to the long-term maintenance mechanisms in the embryo proper.

A second and unique mechanism has been found at the *H19/Igf2* domain and involves a genomic insulator that binds the CCCTC-binding factor on the unmethylated maternal allele [62,63]. This interferes in the communication between the *Igf2* promoter and specific enhancers shared with *H19*, abrogating maternal *Igf2* expression from that allele, and also has the effect of activating the *H19* promoter. The paternally methylated version of the insulator cannot bind *Ctcf*, thus allowing contact between the *Igf2* promoter and the enhancers and resulting in active paternal expression (Figure 4(B)). The *H19* DMR is the only genetically defined insulator in mammals and has been a prime model for mechanistic studies on these elements [64–66].

It is interesting to note that in the four currently known cases of imprinted domains with paternal methylation, the ICR is located in intergenic regions, which may be meaningful somehow for chromatin dynamics during spermatogenesis. In contrast, in the female germ line, methylation is placed on promoters, resulting in maternal repression of transcription. There is evidence that these maternally repressed genes are actively transcribed during oogenesis, and transcription has been suggested to be a requirement for methylation [67]. Although this seems rather contradictory, it is possible that this ensures that the germ cell program is not activated after fertilization. Perhaps there is an oocyte-specific mechanism of repression for ncRNAs such as *Kcnq1ot1* that linger in their sites of transcription.

The examples above do not preclude the existence of other mechanisms whereby methylation can lead to monoallelic or biased expression. For example, in the case of predominantly intergenic imprints, other regulatory elements, in addition to insulators, such as enhancers and silencers, will likely be involved in imprinted gene expression. Posttranscriptional mechanisms could also be involved. Interestingly, in most imprinted domains, the transcriptional consequences of methylation of a single DNA element are far-reaching, affecting clusters of genes, and have provided important mechanistic insights into long-range regulation in general.

6. IMPRINTING AND HUMAN DISEASE

Clinical management of genetic diseases requires considering whether they are biallelic or monoallelic, and of the latter, whether the copy that is silenced is random or parent-of-origin dependent. Familial diseases due to imprinting defects present an unusual inheritance pattern, which can be misleading. Further, the fact that in all or most expressing cell types, imprinted genes are only active from one allele means that any perturbation in their expression is dominant. Indeed, proper imprinted gene regulation is vital and its alteration can lead to both genetic diseases and cancer. A comprehensive list of some of the best described imprinting disorders can be found in several recent reviews [68].

Imprinting diseases can be caused by a variety of mechanisms, all of which inflict an alteration of dosage of imprinted genes, with either doubling of the normal amount of protein or no protein expression at all (Figure 3(B and C)) [69]. Among other etiologies, there are chromosomal duplications, deletions, and uniparental disomies, in which the two copies of a chromosome have a single parental origin; additionally, there are cases due to failure to establish, erase, or maintain imprints. These diseases, though rare, are severe, since they affect genes regulating early development. In addition, loss of imprinting resulting from aberrant DNA methylation states can occur in later stages, and much effort is being invested to elucidate its effects in aging and cancer.

Since many imprinted genes regulate embryonic growth, it is not surprising that mutations or epimutations cause overgrowth or undergrowth syndromes. Examples are Beckwith–Wiedemann syndrome (BWS) and Silver–Russell syndrome (SRS) [70]. BWS is frequently associated with loss of imprinting at the *Kcnq1ot1* ICR, leading to silencing of both alleles of *Cdkn1c*, a gene encoding a cell cycle inhibitor. Predisposition to several pediatric cancers is characteristic of these patients. On the other hand, SRS patients exhibit the opposite phenotype, with a smaller than average size throughout development, among other symptoms. Most of these patients exhibit hypomethylation of the *H19* ICR, accompanied by biallelic repression of *Igf2*.

Elegant studies by Keverne and colleagues predicted that some imprinted genes have an important role in brain development [71]. The fate of androgenetic (duplicated paternal genome) and parthenogenetic/gynogenetic (duplicated maternal genome) cells in chimeric embryos was asymmetric in their distribution in the brain. Androgenetic cells were more abundant in the hypothalamus, whereas parthenogenetic cells contributed more to the cortex, striatum, and hippocampus. In fact, patients with reciprocal parental disruption of the *Snrpn* imprinted domain exhibit a variety of neurodevelopmental and behavioral impairments. Angelman syndrome, characterized by severe intellectual delay, speech deficits, and seizures, is caused by genetic and epigenetic mutations that ablate maternal expression of *Ube3A* [72]. On the other hand, Prader–Willi syndrome patients exhibit hypothalamic dysfunction, obesity, and cognitive delay and are characterized by similar anomalies affecting the paternally expressed genes in the *Snrpn* region [73].

Replicating mutations found in patients with growth disorders in mouse models has been very enlightening in understanding imprinting mechanisms. Interestingly, one mutant mouse model at the *H19* DMR anticipated the description of patients with similar mutations, although only the growth phenotype was reproduced [74]. Behavioral disorders are more difficult to model in mice, although some success has been achieved with parental-specific knockouts of *Ube3A*, *Peg1*, and *Peg3*, with recapitulation of some analogous phenotypes.

Lab tests for imprinting diseases range from cytogenetics to molecular or biochemical assays and reflect the frequency of the types of mutations or epimutations for each disorder. For example, for Angelman syndrome, methylation tests will detect more than 75% of cases; but if these are not informative, FISH analysis, uniparental disomy studies for chromosome 15, and sequencing of *Ube3A* will follow sequentially. For BWS, methylation analysis for the *Kcnq1ot1* and *H19* ICRs is the first choice, detecting more than 70% of cases. Additionally, assays to distinguish diseases with overlapping symptoms, carrier testing and/or prenatal diagnosis may be required.

7. FINDING NEW IMPRINTED GENES

Fine mapping of human and mouse traits or diseases exhibiting parent-of-origin bias has been the traditional way of identifying novel imprinted genes, but new sequencing technologies have accelerated those efforts. Parental-specific DMRs have been identified by genome-wide profiling of sodium bisulfite-treated DNA from human and mouse tissues and have uncovered associated genes with parental expression biases [75,76]. Transcriptome-wide approaches detecting parent-of-origin expression biases have also yielded new imprinted genes [77,78]. Genome-wide methods for imprinted gene discovery require complicated bioinformatics analysis and ultimately, validation by independent methods at the bench. All in all, and as originally predicted, the number of imprinted genes is likely not more than 150–200.

8. THE EVOLUTION OF IMPRINTING

8.1 The how

The methylation machinery involved in silencing imprinted genes is essentially the same one that the genome employs to repress transposons and retroviral elements. Based on this observation, some investigators proposed that imprinting had taken over the genome defense system in response to a specific selective pressure [79,80]. Of course, the methylation differences between male and female gametes could be explained by the asymmetric developmental circumstances between oogenesis and spermatogenesis [81].

Although imprinted genes have a high density of repetitive sequences, their presence alone does not seem to be a determinant of imprinting acquisition. Comparative studies of imprinting in eutherians and marsupials have uncovered at least one region—the *Dlk1* domain, imprinted in mammals—in which accumulation of repeats in the marsupial domain did not result in imprinting [82]. Therefore, additional details on how these genes are recognized are still obscure.

8.2 The why

Most of what we have learned about imprinting relates to the mechanisms that result in parent-of-origin dependent monoallelic expression, although it should be noted that there are likely to be other consequences of differential methylation that are unrelated to gene expression [83]. Thus, many hypotheses on why imprinting exists in mammals are based on the phenotypes resulting from disruption of imprinted genes and the information it provides on imprinted gene function [84]. In particular, there has been a focus on what advantages imprinting could bestow that would outweigh the benefits of diploidy, which masks recessive alleles and provides a backup in the case of mutations on one allele.

Establishment and maintenance of imprints is needed for viability and healthy embryogenesis. As mentioned above, many imprinted genes are expressed during development, both in the embryo and placenta. In fact, the appearance of imprinting in mammals goes hand in hand with the emergence of the placenta as an intermediary between the embryo and the mother. Studies from mouse knockouts and from patients affected by imprinting disorders showed that maternally expressed genes are usually growth inhibitors, whereas paternally expressed genes tend to promote growth. In both cases, the gene products can act directly in the embryo or indirectly by influencing placental function. For example, *Igf2r*, a decoy receptor for the paternally expressed fetal growth factor *Igf2*, is maternally expressed. *Grb10* is also a maternally expressed growth inhibitor. Examples of paternally expressed genes that enhance embryonic growth are *Igf2*, *Rasgrf1*, and *Dlk1*. Some imprinted genes required for placental health and that indirectly affect embryonic development are *Ascl2* and *Phlda2*.

The kinship theory [85] is the most widely cited and posits that there is a conflict between parental interests for allocation of maternal resources during embryogenesis and in early postnatal development. Thus, the paternally expressed imprinted genes favor embryonic growth, increasing the fitness of an individual bearing the paternal genome. In contrast, maternally expressed imprinted genes suppress growth, heightening the fitness of the mother, and allowing greater distribution of the maternal genome. Some of the imprinted genes fit the predictions of this hypothesis, but it is interesting to note that not all mammals have the same repertoire of imprinted genes, suggesting that each domain acquired this regulatory system at different times.

Another interesting idea put forth is the "ovarian time bomb" hypothesis [86], in which imprinting is proposed to be a surveillance mechanism to avoid parthenogenetic events that could lead to ovarian teratomas, or other aneuploidies that could trigger cancerous growth. A logical consequence for this scenario, in which growth inhibitors are maternally expressed, could be that the paternal counterparts were to be downregulated as a compensatory mechanism.

The maternal–fetal coadaptation hypothesis postulates that, in the presence of allelic variability of genes involved in mother–offspring interactions, equalizing the alleles expressed from both parts can avoid incompatibilities arising from mismatched alleles [87]. Support for this view was recently provided by experiments targeting the *Grb10* gene, a growth inhibitor expressed maternally in lactating females and in fetal and neonatal mice [88]. Deletion of *Grb10* in the offspring results in overgrowth, whereas the opposite effect is observed upon deletion in the mother, i.e., delivery of smaller than normal mice. Strikingly, deletion of *Grb10* in both mother and progeny produced normally sized mice, strongly suggesting that this cooperative interaction was either selected for after the appearance of imprinting or it drove its acquisition [89].

Several authors have proposed that imprinting was derived from the X-inactivation mechanism [90,91], a hypothesis that makes sense of many of the molecular links between the two, and comparative genomics of dosage compensation mechanisms are expected to provide more information on this possible association.

Pardo-Manuel de Villena et al. take a completely different approach by proposing that selection is directed toward the distinction between parental homologues and sister chromatids during meiosis, so that recombination occurs only between chromosomes coming from different germ cells [92]. This would predict that sex-specific differential modifications will eventually be found on the chromosomes on which no imprinted regions have been reported (three, four, and eight in mice), perhaps because gene expression has not been affected. Further, imprinted regions are predicted to be hotspots for recombination [93,94].

Based on the currently cataloged imprinted genes, most of the gametic methylation marks at DMRs affecting gene expression originate in oocytes. Most of the imprinted genes are clustered. Imprinted genes also tend to have shorter introns [95]. However, until we have a complete picture of imprinted regions, it may be premature to predict that these features are specifically relevant to imprinting evolution.

None of the ideas put forth until now explain all the features of genomic imprinting, i.e., differences in imprinting or in the direction of expression bias between species or the restriction of transcriptional imprinting to placental mammals and angiosperms, but most of them have some support from theoretical modeling and address a subset of imprinted genes. Some of the caveats raised by models might be due to incomplete data or because imprinting is still in the process of evolving. Because some genes are imprinted in one species and not others, it is possible that different imprinted domains arose as a result of different selective pressures (as predicted by Latham) [13], assuming that the processes leading to an initial imprinting polymorphism are relatively simple and arise frequently. If there is no global model for the function of imprinting and the answers are unique to each individual domain, we still have a long way to go.

Several questions remain open: (1) How can we explain that in several clusters, some genes are paternally while others are maternally expressed, and yet others are biallelic? (2) Why are there more female than male germ line imprinted genes? (3) Whose fitness is increased by imprinting—the mother, the father, or the progeny? (4) Is imprinting still evolving, or has it reached an equilibrium in some species?

9. CONCLUSION

To summarize, imprinted genes remember their history and sexual origin, and epigenetic marks such as DNA methylation function to preserve this memory after fertilization. The study of genomic imprinting has been foundational for the epigenetics field in providing a unique model for transcriptional regulation. Challenges that remain are to understand and explain tissue-specific imprinting, to determine genetic effects that are unrelated to gene expression, and to elucidate how specific sequences are targeted for imprint establishment. Why imprinting exists at all is still under debate, but undoubtedly its existence mandates that both maternal and paternal genomes are required for proper development. Further, disruption of imprinting by both mutations and epimutations poses significant risks to human health and understanding of this phenomenon is vital for clinical management of genetic diseases.

ACKNOWLEDGMENTS

I am grateful to the members of my lab, Bryant Schultz and Gwen Galiccio, for their suggestions, and to Joanne Thorvaldsen and Carmen Sapienza for discussions on the ideas presented here. I apologize to the numerous scientists whose contributions to the field I could not cite because of limits on the number of references.

REFERENCES

[1] McGrath J, Solter D. Completion of mouse embryogenesis requires both the maternal and paternal genomes. Cell 1984;37:179–83.
[2] Surani MAH, Barton SC, Norris ML. Development of reconstituted mouse eggs suggest imprinting of the genome during gametogenesis. Nature 1984;308:548–50.
[3] Cattanach BM, Kirk M. Differential activity of maternally and paternally derived chromosome regions in mice. Nature 1985;315:496–8.
[4] Takagi N, Sasaki M. Preferential inactivation of the paternally derived X chromosome in the extraembryonic membranes of the mouse. Nature 1975;256(5519):640–2.
[5] Ishida M, Moore GE. The role of imprinted genes in humans. Mol Aspects Med 2013;34(4):826–40.
[6] Barlow DP, Stoger R, Herrmann BG, Saito K, Schweifer N. The mouse insulin-like growth factor type-2 receptor is imprinted and closely linked to the *Tme* locus. Nature 1991;349:84–7.
[7] Bartolomei MS, Zemel S, Tilghman SM. Parental imprinting of the mouse *H19* gene. Nature 1991;351:153–5.
[8] Ferguson-Smith AC, Cattanach BM, Barton SC, Beechey CV, Surani MA. Embryological and molecular investigations of parental imprinting on mouse chromosome 7. Nature 1991;351:667–70.
[9] DeChiara TM, Robertson EJ, Efstratiadis A. Parental imprinting of the mouse insulin-like growth factor II gene. Cell 1991;64(4):849–59.
[10] Barlow DP, Bartolomei MS. Genomic imprinting in mammals. Cold Spring Harb Perspect Biol 2014;6(2).
[11] Sasaki H, Matsui Y. Epigenetic events in mammalian germ-cell development: reprogramming and beyond. Nat Rev Genet 2008;9(2):129–40.
[12] Szabo PE, Mann JR. Biallelic expression of imprinted genes in the mouse germ line: implications for erasure, establishment, and mechanisms of genomic imprinting. Genes Dev 1995;9:1857–68.
[13] Latham KE. Stage-specific and cell type-specific aspects of genomic imprinting effects in mammals. Differentiation 1995;59(5):269–82.

[14] Ginsburg M, Snow MH, McLaren A. Primordial germ cells in the mouse embryo during gastrulation. Development 1990;110(2):521–8.

[15] Lawson KA, Hage WJ. Clonal analysis of the origin of primordial germ cells in the mouse. Ciba Found Symp 1994;182:68–84. discussion-91.

[16] Lucifero D, Mertineit C, Clarke HJ, Bestor TH, Trasler JM. Methylation dynamics of imprinted genes in mouse germ cells. Genomics 2002;79(4):530–8.

[17] Davis TL, Trasler JM, Moss SB, Yang GJ, Bartolomei MS. Acquisition of the *H19* methylation imprint occurs differentially on the parental alleles during spermatogenesis. Genomics 1999;58:18–28.

[18] Ueda T, Abe K, Miura A, et al. The paternal methylation imprint of the mouse H19 locus is acquired in the gonocyte stage during foetal testis development. Genes Cells 2000;5(8):649–59.

[19] McEwen KR, Ferguson-Smith AC. Distinguishing epigenetic marks of developmental and imprinting regulation. Epigenetics Chromatin 2010;3(1):2.

[20] Barlow DP. Genomic imprinting: a mammalian epigenetic discovery model. Annu Rev Genet 2011;45:379–403.

[21] Tremblay KD, Duran KL, Bartolomei MS. A 5' 2-kilobase-pair region of the imprinted mouse *H19* gene exhibits exclusive paternal methylation throughout development. Mol Cell Biol 1997;17:4322–9.

[22] Thorvaldsen JL, Duran KL, Bartolomei MS. Deletion of the *H19* differentially methylated domain results in loss of imprinted expression of *H19* and *Igf2*. Genes Dev 1998;12:3693–702.

[23] Wutz A, Smrzka OW, Schweifer N, Schellander K, Wagner EF, Barlow DP. Imprinted expression of the *Igf2r* gene depends on an intronic CpG island. Nature 1997;389:745–9.

[24] Liu J, Chen M, Deng C, et al. Identification of the control region for tissue-specific imprinting of the stimulatory G protein alpha-subunit. Proc Natl Acad Sci USA 2005;102(15):5513–8.

[25] Kantor B, Makedonski K, Green-Finberg Y, Shemer R, Razin A. Control elements within the PWS/AS imprinting box and their function in the imprinting process. Hum Mol Genet 2004;13(7):751–62.

[26] Fitzpatrick GV, Soloway PD, Higgins MJ. Regional loss of imprinting and growth deficiency in mice with a targeted deletion of KvDMR1. Nat Genet 2002;32(3):426–31.

[27] Yoon BJ, Herman H, Sikora A, Smith LT, Plass C, Soloway PD. Regulation of DNA methylation of Rasgrf1. Nat Genet 2002;30(1):92–6.

[28] Takada S, Paulsen M, Tevendale M, et al. Epigenetic analysis of the Dlk1-Gtl2 imprinted domain on mouse chromosome 12: implications for imprinting control from comparison with Igf2-H19. Hum Mol Genet 2002;11(1):77–86.

[29] Coombes C, Arnaud P, Gordon E, et al. Epigenetic properties and identification of an imprint mark in the Nesp-Gnasxl domain of the mouse Gnas imprinted locus. Mol Cell Biol 2003;23(16):5475–88.

[30] Kaneda M, Okano M, Hata K, et al. Essential role for de novo DNA methyltransferase Dnmt3a in paternal and maternal imprinting. Nature 2004;429(6994):900–3.

[31] Hata K, Okano M, Lei H, Li E. Dnmt3L cooperates with the Dnmt3 family of de novo DNA methyltransferases to establish maternal imprints in mice. Development 2002;129(8):1983–93.

[32] Smallwood SA, Tomizawa S, Krueger F, et al. Dynamic CpG island methylation landscape in oocytes and preimplantation embryos. Nat Genet 2011;43(8):811–4.

[33] Davis TL, Yang GJ, McCarrey JR, Bartolomei MS. The H19 methylation imprint is erased and re-established differentially on the parental alleles during male germ cell development. (In Process Citation) Hum Mol Genet 2000;9(19):2885–94.

[34] Watanabe T, Tomizawa S, Mitsuya K, et al. Role for piRNAs and noncoding RNA in de novo DNA methylation of the imprinted mouse Rasgrf1 locus. Science 2011;332(6031):848–52.

[35] Lee JT. Lessons from X-chromosome inactivation: long ncRNA as guides and tethers to the epigenome. Genes Dev 2009;23(16):1831–42.

[36] Nakamura T, Arai Y, Umehara H, et al. PGC7/Stella protects against DNA demethylation in early embryogenesis. Nat Cell Biol 2007;9(1):64–71.

[37] Li X, Ito M, Zhou F, et al. A maternal-zygotic effect gene, Zfp57, maintains both maternal and paternal imprints. Dev Cell 2008;15(4):547–57.

[38] Ooi SK, O'Donnell AH, Bestor TH. Mammalian cytosine methylation at a glance. J Cell Sci 2009;122(Pt 16):2787–91.

[39] Hermann A, Gowher H, Jeltsch A. Biochemistry and biology of mammalian DNA methyltransferases. Cell Mol Life Sci 2004;61(19–20):2571–87.

[40] Li E, Bestor TH, Jaenisch R. Targeted mutation of the DNA methyltransferase gene results in embryonic lethality. Cell 1992;69:915–26.

[41] Verona RI, Thorvaldsen JL, Reese KJ, Bartolomei MS. The transcriptional status but not the imprinting control region determines allele-specific histone modifications at the imprinted H19 locus. Mol Cell Biol 2008;28(1):71–82.

[42] Hsieh CL. Dynamics of DNA methylation pattern. Curr Opin Genet Dev 2000;10(2):224–8.

[43] Cranston MJ, Spinka TL, Elson DA, Bartolomei MS. Elucidation of the minimal sequence required to imprint h19 transgenes. Genomics 2001;73(1):98–107.

[44] Schoenherr CJ, Levorse JM, Tilghman SM. CTCF maintains differential methylation at the Igf2/H19 locus. Nat Genet 2003;33(1):66–9.

[45] Pant V, Kurukuti S, Pugacheva E, et al. Mutation of a single CTCF target site within the H19 imprinting control region leads to loss of Igf2 imprinting and complex patterns of de novo methylation upon maternal inheritance. Mol Cell Biol 2004;24(8):3497–504.

[46] Srivastava M, Hsieh S, Grinberg A, Williams-Simons L, Huang S-P, Pfeifer K. *H19* and *Igf2* monoallelic expression is regulated in two distinct ways by a shared cis acting regulatory region upstream of *H19*. Genes Dev 2000;14:1186–95.

[47] Chaillet JR, Vogt TF, Beier DR, Leder P. Parental-specific methylation of an imprinted transgene is established during gametogenesis and progressively changes during embryogenesis. Cell 1991;66:77–83.

[48] Paoloni-Giacobino A, D'Aiuto L, Cirio MC, Reinhart B, Chaillet JR. Conserved features of imprinted differentially methylated domains. Gene 2007;399.

[49] Hutter B, Helms V, Paulsen M. Tandem repeats in the CpG islands of imprinted genes. Genomics 2006;88.

[50] Bestor TH, Bourc'his D. Transposon silencing and imprint establishment in mammalian germ cells. Cold Spring Harb Symp Quant Biol 2004;69:381–7.

[51] Sleutels F, Zwart R, Barlow DP. The non-coding air RNA is required for silencing autosomal imprinted genes. Nature 2002;415(6873):810–3.

[52] Smilinich NJ, Day CD, Fitzpatrick GV, et al. A maternally methylated CpG island in KvLQT1 is associated with an antisense paternal transcript and loss of imprinting in Beckwith-Wiedemann syndrome. Proc Natl Acad Sci USA 1999;96(14):8064–9.

[53] Wroe SF, Kelsey G, Skinner JA, et al. An imprinted transcript, antisense to Nesp, adds complexity to the cluster of imprinted genes at the mouse Gnas locus. Proc Natl Acad Sci USA 2000;97(7):3342–6.

[54] Runte M, Huttenhofer A, Gross S, Kiefmann M, Horsthemke B, Buiting K. The IC-SNURF-SNRPN transcript serves as a host for multiple small nucleolar RNA species and as an antisense RNA for UBE3A. Hum Mol Genet 2001;10(23):2687–700.

[55] Wood AJ, Schulz R, Woodfine K, et al. Regulation of alternative polyadenylation by genomic imprinting. Genes Dev 2008;22(9):1141–6.

[56] Cowley M, Wood AJ, Bohm S, Schulz R, Oakey RJ. Epigenetic control of alternative mRNA processing at the imprinted Herc3/Nap1l5 locus. Nucleic Acids Res 2012;40(18):8917–26.

[57] Froberg JE, Yang L, Lee JT. Guided by RNAs: X-inactivation as a model for lncRNA function. J Mol Biol 2013;425(19):3698–706.

[58] Latos PA, Pauler FM, Koerner MV, et al. Airn transcriptional overlap, but not its lncRNA products, induces imprinted Igf2r silencing. Science 2013;338(6113):1469–72.

[59] Golding MC, Magri LS, Zhang L, Lalone SA, Higgins MJ, Mann MR. Depletion of Kcnq1ot1 non-coding RNA does not affect imprinting maintenance in stem cells. Development 2011;138(17):3667–78.

[60] Gould TD, Pfeifer K. Imprinting of mouse *Kvlqt1* is developmentally regulated. Hum Mol Genet 1998;7:483–7.

[61] Korostowski L, Raval A, Breuer G, Engel N. Enhancer-driven chromatin interactions during development promote escape from silencing by a long non-coding RNA. Epigenetics Chromatin 2011;4:21–32.

[62] Bell AC, Felsenfeld G. Methylation of a CTCF-dependent boundary controls imprinted expression of the *Igf2* gene. Nature 2000;2000:482–5.

[63] Hark AT, Schoenherr CJ, Katz DJ, Ingram RS, Levorse JM, Tilghman SM. CTCF mediates methylation-sensitive enhancer-blocking activity at the *H19/Igf2* locus. Nature 2000;405:486–9.

[64] Murrell A, Heeson S, Reik W. Interaction between differentially methylated regions partitions the imprinted genes Igf2 and H19 into parent-specific chromatin loops. Nat Genet 2004;36(8):889–93.

[65] Engel N, Raval AK, Thorvaldsen JL, Bartolomei SM. Three-dimensional conformation at the H19/Igf2 locus supports a model of enhancer tracking. Hum Mol Genet 2008;17(19):3021–9.

[66] Essien K, Vigneau S, Apreleva S, Singh LN, Bartolomei MS, Hannenhalli S. CTCF binding site classes exhibit distinct evolutionary, genomic, epigenomic and transcriptomic features. Genome Biol 2009;10(11):R131.

[67] Chotalia M, Smallwood SA, Ruf N, et al. Transcription is required for establishment of germline methylation marks at imprinted genes. Genes Dev 2009;23(1):105–17.

[68] Bartolomei MS, Ferguson-Smith AC. Mammalian genomic imprinting. Cold Spring Harb Perspect Biol 2011;3(7).

[69] Weksberg R. Imprinted genes and human disease. Am J Med Genet C Semin Med Genet 2010; 154C(3):317–20.

[70] Jacob KJ, Robinson WP, Lefebvre L. Beckwith-Wiedemann and Silver-Russell syndromes: opposite developmental imbalances in imprinted regulators of placental function and embryonic growth. Clin Genet 2013;84(4):326–34.

[71] Keverne EB, Fundele R, Narasimha M, Barton SC, Surani MA. Genomic imprinting and the differential roles of parental genomes in brain development. Brain Res Dev Brain Res 1996;92(1):91–100.

[72] Kishino T, Lalande M, Wagstaff J. *UBE3A*/E6-AP mutations cause Angelman syndrome. Nat Genet 1997;15:70–3.

[73] Ohta T, Gray TA, Rogan PK, et al. Imprinting-mutation mechanism in Prader-Willi syndrome. Am J Hum Genet 1999;64:397–413.

[74] Engel N, West AG, Felsenfeld G, Bartolomei MS. Antagonism between DNA hypermethylation and enhancer-blocking activity at the H19 DMD is uncovered by CpG mutations. Nat Genet 2004;36(8):883–8.

[75] Choufani S, Shapiro JS, Susiarjo M, et al. A novel approach identifies new differentially methylated regions (DMRs) associated with imprinted genes. Genome Res 2011;21(3):465–76.

[76] Xie W, Barr CL, Kim A, et al. Base-resolution analyses of sequence and parent-of-origin dependent DNA methylation in the mouse genome. Cell 2012;148(4):816–31.

[77] Babak T, Deveale B, Armour C, et al. Global survey of genomic imprinting by transcriptome sequencing. Curr Biol 2008;18(22):1735–41.

[78] Wang X, Sun Q, McGrath SD, Mardis ER, Soloway PD, Clark AG. Transcriptome-wide identification of novel imprinted genes in neonatal mouse brain. PLoS One 2008;3(12):e3839.

[79] Bestor TH. DNA methylation: evolution of a bacterial immune function into a regulator of gene expression and genome structure in higher eukaryotes. Philos Trans R Soc Lond B Biol Sci 1990;326(1235):179–87.

[80] Barlow DP. Methylation and imprinting: from host defense to gene regulation? Science 1993;260(5106):309–10.

[81] Bourc'his D, Bestor TH. Origins of extreme sexual dimorphism in genomic imprinting. Cytogenet Genome Res 2006;113(1–4):36–40.

[82] Pask A. Insights on imprinting from beyond mice and men. Methods Mol Biol 2012;925:263–75.

[83] de la Casa-Esperon E, Pardo-Manuel de Villena F, Verner AE, et al. Sex-of-offspring-specific transmission ratio distortion on mouse chromosome X. Genetics 2000;154(1):343–50.

[84] Spencer HG, Clark AG. Non-conflict theories for the evolution of genomic imprinting. Heredity (Edinb) 2014;113(2):93–5.

[85] Moore T, Haig D. Genomic imprinting in mammalian development: a parental tug-of-war. Trends Genet 1991;7(2):45–9.

[86] Varmuza S, Mann M. Genomic imprinting–defusing the ovarian time bomb. Trends Genet 1994;10(4):118–23.

[87] Wolf JB, Hager R. A maternal-offspring coadaptation theory for the evolution of genomic imprinting. PLoS Biol 2006;4(12):e380.

[88] Cowley M, Garfield AS, Madon-Simon M, et al. Developmental programming mediated by complementary roles of imprinted Grb10 in mother and pup. PLoS Biol 2014;12(2):e1001799.

[89] Wilkins JF. Genomic imprinting of Grb10: coadaptation or conflict? PLoS Biol 2014;12(2):e1001800.

[90] Lyon MF. Imprinting and X-chromosome inactivation. Results Probl Cell Differ 1999;25:73–90.

[91] Lee JT. Molecular links between X-inactivation and autosomal imprinting: X-inactivation as a driving force for the evolution of imprinting? Curr Biol 2003;13(6):R242–54.

[92] Pardo-Manuel de Villena F, de la Casa-Esperon E, Sapienza C. Natural selection and the function of genome imprinting: beyond the silenced minority. Trends Genet 2000;16(12):573–9.

[93] Sandovici I, Kassovska-Bratinova S, Vaughan JE, Stewart R, Leppert M, Sapienza C. Human imprinted chromosomal regions are historical hot-spots of recombination. PLoS Genet 2006; 2(7):e101.

[94] Lercher MJ, Hurst LD. Imprinted chromosomal regions of the human genome have unusually high recombination rates. Genetics 2003;165(3):1629–32.

[95] Hurst LD, McVean G, Moore T. Imprinted genes have few and small introns. Nat Genet 1996;12(3):234–7.

CHAPTER 4

Polycomb and Trithorax factors in transcriptional and epigenetic regulation

Priscilla Nga Ieng Lau, Chi Wai Eric So
Leukemia and Stem Cell Biology Group, Department of Haematological Medicine, King's College London, London, UK

Contents

1. INTRODUCTION

Nearly all cells in a multicellular organism contain the same genetic information encoded in their DNA. Nevertheless, differential expression of genes in different cell types confers specialized functions and unique identity to these cells. These cell type–specific transcriptional programs are established early during development, and maintenance of cellular identity requires stable transmission of gene expression patterns through cell

Epigenetic Gene Expression and Regulation
http://dx.doi.org/10.1016/B978-0-12-799958-6.00004-4

divisions. Epigenetics, which refers to heritage changes in gene expression without alterations in DNA sequence, plays a critical role in the establishment and maintenance of cellular memory. Epigenetic mechanisms mainly regulate gene expression through modulating chromatin structure and include covalent modifications on DNA and histones, ATP-dependent chromatin remodeling, incorporation of histone variants as well as noncoding RNAs [1]. In addition to mediating epigenetic memory, these mechanisms also allow cells to alter their gene expression patterns in response to environmental cues and developmental signals.

The organization of DNA into chromatin in eukaryotic cells was originally thought to have a function only in packaging lengthy DNA into the tiny nucleus. It is now clear that this organization also plays critical roles in all DNA-templated processes such as gene transcription, DNA replication, DNA repair, and recombination. The basic unit of chromatin is the nucleosome, which consists of 146 base pair (bp) of DNA wrapped around a histone octamer containing two copies each of core histones H2A, H2B, H3, and H4 [2]. This packaging of DNA into nucleosomes restricts access of cellular machineries to DNA, thereby controlling gene expression and other cellular activities.

One mechanism by which cells regulate the accessibility of DNA is through the addition of covalent posttranslational modifications (PTMs) on histones, particularly on their unstructured N-terminal tails [3]. The most common and best-studied PTMs on histones include phosphorylation (ph), acetylation (ac), methylation (me), and ubiquitylation (ub). These PTMs have been correlated with specific transcriptional states, and are also implicated in various biological processes including mitosis, DNA repair, and DNA replication. They mainly function by inducing structural changes to chromatin and/or by recruiting downstream effector proteins to chromatin.

Importantly, most (if not all) histone PTMs are reversible, and they respond dynamically to changes in cellular conditions and during development. The stability and dynamics of histone PTMs are controlled through the delicate balance between two groups of functionally opposing enzymes, which add or remove PTMs from specific amino acids on histones in a highly regulated manner. These two classes of antagonizing histone-modifying enzymes determine the levels of histone PTMs on chromatin and are sometimes referred to as writers and erasers of histone marks [4]. The identification of cellular factors or protein domains that specifically recognize modified residues on histones has greatly contributed to our understanding of how histone PTMs mediate downstream functions [5]. Conserved binding domains that recognize histone methylation, acetylation, and phosphorylation have been well characterized. These PTM-recognition domains are present in a large number of chromatin regulators including histone-modifying enzymes, ATP-dependent chromatin-remodeling enzymes, as well as other cellular machineries involved in transcription, DNA repair, and RNA processing. Histone PTMs therefore play a crucial role in recruiting these effector proteins (also known as readers of histone marks) to the appropriate chromatin locations to mediate downstream functions.

It is now clear that cancer is not only a genetic disease, but also involves epigenetic alterations and deregulation of the epigenetic machineries [6]. Genome-wide studies of DNA methylation and histone PTM patterns indicate that aberrant epigenetic patterns are common in the cancer genome. Mutations and altered expression of epigenetic regulators are often found in patient samples and are pivotal in cancer initiation and progression. As a result, the epigenetic landscape in cancer cells is altered, leading to aberrant transcription of oncogenes and transcriptional silencing of tumor suppressors. In addition to having a role in tumorigenesis, recent studies have also shed light on the role of epigenetic alterations in other human diseases, such as neurological disorders and autoimmune diseases [7]. In this chapter, we will focus on two groups of epigenetic regulators: Polycomb group (PcG, *Drosophila melanogaster*) and Trithorax group (TrxG, *D. melanogaster*) proteins. Extensive studies in the last couple of decades have expanded our knowledge on these two groups of proteins and provided strong evidence for their involvement in development, stem cell differentiation, and cancer [8–10]. Their functions and mode of actions exemplify how chromatin dynamics and histone PTMs regulate chromatin structure, gene expression, and resultant biological processes.

2. REGULATORS OF HOMEOTIC GENES—PcG AND TrxG PROTEINS

PcG and TrxG proteins were first identified in genetic screens in *D. melanogaster* as regulators of homeotic (*Hox*) genes (reviewed in Ref. [11]). *Hox* genes are transcription factors that determine patterning and cell identities along the anterior–posterior body axis of invertebrates and vertebrates. Their expression at specific body segments is tightly regulated and is essential for normal embryonic development. Altered expression of *Hox* genes, either temporally or spatially, will lead to homeotic transformation, which refers to the transformation of one entire body segment to another segment. These homeotic phenotypes are easily recognized in *Drosophila* and have formed the basis for many genetic screens that led to the discovery of *PcG* and *TrxG* genes.

Flies with mutations in PcG proteins exhibited developmental defects in body patterning and segmentation. Careful examination of these homeotic phenotypes indicates that *PcG* mutants resembled mutants with aberrant expression of *Hox* genes, thus suggesting that PcG proteins are crucial for repression of *Hox* genes [12]. Subsequent genetic screens for suppressors of *PcG* mutations identified TrxG as positive regulators of *Hox* genes [13]. Together, PcG and TrxG proteins act antagonistically to ensure proper *Hox* gene expression in the appropriate body segments. Such functions are also evolutionarily conserved in mammals; mice lacking PcG and TrxG proteins also display homeotic transformation of the axial skeleton [14–16].

PcG and TrxG proteins are highly conserved across species, from *Arabidopsis* to *Drosophila* to human. They often function in the context of multi-subunit complexes, and molecular studies in the last two decades indicate that both families of proteins are chromatin regulators that can covalently modify histone tails and remodel nucleosomes on chromatin. PcG complexes catalyze the methylation of histone H3 on K27 and monoubiquitylation of histone H2A on K119, both of which are linked to transcriptional silencing [10]. TrxG complexes, on the other hand, methylate histone H3 on K4, and this mark is associated with transcriptionally active regions of the genome [17]. Therefore, in general, PcG proteins function as transcriptional repressors, whereas TrxG proteins are transcriptional activators. Other noncatalytic subunits within the complexes often contain chromatin/PTM-binding domains and play critical roles in controlling enzyme recruitment, activity, and substrate specificity. In addition to regulating the originally identified *Hox* genes, genome-wide expression and location analyses reveal that PcG and TrxG proteins are also required for expression of key developmental regulators during differentiation and development [18,19]. Importantly, the regulation of PcG and TrxG target genes can be highly dynamic. Their expression status is not as static as previously suggested, and can be altered in response to environmental and developmental signals.

3. PcG-MEDIATED GENE REPRESSION

3.1 Polycomb complexes

PcG proteins mainly reside in two core multiprotein complexes: Polycomb repressive complex 1 (PRC1) and Polycomb repressive complex 2 (PRC2). The *Drosophila* PRC2 contains Enhancer of Zeste (E(z)), Suppressor of Zeste-12 (Su(z)12), Extra Sex Combs (Esc), and Chromatin Assembly Factor 1 (Caf1 or Nurf55), (Figure 1(A)). Homologs of each component have been identified in the mammalian PRC2 complex and include either Enhancer of Zeste homolog 1or 2 (EZH1 or EZH2, humans) as the enzymatic subunit and three noncatalytic subunits SUZ12, Embryonic Ectoderm Development (EED1/2/3/4, humans), and Retinoblastoma Binding Protein (RBBP4/7, humans) [20]. The E(z)/EZH component catalyzes mono-, di-, and trimethylation of H3K27 (H3K27me1/2/3), and this histone methyltransferases (HMT) activity is carried out by the evolutionarily conserved SET domain, which was named after the three founding members in *Drosophila* that share this domain: Suppressor of variegation 3–9 (Su(var)3–9), *E(z)*, and *Trithorax* (Trx). Methylated H3K27 in turn serves as a docking site for the chromodomain-containing Polycomb (Pc, D. melanogaster)/Chromobox Protein Homolog (Cbx in mouse; CBX in humans) family of proteins in PRC1 and recruits the PRC1 complex to target genes. Other core PRC1 members in *Drosophila* include Sex Combs Extra (Sce, dRING, or RING1), Posterior sex combs (Psc), and Polyhomeotic (Ph) (Figure 1(B)). This second complex carries another histone-modifying activity—monoubiquitylation of histone H2A at K119 (H2AK119ub1), and this is mediated by the E3 ubiquitin ligase dRING in the

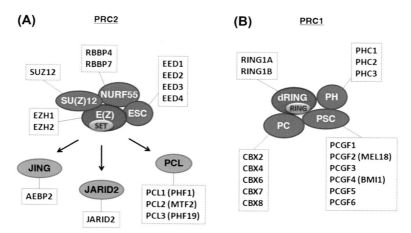

Figure 1 *Subunit composition of* Drosophila *and mammalian PcG complexes.* Subunits of the *Drosophila* (A) PRC2 and (B) PRC1 complexes are depicted. The enzymatic subunits and catalytic domains are highlighted in red. Core components are in blue, and accessory subunits of PRC2 are shown in green. The mammalian homologs of each subunit are indicated in the attached boxes. Incorporation of different paralogs in mammals allows the generation of multiple forms of PRC1 and PRC2 complexes.

complex. Multiple homologs for each of the *Drosophila* PRC1 component are present in the mammalian genomes, and the core mammalian PRC1 complex contains one member each of the CBX (CBX2/4/6/7/8), Polycomb Group Ring Finger (PCGF1–6, humans), Polyhomeotic Homolog (PHC1–3, humans), and Ring Finger Protein 1 (RING1A/B) family (Figure 1(B)).

Due to the presence of large number of paralogs, distinct combinations can generate multiple forms of PRC1 and PRC2 complexes, whose functions are defined by the activity of the specific paralog present in the complex. These paralogs can have redundant as well as complementary roles. Composition of these complexes is context-dependent and often varies according to cell types and developmental status. This is best illustrated by EZH2 and its close homolog EZH1, the two methyltransferases of PRC2. EZH1 and EZH2 form mutually exclusive complexes with other PRC2 components, and in comparison, PRC2-EZH1 has intrinsically lower methyltransferase activity than PRC2-EZH2 [21]. Nevertheless, only PRC2-EZH1 can directly interact with nucleosomes and induce chromatin compaction independent of its HMT activity [21,22]. Expression profiles of EZH1 and EZH2 are also different: EZH2 is only found in proliferative cells, whereas EZH1 is ubiquitously expressed and is found in nondividing differentiated cells as well. Studies in mouse embryonic stem cells (mESCs) indicate that *Ezh2* is the major H3K27 methyltransferase in pluripotent cells, and $Ezh2^{-/-}$ cells have significantly reduced H3K27me2/3, but still retain robust H3K27me1 levels; knockdown of Ezh1 has no impact on global H3K27 methylation, and only modest reduction of H3K27me3 at PcG target genes

[23]. However, further knockdown of Ezh1 in $Ezh2^{-/-}$ cells abolishes all three forms of H3K27 methylation and phenotypically resembles $Eed^{-/-}$ cells, suggesting that Ezh1 has complementary but nonredundant roles in H3K27 methylation and maintenance of ES cell function. By contrast, when Ezh1 and Ezh2 were studied in a mouse skin epidermis model, they were found to be functionally redundant. Defects in skin homeostasis were observed only in double knockout cells [24]. Ezh1 and Ezh2 may therefore have different functional relationships in pluripotent ES cells and tissue-specific stem cells.

Within the PRC2 complex, the core noncatalytic subunits and other newly identified auxiliary subunits are often crucial in modulating enzymatic activity, nucleosome binding, and recruitment to target genes. Binding of EED to H3K27me3 histone tails allosterically activates the HMT activity of EZH2 [25]. Such binding of PRC2 to the same PTM that it deposits suggests a way to propagate this repressive mark during DNA replication, thereby allowing inheritance of gene expression status. Binding of SUZ12 to a portion of H3 tail (residues 31–42) also stimulates PRC2 activity, thus allowing PRC2 to sense the local chromatin status and preferentially methylate densely compacted chromatin [26]. A number of PRC2 accessory proteins, including AE Binding Protein 2 (AEBP2, humans), Jumonji, AT Rich Interactive Domain 2 (JARID2, humans), and polycomblike proteins (PCL1-3, humans), are also found to enhance the enzymatic activity of PRC2 core complex [27–30]. In particular, PCL1 (as well as its *Drosophila* homolog, Pcl) can specifically stimulate EZH2 to catalyze the conversion of H3K27me2 to H3K27me3. The role of these auxiliary proteins in PRC2 recruitment will be discussed in more details in Section 5: *Targeting of PcG/TrxG complexes to chromatin*. In addition to the above-described canonical PRC1, recent biochemical studies have discovered some noncanonical PRC1, which contain the RING1-PCGF core without CBX proteins, and incorporate other non-PcG proteins instead. This gives rise to functionally distinct PRC1 complexes, and for more detailed description and functions of variant PcG complexes, readers are referred to other recent reviews on this subject (see Refs [31,32]).

The classical PcG-mediated silencing model suggests that PRC2 recruits PRC1 to chromatin through the binding of the PRC1 component CBX to the H3K27me3 mark established by PRC2, and together, the two complexes function to repress polycomb target genes. Several recent findings have challenged this hierarchical recruitment model: some genes are redundantly repressed by the two PRC complexes, and they do not always cooccupy the same set of target genes, suggesting that PRC1 and PRC2 can act independently of each other [33,34]. Studies in *Eed* knockout mESCs illustrate that PRC1 and H2Aub activity can be recruited in a H3K27me3-independent manner [35,36]. These cells are devoid of H3K27me3, but the levels of H2AK119ub1 are largely unaffected, both at the global level and at PcG target gene loci. RING1B is also retained at the majority of PcG target sites.

Contrary to the hierarchical recruitment model, two recent studies indicate that variant PRC1 complexes with KDM2B (Lysine (K)-specific Demethylase 2B) and PCGF1/3/5/6, but not PCGF2/4, can also lead to the formation of polycomb domains, in which deposition of H2AK119ub1 precedes the H3K27me3 mark and recruitment of PRC2 components [37,38]. This is further supported by an in vitro study that shows that JARID2 and AEBP2, both accessory proteins of PRC2, are enriched on nucleosomes with ubiquitylated-H2A; and in the presence of JARID2 and AEBP2, PRC2 also preferentially methylates H2Aub nucleosomes [39]. This suggests the existence of a positive feedback loop in which H2Aub (mediated by noncanonical PRC1 complexes) enhances PRC2 binding as well as H3K27 methylation, and this in turn recruits canonical PRC1 to chromatin through the chromodomain of the CBX component. Current evidence therefore suggests it is possible that canonical and noncanonical PRC1 complexes are recruited to chromatin through different mechanisms.

3.2 Mechanism of action of PcG proteins

H3K27me3 and H2AK119ub1 are hallmarks of polycomb silencing. Functional roles of these two repressive marks and the molecular mechanisms of how PcG proteins mediate transcriptional repression are beginning to be understood. Transcriptional repression can be mediated through multiple mechanisms. These different mechanisms are not mutually exclusive and may function together to regulate gene expression. One mechanism by which PcG components maintain gene repression is through altering chromatin conformation and physically compacting chromatin. As visualized by electron microscopy, recombinant PRC1 complex condenses reconstituted nucleosomal arrays, resulting in the formation of DNA loops and nucleosome clusters [40]. This is mediated through interaction with nucleosomes and does not require the presence of histone tails. A high-resolution map of chromosomal contacts in *Drosophila* embryos identifies long-range chromosomal contacts between polycomb-repressed domains, suggesting that PcG proteins are involved in higher order chromatin organization [41]. A study using $Ring1b^{-/-}$ mESCs indicates that RING1B is required for chromatin compaction [42]. Using fluorescence in situ hybridization to assay chromatin compaction in vivo, the authors showed that significant chromatin decompaction at *Hox* loci is observed in $Ring1b^{-/-}$ cells. Expression of wild-type (WT) Ring1b, but not a catalytically inactive mutant, in $Ring1b^{-/-}$ cells rescues H2Aub levels; however, both WT and mutant Ring1b can rescue *Hox* chromatin compaction, suggesting that this is independent of its histone ubiquitylation activity. In addition to RING1B, L3MBTL2 (an MBT domain-containing protein in noncanonical PRC1 (PRC1L4)) is also able to compact chromatin in vitro in a histone methylation-independent manner [43].

PcG complexes also directly influence steps in the transcription process and interfere with productive gene expression. Chromatin immunoprecipitation sequencing (ChIP-seq) and global run-on sequencing (GRO-seq) assays performed in *esc* mutant *Drosophila* embryos indicate that loss of PRC2 and H3K27me3 is associated with increased levels of paused RNA polymerase II (RNAP II) [44]. This suggests that PRC2/H3K27me3 blocks RNAP II binding and/or initiation at target gene promoters. PRC1-mediated H2A ubiquitylation also restrains poised RNAP II at bivalent genes in mESCs, and inhibits eviction of H2A-H2B dimers from nucleosomes, thus preventing promoter escape and transcriptional elongation [45,46].

H3K27 methylation and acetylation are mutually exclusive. Presence of H3K27me3 therefore prevents accumulation of the activating H3K27ac mark, which is associated with promoter and enhancer of active genes [47]. A *Drosophila* mutant containing the H3K27R (lysine 27 of H3 mutated to arginine) point mutation phenocopies the E(z) mutant, thereby confirming that H3K27 is the physiological relevant substrate for PRC2-mediated repression [48]. Nonetheless, it is now increasingly evident that histone PTMs do not function in isolation, instead, chromatin modifiers often work together to coordinately mediate downstream processes. PcG proteins cooperate with a number of other repressive complexes such as histone deacetylases (HDAC) and DNA methyltransferases (DNMT) to reinforce the establishment of a repressed chromatin state at the target genes [49]. This role is mainly achieved through the physical interaction with different repressors. Thus, both enzymes are recruited to target genes in a complex and to comodify the nucleosomes. In addition, PRC complexes are also found to interact with different histone demethylases: PRC2 interacts with the H3K4 demethylase JARID1A/RBP2 (Jumonji, AT rich Interactive Domain 1A/Retinol-Binding Protein 2) and RING1A/B with H3K36 demethylase KDM2B [50]. This mechanism couples the deposition of polycomb repressive marks (H3K27me3 and H2AK119ub1) with removal of active histone marks (H3K4me3 and H3K36me3), and is a self-reinforcing mechanism to ensure efficient repression of gene expression.

In addition to methylating histone H3K27, EZH2 also methylates nonhistone substrates. In a cardiomyocyte model, EZH2 interacts with cardiac transcription factor GATA4 (GATA binding protein 4) and methylates residue K299 [51]. This methylation attenuates GATA4 transcriptional activity by reducing its interaction with histone acetyltransferase p300 and subsequent acetylation. Knockdown of PRC2 components has no effect on GATA4 binding to target genes, but strongly augmented p300 binding and expression of GATA4 target genes. Thus, it suggests that EZH2 can also mediate transcriptional repression through modifying other cellular factors.

3.3 Polycomb-regulated cellular processes

PcG proteins play key roles in early steps of embryogenesis. Knockout of core PRC2 components (Ezh2, Eed, or Suz12) or Ring1b results in early embryonic lethality [10]. Interestingly, loss of the closely related homologs Ezh1 or Ring1a is viable and is associated with a less severe phenotype, suggesting that they may have restricted functions and cannot compensate for their homologous proteins during embryonic development. In ESCs, PRCs bind to key developmental regulators, whose expression would otherwise promote differentiation [18,19]. PRC2 components are dispensable for self-renewal of ESCs, and cells lacking Eed, Ezh2, or Suz12 remain pluripotent. However, these cells have differentiation defects, partially due to inappropriate expression of pluripotency factors during differentiation [23,52,53]. On the other hand, Ring1a/b double knockout ESCs exhibit spontaneous differentiation, indicating that PRC1 is required for ESCs self-renewal [54]. In contrast to the situation in ESCs, tissue-specific knockout studies indicate that both PRC1 and PRC2 components are often required for self-renewal of somatic stem and progenitor cells, such as hematopoietic stem cells and neural stem cells, reinforcing their critical functions in different stages of development [10].

One of the main functions of PcG proteins is the regulation of cellular senescence. This is mediated through suppression of *CDKN2A/2B* loci, which encode for the cyclin-dependent kinase (CDK) inhibitors p14^{ARF} (p19^{ARF} in mouse), p15^{INK4B}, and p16^{INK4A} [55]. These CDK inhibitors function as tumor suppressors and prevent CDK4-dependent phosphorylation of RB (retinoblastoma protein), and this in turn prevents E2F-mediated cell cycle progression. Repression of these loci by PRC1/2 therefore inhibits senescence and enhances cell proliferation. In a similar way, PcG proteins play an important role in oncogene-induced senescence, a mechanism by which normal primary cells resist oncogene-mediated transformation. Activated oncogenes trigger senescence by inducing expression of *CDKN2A/B*. This process is accomplished by the loss of PcG binding and the simultaneous recruitment of H3K27 demethylase JMJD3/KDM6B (Jumonji Domain Containing Protein 3 / Lysine (K)-Specific Demethylase 6B) as well as other TrxG proteins. In this way, repressive H3K27me3 is replaced by activating H3K4me3 at *CDKN2A/B*, leading to cell cycle arrest and tumor suppression. On the other hand, PcG proteins have long been associated with sex chromosome inactivation for dosage compensation in female mammals [56]. PRC1/2 are recruited to the inactive X chromosome by the noncoding RNA *Xist*, and then spread along the chromosome to enable chromosome-wide silencing. PcG-mediated repressive marks H3K27me3 and H2AK119ub1 are enriched on the inactive X chromosome, and are important for both formation and maintenance of this transcriptionally silenced facultative heterochromatin.

Given the crucial roles of PcG proteins in regulating senescence, cell fate, and differentiation, it is therefore not surprising that they are important players in tumorigenesis. PRC1/2 components are often expressed or mutated in a broad spectrum of human cancers, and these mutations can affect enzymatic activity, complex formation, and/or chromatin targeting [57]. Repression of *CDKN2A/B* loci is a major mechanism for PcG-mediated tumorigenesis, and among the PcG members, the roles of EZH2 and

BMI1 (B Lymphoma Mo–MLV Insertion Region 1) in cancer are the best studied. EZH2 is overexpressed or amplified in several solid tumors, such as breast, prostate, and colorectal cancers [58]. High expression level of EZH2 is associated with the most aggressive tumors and worst prognosis; suppression of EZH2, on the other hand, inhibits tumor growth in breast and prostate tumor xenograft models. In follicular and diffuse large B-cell lymphoma (DLBCL), a somatic point mutation has been identified in the SET domain of EZH2 (Y641). This dominant gain-of-function mutation has led to increased levels of H3K27me3, as a result of altered substrate preference of the mutated enzyme [59]. On the other hand, inactivating mutations and deletions of EZH2 can be found in a number of hematological malignancies, such as myelodysplastic syndromes (MDS), myeloproliferative neoplasia (MPN), acute myeloid leukemia (AML), and T-acute lymphoblastic leukemia (T-ALL). Consistently, one of the Ezh2 knockout models develops MDS/MPN, whereas the other induces T-ALL, suggesting that Ezh2 functions as a tumor suppressor in these settings [60,61]. Recently, a recurrent K27M mutation in histone variant H3.3 was identified in pediatric glioblastoma. This mutation dominant negatively inhibits PRC2 activity and leads to a global loss of H3K27me3 [62]. It therefore appears that EZH2 can have both pro-oncogenic and antitumor activities, and this depends highly on the context of specific diseases. In addition to EZH2, the PRC1 component BMI1 is actually the first PcG protein implicated in cancer development, in which it cooperates with c-myc (v-myc Avian Myelocytomatosis Viral Oncogene Homolog) in lymphomagenesis in the Eμ-*myc* transgenic mouse model. This is achieved through the inhibition of c-Myc-induced apoptosis by regulating $p16^{INK4A}$/$p19^{ARF}$ expression [63]. Studies in different models have now confirmed that BMI1 is crucial for self-renewal of both normal and cancer stem cells. We refer readers to several recent reviews for more detailed description of the roles of PcG proteins in cancer [10,57].

3.4 Noncanonical role of EZH2

EZH2 is best known to function as the H3K27 methyltransferase within PRC2 in polycomb-mediated silencing. Interestingly, recent studies have highlighted some noncanonical roles of EZH2, indicating that EZH2 can act as a transcriptional activator instead. In addition, EZH2 can function independently of PRC2 and, in some cases, independently of its HMT activity. ChIP-seq analysis of EZH2 in castration-resistant prostate cancer (CRPC) cells revealed that a subset of EZH2-bound genes lacks H3K27me3 as well as the PRC2 subunit SUZ12 [64]. Intriguingly, many of these solo genes are down regulated upon EZH2 knockdown, indicating that EZH2 functions as an activator at these genes. AKT-mediated phosphorylation of EZH2 at S21 and the HMT activity of EZH2 are both required for expression of solo genes and androgen-independent growth. More detailed analysis showed that phosphorylated EZH2 interacts with androgen receptor (AR), and EZH2 depletion reduces lysine methylation on

AR, which was previously proposed to enhance AR transcriptional activity. Methylation of nonhistone proteins (AR or AR-associated proteins in CRPC) by EZH2 can therefore be a potential mechanism for EZH2-mediated gene activation.

A number of reports have indicated that EZH2 can also function as a co-activator in breast cancer cells [65–67]. In luminal-like estrogen receptor (ER) positive cells, EZH2 physically links estrogen and the Wnt signaling pathway through its interaction with ERα and β-catenin [65]. These proteins are recruited to the *c-myc* and *cyclin D1* promoters and induce expression of these genes, thereby promoting cell cycle progression and cell proliferation. This transactivation function is independent of the SET domain of EZH2, but involves an interaction with the Mediator complex, which can be a potential mechanism for EZH2-mediated transcriptional activation. In basal-like triple-negative cells, EZH2 forms a ternary complex with NFκB (Nuclear Factor kappa-light-chain-enhancer of Activated B cells) components RelA/RelB (v-rel reticuloendotheliosis viral oncogene homolog A/B) to induce expression of NFκB target genes such as IL-6 and TNF [66]. This complex is also present at the *NOTCH1* promoter and EZH2 is required for *NOTCH1* expression as well as the expansion of breast tumor initiating cells [67]. In both cases, EZH2 functions independently of its HMT activity and other PRC2 components. These studies suggested that the oncogenic effect of EZH2 could be a result of transcriptional activation of its target genes and independent of its gene silencing function. Further studies will help uncover the situations under which EZH2 can switch from a repressor in PRC2 to a transcriptional co-activator, and whether other PcG proteins can function as an activator as well.

4. TrxG-MEDIATED TRANSCRIPTIONAL REGULATION

TrxG proteins counteract the function of PcG proteins and mainly act as a positive regulator of gene expression. Similar to PcGs, TrxG proteins also function in the context of multi-protein complexes. In comparison, TrxG proteins form a more heterogeneous group than the PcG proteins and can be divided into two classes: histone modifiers and chromatin remodelers [9].

4.1 Histone-modifying complexes

The founding member of the TrxG family is the *Drosophila melanogaster* gene *Trithorax* (*Trx*), and it was first discovered for its role in regulating homeotic gene expression during development. The human *mixed lineage leukemia* (*MLL1*) gene, originally cloned as the gene involved in 11q23 chromosomal translocations in acute leukemias, is the mammalian homolog of *Trx*. Despite their importance in development and hematological malignancies, the molecular properties and biochemical functions of TrxG proteins were not known until the discovery and studies of its

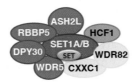

Figure 2 *Subunit composition of mammalian COMPASS-like complexes.* Mammals contain six COM-PASS-like complexes, and they can be divided into three subbranches: SET1A/B, MLL1/2, and MLL3/4 complexes. SET domain-containing proteins are the catalytic subunits of each complex (highlighted in red) and methylate histone H3 on K4. All complexes share four common subunits, which are shown in blue. Complex-specific subunits are also indicated.

Saccharomyces cerevisiae homolog *SET1.* Yeast SET1 exists in a protein complex known as COMPASS and is the first identified H3K4 methyltransferase [17]. In addition to Trx and MLL1, multiple homologs of the SET1 protein exist in metazoans. There are three SET1 homologs in *Drosophila*: dSet1, Trx, and Trr (Trithorax-related); whereas mammals have six: SET1A/B (counterpart of dSet1), MLL1/2 (counterpart of Trx), and MLL3/4 (counterpart of Trr). These SET-domain containing proteins all form COMPASS-like complexes and catalyze all three degrees of methylation on H3K4 (H3K4me1/2/3) (Figure 2).

Besides containing the catalytic component, all six mammalian COMPASS-like complexes share some common subunits, which include WDR5 (WD repeat-containing protein 5), RBBP5 (retinoblastoma binding protein 5), ASH2L (ash2 (absent, small, or homeotic discs 2)-like), and DPY30. These core members (known as the WRAD module) are highly conserved in yeast and flies, and in the absence of any of these subunits, the HMT activities and biological functions of these TrxG complexes are severely compromised [68]. Structural and biochemical studies further indicate that RBBP5 and ASH2L directly stimulate MLL1 enzymatic activity in vitro, whereas WDR5 is required for substrate binding and complex integrity [68,69]. In addition to the common subunits, these H3K4 HMT complexes also contain a number of complex-specific components: for example, WDR82 (WD repeat domain 82) and CXXC1 (CXXC finger protein 1) in SET1A/B complexes; tumor suppressor Menin in MLL1/2 complexes; and PA1, PTIP, NCOA6, and H3K27 demethylase UTX (lysine-specific demethylase 6A) in

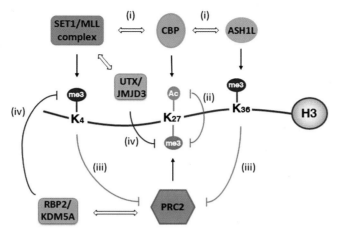

Figure 3 *PcG and TrxG proteins act antagonistically to regulate histone PTMs on H3.* (i) TrxG proteins MLL (Trx in *Drosophila*) and ASH1L (Ash1 in *Drosophila*) methylate H3K4 and H3K36, respectively. They interact with histone acetyltransferase CBP to facilitate acetylation of H3K27. (ii) Acetylation and methylation of H3K27 are mutually exclusive, and the presence of H3K27ac prevents the deposition of methyl marks on K27 by PRC2. (iii) Active H3K4me3 and H3K36me3 marks also inhibit PRC2 activity and methylation of H3K27. (iv) PcG and TrxG proteins cooperate with histone demethylases to couple histone methylation and demethylation—PRC2 interacts with RBP2/KDM5A to catalyze H3K4 demethylation, whereas MLL associates with UTX/JMJD3 to remove the repressive H3K27me3 mark.

MLL3/4 complexes (Figure 2). These unique subunits might confer different biological functions to the individual complexes due to their roles in regulating recruitment to target genes and interaction with other cellular proteins.

An increasing number of histone-modifying complexes are found to contain more than one enzymatic activity, and this is commonly observed in TrxG complexes. For example, MLL1 co-purifies with H4K16 acetyltransferase MOF (males-absent on the first protein) and together, they are required for optimal transcription both in vitro and in vivo [70]. Besides forming a COMPASS-like complex, *Drosophila* Trx associates with histone acetyltransferase CBP a well-studied transcriptional co-activator known to acetylate H3K27 as well as other lysine residues on H3/H4, to form the TAC1 complex (Figure 3(i)) [71]. CBP also interacts with another SET-domain containing TrxG protein ASH1 (ASH1L in mammals), which methylates H3 on K36 instead of K4 [72]. Consequently, histone acetylation is coupled to the methylation of H3K4 or H3K36. This results in histone acetylation being recruited to TrxG target genes to coordinately activate gene expression.

4.2 Chromatin remodelers

The first indication that TrxG proteins can also regulate gene expression by remodeling nucleosomes came from the realization that the *Drosophila* TrxG member Brahma (BRM) was homologous to the yeast SWI2/SNF2 protein, the catalytic subunit of the first discovered chromatin remodeling complex SWI/SNF. ATP-dependent chromatin remodeling complexes each contain a catalytic ATPase subunit that belongs to the SF2 superfamily of helicases (reviewed in Ref. [73]). Based on the presence of distinct structural domains on the ATPase subunit, chromatin remodeling complexes can be divided into subfamilies: SWI-SNF, ISWI, and CHD family. Using energy from ATP hydrolysis, these remodeling complexes disrupt DNA–histone contacts and alter the position and composition of nucleosomes on chromatin. This can be mediated through the sliding of nucleosomes along DNA, exchange of histone variants, or complete eviction of nucleosomes from chromatin. In this way, these ATP-driven machineries can modulate the accessibility of the underlying DNA to cellular proteins. Although the first chromatin remodeler SWI2/SNF2 was identified as a transcriptional co-activator in yeast, it is important to note that chromatin remodeling complexes can also organize nucleosome positioning to induce gene repression. The two classes of TrxG proteins can also function together to control gene expression. This is best illustrated by the binding of the NURF chromatin remodeling complex subunit BPTF to H3K4me2/3 at the *HOXC8* locus [74]. This binding is mediated through the PHD finger of BPTF and recruits the remodeling complex to MLL-regulated genes, indicating that the two types of chromatin-modifying activities can be physically and functionally coupled to regulate chromatin dynamics and accessibility.

4.3 Mechanism of action of TrxG proteins

TrxG proteins employ diverse strategies to regulate gene expression. One of their main roles is to counteract PcG-mediated gene repression, and this is mainly achieved through TrxG-mediated H3K4me3 and other PTMs associated with TrxG complexes. Methyl marks deposited by TrxG proteins (H3K4me3 and H3K36me2/3) directly antagonize polycomb silencing by inhibiting PRC2-mediated methylation of H3K27 (Figure 3(iii)) [75,76]. In vitro HMT assays show that H3K27 methylation by PRC2 is substantially inhibited on H3K4me3-modified or H3K36me2/3-modified nucleosomes compared to unmodified nucleosomes. This is not due to impaired binding of PRC2 to H3K4me3-modified nucleosomes, but is a result of allosteric inhibition of the PRC2 HMTase activity. These active marks therefore limit the enzymatic activity of PcG complexes and prevent the establishment of a repressive state in actively transcribed regions of the genome. As discussed in the previous section, a number of enzymes that act on H3K27 are found in complex with TrxG proteins. Histone demethylases UTX and JMJD3 interact with MLL3/4 and can actively remove the repressive H3K27me2/3 marks on TrxG target genes (Figure 3(iv)) [77,78]. Association of CBP with TrxG proteins also allows concomitant delivery of H3K27 acetylation and H3K4/K36 methylation (Figure 3(i)) [71,72]. This addition of an acetyl group on H3K27 physically blocks further addition of a repressive methyl mark on the same residue by PcG proteins (Figure 3(ii)). As a result, TrxG complexes can oppose H3K27me3-mediated silencing.

Another mechanism by which TrxG proteins regulate gene expression is through the direct recruitment of downstream effectors proteins (readers) to the H3K4me3 mark (Figure 4) [79]. Two binding domains that specifically recognize this PTM include PHD finger and chromodomain, and they can also cooperate with other binding domains to recognize combinatorial PTMs on nucleosomes. Multiple H3K4me3 readers have now been identified and they are often components of various chromatin-modifying complexes or are involved in different steps of the transcription process. ATP-dependent chromatin-remodeling complexes, such as CHD1 and NURF, can be targeted to selective chromatin regions through H3K4me3. This mobilizes and repositions nucleosomes to change accessibility of the chromatin fiber, thus allowing the binding of transcription factors and transcription machinery. Binding of CHD1 to H3K4me3 requires its tandem chromodomains, whereas binding of NURF subunit BPTF occurs through its PHD finger [74,80]. Importantly, comodification of nucleosomes by MLL1–MOF complex

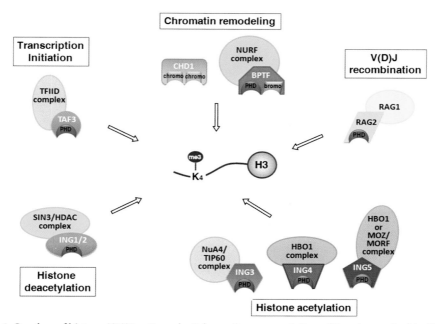

Figure 4 Readers of histone H3K4me3 mark. Schematic representation of the chromatin-binding proteins (readers) that specifically recognize H3K4me3, as well as the molecular or biological functions associated with these complexes. This binding is mainly mediated by two conserved PTM-recognition domains: chromodomains (chromo) and PHD fingers (PHD). These readers are often components of multi-subunit chromatin-modifying complexes, and play important roles in recruiting various enzymatic activities and biological processes to chromatin. As an example of combinatorial readout of histone PTMs, binding of BPTF-PHD finger to H3K4me3 is enhanced by acetylation of H4K16, and this is achieved through its paired PHD finger-bromodomain (bromo) module; H3R2me2s also enhances the binding of RAG2 to chromatin.

generates the H3K4me3/H4K16ac dimodified mark, and this further enhances the binding of BPTF through its paired PHD finger–bromodomain modules [81]. Such multivalent binding of histone PTMs allows for cooperative and more specific binding, and has been observed in numerous chromatin-binding proteins. H3K4me3 can also recruit other histone-modifying enzymes to chromatin. The ING family of tumor suppressor proteins all contain a C-terminal PHD finger that binds H3K4me3 [82]. Interestingly, some members (ING1 and ING2) associate with HDACs, while others (ING3, ING4, and ING5) are components of HAT complexes. Changes in histone acetylation levels at H3K4me3-marked genes can therefore be achieved through the recruitment of different ING proteins, and this would have distinct effects on gene expression. H3K4me3 also plays a role in transcription initiation through its recruitment of basal transcription machinery to active promoters. TFIID subunit TAF3 binds to H3K4me3 through its PHD finger, and this binding of TFIID is further enhanced by increased acetylation levels on H3K9 and H3K14 [83]. Multiple lines of evidence suggest that TrxG proteins are involved in post-initiation events. Besides being a component of chromatin remodeling complex, H3K4me3-binding protein CHD1 is found to co-purify with components of spliceosome as well as PAF and FACT elongation complexes, and this is essential for efficient pre-mRNA splicing [84]. In *Drosophila*, recruitment of FACT also depends on TAC1 complex [85]. These indicate that TrxG proteins can regulate gene expression by directly facilitating transcription elongation.

4.4 Trithorax-regulated cellular processes

Similar to PcG proteins, most SET1/MLL complex components (such as MLL1, MLL2, RBBP5, and DPY30) are not required for self-renewal of ESCs, with the exception of ASH2L and WDR5 [86]. ASH2L is required for maintaining open chromatin in mouse ESCs [87]. Knockdown of ASH2L reduces H3K4 methylation levels and causes ESCs to acquire a silenced chromatin state. On the other hand, WDR5 directly interacts with pluripotency transcription factor OCT4, and together with SOX2 and NANOG, activate expression of genes necessary for pluripotency and self-renewal [88]. Similarly, SWI/SNF subunit BRG1 cooccupy target genes of ES cell master regulators OCT4, SOX2, and NANOG, and is an essential component of the core pluripotency transcriptional network [89].

One striking feature in ES cells is that activating (K4) and repressive (K27) methyl marks often exist on the same loci. Bivalent domains, which consist of peaks of H3K4me3 within broader regions of H3K27me3, are frequently found at promoters of key developmental genes [90]. It has been proposed that this bivalent chromatin signature keeps these loci at a repressed state in pluripotent cells, yet poised for lineage-specific transcription in lineage-committed cells. Most of these bivalent domains are resolved into H3K4me3–only or H3K27me3–only regions in differentiated cells, in a manner consistent with their subsequent expression status. Most studies on bivalent domains have focused on PRC complexes, and the COMPASS family member responsible for H3K4me3 at these genes was not known until recently [91,92]. Knockdown of *MLL2*, and not *MLL3*, reduces H3K4me3 levels specifically at bivalent promoters [91]. Similar results are observed in *Mll2*$^{-/-}$ but not *Mll1*$^{-/-}$ ESCs, despite the high similarity between these two proteins, confirming that MLL2 is the major enzyme responsible for H3K4me3 at bivalent genes [92]. Surprisingly, loss of MLL2 and H3K4me3 at these genes has no substantial effect on gene activation upon retinoic acid-induced differentiation, thus questioning the previously proposed priming model for bivalency. Further work will be needed to evaluate the functional relevance of bivalent domains in ESCs.

Even though RBBP5 and DPY30 are not required for self-renewal in ESCs, they are required for differentiation of ESCs into neuronal progenitor cells [93]. Knockdown of RBBP5 or DPY30 strongly reduces global and gene-specific H3K4me3 levels and inhibits transcription of genes normally induced during neuronal differentiation. Studies showed that ASH2L and WDR5 also play important roles during other differentiation processes. For example, during myogenic differentiation, phosphorylation of muscle-specific transcription factor MEF2D by p38α enhances its interaction with MLL2/ASH2L complex and recruits it to muscle-specific genes such as myogenin and creatine kinase [94]. Paired-box transcription factor PAX7 also recruits MLL2/WDR5/ASH2L complex to myogenic gene promoters, inducing H3K4me3 levels and transcription of these target genes [95]. On the other hand, WDR5 was found to cooperate with canonical Wnt pathway to induce osteoblast and chondrocyte differentiation [96]. It is also involved in epithelial–mesenchymal transition, which is an important step for tumor metastasis [97]. Knockdown of WDR5 abolishes mesenchymal gene expression and decreases in vitro invasion/migration activity, suggesting an oncogenic role for WDR5.

Interestingly, a recent study in skeletal muscle myoblasts indicates that MLL3/4-mediated H3K4me1 at promoters can also have a role in gene repression [98]. Genes marked by H3K4me1 lack active marks such as H3K4me3, but are associated with H3K27me3 and are transcribed at low levels. It is suggested that this prevents the binding of H3K4me3 readers to target promoters and therefore, inhibit gene expression. During muscle differentiation, MLL3/4 are replaced by SET1/MLL1, and this is associated with the conversion of H3K4me1 to H3K4me3 as well as transcriptional activation of these muscle-specific genes. Monomethylation of H3K4 by MLL3/4 was previously associated with enhancers [99,100]. H3K4me1 together with H3K27ac and HAT p300 are currently used as a marker for enhancers. This study indicates that H3K4me1 likely serves different functions at distal and proximal regulatory elements.

Most studies involving the role of TrxG proteins in cancer had focused on MLL1, mainly due to its frequent involvement in human leukemias. Mutations in *MLL1* gene are associated with poor prognosis and four different kinds of chromosomal aberrations can occur: chromosomal translocations, partial tandem duplication, gene amplification, and internal deletion [101]. MLL chromosomal translocations are the most frequent

11q23 abnormalities in acute leukemia. It involves the in-frame fusion of N-terminus of MLL to C-terminus of over 60 different partners. The translocation breakpoint mainly occurs at the breakpoint cluster region, which is located between the CXXC DNA-binding domain and the central PHD fingers. As a result, the MLL fusion proteins retain DNA-binding motifs, but exclude the enzymatic SET domain. Some common MLL fusion partners include AF4, AF6, AF9, AF10, ENL, and ELL. Interestingly, many of these fusion partners belong to the Super Elongation Complex (SEC) or other related complexes such as EAP, AEP, and DotCom. These complexes contain elongation factors including PAF1C and pTEFb, and are aberrantly recruited to MLL target genes. Therefore, it is suggested that deregulation of transcriptional elongation is a mechanism employed by MLL-based leukemogenesis. In addition, aberrant H3K4me3 and H3K79me2 levels are often found at MLL fusion target genes [102]. This is due to aberrant recruitment of DOT1L, which methylates H3K79 and is the only non-SET domain-containing HMT identified to date. Another histone-modifying enzyme, protein arginine methyltransferase PRMT1, asymmetrically dimethylates H4R3 and is also required for a subset of MLL-fusion leukemias [103].

In contrast to the gain-of-function mutations of MLL1 in leukemia, loss-of-function mutations in MLL2, MLL3, and MLL4 are observed in human cancers. MLL3 was recently identified as a haploinsufficient tumor suppressor in human-7/del (7q) AML [104]. MLL3 suppression cooperates with other lesions (including NF1 and TP53 deletions) to drive the development of myeloid leukemia and impairs the differentiation of hematopoietic stem and progenitor cells. TrxG proteins of the chromatin-remodeling group such as BRG1, BRM, CHD5, and SNF5 also function as tumor suppressors [8]. They are involved in modulating transcription of the *CDKN2A/B* locus, and therefore, play a role in regulating cell cycle arrest and senescence. SWI/SNF subunit SNF5 recruits BRG1 to *CDKN2A/B* locus, and this coincides with removal of PRC1/2 and DNMT3B [105]. MLL1 is simultaneously recruited, leading to deposition of active histone marks and expression of *p15^{INK4B}* and *p16^{INK4A}*. Loss of SNF5 in humans results in malignant rhabdoid tumor (MRT), an aggressive form of childhood cancer. In mouse models, mice heterozygous for *Snf5* are tumor prone, and homozygous conditional inactivation of *Snf5* leads to highly penetrant T-cell lymphoma or rare rhabdoid tumors [106].

5. TARGETING OF PcG/TrxG COMPLEXES TO CHROMATIN

5.1 Polycomb/Trithorax response element

In *Drosophila*, recruitment of PcG and TrxG complexes to target genes mainly occurs through *cis*-regulatory elements known as Polycomb/Trithorax response elements (PRE/TRE). These elements can be found up to tens of kilobases from the genes they regulate. They contain consensus-binding motifs for a number of DNA-binding proteins: Pho, Phol, GAF, Dsp1, Pipsqueak, and Zeste, which in turn recruit PcG complexes to chromatin [107]. Even though PcG and TrxG proteins are highly conserved between *Drosophila* and mammals, the zinc finger protein Pho is the only PRE-binding protein with a mammalian homolog (YY1). There is however little overlap between YY1-binding sites and PcG proteins in mammals, and despite much effort, mammalian PREs have remained elusive until recently, when two groups reported the identification of putative PREs in mouse and human cells.

The first identified vertebrate PRE-like element, termed PRE-*kr*, regulates expression of *MafB* (v-Maf Avian Musculoaponeurotic Fibrosarcoma Oncogene Homolog B) in the mouse hindbrain [108]. This element displays properties that resemble *Drosophila* PREs, and induces repression of reporter genes in both flies and mouse embryos. Importantly, it binds components of PRC1/2 and represses gene expression in a PcG-dependent manner. Another PRE-like element, called *D11.12*, is located at the human *HOXD* locus, in the intergenic region between *HOXD11* and *HOXD12* [109]. Several Pho/YY1-binding sites are present at *D11.12*, and ChIP assays confirmed the presence of YY1 and PcG proteins at *D11.12* in human ESCs and other cells types. These sites are important for D11.12 function, and disruption of YY1 sites compromised BMI1 binding and D11.12-mediated repression. PRE-*kr* also contains Pho/YY1-binding sites, but their importance at this element has not been tested.

Even though PRE elements are generally lacking in the mammalian genome, it is suggested that another feature in the DNA sequence—unmethylated CpG islands that are devoid of transcriptional activator binding sites—is sufficient to recruit PcG/TrxG complexes to chromatin. This is supported by genome-wide sequencing analysis, which shows that there is a significant overlap between localization of PRC2 components and CpG islands [33]. In addition, mouse-derived GC-rich elements inserted into the mouse ESCs genome are also capable of recruiting PRC2 [110]. Since core PcG components do not contain DNA-binding domains, the mechanism by which PcG complexes is recruited to CpG islands is not clear and likely involves additional PRC accessory subunits. A number of TrxG proteins, including MLL1, MLL2, and CXXC1 (a component of mammalian SETD1A/B complexes), have a CXXC domain that can recognize unmethylated CpG, and they play a role in recruiting MLL complexes and MLL-fusions to GC-rich elements [111,112]. Since most PcG/TrxG target genes in mammals lack consensus PRE-like sequences in their regulatory regions, mammalian cells have employed multiple strategies for recruiting PcG and TrxG complexes to chromatin.

5.2 Accessory subunits of PcG/TrxG complexes

Recent biochemical characterizations of PcG complexes have identified additional auxiliary subunits that bind to PRC complexes at near-stoichiometric levels and are suggested to play a role in recruiting PcG complexes to target genes. Two accessory components of PRC2, JARID2 and AEBP2, have DNA-binding domains and are found to have weak CpG-rich DNA binding capacity [113,114]. JARID2 was identified as a

PRC2-associated protein in ESCs, and loss of JARID2 impaired PRC2 recruitment to PcG target genes, suggesting that it has a prominent role in recruiting PRC2 to chromatin. AEBP2 also binds to a DNA-binding motif with an unusual bipartite structure, and ChIP assays show that AEBP2 colocalizes with SUZ12 at target loci. Besides binding to specific DNA sequences, other PcG/TrxG subunits can be recruited to chromatin through preestablished histone PTMs. For example, polycomb-like (PCL) proteins bind specifically to H3K36me3 through the Tudor domain and may serve to recruit PRC2 to previously active genes, which contain the H3K36me3 elongation mark [115]. PHD finger domain of MLL binds to H3K4me2/3, and is essential for the recruitment of both WT MLL and MLL-fusions to the *HOXA9* locus [111]. This recruitment of MLL is further enhanced by its interaction with PAF elongation complex and CpG-rich sequences via its CXXC domain.

5.3 Sequence-specific transcription factors

PcG complexes are also recruited indirectly to chromatin through sequence-specific transcription factors. BMI1 interacts with transcription factor Runx1/CBFβ, and genome-wide studies show that PRC1 component RING1 and Runx1 largely occupy the same sites on chromatin in megakaryocytic cells [116]. Binding of PRC1 to Runx1-binding sites depends on Runx1, but is independent of PRC2. Other transcription repressors such as SNAIL, REST, and PLZF have also been proposed to regulate recruitment of PRC complexes to target genes [117,118]. Oncogenic RARα fusions, which are involved in acute promyelocytic leukemia (APL), are also involved in PRC recruitment. PML-RARα directly associates with PRC2, and loss of PRC2 in APL mouse models results in differentiation of leukemic cells [119]. On the other hand, PLZF–RARα recruits both PRC2 and PRC1 to retinoic acid target genes. All-*trans* retinoic acid treatment results in loss of PRC2, but not PRC1, recruitment, and as a result, PLZF–RARα cells are insensitive to all-*trans* retinoic acid treatment [120]. A recent study suggests that MLL1 can also be recruited to chromatin through the binding with p65, the transactivation subunit of NFκB [121].

5.4 Non-coding RNA

Several studies indicated that long noncoding RNAs (lncRNAs) play an important role in recruiting PRC and MLL complexes to chromatin [122]. The involvement of lncRNAs in targeting PRC complexes was originally illustrated in the mammalian X chromosome inactivation model. *Xist* RNA interacts with EZH2 in PRC2 and is required for the accumulation of H3K27me3 and PRC2 components on the inactive X chromosome. Other lncRNAs are also reported to recruit chromatin regulators to *HOX* gene loci. *HOTAIR*, which is transcribed from the *HOXC* locus, also interacts with PRC2, and is required for PRC2 occupancy and H3K27me3 at *HOXD* locus [123]. The first indication that lncRNA might also function in recruiting TrxG complexes comes from the study of lncRNA *HOTTIP*. This lncRNA from the 5′ end of *HOXA* locus interacts with WDR5 and recruits MLL complex to *HOXA* locus, leading to H3K4 methylation and gene activation [124]. *Mistral* also recruits MLL1 and activates transcription of *Hoxa6/7* in mouse ESCs [125].

6. CROSSTALK WITH OTHER PTMs

One established theme about histone PTMs is the existence of crosstalk among histone PTMs. A specific histone PTM can dictate the establishment of another mark by other histone-modifying enzymes and can also regulate the binding of chromatin readers to specific modified histone residue. Multiple examples of histone crosstalk involving H3K4 methylation and MLL complexes have been described [126]. PTM crosstalk can be mediated through multiple mechanisms. One involves a direct influence of the established PTM on enzymatic activity. For example, activity of the MLL-SET domain is stimulated by acetylation on the H3 tail; more strikingly, phospho-acetylation (H3K9ac/S10ph/K14ac) of the substrate further enhances its activity in HMT assays [127,128]. An established mark can also regulate the binding of another histone-modifying enzyme. For instance, H3R2me2a by PRMT6 inhibits H3K4 methylation by preventing the MLL subunit WDR5 from interacting with H3 [129,130]. In contrast to the asymmetrically dimethylated form, symmetric dimethylation of the same residue (H3R2me2s) facilitates binding with WDR5 and enhances H3K4 methylation. H3R2me2s overlaps genome-wide with H3K4me3 in mouse, *Xenopus*, *Saccharomyces cerevisiae*, and *Drosophila*, and this dimodified mark (H3R2me2s/K4me3) enhances the binding of the PHD finger of RAG2 (Recombination Activating Gene 2), a subunit of the V(D)J recombinase, at the antigen receptor loci (Figure 4) [131,132].

Crosstalk between histone phosphorylation and H3K4/K27 methylation has also been observed. Phosphorylation of H3T3 by haspin prevents methylation of H3K4me3 by MLL1, possibly due to the presence of the bulky, negatively charged phosphate group on H3T3 [69]. Phosphorylation of a serine/threonine next to a methylated lysine often has a negative impact on the binding of a methyl-reader and is referred to as the phos/methyl switch. TFIID subunit TAF3 binds to H3K4me3 through its PHD finger, but its binding to mitotic chromatin is decreased as a result of H3T3 phosphorylation during mitosis [133]. PRC components are also displaced from chromatin upon phosphorylation of H3S28, resulting in the activation of PcG target genes [134,135]. In contrast, phosphorylation of H3T11 by PKN1 during AR signaling has a positive effect on H3K4 methylation [136]. WDR5 binds to H3T11ph and recruits MLL1 complex to AR target genes, thereby inducing H3K4me3 and promoting prostate cancer cell proliferation.

In addition to affecting modifications on the same histone tail, a *trans*-tail crosstalk between H2B and H3 was also described. H3K4 methylation by Set1 and H3K79 methylation by Dot1 are dependent on monoubiquitylation of H2B at the C-terminus [17].

This process was originally discovered in yeast, but is now shown to be conserved in mammalian cells as well. All these examples of synergistic and antagonistic crosstalk strongly indicate that histone PTMs mediated by TrxG and PcG proteins do not function in isolation. Instead, combinations of PTMs are often involved in regulating downstream processes.

7. REGULATION OF PcG/TrxG FUNCTION BY UPSTREAM SIGNALING PATHWAYS

Regulation of epigenetic machineries and chromatin by signaling cascades has received much attention in recent years. Histone-modifying enzymes, and their resultant PTMs, can be regarded as an extension of intracellular signal transduction. They function to transmit intrinsic or extrinsic signals to chromatin, which then translates the cellular inputs into appropriate nuclear responses. In response to diverse cellular cues such as growth factors or stress signals, a number of signaling kinases (e.g., AKT, MAPKs, IKKα) have been reported to target chromatin directly, phosphorylating histone H3 and regulating expression of downstream targets [137]. In addition to phosphorylating histones, many PcG and TrxG proteins are themselves subjected to phosphorylation and other PTMs in response to cellular signaling, regulating the activity or recruitment of these chromatin regulators.

PI3K-AKT and MAPK pathways are two well-studied pathways involved in cell growth and oncogenesis, and both pathways have been reported to directly phosphorylate PcG and TrxG proteins. Phosphorylation of EZH2 at S21 impairs its binding to histone H3 and decreases its enzymatic activity. This reduces H3K27me3 levels, leading to the derepression of PRC2 target genes [138]. The same phosphorylation event also occurs in CRPC and is required for methylation of AR and expression of AR-target genes [64]. In addition to EZH2, Akt also phosphorylates the PRC1 component BMI1 [139]. This enhances the ubiquitylation activity of PRC1 in vitro and at sites of double-strand breaks, and is important for mediating DNA repair by homologous recombination. Phosphorylation of BMI1 paralog, MEL-18, also similarly increases the ubiquitylation activity of PRC1, but the MEL-18 kinase has yet to be identified [140].

p38 MAPK signaling plays a key role in myogenesis. Differentiation-activated p38α phosphorylates BAF60c, a structural component of SWI/SNF remodeling complex, as well as EZH2, but with different effects on gene expression [141,142]. During muscle regeneration, muscle satellite cells are stimulated to differentiate but inhibit proliferation. EZH2-T372 phosphorylation by TNFα-activated p38 in these cells promotes its interaction with YY1 and the subsequent recruitment of PRC2 to the promoter of lineage-specific gene *Pax7*. This represses *Pax7* expression and inhibits proliferation of muscle satellite cells [141]. In contrast, phosphorylation of BAF60c by p38 promotes the recruitment of BRG1-containing SWI/SNF to MyoD target genes such as myogenin, and promotes myogenic differentiation [142]. Phosphorylation of histone H3 by MAPK pathway also regulates gene expression by modulating the binding of TrxG and PcG proteins to chromatin. MAPK-induced phosphorylation of H3S10 and H3S28 at promoters of immediate early genes recruits BRG1 to facilitate nucleosome remodeling and to induce gene expression [143]. On the other hand, phosphorylation of H3S28 at PcG target genes results in eviction of PRC complexes from chromatin and antagonizes polycomb silencing [134,135].

In addition to signaling pathways, PcG/TrxG proteins can also be regulated by the cell cycle. Phosphorylation of EZH2 by cyclin-dependent kinases CDK1/2 at two different threonine residues has different outcomes on PRC2 activity [144–146]. Phosphorylation of EZH2 at T345 (T350 in human) enhances the binding of EZH2 to ncRNAs, such as *HOTAIR* and *XIST*, and facilitates PRC2 recruitment to target genes [145]. This enhances H3K27 methylation and suppresses downstream target genes such as *HOXA9*, and is important for cell proliferation and cell migration [144]. In contrast, phosphorylation of EZH2 at T487 disrupts its interaction with other PRC2 components EED and SUZ12, and has a negative impact on PRC2 activity [146]. This phosphorylation event is crucial for osteogenic differentiation of human mesenchymal stem cells and suggests a way in which PRC activity can be regulated by the cell cycle machinery. A number of PcG/TrxG proteins are also phosphorylated during mitosis, resulting in their dissociation from condensed mitotic chromosomes. These include BRM and BRG1, which are components of the human SWI/SNF complex, as well as members of the PCGF family (BMI1, MEL-18, PCGF6).

Besides phosphorylation, other PTMs can also play a role in regulating the recruitment and activity of chromatin regulators. A recent study suggests that recruitment of MLL complex to target genes can be regulated by sumoylation of its RBBP5 subunit [147]. SUMO-specific isopeptidase SENP3 desumolates RBBP5 and is required for association of MLL subunits MENIN and ASH2L at the developmental regulator *DLX3* gene locus. Knockdown of SENP3 decreased H3K4 methylation and expression of DLX3, impairing osteogenic differentiation of human dental follicle stem cells. These signaling-induced changes on chromatin and chromatin modifiers are beginning to be understood. Such dynamic regulation of PcG and TrxG proteins therefore allows cells to be responsive to environmental cues and regulate gene expression accordingly.

8. CONCLUDING REMARKS AND FUTURE PERSPECTIVE

TrxG and PcG proteins were first identified as regulators of HOX gene expression and as important factors during organism development in the 1970s and 1980s. Since then, we have gained an extensive knowledge on the molecular and biological functions of these important chromatin regulators. With advances in genome-wide profiling technologies, we have significantly expanded the repertoire of genes and processes regulated by TrxG and PcG proteins, and they are now recognized as key regulators of cell fate decisions. In spite of this, a number of key questions regarding these chromatin regulators remain to be understood. Studies in the past decade have led to a better biochemical characterization of PcG and TrxG complexes in both *Drosophila* and mammalian systems. The presence of large number of paralogs in mammals results in the formation of multiple forms of PRC1/2 and COMPASS-like complexes, and a number of noncanonical variant complexes with alternative subunits have also been identified. We are now only beginning to understand the molecular and functional roles of these various complexes, and it appears that they can be involved in regulating different subsets of target genes and/or different cellular processes. Interestingly, in contrary to the proposed hierarchical recruitment of canonical PRC1 to H3K27me3 regions, a few recent studies found that a variant PRC1 complex and its associated H2AK119ub1 is sufficient to recruit PRC2 and H3K27me3 for polycomb domain formation [37,38]. A good characterization of various canonical and noncanonical complexes in different cell types would help uncover their precise roles and gene targets, as well as their mechanism of action. It is also likely that different targeting mechanisms are used for their recruitment to chromatin. Future work will be required to clarify the functional relationship between these complexes.

It is now clear that PcG/TrxG proteins are involved in both short-term responses to transient signals as well as long-term maintenance of gene expression patterns during development. Numerous upstream signaling pathways are found to directly act on TrxG/PcG proteins and their target genes, indicating that these genes are more dynamically regulated than previously anticipated. Such dynamic regulation of PcG/TrxG proteins enables cells to be responsive to cellular conditions and transmit cellular signals to changes in gene expression. At the same time, TrxG/PcG proteins play crucial roles in maintaining cellular memory and identity, and this requires the inheritance of epigenetic memory through cell generations. This is particularly challenging because dramatic changes in chromosome structure and binding of chromatin proteins occur at two stages of the cell cycle—during DNA replication and fork passage in S phase and during chromosome condensation and segregation in mitosis. We are only beginning to understand the mechanisms for such epigenetic inheritance, and further studies are required to distinguish the roles and mechanisms of TrxG/PcG proteins in stable inheritance of expression states from their roles in dynamic regulation of gene expression [148].

PcG and TrxG proteins are known to act antagonistically in regulating gene expression. A number of recent studies, however, suggest that PcG and TrxG proteins can cooperate in leukemogenic transformation by MLL-fusions. Using MLL-AF9 leukemia as a model, BMI1 and MLL/HOX were found to cooperate in suppressing *Cdkn2a* locus to promote oncogenic transformation [149]. Repression of the same locus by PRC2 components EED and SUZ12 is also required to prevent differentiation and support self-renewal in an MLL-AF9:Nras^{G12D} AML model [150]. On the other hand, PRC1 component CBX8 directly interacts with MLL-AF9 and histone acetyltransferase TIP60, and together they are recruited to activate the MLL-AF9 target gene *HOXA9*. In this case, CBX8 functions independently of other PRC1 components, and is required for MLL-fusion-induced transcriptional activation and leukemogenesis [151]. In addition, recent studies showed that MLL3/4-mediated H3K4me1 has a repressive role at promoters, whereas EZH2 can also function as a transcriptional activator [64–67,98]. These point to the existence of noncanonical functions for TrxG/PcG proteins. Additional studies will be required to examine the prevalence and importance of these noncanonical functions, and the mechanisms by which these proteins function in an opposite way to their classical functions.

Extensive studies in mouse models and patient samples clearly indicate that TrxG and PcG proteins are key mediators of tumorigenesis. They play critical roles in driving oncogenesis, from tumor initiation to metastasis, and aberrant expression of these factors has often been associated with poor prognosis. Recent years have seen the development of small molecule inhibitors that target PcG/TrxG pathways as well as other epigenetic machineries in cancer [152]. Most of these epigenetic drugs target the enzymes that write or erase the epigenetic marks, and there are other inhibitors that target the binding of readers to chromatin. For instance, several EZH2 inhibitors have been developed, and they are effective in selectively inhibiting the growth of DLBCL cell lines with EZH2-Y641 activating mutation as well as in DLBCL mouse xenograft models. DOT1L inhibitors also selectively kill MLL-rearranged leukemias, but have no effect on non-MLL rearranged cells. These HMT inhibitors have now entered into early phases of clinical trials. In addition, inhibitors that interfere with the binding of bromodomain-containing proteins to acetylated histones (e.g., JQ1 and I-BET151, which inhibit the BET (bromodomain and extraterminal) family of proteins) have shown encouraging results in a number of hematological malignancies, indicating that histone reader modules can be promising therapeutic targets [153]. Discovery of the first small molecule probe that targets the methyl–lysine reader domain MBT has recently been reported, but inhibitors for chromodomains or PHD fingers (readers for methylated H3K4 and H3K27) are yet to be identified. Further studies into the molecular and cellular functions of TrxG and PcG factors will provide a better insight into how their dysregulation contributes to cancer and other human diseases, and will further provide opportunities for intervention in the clinic.

ACKNOWLEDGMENTS

This work is supported by Cancer Research UK (CRUK), Leukemia and Lymphoma Research (LLR), the Association for International Cancer Research (AICR), Medical Research Council (MRC), and Kay Kendall Leukemia Fund (KKLF).

REFERENCES

[1] Felsenfeld G, Groudine M. Controlling the double helix. Nature 2003;421(6921):448–53.
[2] Luger K, Mader AW, Richmond RK, Sargent DF, Richmond TJ. Crystal structure of the nucleosome core particle at 2.8 A resolution. Nature 1997;389(6648):251–60.
[3] Bannister AJ, Kouzarides T. Regulation of chromatin by histone modifications. Cell Res 2011;21(3):381–95.
[4] Gardner KE, Allis CD, Strahl BD. Operating on chromatin, a colorful language where context matters. J Mol Biol 2011;409(1):36–46.
[5] Musselman CA, Lalonde ME, Cote J, Kutateladze TG. Perceiving the epigenetic landscape through histone readers. Nat Struct Mol Biol 2012;19(12):1218–27.
[6] Plass C, Pfister SM, Lindroth AM, Bogatyrova O, Claus R, Lichter P. Mutations in regulators of the epigenome and their connections to global chromatin patterns in cancer. Nat Rev Genet 2013;14(11):765–80.
[7] Portela A, Esteller M. Epigenetic modifications and human disease. Nat Biotechnol 2010;28(10):1057–68.
[8] Mills AA. Throwing the cancer switch: reciprocal roles of polycomb and trithorax proteins. Nat Rev Cancer 2010;10(10):669–82.
[9] Schuettengruber B, Martinez AM, Iovino N, Cavalli G. Trithorax group proteins: switching genes on and keeping them active. Nat Rev Mol Cell Biol 2011;12(12):799–814.
[10] Laugesen A, Helin K. Chromatin repressive complexes in stem cells, development, and cancer. Cell Stem Cell 2014;14(6):735–51.
[11] Kennison JA. The Polycomb and trithorax group proteins of Drosophila: trans-regulators of homeotic gene function. Annu Rev Genet 1995;29:289–303.
[12] Lewis EB. A gene complex controlling segmentation in Drosophila. Nature 1978;276(5688):565–70.
[13] Ingham PW, Whittle R. Trithorax: a new homoeotic mutation of Drosophila melanogaster causing transformations of abdominal and thoracic imaginal segments. Mol Gen Genet MGG 1980;179:607–14.
[14] van der Lugt NM, Domen J, Linders K, van Roon M, Robanus-Maandag E, te Riele H, et al. Posterior transformation, neurological abnormalities, and severe hematopoietic defects in mice with a targeted deletion of the bmi-1 proto-oncogene. Genes Dev 1994;8(7):757–69.
[15] Yu BD, Hess JL, Horning SE, Brown GA, Korsmeyer SJ. Altered Hox expression and segmental identity in Mll-mutant mice. Nature 1995;378(6556):505–8.
[16] del Mar Lorente M, Marcos-Gutierrez C, Perez C, Schoorlemmer J, Ramirez A, Magin T, et al. Loss-and gain-of-function mutations show a polycomb group function for Ring1A in mice. Development 2000;127(23):5093–100.
[17] Shilatifard A. The COMPASS family of histone H3K4 methylases: mechanisms of regulation in development and disease pathogenesis. Annu Rev Biochem 2012;81:65–95.
[18] Bracken AP, Dietrich N, Pasini D, Hansen KH, Helin K. Genome-wide mapping of Polycomb target genes unravels their roles in cell fate transitions. Genes Dev 2006;20(9):1123–36.
[19] Lee TI, Jenner RG, Boyer LA, Guenther MG, Levine SS, Kumar RM, et al. Control of developmental regulators by Polycomb in human embryonic stem cells. Cell 2006;125(2):301–13.
[20] O'Meara MM, Simon JA. Inner workings and regulatory inputs that control Polycomb repressive complex 2. Chromosoma 2012;121(3):221–34.
[21] Margueron R, Li G, Sarma K, Blais A, Zavadil J, Woodcock CL, et al. Ezh1 and Ezh2 maintain repressive chromatin through different mechanisms. Mol Cell 2008;32(4):503–18.
[22] Son J, Shen SS, Margueron R, Reinberg D. Nucleosome-binding activities within JARID2 and EZH1 regulate the function of PRC2 on chromatin. Genes Dev 2013;27(24):2663–77.
[23] Shen X, Liu Y, Hsu YJ, Fujiwara Y, Kim J, Mao X, et al. EZH1 mediates methylation on histone H3 lysine 27 and complements EZH2 in maintaining stem cell identity and executing pluripotency. Mol Cell 2008;32(4):491–502.
[24] Ezhkova E, Lien WH, Stokes N, Pasolli HA, Silva JM, Fuchs E. EZH1 and EZH2 cogovern histone H3K27 trimethylation and are essential for hair follicle homeostasis and wound repair. Genes Dev 2011;25(5):485–98.

[25] Margueron R, Justin N, Ohno K, Sharpe ML, Son J, Drury 3rd WJ, et al. Role of the polycomb protein EED in the propagation of repressive histone marks. Nature 2009;461(7265):762–7.

[26] Yuan W, Wu T, Fu H, Dai C, Wu H, Liu N, et al. Dense chromatin activates Polycomb repressive complex 2 to regulate H3 lysine 27 methylation. Science 2012;337(6097):971–5.

[27] Cao R, Zhang Y. SUZ12 is required for both the histone methyltransferase activity and the silencing function of the EED-EZH2 complex. Mol Cell 2004;15(1):57–67.

[28] Herz HM, Shilatifard A. The JARID2-PRC2 duality. Genes Dev 2010;24(9):857–61.

[29] Nekrasov M, Klymenko T, Fraterman S, Papp B, Oktaba K, Kocher T, et al. Pcl-PRC2 is needed to generate high levels of H3-K27 trimethylation at Polycomb target genes. EMBO J 2007;26(18): 4078–88.

[30] Sarma K, Margueron R, Ivanov A, Pirrotta V, Reinberg D. Ezh2 requires PHF1 to efficiently catalyze H3 lysine 27 trimethylation in vivo. Mol Cell Biol 2008;28(8):2718–31.

[31] Schwartz YB, Pirrotta V. A new world of Polycombs: unexpected partnerships and emerging functions. Nat Rev Genet 2013;14(12):853–64.

[32] Di Croce L, Helin K. Transcriptional regulation by Polycomb group proteins. Nat Struct Mol Biol 2013;20(10):1147–55.

[33] Ku M, Koche RP, Rheinbay E, Mendenhall EM, Endoh M, Mikkelsen TS, et al. Genomewide analysis of PRC1 and PRC2 occupancy identifies two classes of bivalent domains. PLoS Genet 2008;4(10):e1000242.

[34] Leeb M, Pasini D, Novatchkova M, Jaritz M, Helin K, Wutz A. Polycomb complexes act redundantly to repress genomic repeats and genes. Genes Dev 2010;24(3):265–76.

[35] Schoeftner S, Sengupta AK, Kubicek S, Mechtler K, Spahn L, Koseki H, et al. Recruitment of PRC1 function at the initiation of X inactivation independent of PRC2 and silencing. EMBO J 2006;25(13): 3110–22.

[36] Tavares L, Dimitrova E, Oxley D, Webster J, Poot R, Demmers J, et al. RYBP-PRC1 complexes mediate H2A ubiquitylation at polycomb target sites independently of PRC2 and H3K27me3. Cell 2012;148(4):664–78.

[37] Cooper S, Dienstbier M, Hassan R, Schermelleh L, Sharif J, Blackledge NP, et al. Targeting polycomb to pericentric heterochromatin in embryonic stem cells reveals a role for H2AK119u1 in PRC2 recruitment. Cell Rep 2014;7(5):1456–70.

[38] Blackledge NP, Farcas AM, Kondo T, King HW, McGouran JF, Hanssen LL, et al. Variant PRC1 complex-dependent H2A ubiquitylation drives PRC2 recruitment and polycomb domain formation. Cell 2014;157(6):1445–59.

[39] Kalb R, Latwiel S, Baymaz HI, Jansen PW, Muller CW, Vermeulen M, et al. Histone H2A monoubiquitination promotes histone H3 methylation in Polycomb repression. Nat Struct Mol Biol 2014.

[40] Francis NJ, Kingston RE, Woodcock CL. Chromatin compaction by a polycomb group protein complex. Science 2004;306(5701):1574–7.

[41] Sexton T, Yaffe E, Kenigsberg E, Bantignies F, Leblanc B, Hoichman M, et al. Three-dimensional folding and functional organization principles of the Drosophila genome. Cell 2012;148(3): 458–72.

[42] Eskeland R, Leeb M, Grimes GR, Kress C, Boyle S, Sproul D, et al. Ring1B compacts chromatin structure and represses gene expression independent of histone ubiquitination. Mol Cell 2010;38(3):452–64.

[43] Trojer P, Cao AR, Gao Z, Li Y, Zhang J, Xu X, et al. L3MBTL2 protein acts in concert with PcG protein-mediated monoubiquitination of H2A to establish a repressive chromatin structure. Mol Cell 2011;42(4):438–50.

[44] Chopra VS, Hendrix DA, Core LJ, Tsui C, Lis JT, Levine M. The polycomb group mutant esc leads to augmented levels of paused Pol II in the Drosophila embryo. Mol Cell 2011;42(6):837–44.

[45] Stock JK, Giadrossi S, Casanova M, Brookes E, Vidal M, Koseki H, et al. Ring1-mediated ubiquitination of H2A restrains poised RNA polymerase II at bivalent genes in mouse ES cells. Nat Cell Biol 2007;9(12):1428–35.

[46] Zhou W, Zhu P, Wang J, Pascual G, Ohgi KA, Lozach J, et al. Histone H2A monoubiquitination represses transcription by inhibiting RNA polymerase II transcriptional elongation. Mol Cell 2008;29(1):69–80.

[47] Pasini D, Malatesta M, Jung HR, Walfridsson J, Willer A, Olsson L, et al. Characterization of an antagonistic switch between histone H3 lysine 27 methylation and acetylation in the transcriptional regulation of Polycomb group target genes. Nucleic Acids Res 2010;38(15):4958–69.

[48] Pengelly AR, Copur O, Jackle H, Herzig A, Muller J. A histone mutant reproduces the phenotype caused by loss of histone-modifying factor Polycomb. Science 2013;339(6120):698–9.

[49] Vire E, Brenner C, Deplus R, Blanchon L, Fraga M, Didelot C, et al. The Polycomb group protein EZH2 directly controls DNA methylation. Nature 2006;439(7078):871–4.

[50] Pasini D, Hansen KH, Christensen J, Agger K, Cloos PA, Helin K. Coordinated regulation of transcriptional repression by the RBP2 H3K4 demethylase and Polycomb-Repressive Complex 2. Genes Dev 2008;22(10):1345–55.

[51] He A, Shen X, Ma Q, Cao J, von Gise A, Zhou P, et al. PRC2 directly methylates GATA4 and represses its transcriptional activity. Genes Dev 2012;26(1):37–42.

[52] Pasini D, Bracken AP, Hansen JB, Capillo M, Helin K. The polycomb group protein Suz12 is required for embryonic stem cell differentiation. Mol Cell Biol 2007;27(10):3769–79.

[53] Chamberlain SJ, Yee D, Magnuson T. Polycomb repressive complex 2 is dispensable for maintenance of embryonic stem cell pluripotency. Stem Cells 2008;26(6):1496–505.

[54] Endoh M, Endo TA, Endoh T, Fujimura Y, Ohara O, Toyoda T, et al. Polycomb group proteins Ring1A/B are functionally linked to the core transcriptional regulatory circuitry to maintain ES cell identity. Development 2008;135(8):1513–24.

[55] Popov N, Gil J. Epigenetic regulation of the INK4b-ARF-INK4a locus: in sickness and in health. Epigenet Off J DNA Methylation Soc 2010;5(8):685–90.

[56] Brockdorff N. Chromosome silencing mechanisms in X-chromosome inactivation: unknown unknowns. Development 2011;138(23):5057–65.

[57] Sauvageau M, Sauvageau G. Polycomb group proteins: multi-faceted regulators of somatic stem cells and cancer. Cell Stem Cell 2010;7(3):299–313.

[58] Crea F, Paolicchi E, Marquez VE, Danesi R. Polycomb genes and cancer: time for clinical application? Crit Rev Oncol/Hematol 2012;83(2):184–93.

[59] Yap DB, Chu J, Berg T, Schapira M, Cheng SW, Moradian A, et al. Somatic mutations at EZH2 Y641 act dominantly through a mechanism of selectively altered PRC2 catalytic activity, to increase H3K27 trimethylation. Blood 2011;117(8):2451–9.

[60] Muto T, Sashida G, Oshima M, Wendt GR, Mochizuki-Kashio M, Nagata Y, et al. Concurrent loss of Ezh2 and Tet2 cooperates in the pathogenesis of myelodysplastic disorders. J Exp Med 2013;210(12): 2627–39.

[61] Simon C, Chagraoui J, Krosl J, Gendron P, Wilhelm B, Lemieux S, et al. A key role for EZH2 and associated genes in mouse and human adult T-cell acute leukemia. Genes Dev 2012;26(7):651–6.

[62] Lewis PW, Muller MM, Koletsky MS, Cordero F, Lin S, Banaszynski LA, et al. Inhibition of PRC2 activity by a gain-of-function H3 mutation found in pediatric glioblastoma. Science 2013;340(6134): 857–61.

[63] Jacobs JJ, Scheijen B, Voncken JW, Kieboom K, Berns A, van Lohuizen M. Bmi-1 collaborates with c-Myc in tumorigenesis by inhibiting c-Myc-induced apoptosis via INK4a/ARF. Genes Dev 1999;13(20):2678–90.

[64] Xu K, Wu ZJ, Groner AC, He HH, Cai C, Lis RT, et al. EZH2 oncogenic activity in castration-resistant prostate cancer cells is Polycomb-independent. Science 2012;338(6113):1465–9.

[65] Shi B, Liang J, Yang X, Wang Y, Zhao Y, Wu H, et al. Integration of estrogen and Wnt signaling circuits by the polycomb group protein EZH2 in breast cancer cells. Mol Cell Biol 2007;27(14):5105–19.

[66] Lee ST, Li ZM, Wu ZL, Aau M, Guan PY, Karuturi RKM, et al. Context-specific regulation of NF-kappa B target gene expression by EZH2 in breast cancers. Mol Cell 2011;43(5):798–810.

[67] Gonzalez ME, Moore HM, Li X, Toy KA, Huang W, Sabel MS, et al. EZH2 expands breast stem cells through activation of NOTCH1 signaling. Proc Natl Acad Sci U S A 2014;111(8):3098–103.

[68] Dou Y, Milne TA, Ruthenburg AJ, Lee S, Lee JW, Verdine GL, et al. Regulation of MLL1 H3K4 methyltransferase activity by its core components. Nat Struct Mol Biol 2006;13(8):713–9.

[69] Southall SM, Wong PS, Odho Z, Roe SM, Wilson JR. Structural basis for the requirement of additional factors for MLL1 SET domain activity and recognition of epigenetic marks. Mol Cell 2009;33(2):181–91.

[70] Dou Y, Milne TA, Tackett AJ, Smith ER, Fukuda A, Wysocka J, et al. Physical association and coordinate function of the H3 K4 methyltransferase MLL1 and the H4 K16 acetyltransferase MOF. Cell 2005;121(6):873–85.

[71] Petruk S, Sedkov Y, Smith S, Tillib S, Kraevski V, Nakamura T, et al. Trithorax and dCBP acting in a complex to maintain expression of a homeotic gene. Science 2001;294(5545):1331–4.

[72] Bantignies F, Goodman RH, Smolik SM. Functional interaction between the coactivator *Drosophila* CREB-binding protein and ASH1, a member of the trithorax group of chromatin modifiers. Mol Cell Biol 2000;20(24):9317–30.

[73] Hargreaves DC, Crabtree GR. ATP-dependent chromatin remodeling: genetics, genomics and mechanisms. Cell Res 2011;21(3):396–420.

[74] Wysocka J, Swigut T, Xiao H, Milne TA, Kwon SY, Landry J, et al. A PHD finger of NURF couples histone H3 lysine 4 trimethylation with chromatin remodelling. Nature 2006;442(7098):86–90.

[75] Schmitges FW, Prusty AB, Faty M, Stutzer A, Lingaraju GM, Aiwazian J, et al. Histone methylation by PRC2 is inhibited by active chromatin marks. Mol Cell 2011;42(3):330–41.

[76] Yuan W, Xu M, Huang C, Liu N, Chen S, Zhu B. H3K36 methylation antagonizes PRC2-mediated H3K27 methylation. J Biol Chem 2011;286(10):7983–9.

[77] Lee MG, Villa R, Trojer P, Norman J, Yan KP, Reinberg D, et al. Demethylation of H3K27 regulates polycomb recruitment and H2A ubiquitination. Science 2007;318(5849):447–50.

[78] De Santa F, Totaro MG, Prosperini E, Notarbartolo S, Testa G, Natoli G. The histone H3 lysine-27 demethylase Jmjd3 links inflammation to inhibition of polycomb-mediated gene silencing. Cell 2007;130(6):1083–94.

[79] Ruthenburg AJ, Allis CD, Wysocka J. Methylation of lysine 4 on histone H3: intricacy of writing and reading a single epigenetic mark. Mol Cell 2007;25(1):15–30.

[80] Sims 3rd RJ, Chen CF, Santos-Rosa H, Kouzarides T, Patel SS, Reinberg D. Human but not yeast CHD1 binds directly and selectively to histone H3 methylated at lysine 4 via its tandem chromodomains. J Biol Chem 2005;280(51):41789–92.

[81] Ruthenburg AJ, Li H, Milne TA, Dewell S, McGinty RK, Yuen M, et al. Recognition of a mononucleosomal histone modification pattern by BPTF via multivalent interactions. Cell 2011;145(5):692–706.

[82] Shi X, Hong T, Walter KL, Ewalt M, Michishita E, Hung T, et al. ING2 PHD domain links histone H3 lysine 4 methylation to active gene repression. Nature 2006;442(7098):96–9.

[83] Vermeulen M, Mulder KW, Denissov S, Pijnappel WW, van Schaik FM, Varier RA, et al. Selective anchoring of TFIID to nucleosomes by trimethylation of histone H3 lysine 4. Cell 2007;131(1):58–69.

[84] Sims 3rd RJ, Millhouse S, Chen CF, Lewis BA, Erdjument-Bromage H, Tempst P, et al. Recognition of trimethylated histone H3 lysine 4 facilitates the recruitment of transcription postinitiation factors and pre-mRNA splicing. Mol Cell 2007;28(4):665–76.

[85] Petruk S, Sedkov Y, Riley KM, Hodgson J, Schweisguth F, Hirose S, et al. Transcription of bxd noncoding RNAs promoted by trithorax represses Ubx in cis by transcriptional interference. Cell 2006;127(6):1209–21.

[86] Gu B, Lee MG. Histone H3 lysine 4 methyltransferases and demethylases in self-renewal and differentiation of stem cells. Cell Biosci 2013;3(1):39.

[87] Wan M, Liang J, Xiong Y, Shi F, Zhang Y, Lu W, et al. The trithorax group protein Ash2l is essential for pluripotency and maintaining open chromatin in embryonic stem cells. J Biol Chem 2013;288(7):5039–48.

[88] Ang YS, Tsai SY, Lee DF, Monk J, Su J, Ratnakumar K, et al. Wdr5 mediates self-renewal and reprogramming via the embryonic stem cell core transcriptional network. Cell 2011;145(2):183–97.

[89] Ho L, Jothi R, Ronan JL, Cui K, Zhao K, Crabtree GR. An embryonic stem cell chromatin remodeling complex, esBAF, is an essential component of the core pluripotency transcriptional network. Proc Natl Acad Sci U S A 2009;106(13):5187–91.

[90] Bernstein BE, Mikkelsen TS, Xie X, Kamal M, Huebert DJ, Cuff J, et al. A bivalent chromatin structure marks key developmental genes in embryonic stem cells. Cell 2006;125(2):315–26.

[91] Hu D, Garruss AS, Gao X, Morgan MA, Cook M, Smith ER, et al. The Mll2 branch of the COMPASS family regulates bivalent promoters in mouse embryonic stem cells. Nat Struct Mol Biol 2013;20(9):1093–7.

[92] Denissov S, Hofemeister H, Marks H, Kranz A, Ciotta G, Singh S, et al. Mll2 is required for H3K4 trimethylation on bivalent promoters in embryonic stem cells, whereas Mll1 is redundant. Development 2014;141(3):526–37.

[93] Jiang H, Shukla A, Wang X, Chen WY, Bernstein BE, Roeder RG. Role for Dpy-30 in ES cell-fate specification by regulation of H3K4 methylation within bivalent domains. Cell 2011;144(4):513–25.

[94] Rampalli S, Li L, Mak E, Ge K, Brand M, Tapscott SJ, et al. p38 MAPK signaling regulates recruitment of Ash2L-containing methyltransferase complexes to specific genes during differentiation. Nat Struct Mol Biol 2007;14(12):1150–6.

[95] McKinnell IW, Ishibashi J, Le Grand F, Punch VG, Addicks GC, Greenblatt JF, et al. Pax7 activates myogenic genes by recruitment of a histone methyltransferase complex. Nat Cell Biol 2008;10(1):77–84.

[96] Gori F, Friedman LG, Demay MB. Wdr5, a WD-40 protein, regulates osteoblast differentiation during embryonic bone development. Dev Biol 2006;295(2):498–506.

[97] Wu MZ, Tsai YP, Yang MH, Huang CH, Chang SY, Chang CC, et al. Interplay between HDAC3 and WDR5 is essential for hypoxia-induced epithelial-mesenchymal transition. Mol Cell 2011;43(5):811–22.

[98] Cheng J, Blum R, Bowman C, Hu D, Shilatifard A, Shen S, et al. A role for H3K4 monomethylation in gene repression and partitioning of chromatin readers. Mol Cell 2014;53(6):979–92.

[99] Herz HM, Mohan M, Garruss AS, Liang K, Takahashi YH, Mickey K, et al. Enhancer-associated H3K4 monomethylation by Trithorax-related, the *Drosophila* homolog of mammalian Mll3/Mll4. Genes Dev 2012;26(23):2604–20.

[100] Hu D, Gao X, Morgan MA, Herz HM, Smith ER, Shilatifard A. The MLL3/MLL4 branches of the COMPASS family function as major histone H3K4 monomethylases at enhancers. Mol Cell Biol 2013;33(23):4745–54.

[101] Yip BH, So CW. Mixed lineage leukemia protein in normal and leukemic stem cells. Exp Biol Med 2013;238(3):315–23.

[102] Bernt KM, Zhu N, Sinha AU, Vempati S, Faber J, Krivtsov AV, et al. MLL-rearranged leukemia is dependent on aberrant H3K79 methylation by DOT1L. Cancer Cell 2011;20(1):66–78.

[103] Cheung N, Chan LC, Thompson A, Cleary ML, So CW. Protein arginine-methyltransferase-dependent oncogenesis. Nat Cell Biol 2007;9(10):1208–15.

[104] Chen C, Liu Y, Rappaport AR, Kitzing T, Schultz N, Zhao Z, et al. MLL3 is a haploinsufficient 7q tumor suppressor in acute myeloid leukemia. Cancer Cell 2014;25(5):652–65.

[105] Kia SK, Gorski MM, Giannakopoulos S, Verrijzer CP. SWI/SNF mediates polycomb eviction and epigenetic reprogramming of the INK4b-ARF-INK4a locus. Mol Cell Biol 2008;28(10):3457–64.

[106] Roberts CW, Leroux MM, Fleming MD, Orkin SH. Highly penetrant, rapid tumorigenesis through conditional inversion of the tumor suppressor gene Snf5. Cancer Cell 2002;2(5):415–25.

[107] Schuettengruber B, Cavalli G. Recruitment of polycomb group complexes and their role in the dynamic regulation of cell fate choice. Development 2009;136(21):3531–42.

[108] Sing A, Pannell D, Karaiskakis A, Sturgeon K, Djabali M, Ellis J, et al. A vertebrate Polycomb response element governs segmentation of the posterior hindbrain. Cell 2009;138(5):885–97.

[109] Woo CJ, Kharchenko PV, Daheron L, Park PJ, Kingston RE. A region of the human HOXD cluster that confers polycomb-group responsiveness. Cell 2010;140(1):99–110.

[110] Mendenhall EM, Koche RP, Truong T, Zhou VW, Issac B, Chi AS, et al. GC-rich sequence elements recruit PRC2 in mammalian ES cells. PLoS Genet 2010;6(12):e1001244.

[111] Milne TA, Kim J, Wang GG, Stadler SC, Basrur V, Whitcomb SJ, et al. Multiple interactions recruit MLL1 and MLL1 fusion proteins to the HOXA9 locus in leukemogenesis. Mol Cell 2010;38(6):853–63.

[112] Bina M, Wyss P, Novorolsky E, Zulkelfi N, Xue J, Price R, et al. Discovery of MLL1 binding units, their localization to CpG Islands, and their potential function in mitotic chromatin. BMC Genomics 2013;14:927.

[113] Li G, Margueron R, Ku M, Chambon P, Bernstein BE, Reinberg D. Jarid2 and PRC2, partners in regulating gene expression. Genes Dev 2010;24(4):368–80.

[114] Kim H, Kang K, Kim J. AEBP2 as a potential targeting protein for Polycomb Repression Complex PRC2. Nucleic Acids Res 2009;37(9):2940–50.

[115] Abed JA, Jones RS. H3K36me3 key to Polycomb-mediated gene silencing in lineage specification. Nat Struct Mol Biol 2012;19(12):1214–5.

[116] Yu M, Mazor T, Huang H, Huang HT, Kathrein KL, Woo AJ, et al. Direct recruitment of polycomb repressive complex 1 to chromatin by core binding transcription factors. Mol Cell 2012;45(3):330–43.

[117] Arnold P, Scholer A, Pachkov M, Balwierz PJ, Jorgensen H, Stadler MB, et al. Modeling of epigenome dynamics identifies transcription factors that mediate Polycomb targeting. Genome Res 2013;23(1):60–73.

[118] Dietrich N, Lerdrup M, Landt E, Agrawal-Singh S, Bak M, Tommerup N, et al. REST-mediated recruitment of polycomb repressor complexes in mammalian cells. PLoS Genet 2012;8(3):e1002494.

[119] Villa R, Pasini D, Gutierrez A, Morey L, Occhionorelli M, Vire E, et al. Role of the polycomb repressive complex 2 in acute promyelocytic leukemia. Cancer Cell 2007;11(6):513–25.

[120] Boukarabila H, Saurin AJ, Batsche E, Mossadegh N, van Lohuizen M, Otte AP, et al. The PRC1 Polycomb group complex interacts with PLZF/RARA to mediate leukemic transformation. Genes Dev 2009;23(10):1195–206.

[121] Wang X, Zhu K, Li S, Liao Y, Du R, Zhang X, et al. MLL1, a H3K4 methyltransferase, regulates the TNFalpha-stimulated activation of genes downstream of NF-kappaB. J Cell Sci 2012;125(Pt 17): 4058–66.

[122] Brockdorff N. Noncoding RNA and Polycomb recruitment. RNA 2013;19(4):429–42.

[123] Rinn JL, Kertesz M, Wang JK, Squazzo SL, Xu X, Brugmann SA, et al. Functional demarcation of active and silent chromatin domains in human HOX loci by noncoding RNAs. Cell 2007;129(7): 1311–23.

[124] Wang KC, Yang YW, Liu B, Sanyal A, Corces-Zimmerman R, Chen Y, et al. A long noncoding RNA maintains active chromatin to coordinate homeotic gene expression. Nature 2011;472(7341):120–4.

[125] Bertani S, Sauer S, Bolotin E, Sauer F. The noncoding RNA Mistral activates Hoxa6 and Hoxa7 expression and stem cell differentiation by recruiting MLL1 to chromatin. Mol Cell 2011;43(6): 1040–6.

[126] Binda O. On your histone mark, SET, methylate! Epigenet Off J DNA Methylation Soc 2013;8(5): 457–63.

[127] Milne TA, Briggs SD, Brock HW, Martin ME, Gibbs D, Allis CD, et al. MLL targets SET domain methyltransferase activity to Hox gene promoters. Mol Cell 2002;10(5):1107–17.

[128] Nightingale KP, Gendreizig S, White DA, Bradbury C, Hollfelder F, Turner BM. Cross-talk between histone modifications in response to histone deacetylase inhibitors: MLL4 links histone H3 acetylation and histone H3K4 methylation. J Biol Chem 2007;282(7):4408–16.

[129] Guccione E, Bassi C, Casadio F, Martinato F, Cesaroni M, Schuchlautz H, et al. Methylation of histone H3R2 by PRMT6 and H3K4 by an MLL complex are mutually exclusive. Nature 2007;449(7164):933–7.

[130] Kirmizis A, Santos-Rosa H, Penkett CJ, Singer MA, Vermeulen M, Mann M, et al. Arginine methylation at histone H3R2 controls deposition of H3K4 trimethylation. Nature 2007;449(7164):928–32.

[131] Migliori V, Muller J, Phalke S, Low D, Bezzi M, Mok WC, et al. Symmetric dimethylation of H3R2 is a newly identified histone mark that supports euchromatin maintenance. Nat Struct Mol Biol 2012;19(2):136–44.

[132] Yuan CC, Matthews AG, Jin Y, Chen CF, Chapman BA, Ohsumi TK, et al. Histone H3R2 symmetric dimethylation and histone H3K4 trimethylation are tightly correlated in eukaryotic genomes. Cell Rep 2012;1(2):83–90.

[133] Varier RA, Outchkourov NS, de Graaf P, van Schaik FM, Ensing HJ, Wang F, et al. A phospho/methyl switch at histone H3 regulates TFIID association with mitotic chromosomes. EMBO J 2010;29(23): 3967–78.

[134] Gehani SS, Agrawal-Singh S, Dietrich N, Christophersen NS, Helin K, Hansen K. Polycomb group protein displacement and gene activation through MSK-dependent H3K27me3S28 phosphorylation. Mol Cell 2010;39(6):886–900.

[135] Lau PN, Cheung P. Histone code pathway involving H3 S28 phosphorylation and K27 acetylation activates transcription and antagonizes polycomb silencing. Proc Natl Acad Sci U S A 2011;108(7): 2801–6.

[136] Kim JY, Banerjee T, Vinckevicius A, Luo Q, Parker JB, Baker MR, et al. A role for WDR5 in integrating threonine 11 phosphorylation to lysine 4 methylation on histone H3 during androgen signaling and in prostate cancer. Mol Cell 2014.

[137] Lau PN, Cheung P. Histone phosphorylation: chromatin modifications that link cell signaling pathways to nuclear function regulation. In: Bradshaw RA, Dennis EA, editors. Handbook of cell signaling. 2nd ed. Oxford: Academic Press; 2009. p. 2399–408.

[138] Cha TL, Zhou BP, Xia W, Wu Y, Yang CC, Chen CT, et al. Akt-mediated phosphorylation of EZH2 suppresses methylation of lysine 27 in histone H3. Science 2005;310(5746):306–10.

[139] Nacerddine K, Beaudry JB, Ginjala V, Westerman B, Mattiroli F, Song JY, et al. Akt-mediated phosphorylation of Bmi1 modulates its oncogenic potential, E3 ligase activity, and DNA damage repair activity in mouse prostate cancer. J Clin Invest 2012;122(5):1920–32.

[140] Elderkin S, Maertens GN, Endoh M, Mallery DL, Morrice N, Koseki H, et al. A phosphorylated form of Mel-18 targets the Ring1B histone H2A ubiquitin ligase to chromatin. Mol Cell 2007;28(1):107–20.

[141] Palacios D, Mozzetta C, Consalvi S, Caretti G, Saccone V, Proserpio V, et al. TNF/p38alpha/polycomb signaling to Pax7 locus in satellite cells links inflammation to the epigenetic control of muscle regeneration. Cell Stem Cell 2010;7(4):455–69.

[142] Forcales SV, Albini S, Giordani L, Malecova B, Cignolo L, Chernov A, et al. Signal-dependent incorporation of MyoD-BAF60c into Brg1-based SWI/SNF chromatin-remodelling complex. EMBO J 2012;31(2):301–16.

[143] Drobic B, Perez-Cadahia B, Yu J, Kung SK, Davie JR. Promoter chromatin remodeling of immediate-early genes is mediated through H3 phosphorylation at either serine 28 or 10 by the MSK1 multiprotein complex. Nucleic Acids Res 2010;38(10):3196–208.

[144] Chen S, Bohrer LR, Rai AN, Pan Y, Gan L, Zhou X, et al. Cyclin-dependent kinases regulate epigenetic gene silencing through phosphorylation of EZH2. Nat Cell Biol 2010;12(11):1108–14.

[145] Kaneko S, Li G, Son J, Xu CF, Margueron R, Neubert TA, et al. Phosphorylation of the PRC2 component Ezh2 is cell cycle-regulated and up-regulates its binding to ncRNA. Genes Dev 2010;24(23):2615–20.

[146] Wei Y, Chen YH, Li LY, Lang J, Yeh SP, Shi B, et al. CDK1-dependent phosphorylation of EZH2 suppresses methylation of H3K27 and promotes osteogenic differentiation of human mesenchymal stem cells. Nat Cell Biol 2011;13(1):87–94.

[147] Nayak A, Viale-Bouroncle S, Morsczeck C, Muller S. The SUMO-specific isopeptidase SENP3 regulates MLL1/MLL2 methyltransferase complexes and controls osteogenic differentiation. Mol Cell 2014.

[148] Steffen PA, Ringrose L. What are memories made of? How Polycomb and Trithorax proteins mediate epigenetic memory. Nat Rev Mol Cell Biol 2014;15(5):340–56.

[149] Smith LL, Yeung J, Zeisig BB, Popov N, Huijbers I, Barnes J, et al. Functional crosstalk between Bmi1 and MLL/Hoxa9 axis in establishment of normal hematopoietic and leukemic stem cells. Cell Stem Cell 2011;8(6):649–62.

[150] Shi J, Wang E, Zuber J, Rappaport A, Taylor M, Johns C, et al. The Polycomb complex PRC2 supports aberrant self-renewal in a mouse model of MLL-AF9;Nras(G12D) acute myeloid leukemia. Oncogene 2013;32(7):930–8.

[151] Tan J, Jones M, Koseki H, Nakayama M, Muntean AG, Maillard I, et al. CBX8, a polycomb group protein, is essential for MLL-AF9-induced leukemogenesis. Cancer Cell 2011;20(5):563–75.

[152] Campbell RM, Tummino PJ. Cancer epigenetics drug discovery and development: the challenge of hitting the mark. J Clin Invest 2014;124(1):64–9.

[153] Dawson MA, Kouzarides T, Huntly BJ. Targeting epigenetic readers in cancer. N Engl J Med 2012;367(7):647–57.

CHAPTER 5

Chromatin dynamics and genome organization in development and disease

Changwang Deng[1,*], Bhavita Patel[2,*], Xiumei Lin[3], Yangqiu Li[4], Suming Huang[1]

[1]Department of Biochemistry and Molecular Biology, University of Florida College of Medicine, Gainesville, FL, USA;
[2]Department of Medicine, University of Florida College of Medicine, Gainesville, FL, USA; [3]Department of Hematology,
Guangzhou First People's Hospital, Guangzhou, China; [4]Institute of Hematology, Jinan University Medical College,
Guangzhou, China

Contents

1. INTRODUCTION

Eukaryotic genomes are packaged and confined into the minute space of the nucleus. This genomic information needs to be readily available for duplication, transmission, and expression. All of these processes require quick and easy transitions between the relatively uncompacted state and highly compacted state during cell cycle while assuring successful transmission of genetic and epigenetic information from mother to daughter cells. Previously, due to limitations in light microscopic techniques, the question of how the genome is arranged and organized in the nucleus was obscured. The development of 3C (Chromosome Conformation Capture) and its derivatives have made it possible to study

★ These authors contribute equally.

Epigenetic Gene Expression and Regulation
http://dx.doi.org/10.1016/B978-0-12-799958-6.00005-6

the relationship between nuclear organization and gene expression. One of the driving forces in the development of these techniques has been the question: *how does an enhancer, located long distances from its target promoter, activate a gene?* With this in mind, various studies have highlighted the presence of intra- and interchromosomal interactions that occur over large distances throughout the genomes. Though most of these interactions are functionally important in gene expression, it is difficult to assess whether the conformation effects are a cause or consequence of gene regulation. However, supportive epigenetic evidence together with loss-of-function studies of essential factors has provided a functional role for chromatin interactions. Furthermore, the invention of the Hi-C (3C extension allowing unbiased identification of long-range interactions) and the 5C (Chromosome Conformation Capture Carbon Copy) platforms allowed the examination of higher-order chromatin structure on a genome-wide scale. These studies revealed that the genome is organized into highly self-interacting regions associating termed topologically associating domains (TADs), which are highly conserved among mammals. Together, these findings have aided the development of our understanding of the roles that the higher-order nuclear chromatin organization plays in regulating gene expression.

2. TECHNOLOGY ADVANCES IN UNRAVELING CHROMOSOME ARCHITECTURE AND GENOME ORGANIZATION

The theory of chromosomal territories was proposed as early as 1982, using traditional light microscopic techniques [1,2]. Later, these findings were confirmed by DNA fluorescence in situ hybridization (FISH) analysis, notably at the mouse beta-globin gene cluster [3]. The power of FISH and microscopic techniques lies in their ability to do single-cell analysis. However, this is limited by poor resolution and a lack of an overall cell population analysis. These limitations were overcome by the introduction of 3C and its variants. One of the original supports for direct interaction between a gene promoter and its enhancer was carried out by Cullen et al. in 1993, where they showed interaction between an enhancer and promoter after stimulation of the estrogen receptor [4]. With the advancement of 3C-based techniques, it is now possible to map long-range *cis* and *trans* chromosomal interactions. This information allows us to generate three-dimensional (3D) maps of the genome in the cell nucleus and to understand the relationship between genome organization, nuclear structure, and function.

2.1 Mapping long-range enhancer and promoter interactions using chromosome conformation capture (3C) and circular chromosome conformation capture (4C) assays

First described by the Job Dekker group in 2002, 3C technology was used to demonstrate interactions between the telomere and centromere in yeast cells, confirming earlier microscopic studies [5]. Later, it was successfully used to identify long-range interactions between the mouse *β−globin* genes and the LCR (locus control region) in a developmental specific phenomena [6–8]. Since then, 3C has been successfully utilized to identify long-range chromatin interactions at various gene loci including, *IGF2R* (insulin-like growth factor 2 receptor), *MYB* (myeloblastosis oncogene), *TAL1* (T-cell acute lymphocytic leukemia 1), and *alpha-globin* locus [9–11]. In a typical 3C assay, interactions among genomic loci are captured by cross-linking with formaldehyde, followed by digestion with a restriction enzyme of choice and ligation in very dilute conditions to allow intramolecular ligation of the cross-linked fragments. This generates a genome-wide 3C interaction library, where interactions between two known loci are detected by traditionally semiquantitative polymerase chain reaction (PCR) using primers in the bait and test sequence. The interaction data, thus obtained, are cell population average representative and are limited to the scope of studying only one-to-one interactions.

To overcome the limitations of the 3C assay, 4C (circular chromosome conformation capture) technology was derived with the power to analyze relatively all possible interactions mediated by a selected region or anchor of interest; in other words, a one-to-all scenario [12,13]. As the name suggests, in 4C, inverse PCR is used to amplify all fragments ligated to a single anchor fragment involving a circularization step. The PCR products are then used to generate the 4C library, which is then subjected to high-throughput microarray or next generation sequencing platforms to obtain a genome-wide interaction profile for that anchor locus. 4C technology was first applied to investigate genomic interactions by comparing the tissue specific β-globin gene with the housekeeping *RAD23A* (RAD23 homolog A, repairosome nucleotide excision repair protein [ubiquitin-like protein]) gene [12]. While *RAD23A* predominantly interacts with active gene clusters in different tissues, the tissue-specific expressed *β-globin* gene had different interaction profiles based on its expression status. Thus, the data suggest that the chromosome is organized into distinct active and inactive domains, dependent on expression status [12]. Consistent with these findings, it was shown that the inactive and active X-chromosomes adopt distinct topologies [14]. During X-inactivation, the long noncoding RNA *Xist* (X-inactive specific transcript) establishes and spreads silencing effects by exploiting 3D repressive chromatin domains [15]. Furthermore, the connection between chromatin loops and gene transcription was demonstrated in the study of the expression of olfactory receptors (OR) in sensory neurons [16]. A single *trans*-acting enhancer, *H* element, interacts with different *OR* gene promoters on different chromosomes, allowing each olfactory sensory neuron to specifically express only one of thousands of *OR* genes [16].

2.2 Spatial organization of topological domains for genome-wide gene regulation

With the emerge of Hi-C technique in 2009, it has become possible to determine chromatin interactions in a truly unbiased and genome-wide manner without needing to limit the analysis to one selected anchor or group of them. The Hi-C technology is also based on 3C, but includes a step before ligation in which the staggered ends of the restriction fragments are filled in with biotinylated nucleotides, followed by blunt-end ligation and biotin-mediated pulldown. The purified DNA is then subjected to pair-end high-throughput genome sequencing, and the reads are then mapped back to the genome. When a pair is found on two different restriction fragments, this pair is scored as an interaction between two fragments [17]. Hi-C has been used to study the overall folding of genomes. For large genomes such as those of humans and mice, Hi-C analysis will produce an interaction map with a resolution of ~0.1–1 Mb. One of the first observations of 5C studies was that the long-range interaction landscape in mammalian genomes comprised highly interacting topologically associating domains [18–20]. Interphase chromosomes occupy individual territories and are compartmentalized at distinct levels, from large active and inactive multi-megabases compartments to smaller than 1 Mb TADs [17]. TADs are invariant across cell types, suggesting an overall conserved domain structure of the genome [18]. Organizing the genome into TADs clearly has regulator implications, because regions within TADS are in close proximity while the regions between TADs rarely interact. The transition regions between TADs are bound by CTCF (CCCTC-DNA binding factor) and cohesin.

Hi-C allows studying genome–interaction profiles on a single-cell level. Nagano and colleagues utilized single-cell Hi-C to analyze individual T-helper cells, where they reported striking cell-to-cell variability in interactions between and within chromosomes [21]. Combining Hi-C and ChIP-seq (Chromatin Immunoprecipitation coupled with massively parallel DNA sequencing) data has revealed that chromatin insulators or an insulator-binding protein, such as CTCF, are generally enriched at the boundaries of TADs, suggesting that CTCF and transcription factors play an important role in organizing the genome structure [18]. Thus, deletion of CTCF and cohesin-bound boundary regions was found to disturb neighboring TADs [19].

In order to further understand protein-dependent DNA interactions and organization in the eukaryotic genome, the Chromatin Interaction Analysis by Paired-End Tag Sequencing (ChIA-PET) technique was developed [22,23]. In addition to 3D genome organization, the precise coordinate of transcription depends on the binding of tissue-specific transcription factors to the regulatory genomic elements. Comprehensive analysis of the estrogen receptor-α (ER-α)-bound chromatin interactome in the human genome demonstrated that ER-α mediates genome-wide chromatin interaction by bringing ER-α–dependent gene loci into close proximity with gene promoters to coordinate transcription [24].

Furthermore, CTCF has been shown to be extensively involved in modulating long-range chromatin interactions in pluripotent stem cells where CTCF-dependent loops demarcate chromatin–nuclear membrane attachments and transcriptional active gene cluster through extensive crosstalk between promoters and genomic regulatory elements. Thus, CTCF-mediated genome-wide chromatin topology plays an important role in gene regulation and stem cell function [25]. The most direct evidence that long-range interactions between promoters and distal regulatory elements are involved in transcription comes from the recent genome-wide ChIA-PET analysis of the RNA polymerase II-bound chromatin interaction networks [26]. In this study, authors demonstrated that promoter-directed interactions usually involve transcriptionally active genes, where the actively transcribed genes with similar functions were found to be physically clustered in close proximity [26,27].

3. PROTEINS INVOLVED IN GENOME ORGANIZATION

3.1 CTCF-mediated genome organization

Insulator proteins are characterized as a specific class of DNA-binding proteins that function to insulate the aberrant activation or repression of genes by restricting enhancer and silencer functions [28,29]. Due to the advancements in 3C-based techniques, a special class of architectural proteins has recently been recognized, which includes some known insulator proteins and is responsible for global genome organization as well as local enhancer–promoter interactions [30–33]. The CTCF-was the first described vertebrate insulator protein that binds to a DNA insulator element, 5′HS4, which was located at the 5′end of the chicken *β-globin* locus [28]. Later, it was shown that CTCF would bind to the mouse *β-globin* locus and mediate long-range chromatin interactions and histone modifications in the locus [34,35]. CTCF is ubiquitously expressed, highly conserved across species, and required for early development of the mouse embryo [36,37]. CTCF is required for genome imprinting, X-chromosome inactivation, and higher-order chromatin organization. This diversity in its role is, in part, due to the ability of CTCF to homodimerize with itself and to heterodimerize with other proteins [38,39]. Genome-wide binding analysis by ChIP-seq in various tissues has revealed that CTCF was specifically enriched at gene-dense regions, localizes to Dnase I hypersensitive sites, and was an open chromatin determinant; therefore, CTCF demarcated active and inactive chromatin or gene boundaries [40–44].

There has been an increasing amount of data validating the role of CTCF in global genome organization. Indeed, CTCF was involved in tethering chromosomal DNA to nuclear lamina by binding outside lamina-associated domains [45]. Furthermore, CTCF can mediate both intra- and interchromosomal interactions and can influence gene transcription both positively and negatively. To date, there are several gene loci where the function of CTCF in chromatin looping and gene transcription has been well characterized. One of the best-characterized loci is the *IGF2-H19*, where the binding of CTCF onto the *H19* imprinting control regions results in the differential expression of the paternal Insulin-like Growth Factor 2 (*IGF2*) allele and maternal *H19* allele through physical chromatin interactions with the differentially methylated regions upstream of the *IGF2* promoter [46–50].

CTCF is also implicated in subnuclear localization and mediating inactive Xist interactions during X-chromosome inactivation [51,52]. Recently, CTCF was reported to organize the *TAL1* locus differentially to allow the *TAL1* enhancer–promoter interaction in normal hematopoietic progenitors, while blocking the same interaction in T-cell leukemia patients [11]. Beyond these well-characterized interactions, additional CTCF-dependent inter- and intrachromosomal interactions have been identified using 4C [13].

Genome-wide analysis of CTCF-associated chromatin loops by the ChIA-PET technique in pluripotent mouse embryonic stem cells (mESCs) implied that CTCF has a multifaceted role in genome organization. It creates local chromatin hubs that allow coordinated expression within gene clusters, mediates interactions between distal enhancers and promoters, and specifies boundaries between active and repressive chromatin domains [25]. In the study by Handoko et al. in 2011, a total of 1816 high-confidence interactions mediated by CTCF were defined, 1480 of which were intrachromosomal interactions and 336 were interchromosomal interactions. These interactions were further confirmed by 4C analysis for selected anchor points and DNA FISH techniques. Furthermore, the CTCF-associated chromatin interactome was defined into five distinct chromatin domains based on the distribution of seven distinct histone modifications. The interaction map was then overlapped with genome-wide RNA polymerase II (RNAPII) and p300 (E1A binding protein p300) binding. The overlap revealed that 28% of the genes, whose promoters are located far away from enhancers and bound by RNAPII, are brought into close proximity to the p300-bound enhancers through CTCF-mediated looping and are upregulated in mESCs compared with neuronal stem cells [25], suggesting that CTCF-mediated enhancer/promoter long-range interactions coordinate transcriptional activation. In accordance with previous findings of CTCF as a boundary element, genome-wide Hi-C analysis in mESCs and human fibroblasts identified CTCF as a prominent boundary element between topological domains [18]. These boundaries are also enriched with tRNAs and SINE (short interspersed nuclear elements), both of which are postulated to function as insulators [53,54]. Given the available information, CTCF has been proposed as one of the leading candidates to act as a global genome organizer to coordinate higher-order chromatin structures and regulate gene expression [55].

3.2 Cohesin-mediated interactions

Cohesin functions to tether chromosomal DNA during mitosis and meiosis to allow proper DNA segregation into daughter cells. It is a macromolecular complex that comprised two long coiled-coil molecules, SMC1 (structural maintenance of chromosome protein 1) and SMC3 (structural maintenance of chromosome protein 3) that form an open-ended heterodimer, RAD21 (mitotic cohesin complex, non-SMC subunit Rad21) that bridges the open end, and SA1 (stromal antigen 1, cohesin subunit SA-1) and SA2 (stromal antigen 2, cohesin subunit SA-2) proteins that interact with RAD21 on the outside of the SMC1-SMC3-RAD21 trimer. This complex is believed to cause chromatin cohesion by trapping DNA into its ring-like structure, thereby influencing chromatin structure and probably forming or stabilizing chromatin loops. Genome-wide ChIP-seq analysis for CTCF and cohesin revealed that these two proteins mostly overlap in their binding patterns [36,56]. Furthermore, CTCF coimmunoprecipitates with cohesin, and in vitro studies revealed a direct interaction between CTCF and SA2 [57]. Another molecule, DEAD-box RNA helicase p68 (DEAD [Asp-Glu-Ala-Asp] box helicase 5, in humans), interacts with both CTCF and cohesin. Depletion of p68 or SRA (steroid receptor RNA activator 1) results in loss of cohesin binding to CTCF [58]. Zuin et al. in 2014 showed that depletion of cohesin resulted in reduced intradomain interactions in TADs, while CTCF depletion led to increased interdomain interactions [59]. However, not all sites occupied by cohesin are CTCF dependent. In CTCF-depleted cells, most of the cohesin-bound sites are unperturbed, indicating overlapping as well as distinct functions of CTCF and cohesin in genome organization [36,60,61]. Interestingly, cohesin copurifies and colocalizes with mediator complex, which is recruited mostly to activate gene promoters and enhancers [62,63]. Mediator is a large multisubunit complex comprising approximately 30 subunits and is highly conserved from yeast to higher organisms [64]. It was recently shown that both cohesin and mediator components are involved in cell-type-specific enhancer–promoter interactions. Depletion of either component results in partial loss of these interactions [65]. In *Drosophila*, cohesin also functionally interacts with the polycomb group (PcG) proteins at both inactive and actively transcribing genes. It recruits PcG to many active genes, but their binding at inactive genes is mutually exclusive. Furthermore, depletion of cohesin disrupts long-range interaction between PcG binding elements [66]. Additionally, cohesin also colocalizes with activator complexes and functions to promote reestablishment of transcription factor clusters after S-phase and chromatin condensation [67]. Thus, cohesin is considered to be important during M-phase, and also emerged as an essential player in roles involved in chromatin and genome organization during G1 and S phases.

3.3 SATB1 and nuclear matrix-associated genome organization

Chromatin, at a higher order of compaction, is organized into looped domains, which are tethered and maintained by interaction with the nuclear matrix. These DNA elements, which span several hundred bases are mostly AT-rich and are termed matrix-associated/attachment regions (MARs), scaffold-associated/attachment regions (SARs), or combined as SMARs (S/MARs). MARs determine chromatin structure and accessibility, and therefore, play an essential role in gene transcription, replication, and repair. Special AT-rich binding protein 1 (SATB1) is the most well-characterized MAR-binding protein, and aids in the maintenance and compaction of chromatin architecture within the nucleus. In thymocyte nuclei, SATB1 forms a cage-like network around heterochromatin and further tethers specialized DNA sequences containing ATC (AT-rich sequence with either C or G exclusively on one strand) onto its network. SATB1 acts to control transcriptional activity of this network by establishing active histone modification across the network [68]. Working along the same lines of thought in 2007, Kumar et al. used SATB1 RNAi-mediated knockdown to show that SATB1 was involved in overall chromatin loop organization of the MHC-1 locus (myosin heavy chain-1) [69]. Furthermore, they identified the promyelocytic leukemia oncoprotein (PML) as a SATB1 interacting partner. PML has been proposed as a site for the regulation of transcriptional activity for specific loci via higher-order chromatin organization [70].

SATB1 is also important for fetal hemoglobin switching by binding to the HS2 (DNase I hypersensitive site 2) region of the *β-globin* LCR and promoting hypomethylation and hyperacetylation of the histones at the *ε-globin* gene promoter, and hypermethylation and hypoacetylation at the *γ-globin* gene promoter [71]. In 2009, Agrelo et al. showed that SATB1 was essential for initiation of X-chromosome inactivation in lymphoma cells and in embryonic stem cells [72]. However, later it was reported that both SATB1 and SATB2 (special AT-rich sequence-binding protein 2) proteins are dispensable for X-chromosome inactivation, since fibroblasts derived from $SATB1^{-/-}$ and $SATB2^{-/-}$ females did not show any upregulation of X-linked genes and contained proper Barr bodies [73]. Apart from its role in tethering AT-rich DNA sequence to MARs and higher-order chromatin organization, SATB1 also serves as a docking site for histone-modifying enzymes such as HDAC1 (histone deacetylase 1), ACF (ATP-dependent chromatin assembly factor, large subunit), and the chromatin remodeler known as ISWI (imitation switching) [74,75].

3.4 Nuclear pore interactions

Nuclear pore complex (NPC) is primarily a site in the nuclear envelope, which regulates entry and exit of cellular material including proteins and RNA. It comprises at least 30 proteins called nucleoporins [76]. Previously, electron micrographs of the nuclear periphery have revealed that euchromatin tends to localize near NPCs, while heterochromatin lies close to the nuclear periphery and lamina [77,78]. Not surprisingly, various studies have shown an association between active genes and NPCs [79,80]. The role of NPCs in genome organization and transcription is well established in yeast; however, its role in higher eukaryotes is just starting to evolve. In yeast, highly expressed genes such as those involved in glycolysis and in ribosomal protein synthesis, as well as genes induced by environmental stimuli, are associated with NPCs constitutively or conditionally, respectively [79,81,82]. The histone acetyltransferase complex, SAGA (Spt-Ada-Gcn5 acetyltransferase) plays an essential role in modulating the interaction of genes with NPC. In *Drosophila*, silencing of SAGA components delocalizes the *HSP70* (Heat Shock Protein 70) locus from the nuclear periphery [83], hyperactivates the X-chromosome in males localized at nuclear periphery, and is associated with nuclear pore components [84]. Despite the advances in our current knowledge, one of the prominent questions still holds, and that is: *what targets the transcribing DNA to the NPCs, and do NPCs or transport machinery directly interact with transcription factors to mediate or regulate this process?*

4. INTRA- AND INTERCHROMOSOMAL INTERACTIONS DURING DEVELOPMENT

Nuclear organization of the genome is a complex and dynamic process that compartmentalizes molecular machines to regulate nuclear processes including transcription, replication, and DNA repair in nuclei [85]. The proper nuclear organization is particularly important for expression of many developmentally regulated complex gene loci in which long-range chromatin topologies coordinate the transcription at specific developmental stages or in specific tissues (Figure 1). The most famous and best characterized example is the human and mouse *β-globin* loci. The *β-globin* locus contains several developmentally regulated globin genes, an upstream LCR, and a number of additional regulatory elements that are embedded in a large region of the inactive *OR* genes. During erythropoiesis, the *β-globin* locus adopts nonuniform histone modification patterns with methylation of H4 at arginine-3 (H4R3) and hyperacetylation of H3 and H4 with enrichment at the LCR and at the transcribed *β-globin* gene [8,86,87]. Both the LCR and the β^{maj}-promoter play important roles in generating an active chromatin territory hub (ACH), for maximal activity [7,88,89]. The LCR can serve as a primary site to recruit transcription factors and chromatin modifying as well as remodeling factors, and to dynamically alter topology of the *β-globin* locus in erythrocytes [90]. In erythroid cells, active genes including the globin genes from discrete loci are frequently organized into

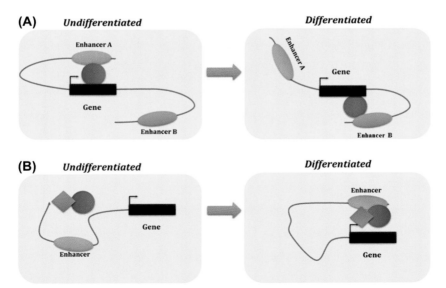

Figure 1 The model illustrates that genes are regulated by transcription factor-and cofactor-mediated long-range chromatin loops during development. (A) Different enhancers differentially interact with a gene for transcriptional activation during cellular differentiation. (B) Enhancer is brought within close proximity of the gene upon differentiation.

shared nuclear compartments with ongoing transcription [7]. Analysis of dynamic chromatin organization in the *β-globin* locus revealed that the LCR forms a conserved ACH with upstream 5′HS (DNase I-hypersensitive site) elements and downstream 3′HS1 (DNase I-hypersensitive site 1) elements, which then serves as a nucleation site to differentially interact with active globin genes during development and erythroid differentiation [91]. The switch of interactions between the ACH and developmentally activated globin genes positively correlates with the switch in their transcriptional activity [6,91]. In erythroid cells, the LCR and *β-globin* gene interaction is mediated by the GATA1/TAL1 (GATA, Globin transcription factor 1/T-cell acute lymphocytic leukemia 1) complex and its associated cofactor Ldb1 (*lim* Domain Binding 1). Although long-range enhancer and promoter interaction has been implicated in transcriptional regulation, the direct evidence that the enhancer/promoter long-range loop is required for gene activation comes from the study of zinc finger (ZF)-Ldb1 fusion-mediated enhancer/promoter dimerization [92]. In this artificial system, the ZF that recognizes the adult *β-globin* promoter was fused with Ldb1 to tether the LCR to the *β-globin* promoter in the *GATA1* null G1E (GATA-1- erythroid) erythroblasts in which the GATA-1 mediated LCR/*β-globin* promoter interaction is disrupted, and the *β-globin* gene is silent. Remarkably, when the long-range loop juxtaposes the LCR, the *β-globin* promoter substantially increases recruitment of phosphorylated RNAPII and activates *β-globin* transcription [92]. Significantly, this mechanism addresses the previous challenge regarding the cause and effect relationship between chromatin looping and gene regulation, which may also support that intra- and interchromosomal interactions play a direct role in a general cell-type-specific gene regulation during development.

During B-lymphocyte development, the pro-B to pre-B cells transition requires the rearrangement of immunoglobin heavy chain (*IGH*) locus [93,94]. PAX5 (paired box 5) is one of the key regulatory factors binding to the sequences that are interspersed in the distal V_H (Ig heavy [H] chain variable [V]) gene region in the *IGH* locus, called the PAX5-activated intergenic repeat (PAIR) elements [95], from which antisense noncoding RNAs (as-ncRNAs) are transcribed in progenitor B-cells (pro-B cells). It was shown that ncRNA transcripts generated within this region are required for *IGH* rearrangement by tethering the distal V_H gene region in close proximity with *Eμ* (IGH intronic enhancer) at the pro-B cell stage [95–98]. During pro-B cell to pre-B (precursor B-cell) cell differentiation stage, the rearranging *IGH* locus contraction requires formation of two distinct multilooped rosette-like domains in the pro-B cells. The first multiple looped domain involves the interactions of the *Eμ* enhancer with 5′DFL and 3′-Regulatory Regions (3′RR) in the 3′ end of the rearranging locus that initiates the $D_H J_H$ (Ig heavy [H] chain variable [D], Ig heavy [H] chain variable [J]; where D = diversity and J = joining) rearrangement. In the second multiple-looped domain, the *Eμ* enhancer brings the $D_H J_H$ region in close proximity to both the distal and proximal parts of the V_H gene regions, which are spread over large megabase distances within the *IGH* locus [99,100].

In addition, antisense transcripts from PAIR4 (PAX4 [Paired Box 4]-activated intergenic repeat element) and PAIR6 (PAX6 [Paired Box 6]-activated intergenic repeat elements) in the distal region of V_H genes are critical to form the $E\mu$-PAIR interactions that also facilitate locus compaction, and allow distal V_H genes to undergo efficient rearrangement [97].

Dynamic regulation of HOX (homeobox) genes plays an important role in the limb formation that includes specification, outgrowth, and patterning of the limb field [101–103]. In vertebrates, temporal and spatial expression of HOX genes during embryonic development relies on the long-range chromatin interactions between the HOX locus and distal enhancers [104]. By using high-resolution 3C methodology, Noordermeer et al. revealed a significant change in spatial configuration of all four HOX loci during mammalian development. When the HOX cluster remains transcriptionally silent, HOX genes are orchestrated into a single tight 3D structure. Once active, the clusters reorganize into separated 3D domains with the activated genes being progressively clustered into a transcriptionally active compartment [105]. This configuration transition coincides with the dynamics of chromatin modification marks, which reflect the progression from silent clusters to a transcriptional active status [106]. In the HOXD locus, the early phase of gene transcription is important for the prospective arm and the forearm formation and includes HOXD3~HOXD11 genes. A regulatory element, located in the telomeric-end gene desert, is required for maintaining the transcriptional activity by long-range interaction with this locus. Deletion of a certain part of this gene desert causes HoxD gene transcription levels, as well as the histone marks, to be abolished. Conversely, although the HOXD9~HOXD11 genes are also activated in the late phase for shaping the digits, a much different element regulates this process. In this latter stage, a centromeric enhancer comes to contact the locus instead of the telomeric enhancer. This fine-tuned switch of chromatin interactions plays a key role in regulating the balance of distinct HoxD gene expression, thus ensuring proper limb patterning and development [107].

5. INTRA- AND INTERCHROMOSOMAL INTERACTIONS DURING CARCINOGENESIS

Disease initiation and progress requires gene expression undergoing large-scale reprogramming in a disease prone manner that results from the alteration in the genetic and the epigenetic status of cells. To facilitate the establishment of this unique cancer prone chromatin signature, aberrant long-range gene interactions could anchor hundreds or thousands of genes together to alter genome expression patterns, thus leading to cancer-related processes (Figure 2) [108,109]. A recent 4C analysis of the breast cancer related IGFBP3 (insulin-like growth factor-binding protein 3) locus interaction network revealed substantial alterations in the long-range intra- or interchromasomal interactions in breast cancer cells. Cancer cells lost many of the chromatin interactions seen in normal breast epithelial cells and gained novel interactions (Figure 2). In breast cancer cells, *IGFBP3* comes in close contact with chromatin regions that are prone to chromosomal translocation. Thus, these data suggest that long-range interactions might favor chromosome rearrangements during tumorigenesis [110]. A ChIA-PET study of the ER-α-bound chromatin interactome in human breast adenocarcinoma cells, treated with estrogen, demonstrated that ER-α mediates chromatin loops, and compartmentalizes and concentrates ER-α cofactors and transcription-related protein complexes with its target genes for coordinated transcriptional regulation [24].

Interestingly, SATB1, a protein that binds exclusively to BURs (base unpairing regions) and plays an important role in genome organization for coordinated gene expression [68,111,112], is highly expressed in aggressive breast cancer cells, and its expression correlates with poor prognoses of breast cancer [113]. SATB1 organizes active chromatin architecture by anchoring numerous gene loci and by recruiting chromatin modifying and remodeling enzymes to control gene expression on a large scale. Thus, elevated SATB1 in aggressive breast cancer cells might promote tumor metastasis by anchoring and regulating thousands of genes. More importantly, these genes are tethered together in the SATB1-organized nuclear architecture by attaching to BURs for coordinated transcriptional regulation, which enables cells to alter their function [113].

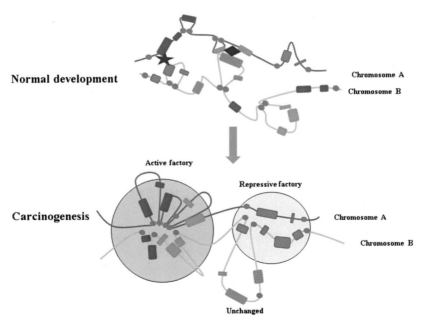

Figure 2 The model depicts that an aberrant alteration in genome organization may lead to drastic changes in global gene expression patterns causing a perturbation of normal cellular function and malignant transformation. Green dots represent chromatin organizers.

T-cell acute lymphoblastic leukemia (T-ALL) is a malignant blood disease, which is mostly characterized by genetic abnormalities that are crucial for T-cell pathogenesis [114]. Aberrant activation of TAL1 is found in 40–60% of T-ALL patients, resulting from chromosomal translocation (4–5%), interstitial chromosome deletion (25–30%), or by an undefined mechanism (60%) [115–117]. Despite its role in T-cell leukemia, TAL1 is a master hematopoietic regulator and is essential for normal development of all blood lineages [118,119]. The human *TAL1* gene is located on chromosome 1p32 and is tightly regulated by various *cis*-regulatory elements spanning ~199 kb on the genome, including distinct *TAL1* enhancers and CTCF-bound elements that together control *TAL1* expression levels in different hematopoietic lineages and stages [120–123].

How do these enhancers and promoters coordinate to differentially regulate TAL1 gene expression in different developmental stages and malignant leukemia? The answer to whether CTCF is directly involved in genome organization to control the enhancer/promoter interactions remains elusive. Two publications in 2014 [11] and 2013 [124] demonstrated a CTCF-mediated genome organization of chromatin architecture in the *TAL1* locus that modulates accessibility of enhancers to the *TAL1* promoter and its expression. Oncogenic *TAL1* expression is regulated by different intra- and interchromosomal loops in normal hematopoietic and leukemic cells, respectively. These intra- and interchromosomal loops alter the cell-type-specific enhancers that interact with the *TAL1* promoter. One study showed that human SET1 (hSET1; Su(var)3–9, Enhancer of Zeste [EHZ] and Trithorax [TRX]) complex-mediated H3K4 methylations promoted a long-range chromatin loop that brought the +51 enhancer in close proximity to the *TAL1* promoter in erythroid cells. CTCF facilitates this long-range enhancer/promoter interaction of the *TAL1* locus in erythroid cells while blocking the same enhancer/promoter interaction of the *TAL1* locus in human T-cell leukemia. In human T-ALL, a T-cell-specific transcription factor c-Maf (cellular v-maf avian musculoaponeurotic fibrosarcoma oncogene homolog)-mediated interchromosomal interaction brings the *TAL1* promoter into close proximity with a T-cell-specific regulatory element located on chromosome 16, activating aberrant *TAL1* oncogene expression [11]. Consistent with this report, the study in the mouse locus supports the model that the *TAL1* promoters and enhancers are in close contact within an erythroid transcriptional ACH. And, CTCF-mediated looping between the +53 and −31 element is hub-dependent in a GATA1-mediated manner [124]. However, the *TAL1* promoter also interacts with a *-81/TAL^d* site, which is the common *TAL1/STIL* (SCL/TAL1 interrupting locus) deletion breakpoint in patients with T-ALL [124]. Thus, the *TAL1* promoter loop interaction is preferentially selected for chromosomal rearranged regions in leukemia.

In addition to the above cell-type-specific regulation, *TAL1* expression in cancer cells was also regulated by an activating ncRNA-a3 (ncRNA-activating3), which is located downstream of the *TAL1* locus and functions as an enhancer to activate their neighboring genes via a *cis*-mediated mechanism [125]. In this case, ncRNA-a3 recruits mediators especially the MED1 (mediator complex subunit 1) and MED12 (mediator complex subunit 12) subunits to its target (the *TAL1* gene promoter), and mediates long-range chromatin interactions at the locus. Depletion of mediator subunits MED1 and MED12, or ncRNA-a3, disrupted the chromatin looping between ncRNA-a3 and the *TAL1* locus. This and other studies suggested that the enhancer-like ncRNAs maybe involved in long-range transcriptional activation.

6. CLOSING REMARKS AND FUTURE DIRECTIONS

Current technology advance and genome-wide studies provide compelling evidence that long-range intra- and interchromosomal interactions underlie transcriptional regulation of the mammalian genome. The genome is organized into distinct 3D structures to facilitate its regulation. Molecular mechanisms of global genome organization have just begun to emerge. Increasing evidence support a genome-wide role for CTCF in chromatin interactions and 3D genome organization [55]. However, many important questions remain to be answered. It is still unclear how CTCF acts to demarcate independent chromatin domains to regulate gene expression. Although dynamic genomic chromatin interactions in the nucleus are implicated in regulating gene expression, a central question is *whether long-range chromatin interaction is a direct cause of transcription*. A recent study in the mouse *β-globin* locus provided the first evidence that enforced looping between the LCR and the adult *β-globin* promoter stimulates globin transcription [92]. However, it still remains largely unknown how chromatin loops coordinate with different histone modification marks and whether they work together to modulate gene transcription.

During lineage commitment and differentiation, these processes require the coordination of external stimuli, signal transduction pathways, and lineage-restricted transcription factors for lineage fate determination. Whether and how enhancers, promoters, and regulatory factors are gathered to the same or to distinct chromatin domains are unknown. It is even important to understand how changes in these factors alter the genome organization and lead to a disease state. All information and knowledge that we have thus far gathered suggests that many, if not all, transcription-related events within the nucleus take place at clusters of the genome, enriched with regulatory factors that facilitate long-range chromatin interactions between genes and their regulatory sites.

ACKNOWLEDGMENTS

We thank our colleague in the Huang and Li Labs for their comments on this manuscript. This work was supported by the grants from the National Institute of Health (HL091929 and HL090589 to S.H.) and from Natural Science Foundation of China (91129720 and 81270604 to Y.L.).

REFERENCES

[1] Cremer T, Cremer C, Schneider T, Baumann H, Hens L, Kirsch-Volders M. Analysis of chromosome positions in the interphase nucleus of Chinese hamster cells by laser-UV-microirradiation experiments. Hum Genet 1982;62(3):201–9.

[2] Cremer T, Cremer M, Dietzel S, Muller S, Solovei I, Fakan S. Chromosome territories–a functional nuclear landscape. Curr Opin Cell Biol June 2006;18(3):307–16.

[3] Noordermeer D, Branco MR, Splinter E, Klous P, van Ijcken W, Swagemakers S, et al. Transcription and chromatin organization of a housekeeping gene cluster containing an integrated beta-globin locus control region. PLoS Genet March 2008;4(3):e1000016.

[4] Cullen KE, Kladde MP, Seyfred MA. Interaction between transcription regulatory regions of prolactin chromatin. Science July 9, 1993;261(5118):203–6.

[5] Dekker J, Rippe K, Dekker M, Kleckner N. Capturing chromosome conformation. Science February 15, 2002;295(5558):1306–11.

[6] Tolhuis B, Palstra RJ, Splinter E, Grosveld F, de Laat W. Looping and interaction between hypersensitive sites in the active beta-globin locus. Mol Cell December 2002;10(6):1453–65.

[7] Osborne CS, Chakalova L, Brown KE, Carter D, Horton A, Debrand E, et al. Active genes dynamically colocalize to shared sites of ongoing transcription. Nat Genet October 2004;36(10):1065–71.

[8] Li X, Hu X, Patel B, Zhou Z, Liang S, Ybarra R, et al. H4R3 methylation facilitates beta-globin transcription by regulating histone acetyltransferase binding and H3 acetylation. Blood March 11, 2010;115(10):2028–37. [Research Support, N.I.H., Extramural Research Support, Non-U.S. Gov't].

[9] Vernimmen D, Marques-Kranc F, Sharpe JA, Sloane-Stanley JA, Wood WG, Wallace HA, et al. Chromosome looping at the human alpha-globin locus is mediated via the major upstream regulatory element (HS -40). Blood November 5, 2009;114(19):4253–60.

[10] Stadhouders R, Thongjuea S, Andrieu-Soler C, Palstra RJ, Bryne JC, van den Heuvel A, et al. Dynamic long-range chromatin interactions control Myb proto-oncogene transcription during erythroid development. EMBO J February 15, 2012;31(4):986–99.

[11] Patel B, Kang Y, Cui K, Litt M, Riberio MS, Deng C, et al. Aberrant TAL1 activation is mediated by an interchromosomal interaction in human T-cell acute lymphoblastic leukemia. Leukemia February 2014;28(2):349–61. [Research Support, American Recovery and Reinvestment Act Research Support, N.I.H., Extramural Research Support, N.I.H., Intramural].

[12] Simonis M, Klous P, Splinter E, Moshkin Y, Willemsen R, de Wit E, et al. Nuclear organization of active and inactive chromatin domains uncovered by chromosome conformation capture-on-chip (4C). Nat Genet November 2006;38(11):1348–54.

[13] Zhao Z, Tavoosidana G, Sjolinder M, Gondor A, Mariano P, Wang S, et al. Circular chromosome conformation capture (4C) uncovers extensive networks of epigenetically regulated intra- and inter-chromosomal interactions. Nat Genet November 2006;38(11):1341–7.

[14] Wutz A, Rasmussen TP, Jaenisch R. Chromosomal silencing and localization are mediated by different domains of Xist RNA. Nat Genet February 2002;30(2):167–74.

[15] Engreitz JM, Pandya-Jones A, McDonel P, Shishkin A, Sirokman K, Surka C, et al. The Xist lncRNA exploits three-dimensional genome architecture to spread across the X chromosome. Science August 16, 2013;341(6147):1237973. [Research Support, N.I.H., Extramural Research Support, Non-U.S. Gov't Research Support, U.S. Gov't, Non-P.H.S.].

[16] Lomvardas S, Barnea G, Pisapia DJ, Mendelsohn M, Kirkland J, Axel R. Interchromosomal interactions and olfactory receptor choice. Cell July 28, 2006;126(2):403–13. [Research Support, N.I.H., Extramural Research Support, Non-U.S. Gov't].

[17] Lieberman-Aiden E, van Berkum NL, Williams L, Imakaev M, Ragoczy T, Telling A, et al. Comprehensive mapping of long-range interactions reveals folding principles of the human genome. Science October 9, 2009;326(5950):289–93.

[18] Dixon JR, Selvaraj S, Yue F, Kim A, Li Y, Shen Y, et al. Topological domains in mammalian genomes identified by analysis of chromatin interactions. Nature May 17, 2012;485(7398):376–80.

[19] Nora EP, Lajoie BR, Schulz EG, Giorgetti L, Okamoto I, Servant N, et al. Spatial partitioning of the regulatory landscape of the X-inactivation centre. Nature May 17, 2012;485(7398):381–5.

[20] Sexton T, Kurukuti S, Mitchell JA, Umlauf D, Nagano T, Fraser P. Sensitive detection of chromatin coassociations using enhanced chromosome conformation capture on chip. Nat Protoc July 2012;7(7):1335–50.

[21] Nagano T, Lubling Y, Stevens TJ, Schoenfelder S, Yaffe E, Dean W, et al. Single-cell Hi-C reveals cell-to-cell variability in chromosome structure. Nature October 3, 2013;502(7469):59–64.

[22] Fullwood MJ, Wei CL, Liu ET, Ruan Y. Next-generation DNA sequencing of paired-end tags (PET) for transcriptome and genome analyses. Genome Res April 2009;19(4):521–32.

[23] Li G, Fullwood MJ, Xu H, Mulawadi FH, Velkov S, Vega V, et al. ChIA-PET tool for comprehensive chromatin interaction analysis with paired-end tag sequencing. Genome Biol 2010;11(2):R22. [Research Support, N.I.H., Extramural Research Support, Non-U.S. Gov't].

[24] Fullwood MJ, Liu MH, Pan YF, Liu J, Xu H, Mohamed YB, et al. An oestrogen-receptor-alpha-bound human chromatin interactome. Nature November 5, 2009;462(7269):58–64. [Research Support, N.I.H., Extramural Research Support, Non-U.S. Gov't].

[25] Handoko L, Xu H, Li G, Ngan CY, Chew E, Schnapp M, et al. CTCF-mediated functional chromatin interactome in pluripotent cells. Nat Genet July 2011;43(7):630–8.

[26] Zhang Y, Wong CH, Birnbaum RY, Li G, Favaro R, Ngan CY, et al. Chromatin connectivity maps reveal dynamic promoter-enhancer long-range associations. Nature December 12, 2013;504(7479): 306–10. [Research Support, N.I.H., Extramural Research Support, Non-U.S. Gov't Research Support, U.S. Gov't, Non-P.H.S.].

[27] Sanyal A, Lajoie BR, Jain G, Dekker J. The long-range interaction landscape of gene promoters. Nature September 6, 2012;489(7414):109–13. [Research Support, N.I.H., Extramural Research Support, Non-U.S. Gov't].

[28] Chung JH, Whiteley M, Felsenfeld GA. 5' Element of the chicken beta-globin domain serves as an insulator in human erythroid cells and protects against position effect in Drosophila. Cell August 13, 1993;74(3):505–14.

[29] Ghirlando R, Giles K, Gowher H, Xiao T, Xu Z, Yao H, et al. Chromatin domains, insulators, and the regulation of gene expression. Biochim Biophys Acta July 2012;1819(7):644–51.

[30] Maksimenko O, Golovnin A, Georgiev P. Enhancer-promoter communication is regulated by insulator pairing in a Drosophila model bigenic locus. Mol Cell Biol September 2008;28(17):5469–77.

[31] Holwerda S, de Laat W. Chromatin loops, gene positioning, and gene expression. Front Genet 2012;3:217.

[32] Gibcus JH, Dekker J. The hierarchy of the 3D genome. Mol Cell March 7, 2013;49(5):773–82.

[33] Kyrchanova O, Georgiev P. Chromatin insulators and long-distance interactions in Drosophila. FEBS Lett January 3, 2014;588(1):8–14.

[34] Farrell CM, West AG, Felsenfeld G. Conserved CTCF insulator elements flank the mouse and human beta-globin loci. Mol Cell Biol June 2002;22(11):3820–31. [Comparative Study In Vitro].

[35] Splinter E, Heath H, Kooren J, Palstra RJ, Klous P, Grosveld F, et al. CTCF mediates long-range chromatin looping and local histone modification in the beta-globin locus. Genes Dev September 1, 2006;20(17):2349–54. [Research Support, Non-U.S. Gov't].

[36] Wendt KS, Yoshida K, Itoh T, Bando M, Koch B, Schirghuber E, et al. Cohesin mediates transcriptional insulation by CCCTC-binding factor. Nature February 14, 2008;451(7180):796–801.

[37] Fedoriw AM, Stein P, Svoboda P, Schultz RM, Bartolomei MS. Transgenic RNAi reveals essential function for CTCF in H19 gene imprinting. Science January 9, 2004;303(5655):238–40.

[38] Zlatanova J, Caiafa P. CTCF and its protein partners: divide and rule? J Cell Sci May 1, 2009;122 (Pt 9):1275–84.

[39] Yusufzai TM, Tagami H, Nakatani Y, Felsenfeld G. CTCF tethers an insulator to subnuclear sites, suggesting shared insulator mechanisms across species. Mol Cell January 30, 2004;13(2):291–8. [Research Support, Non-U.S. Gov't].

[40] Kim TH, Abdullaev ZK, Smith AD, Ching KA, Loukinov DI, Green RD, et al. Analysis of the vertebrate insulator protein CTCF-binding sites in the human genome. Cell March 23, 2007;128(6):1231–45.

[41] Chen H, Tian Y, Shu W, Bo X, Wang S. Comprehensive identification and annotation of cell type-specific and ubiquitous CTCF-binding sites in the human genome. PLoS One 2012;7(7):e41374.

[42] Wang H, Maurano MT, Qu H, Varley KE, Gertz J, Pauli F, et al. Widespread plasticity in CTCF occupancy linked to DNA methylation. Genome Res September 2012;22(9):1680–8.

[43] Lee BK, Iyer VR. Genome-wide studies of CCCTC-binding factor (CTCF) and cohesin provide insight into chromatin structure and regulation. J Biol Chem September 7, 2012;287(37):30906–13.

[44] Cuddapah S, Jothi R, Schones DE, Roh TY, Cui K, Zhao K. Global analysis of the insulator binding protein CTCF in chromatin barrier regions reveals demarcation of active and repressive domains. Genome Res January 2009;19(1):24–32. [Research Support, N.I.H., Intramural].

[45] Guelen L, Pagie L, Brasset E, Meuleman W, Faza MB, Talhout W, et al. Domain organization of human chromosomes revealed by mapping of nuclear lamina interactions. Nature June 12, 2008;453(7197):948–51.

[46] Murrell A, Heeson S, Reik W. Interaction between differentially methylated regions partitions the imprinted genes Igf2 and H19 into parent-specific chromatin loops. Nat Genet August 2004;36(8):889–93.

[47] Yoon YS, Jeong S, Rong Q, Park KY, Chung JH, Pfeifer K. Analysis of the H19ICR insulator. Mol Cell Biol May 2007;27(9):3499–510.

[48] Bell AC, Felsenfeld G. Methylation of a CTCF-dependent boundary controls imprinted expression of the Igf2 gene. Nature May 25, 2000;405(6785):482–5.

[49] Hark AT, Schoenherr CJ, Katz DJ, Ingram RS, Levorse JM, Tilghman SM. CTCF mediates methylation-sensitive enhancer-blocking activity at the H19/Igf2 locus. Nature May 25, 2000;405(6785):486–9. [Research Support, Non-U.S. Gov't Research Support, U.S. Gov't, P.H.S.].

[50] Kanduri C, Pant V, Loukinov D, Pugacheva E, Qi CF, Wolffe A, et al. Functional association of CTCF with the insulator upstream of the H19 gene is parent of origin-specific and methylation-sensitive. Curr Biol July 13, 2000;10(14):853–6. [Research Support, Non-U.S. Gov't Research Support, U.S. Gov't, P.H.S.].

[51] Augui S, Filion GJ, Huart S, Nora E, Guggiari M, Maresca M, et al. Sensing X chromosome pairs before X inactivation via a novel X-pairing region of the Xic. Science December 7, 2007;318(5856):1632–6.

[52] Donohoe ME, Zhang LF, Xu N, Shi Y, Lee JT. Identification of a Ctcf cofactor, Yy1, for the X chromosome binary switch. Mol Cell January 12, 2007;25(1):43–56. [Research Support, N.I.H., Extramural Research Support, Non-U.S. Gov't].

[53] Lunyak VV, Prefontaine GG, Nunez E, Cramer T, Ju BG, Ohgi KA, et al. Developmentally regulated activation of a SINE B2 repeat as a domain boundary in organogenesis. Science July 13, 2007;317(5835):248–51.

[54] Ebersole T, Kim JH, Samoshkin A, Kouprina N, Pavlicek A, White RJ, et al. tRNA genes protect a reporter gene from epigenetic silencing in mouse cells. Cell Cycle August 15, 2011;10(16):2779–91.

[55] Phillips JE, Corces VG. CTCF: master weaver of the genome. Cell June 26, 2009;137(7):1194–211.

[56] Parelho V, Hadjur S, Spivakov M, Leleu M, Sauer S, Gregson HC, et al. Cohesins functionally associate with CTCF on mammalian chromosome arms. Cell February 8, 2008;132(3):422–33.

[57] Xiao T, Wallace J, Felsenfeld G. Specific sites in the C terminus of CTCF interact with the SA2 subunit of the cohesin complex and are required for cohesin-dependent insulation activity. Mol Cell Biol June 2011;31(11):2174–83.

[58] Yao H, Brick K, Evrard Y, Xiao T, Camerini-Otero RD, Felsenfeld G. Mediation of CTCF transcriptional insulation by DEAD-box RNA-binding protein p68 and steroid receptor RNA activator SRA. Genes Dev November 15, 2010;24(22):2543–55.

[59] Zuin J, Dixon JR, van der Reijden MI, Ye Z, Kolovos P, Brouwer RW, et al. Cohesin and CTCF differentially affect chromatin architecture and gene expression in human cells. Proc Natl Acad Sci USA January 21, 2014;111(3):996–1001.

[60] Hadjur S, Williams LM, Ryan NK, Cobb BS, Sexton T, Fraser P, et al. Cohesins form chromosomal cis-interactions at the developmentally regulated IFNG locus. Nature July 16, 2009;460(7253):410–3.

[61] Nativio R, Wendt KS, Ito Y, Huddleston JE, Uribe-Lewis S, Woodfine K, et al. Cohesin is required for higher-order chromatin conformation at the imprinted IGF2-H19 locus. PLoS Genet November 2009;5(11):e1000739.

[62] Ebmeier CC, Taatjes DJ. Activator-mediator binding regulates mediator-cofactor interactions. Proc Natl Acad Sci USA June 22, 2010;107(25):11283–8.

[63] Kagey MH, Newman JJ, Bilodeau S, Zhan Y, Orlando DA, van Berkum NL, et al. Mediator and cohesin connect gene expression and chromatin architecture. Nature September 23, 2010;467(7314): 430–5.

[64] Malik S, Roeder RG. The metazoan mediator co-activator complex as an integrative hub for transcriptional regulation. Nat Rev Genet November 2010;11(11):761–72.

[65] Phillips-Cremins JE, Sauria ME, Sanyal A, Gerasimova TI, Lajoie BR, Bell JS, et al. Architectural protein subclasses shape 3D organization of genomes during lineage commitment. Cell June 6, 2013;153(6):1281–95.

[66] Schaaf CA, Misulovin Z, Gause M, Koenig A, Gohara DW, Watson A, et al. Cohesin and polycomb proteins functionally interact to control transcription at silenced and active genes. PLoS Genet June 2013;9(6):e1003560.

[67] Yan J, Enge M, Whitington T, Dave K, Liu J, Sur I, et al. Transcription factor binding in human cells occurs in dense clusters formed around cohesin anchor sites. Cell August 15, 2013;154(4):801–13.

[68] Cai S, Han HJ, Kohwi-Shigematsu T. Tissue-specific nuclear architecture and gene expression regulated by SATB1. Nat Genet May 2003;34(1):42–51. [Research Support, U.S. Gov't, P.H.S.].

[69] Kumar PP, Bischof O, Purbey PK, Notani D, Urlaub H, Dejean A, et al. Functional interaction between PML and SATB1 regulates chromatin-loop architecture and transcription of the MHC class I locus. Nat Cell Biol January 2007;9(1):45–56.

[70] Wang J, Shiels C, Sasieni P, Wu PJ, Islam SA, Freemont PS, et al. Promyelocytic leukemia nuclear bodies associate with transcriptionally active genomic regions. J Cell Biol February 16, 2004;164(4):515–26.

[71] Wen J, Huang S, Rogers H, Dickinson LA, Kohwi-Shigematsu T, Noguchi CT. SATB1 family protein expressed during early erythroid differentiation modifies globin gene expression. Blood April 15, 2005;105(8):3330–9.

[72] Agrelo R, Souabni A, Novatchkova M, Haslinger C, Leeb M, Komnenovic V, et al. SATB1 defines the developmental context for gene silencing by Xist in lymphoma and embryonic cells. Dev Cell April 2009;16(4):507–16.

[73] Nechanitzky R, Davila A, Savarese F, Fietze S, Grosschedl R. Satb1 and Satb2 are dispensable for X chromosome inactivation in mice. Dev Cell October 16, 2012;23(4):866–71.

[74] Yasui D, Miyano M, Cai S, Varga-Weisz P, Kohwi-Shigematsu T. SATB1 targets chromatin remodelling to regulate genes over long distances. Nature October 10, 2002;419(6907):641–5.

[75] Kumar PP, Purbey PK, Ravi DS, Mitra D, Galande S. Displacement of SATB1-bound histone deacetylase 1 corepressor by the human immunodeficiency virus type 1 transactivator induces expression of interleukin-2 and its receptor in T cells. Mol Cell Biol March 2005;25(5):1620–33.

[76] Alber F, Dokudovskaya S, Veenhoff LM, Zhang W, Kipper J, Devos D, et al. The molecular architecture of the nuclear pore complex. Nature November 29, 2007;450(7170):695–701.

[77] Belmont AS, Zhai Y, Thilenius A. Lamin B distribution and association with peripheral chromatin revealed by optical sectioning and electron microscopy tomography. J Cell Biol December 1993;123(6 Pt 2):1671–85.

[78] Schermelleh L, Carlton PM, Haase S, Shao L, Winoto L, Kner P, et al. Subdiffraction multicolor imaging of the nuclear periphery with 3D structured illumination microscopy. Science June 6, 2008;320(5881):1332–6.

[79] Casolari JM, Brown CR, Komili S, West J, Hieronymus H, Silver PA. Genome-wide localization of the nuclear transport machinery couples transcriptional status and nuclear organization. Cell May 14, 2004;117(4):427–39.

[80] Capelson M, Liang Y, Schulte R, Mair W, Wagner U, Hetzer MW. Chromatin-bound nuclear pore components regulate gene expression in higher eukaryotes. Cell February 5, 2010;140(3):372–83.

[81] Sarma NJ, Haley TM, Barbara KE, Buford TD, Willis KA, Santangelo GM. Glucose-responsive regulators of gene expression in Saccharomyces cerevisiae function at the nuclear periphery via a reverse recruitment mechanism. Genetics March 2007;175(3):1127–35.

[82] Taddei A, Van Houwe G, Hediger F, Kalck V, Cubizolles F, Schober H, et al. Nuclear pore association confers optimal expression levels for an inducible yeast gene. Nature June 8, 2006;441(7094): 774–8.

[83] Kurshakova MM, Krasnov AN, Kopytova DV, Shidlovskii YV, Nikolenko JV, Nabirochkina EN, et al. SAGA and a novel Drosophila export complex anchor efficient transcription and mRNA export to NPC. EMBO J December 12, 2007;26(24):4956–65.

[84] Mendjan S, Taipale M, Kind J, Holz H, Gebhardt P, Schelder M, et al. Nuclear pore components are involved in the transcriptional regulation of dosage compensation in Drosophila. Mol Cell March 17, 2006;21(6):811–23.

[85] Misteli T. Beyond the sequence: cellular organization of genome function. Cell February 23, 2007;128(4):787–800. [Research Support, N.I.H., Intramural Review].

[86] Forsberg EC, Downs KM, Christensen HM, Im H, Nuzzi PA, Bresnick EH. Developmentally dynamic histone acetylation pattern of a tissue-specific chromatin domain. Proc Natl Acad Sci USA December 19, 2000;97(26):14494–9. [Research Support, Non-U.S. Gov't Research Support, U.S. Gov't, P.H.S.].

[87] Schubeler D, Francastel C, Cimbora DM, Reik A, Martin DI, Groudine M. Nuclear localization and histone acetylation: a pathway for chromatin opening and transcriptional activation of the human beta-globin locus. Genes Dev April 15, 2000;14(8):940–50. [Research Support, Non-U.S. Gov't Research Support, U.S. Gov't, P.H.S.].

[88] Carter D, Chakalova L, Osborne CS, Dai YF, Fraser P. Long-range chromatin regulatory interactions in vivo. Nat Genet December 2002;32(4):623–6. [Research Support, Non-U.S. Gov't].

[89] Ragoczy T, Bender MA, Telling A, Byron R, Groudine M. The locus control region is required for association of the murine beta-globin locus with engaged transcription factories during erythroid maturation. Genes Dev June 1, 2006;20(11):1447–57. [Research Support, N.I.H., Extramural Research Support, Non-U.S. Gov't].

[90] Fiering S, Epner E, Robinson K, Zhuang Y, Telling A, Hu M, et al. Targeted deletion of 5'HS2 of the murine beta-globin LCR reveals that it is not essential for proper regulation of the beta-globin locus. Genes Dev September 15, 1995;9(18):2203–13. [Comparative Study Research Support, U.S. Gov't, P.H.S.].

[91] Palstra RJ, Tolhuis B, Splinter E, Nijmeijer R, Grosveld F, de Laat W. The beta-globin nuclear compartment in development and erythroid differentiation. Nat Genet October 2003;35(2):190–4. [Research Support, Non-U.S. Gov't].

[92] Deng W, Lee J, Wang H, Miller J, Reik A, Gregory PD, et al. Controlling long-range genomic interactions at a native locus by targeted tethering of a looping factor. Cell June 8, 2012;149(6):1233–44. [Research Support, N.I.H., Extramural Research Support, N.I.H., Intramural].

[93] Bartholdy B, Matthias P. Transcriptional control of B cell development and function. Gene February 18, 2004;327(1):1–23. [Research Support, Non-U.S. Gov't Review].

[94] Hardy RR, Hayakawa K. B cell development pathways. Annu Rev Immunol 2001;19:595–621. [Review].

[95] Ebert A, McManus S, Tagoh H, Medvedovic J, Salvagiotto G, Novatchkova M, et al. The distal V(H) gene cluster of the Igh locus contains distinct regulatory elements with Pax5 transcription factor-dependent activity in pro-B cells. Immunity February 25, 2011;34(2):175–87. [Research Support, Non-U.S. Gov't].

[96] Bolland DJ, Wood AL, Johnston CM, Bunting SF, Morgan G, Chakalova L, et al. Antisense intergenic transcription in V(D)J recombination. Nat Immunol June 2004;5(6):630–7. [Research Support, Non-U.S. Gov't].

[97] Verma-Gaur J, Torkamani A, Schaffer L, Head SR, Schork NJ, Feeney AJ. Noncoding transcription within the Igh distal V(H) region at PAIR elements affects the 3D structure of the Igh locus in pro-B cells. Proc Natl Acad Sci USA October 16, 2012;109(42):17004–9. [Research Support, N.I.H., Extramural Research Support, Non-U.S. Gov't].

[98] Yancopoulos GD, Alt FW. Developmentally controlled and tissue-specific expression of unrearranged VH gene segments. Cell February 1985;40(2):271–81. [Research Support, Non-U.S. Gov't Research Support, U.S. Gov't, P.H.S.].

[99] Degner-Leisso SC, Feeney AJ. Epigenetic and 3-dimensional regulation of V(D)J rearrangement of immunoglobulin genes. Semin Immunol December 2010;22(6):346–52. [Research Support, N.I.H., Extramural Review].

[100] Guo C, Gerasimova T, Hao H, Ivanova I, Chakraborty T, Selimyan R, et al. Two forms of loops generate the chromatin conformation of the immunoglobulin heavy-chain gene locus. Cell October 14, 2011;147(2):332–43.

[101] Zakany J, Duboule D. The role of Hox genes during vertebrate limb development. Curr Opin Genet Dev August 2007;17(4):359–66. [Review].

[102] Krumlauf R. Hox genes in vertebrate development. Cell July 29, 1994;78(2):191–201. [Review].

[103] Lewis EB. A gene complex controlling segmentation in Drosophila. Nature December 7, 1978;276(5688):565–70. [Research Support, U.S. Gov't, P.H.S.].

[104] Rodrigues AR, Tabin CJ. Developmental biology. Deserts and waves in gene expression. Science June 7, 2013;340(6137):1181–2. [Comment].

[105] Noordermeer D, Leleu M, Splinter E, Rougemont J, De Laat W, Duboule D. The dynamic architecture of Hox gene clusters. Science October 14, 2011;334(6053):222–5. [Research Support, Non-U.S. Gov't].

[106] Soshnikova N, Duboule D. Epigenetic temporal control of mouse Hox genes in vivo. Science June 5, 2009;324(5932):1320–3. [Research Support, Non-U.S. Gov't].

[107] Andrey G, Montavon T, Mascrez B, Gonzalez F, Noordermeer D, Leleu M, et al. A switch between topological domains underlies HoxD genes collinearity in mouse limbs. Science June 7, 2013;340(6137):1234167. [Research Support, Non-U.S. Gov't].

[108] Richon VM. A new path to the cancer epigenome. Nat Biotechnol June 2008;26(6):655–6.

[109] Vu TH, Nguyen AH, Hoffman AR. Loss of IGF2 imprinting is associated with abrogation of long-range intrachromosomal interactions in human cancer cells. Hum Mol Genet March 1, 2010;19(5): 901–19. [Research Support, N.I.H., Extramural Research Support, U.S. Gov't, Non-P.H.S.].

[110] Zeitz MJ, Ay F, Heidmann JD, Lerner PL, Noble WS, Steelman BN, et al. Genomic interaction profiles in breast cancer reveal altered chromatin architecture. PLoS One 2013;8(9):e73974. [Research Support, Non-U.S. Gov't Research Support, U.S. Gov't, Non-P.H.S.].

[111] Cai S, Lee CC, Kohwi-Shigematsu T. SATB1 packages densely looped, transcriptionally active chromatin for coordinated expression of cytokine genes. Nat Genet November 2006;38(11):1278–88. [Research Support, N.I.H., Extramural].

[112] Kohwi-Shigematsu T, Poterlowicz K, Ordinario E, Han HJ, Botchkarev VA, Kohwi Y. Genome organizing function of SATB1 in tumor progression. Semin Cancer Biol April 2013;23(2):72–9. [Research Support, N.I.H., Extramural Research Support, Non-U.S. Gov't Research Support, U.S. Gov't, Non-P.H.S. Review].

[113] Han HJ, Russo J, Kohwi Y, Kohwi-Shigematsu T. SATB1 reprogrammes gene expression to promote breast tumour growth and metastasis. Nature March 13, 2008;452(7184):187–93. [Research Support, N.I.H., Extramural Research Support, Non-U.S. Gov't].

[114] Van Vlierberghe P, Pieters R, Beverloo HB, Meijerink JP. Molecular-genetic insights in paediatric T-cell acute lymphoblastic leukaemia. Br J Haematol October 2008;143(2):153–68. [Review].

[115] Brown L, Cheng JT, Chen Q, Siciliano MJ, Crist W, Buchanan G, et al. Site-specific recombination of the tal-1 gene is a common occurrence in human T cell leukemia. EMBO J October 1990;9(10): 3343–51. [Research Support, Non-U.S. Gov't Research Support, U.S. Gov't, P.H.S.].

[116] Bash RO, Hall S, Timmons CF, Crist WM, Amylon M, Smith RG, et al. Does activation of the TAL1 gene occur in a majority of patients with T-cell acute lymphoblastic leukemia? A pediatric oncology group study. Blood July 15, 1995;86(2):666–76. [Comparative Study Multicenter Study Research Support, Non-U.S. Gov't Research Support, U.S. Gov't, P.H.S.].

[117] Carroll AJ, Crist WM, Link MP, Amylon MD, Pullen DJ, Ragab AH, et al. The t(1;14)(p34;q11) is nonrandom and restricted to T-cell acute lymphoblastic leukemia: a Pediatric Oncology Group study. Blood September 15, 1990;76(6):1220–4. [Research Support, Non-U.S. Gov't Research Support, U.S. Gov't, P.H.S.].

[118] Hu X, Ybarra R, Qiu Y, Bungert J, Huang S. Transcriptional regulation by TAL1: a link between epigenetic modifications and erythropoiesis. Epigenetics August 16, 2009;4(6):357–61. [Research Support, N.I.H., Extramural Research Support, Non-U.S. Gov't].

[119] Porcher C, Swat W, Rockwell K, Fujiwara Y, Alt FW, Orkin SH. The T cell leukemia oncoprotein SCL/tal-1 is essential for development of all hematopoietic lineages. Cell July 12, 1996;86(1):47–57. [Research Support, Non-U.S. Gov't].

[120] Gottgens B, Barton LM, Chapman MA, Sinclair AM, Knudsen B, Grafham D, et al. Transcriptional regulation of the stem cell leukemia gene (SCL)–comparative analysis of five vertebrate SCL loci. Genome Res May 2002;12(5):749–59. [Comparative Study Letter Research Support, Non-U.S. Gov't].

[121] Gottgens B, Barton LM, Gilbert JG, Bench AJ, Sanchez MJ, Bahn S, et al. Analysis of vertebrate SCL loci identifies conserved enhancers. Nat Biotechnol February 2000;18(2):181–6. [Comparative Study].

[122] Delabesse E, Ogilvy S, Chapman MA, Piltz SG, Gottgens B, Green AR. Transcriptional regulation of the SCL locus: identification of an enhancer that targets the primitive erythroid lineage in vivo. Mol Cell Biol June 2005;25(12):5215–25. [Research Support, Non-U.S. Gov't].

[123] Dhami P, Bruce AW, Jim JH, Dillon SC, Hall A, Cooper JL, et al. Genomic approaches uncover increasing complexities in the regulatory landscape at the human SCL (TAL1) locus. PLoS One 2010;5(2):e9059. [Research Support, Non-U.S. Gov't].

[124] Zhou Y, Kurukuti S, Saffrey P, Vukovic M, Michie AM, Strogantsev R, et al. Chromatin looping defines expression of TAL1, its flanking genes, and regulation in T-ALL. Blood December 19, 2013;122(26):4199–209. [Research Support, Non-U.S. Gov't].

[125] Lai F, Orom UA, Cesaroni M, Beringer M, Taatjes DJ, Blobel GA, et al. Activating RNAs associate with mediator to enhance chromatin architecture and transcription. Nature February 28, 2013;494(7438):497–501. [Research Support, N.I.H., Extramural].

CHAPTER 6

ncRNA function in chromatin organization

Keith E. Giles, Jessica L. Woolnough, Blake Atwood
Stem Cell Institute, Department of Biochemistry and Molecular Genetics, University of Alabama at Birmingham, Birmingham, AL, USA

Contents

1. INTRODUCTION

The advent of high-throughput DNA sequencing and its adaptation to sequence RNA through cDNA libraries has uncovered the surprising fact that the vast majority of the human genome is transcribed [1,2]. These numbers vary, but reports indicate as much as 75% of the human genome is transcribed. The expansive transcriptome, as it can be referred, is also a fact for most commonly used model organisms throughout the biomedical research community. This pervasive transcription has led to a series of questions: do these RNAs code for a protein, what is the biological function of these RNAs, if any,

Epigenetic Gene Expression and Regulation
http://dx.doi.org/10.1016/B978-0-12-799958-6.00006-8

and what is the molecular mechanism by which these RNAs execute that function? The majority of these newly discovered RNA molecules do not have any identifiable open reading frame, do not associate with the ribosomes, and many are nuclear localized. Thus, the consensus is that these represent noncoding RNAs (ncRNAs).

In spite of the power of high-throughput sequencing to promote the discovery of newly identified biological molecules, the most well understood of these ncRNAs that affect chromatin structure were discovered without the aid of high-throughput DNA sequencing. These two very different ncRNAs are Xist and siRNAs. Xist was originally discovered in the early 1990s and is shown to be an ncRNA required for maintaining a silent X chromosome in mammals [3–6]. This long ncRNA (17 kb in humans and 15 kb in mice) functions by coating the silent X chromosome and recruiting factors that promote heterochromatin.

The other well-studied ncRNA was discovered using the model systems *Arabidopsis thaliana* and *Caenorhabditis elegans*. Using these genetic models, the observation was made that transgenes containing a high degree of homology to an endogenous gene were being transcriptionally silenced [7–9]. Interestingly, this silencing extended to the endogenous counterpart of the transgene. The mechanism by which these genes were being silenced was ultimately shown to be due to the action of small RNAs, spanning 21–24 nucleotides in length. These were named small interfering RNAs (siRNAs), and the mechanism by which they silenced both the transgene and its endogenous counterpart in *C. elegans* was termed RNA interference (RNAi). The mechanism, as it was originally understood, acted to degrade the messenger RNA (mRNA) of these genes posttranscriptionally. This type of RNAi does not involve changes to chromatin structure. However, later work in another model organism, *Schizosaccharomyces pombe*, identified the RNAi machinery as being critical for the establishment and maintenance of heterochromatin at the centromere [10,11]. Over the past decade, the ability of the RNAi machinery to regulate chromatin structure has been extended to include regulation of genomic regions other than the centromere, and has also been extended to flies, mice, and humans [12]. As might be expected, the precise mechanisms by which the RNAi machinery functions differ in each system.

The identification of novel ncRNAs as a part of the ever-expanding transcriptome has been aided in part by knowledge about the relationship between chromatin modifications and transcriptional regulation. Specifically, it is now known that histone H3 lysine 4 trimethylation (H3K4me3) marks promoters and that H3K36me marks actively transcribed coding regions. The strong correlation of these histone modifications with transcription start sites and active coding regions, respectively, led to a clever method of searching the genome for previously unannotated genes. To wit, unannotated regions of the genome containing an H3K4me3 peak, followed by broad enrichment of H3K36me, were identified bioinformatically, and ultimately led to the discovery of an entire class of RNAs called long intergenic ncRNAs (lincRNAs) [13]. Xist can be considered to be the founding member of this class of RNAs, despite predating the nomenclature by 20 years. Many of these RNAs have been shown, much like Xist, to regulate chromatin structure in both *cis* and *trans* [14].

This chapter presents a hierarchical method to catalog the known ncRNAs. Each ncRNA can be classified as either short (less than 200 nts) or long (greater than 200 nts), activating or repressive, and as acting either in *cis* or in *trans*. The forces driving specific interactions between proteins and RNAs remain murky and are best explored elsewhere. However, information regarding specificity between ncRNA and a specific protein or protein complex will be presented wherever possible. A couple caveats regarding the distinction between *cis*- and *trans*-acting ncRNAs is that many *cis*-acting ncRNAs can be induced to function in *trans* experimentally. Such ncRNA will be considered *cis* acting if this is its predominant mode in its normal biological context. Furthermore, it can often be considered a matter of debate as to whether a given ncRNA functions in *cis* or *trans*. In these gray-area type situations, an ncRNA may function to silence or activate genes other than itself, but these genes may be restricted to a nearby region of the genome. We have defined *cis* regulation as cases where the RNA does not dissociate from the chromatin template and does not an obvious stage of its biogenesis within the soluble nucleoplasm or cytoplasm. These *cis* ncRNAs may act at a distance, but the regulation is dependent on the chromatin context. Additionally, it is unlikely that such ncRNAs would be able to function properly if provided as a transgene.

2. SMALL ncRNAs

2.1 Repressive *cis*-acting small ncRNAs

2.1.1 RNAi and heterochromatin structure at the S. pombe centromere

The discovery of RNAi in 1998 revolutionized thinking toward the central dogma of molecular biology [7,8]. It solidified the notion that RNA was not just a mere carrier of genetic information or structural scaffold, but that it had vast regulatory potential. Although the knowledge of Xist as an ncRNA predated the discovery of RNAi by almost a decade, it did not have the magnitude of impact that was to come from the discovery of RNAi. Initially viewed as an intracellular immunity for invertebrate organisms, RNAi would in a short time be shown to be the mechanism by which microRNAs function to regulate gene expression and also to be utilized for experimental manipulation of gene expression levels [15]. The essence of RNAi constitutes the formation of a long double-stranded RNA (dsRNA), which is processed by Dicer (double-stranded RNA-specific endoribonuclease type III), an RNase III-type enzyme into 21–24 nucleotide siRNAs. The siRNAs remain double stranded only transiently, whereupon one of the two strands is retained and bound by a member of the Argonaute protein family [16]. The other strand is degraded. The bound single-stranded siRNA tethers an Argonaute protein to a complementary region, typically within the 3′ untranslated

cis repression: RNAi control of *S. pombe* centromere

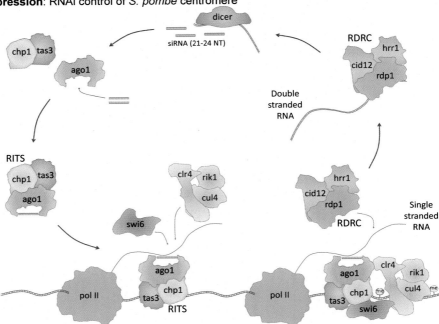

Figure 1 *The control of centromeric heterochromatin structure in* **S. pombe** *by the RNAi machinery.* The centromeric repeat region is transcribed by RNA polymerase II, creating a long, single-stranded RNA (ssRNA, thin red line). This RNA is copied by the RNA-dependent RNA polymerase complex (RDRC) into a long double-stranded RNA (dsRNA, double red line). The long dsRNA is cleaved into short interfering RNAs (siRNAs) by Dicer. These siRNAs are unwound and loaded into the RNA-induced transcriptional silencing (RITS) complex by binding to Argonaute. The complementarity between the siRNA and the nascent RNA, which was originally transcribed by RNA polymerase II, targets the RITS complex to the chromatin. The formation of silent chromatin is facilitated by the recruitment of the CLRC complex by RITS, which facilitates H3K9 methylation via the Clr3 enzyme. The presence of H3K9me enrichment promotes Swi6 binding, the *S. pombe* homolog to HP1. The combination of H3K9me and Swi6 creates the silent chromatin *(figure adapted from [12]).*

region of an mRNA. Argonaute and its cognate siRNAs do not bind to these mRNAs alone, but as part of the repeat induced silencing complex [17] (Figure 1). This binding can induce both translational repression and RNA degradation.

This pathway was initially described as an intracellular immune response for invertebrates. However, the RNAi pathway is also critical for processing genome encoded microRNAs (miRNAs) [18,19]. These RNAs function in an identical pathway as siRNAs, but where siRNAs tend to signal the degradation of a complementary mRNA target, miRNAs typically signal translational repression. The differences largely depend upon the degree of base pairing. The binding of siRNAs to a cognate mRNA results in perfect base pairing, whereas miRNA binding to a target typically involves 3–7 mismatches. The physiological relevance of miRNAs is vast; they have been implicated in practically every aspect of mammalian biology [20].

The ubiquity of miRNAs in the animal kingdom, combined with the ability of the RNAi machinery to be adapted to experimental manipulation, spawned a great deal of interest in the protein components of the RNAi machinery. A fascinating result came in 2002 when genetic screens for novel factors that control the heterochromatin structure at the fission yeast *S. pombe* centromere turned up two genes most central to RNAi: Dicer (*dcr1*) and Argonaute (*ago1*) [10,11]. In addition, an RNA-dependent RNA polymerase (*rdp1*, also *rdr1*; nb, there are two *S. pombe* genes with the name "rdp1." In the original papers, rdp1 referred to the RNA-dependent RNA polymerase involved in RNAi; thus we will use that nomenclature here) was also shown to both interact with these RNAi factors and be critical for heterochromatin formation in *S. pombe* [21]. To date, there is no known RNA-dependent RNA polymerase in metazoans. A follow-up study indicated that *S. pombe* protein Ago1 was found in a nuclear-specific protein complex, RNA-induced transcriptional silencing complex (RITS), along with *trans*-activator of silencing 3 (Tas3) and chromodomain protein 1 (Chp1) [22]. Although their functions were not immediately known, all three of these factors were needed for maintenance of the heterochromatic structure at the *S. pombe* centromere.

The mystery of this discovery was in how could these proteins, known for their cytoplasmic roles in degrading mRNAs, be critical in the regulation of silenced chromatin? Some light was shed on this problem by the discovery that despite it being heterochromatic, the *S. pombe* centromeres are transcribed to a small extent by RNA polymerase II [23]. Furthermore, siRNAs are generated from the Dicer-dependent processing of these heterochromatic transcripts. These transcripts, if not already, are converted into dsRNA by the action of Rdp1. This protein is found in another protein complex, called the RNA-dependent RNA polymerase complex (RDRC), along with Hrr1 (an RNA helicase) and Cid12 (a poly-A polymerase).

Heterochromatin is characterized by a physically condensed chromatin structure, specific posttranslational histone modifications, and the binding of heterochromatin protein 1 (HP1, or its *S. pombe* homolog Swi6) [24]. To facilitate the formation and/or the maintenance of heterochromatin, the RITS and RDRC complexes require interaction with a third protein complex, called Clr4 Complex (CLRC) [25]. This complex contains the proteins Clr4, Rik1, and Rul4. The Rik1 protein has nucleic acid binding properties and is likely the factor responsible for the initial recruitment of the complex to chromatin. The Clr4 protein is an enzyme that is capable of catalyzing the formation of H3K9me (writing) and binding to previously deposited H3K9me (reading). Although final proof remains, it is likely that the RITS complex promotes binding of CLRC to chromatin, and ultimately effects heterochromatin formation. The final component of heterochromatin, Swi6 (the *S. pombe* HP1 homolog) is recruited through its chromodomain to the H3K9me-containing nucleosomes. Similarly, the loop is solidly reinforced by virtue of the chromodomain of Chp1, which facilitates the RITS complex targeting to maintain heterochromatin structure.

The RNAi-mediated regulation of chromatin structure in *S. pombe* can be considered to occur in *cis*. However, ectopic hairpin-generated siRNAs containing *S. pombe* centromere sequence can localize to the centromere in *trans* [26]. This observation is consistent with the widely accepted model that Watson-Crick base pairing between the RITS complex and the heterochromatic transcripts contributes to the stability and specificity of the RITS–chromatin interaction. However, there have been reports that propose the formation of an RNA:DNA hybrid to tether RITS to the centromere [27]. Regardless, base pairing interactions are only partially responsible for the stability of the interaction, as naked RNAs have not been localized to the *S. pombe* centromere.

The notion that transcription of a heterochromatic region is essential for it to remain silent is paradoxical. However, one can resolve the paradox if the processes listed above are considered in the context of the cell cycle. The transcription of the centromeric repeats by RNA polymerase II does not occur throughout, but is limited to early S phase [28]. This transcription is made possible by increases in H3S10p during M phase. This modification evicts Swi6 and allows RNA polymerase II transcription to occur briefly during early S phase. The transcription from these centromeric repeats is acted upon by the RDRC to form dsRNA, followed by Dcr1 to form siRNAs, which ultimately become loaded up into the RITS complex via Ago1 before the completion of S phase [29]. The physical interactions between the RDRC and RITS, and the base pairing between RITS and the nascent transcripts, ultimately target the CLRC complex to the centromeres. The CLRC complex then facilitates heterochromatin formation via histone deacetylation, H3K9 methylation, and the deposition of swi6.

The activity of RNA polymerase II at the centromeric loci during early S phase poses a problem for the DNA replication. There are numerous origins of replications within the centromere regions. Inevitably, a transcribing molecule of RNA polymerase II will become juxtaposed with a replication fork, stalling the replication fork [30]. Since this occurs in S phase, the RNAi machinery is beginning to reestablish a heterochromatic structure in the region, and the deposition of heterochromatin ultimately releases RNA polymerase II from the chromatin and allows replication to continue.

In addition to the centromere, the RNAi machinery has been shown to localize to small regions of heterochromatin distributed throughout the *S. pombe* genome. Many of the RNAi genes themselves are arranged in a head to head fashion, and have been termed convergent genes (CGs) [31]. The continuation of RNA polymerase II transcription downstream of each gene in G1 phase results in the formation of dsRNA in the intergenic region. This dsRNA is recognized by the RNAi machinery and results in the deposition of H3K9me and swi6. Cohesin is next recruited by Swi6 and remains bound throughout the S and G2 phases. The accumulation of cohesin allows normal transcriptional termination, which prevents the formation of the intergenic dsRNAs, alleviates the signal for the recruitment of the RITS complex, and ultimately releases the RNAi machinery from the region. The cycle repeats itself at the onset of mitosis when cohesin is depleted from this region to fulfill its role in chromatin condensation and proper segregation.

Interestingly, this process can be thought of as an efficient evolutionary mechanism that helps to regulate the chromatin structure of the centromeres. As mentioned previously, the centromeres gain H3S10p during mitosis, which displaces Swi6 and allows transcription. This continues throughout G1 and early S phase. The localization of the RITS complex to the CGs during this time allows the centromeric transcripts to accumulate. As cohesin accumulates in the CG intergenic region and normal termination occurs, the need for the RITS complex at the CGs is lost, and it dissociates. By this time, the transcription of the centromeric repeats has accumulated sufficient RNA to allow the reformation of heterochromatin, which requires the newly released RITS complex, and is completed by the onset of G2 phase.

2.1.1.1 RNAi and the exosome

The formation of heterochromatin is not the exclusive domain of the RNAi machinery in *S. pombe*. The exosome is a well-studied RNA quality control machinery that functions as the RNA equivalent to the proteasome [32]. Both complexes resemble a hollow barrel shape, and function by feeding the polymer (RNA or protein) into the lumen of the barrel for degradation. It remains unclear precisely how the exosome can distinguish normal from aberrant RNAs. In *Saccharomyces cerevisiae*, Trf4 is found in a complex with the zinc knuckle protein Air2 and the RNA helicase Mtr4 [33]. Together, these form the TRAMP complex (for **Tr**f4, **A**ir2, **M**tr4, **p**olyadenylation complex). Polyadenylation by TRAMP facilitates the targeting of RNA to the exosome. Although this observation provides some insight into how RNAs are targeted for destruction by the exosome, it remains unclear how TRAMP recognizes its RNA targets.

The role of TRAMP and the exosome was connected to the regulation of chromatin structure in 2008 when a genetic interaction was observed between *cid14*, the *S. pombe* functional homolog of *trf4*, and *ago1* [34]. In *cid14* deletion strains (*cid14Δ*), the composition of RNA molecules that were bound by Ago1 was drastically changed and came to be overwhelmed by rRNA- and tRNA-derived sequences. These RNAs are typically degraded by the exosome, but in the absence of Cid14, they accumulated to such levels that they become targets for the RNAi machinery [35]. Conversely, Mlo3, a protein that interacts with the TRAMP complex, was shown to nucleate heterochromatin in RNAi deletion strains [36]. It has also been suggested that the exosome-dependent turnover of abundant RNAs produces small RNAs called priRNAs [37]. These RNAs are produced independent of Dcr1 and have a range in size from 20 to 30 nts. Once priRNAs are bound by Ago1, they get cleaved by another enzyme called Triman to coincide with the

21–24 nt size range typical of siRNAs [38]. This pathway is believed to represent the initial targeting of the RITS complex to chromatin. Consistent with a functional interaction between these two cellular machineries, Mlo3 physically interacts with both the RITS complex and Clr4. Thus, both the RNAi machinery and the exosome have physical and functional interactions and work together to control both gene expression and chromatin structure.

2.1.1.2 RNAi in metazoans

The roles of the RNAi machinery in controlling heterochromatin in *S. pombe* have been exhaustively studied. Yet, recent connections between the RNAi machinery and the exosome make it clear that much work remains in understanding how ncRNAs function in regulating the *S. pombe* genome. This task is even more daunting in metazoans where the centromeric regions are nearly impossible to study in higher organisms due to their increased length and repetitiveness. The *S. pombe* genome is largely devoid of heterochromatin, with singular peaks occurring over only a handful of loci, such as the centromeres, telomeres, and mating-type loci. The story is different in metazoans. The genomes of both flies and mammals consist more of heterochromatin than euchromatin. In addition to H3K9 methylation, metazoans extensively use H3K27 methylation and DNA methylation of CpG regions as chromatin regulatory modules. These and many other added complexities have made discerning what role, if any, the RNAi machinery plays in metazoans challenging.

2.1.1.2.a *Drosophila melanogaster* Some of the earliest indications that the RNAi machinery was involved in regulating chromatin structure in flies came in 2004 with the observation that mutants in many of the fly homologs to RNAi factors lost the ability to perform position effect variegation [39]. This coincided with a drop in H3K9me2 levels and a mislocalization HP1 (the fly homolog of Swi6). A similar study showed that deletions of fly AGO2 caused embryonic lethality, which was determined to stem from a failure to form a proper centromere [40]. This result is consistent with the observed roles for the RNAi machinery in *S. pombe*.

In addition to the highly repetitive centromeres, the metazoan genome is littered with various types of repeat sequence. So much so in fact, that it can be said that humans are more repeat than gene. In flies, transposable elements (TEs) are transcribed and produced into endogenous siRNAs (endo-siRNAs) by Dicer-2 [41]. These siRNAs were shown to silence the TEs by virtue of heterochromatin formation. In addition to the transcriptional silencing of TEs, the RNAi machinery was shown to regulate the expression of normal RNA polymerase II protein-coding genes. The deletion of RNAi factors altered the genome wide distribution of RNA polymerase II [42]. Supporting a direct role for this, Dicer-2 and Argonaute-2 physically interact with RNA polymerase II and localize to euchromatic regions of the *D. melanogaster* genome [43]. This work indicates a new role for the RNAi machinery in regulating developmental-stage specific gene expression and not just the constitutive silencing of centromeres.

In keeping with a role in regulating gene expression, the function of the RNAi machinery was expanded to include the regulation of insulators [44]. These DNA sequences and their cognate-specific DNA-binding proteins form genomic boundaries that prevent looping interactions if placed between enhancers and promoters. Mutations of *D. melanogaster* Argonaute-2 reduced the functional ability of the *gypsy* insulator. However, a follow-up of this study demonstrated that although Argonaute-2 was bound to *D. melanogaster* insulators, neither binding nor function was dependent on Dicer-1, Dicer-2, or on siRNAs [45]. Interestingly, these Argonaute-2 containing insulator complexes were shown to contain mRNAs [46]. The association of mRNAs with insulators distal from their site of transcription indicates that although the RNAi machinery is important for the regulation of chromatin structure in metazoans, it does appear to have many divergent roles from those originally identified in *S. pombe*.

2.1.1.2.b Mammals The original studies that suggested a similar mechanism as those outlined above could be of functional importance in mammals came in 2004, with the use of a conditional Dicer knockout cell line [47]. This cell line DT40 carried an entire human chromosome. The loss of Dicer coincided with an increase in aneuploidy and a loss of silencing of centromeric transcription from human chromosome 21. A similar study also demonstrated that Dicer knockout mouse ES cells also lost the ability to maintain silenced centromeres [48].

The RNAi machinery is also capable of regulating transcription outside of the centromeric regions in mammals. Mouse Dicer was shown to reside in the nucleus and degrade intergenic transcripts originating from within the beta-globin locus [49]. Although not a mammalian system, a constitutive heterochromatic region within the chicken beta-globin locus was shown to generate heterochromatin-associated transcripts that were processed by Dicer into siRNAs [50]. These siRNAs were loaded into chicken Argonaute-2 and facilitated chromatin localization and the maintenance of heterochromatin. Exogenous siRNAs targeted to the CCR5 HIV co-receptor and to the tumor suppressor gene *RASSF1A* were able to recruit human AGO1 [51]. This recruitment caused transcriptional silencing via the H3K27me3 deposition by the PcG complex. Human AGO1 was also shown to be recruited to actively transcribed coding regions in an siRNA-dependent manner [52]. This facilitated the formation of H3K9me2/3 levels, which was believed to alter the elongation rate of RNA polymerase II and thus alter alternative splicing.

Understanding of the RNAi machinery and its ability to regulate chromatin in *S. pombe* is simplified by the existence of a single Argonaute (Ago1), a single Dicer (Dcr1), and a small number of stable protein complexes that perform a similar function at each locus. The situation in metazoans could not be more different. Mammals do not appear to have a stable protein complex reminiscent of the RITS complex in *S. pombe*. Mammals also lack any obvious RNA-directed RNA polymerase, which is essential for RNAi-directed chromatin modification in *S. pombe*. The *S. Pombe* RDRC is believed to function by amplifying dsRNA levels, which are the targets of Dcr1 cleavage and the source of siRNAs needed for RITS targeting. Perhaps the sheer size and expansiveness of the mammalian transcriptome alleviates the need for such an amplification mechanism.

2.2 Repressive *trans*-acting small RNAs
2.2.1 PIWI-interacting RNAs (piRNA) mediated gene silencing

The family of proteins that can be considered a part of the RNAi machinery is greatly expanded in metazoans. *D. melanogaster* has two Dicer proteins, although mammals have only one. The roles of Dicer-1 and Dicer-2 in *D. melanogaster* are not redundant; Dicer-1 is needed for miRNA biogenesis while Dicer-2 is typically involved in the processing of endo-siRNAs [53]. Even more complex is the Argonaute family of proteins, which can be grouped into the Argonaute and Piwi clades [54]. *D. melanogaster* contains two members of the Argonaute clade (Argonaute-1 and Argonaute-2), and three members of the Piwi clade (Piwi, Aubergine, and Argonaute-3). Mice and humans both have four Argonaute proteins; but mice have three piwi proteins and humans have four.

The levels of redundancy between the various Argonaute proteins remain to be fully understood. Studies have clearly shown roles specific to one or the other, but a general rule or trend has yet to emerge, as is the case for the two Dicer proteins in *D. melanogaster*. However, the variation in function between the Argonaute and Piwi clades is drastic. In a similar manner to what has been discussed regarding the ability of Argonaute to interact with small RNAs, members of the Piwi clade can also interact with small RNAs, termed Piwi-interacting RNAs (piRNA). This pathway can be traced to the discovery of Piwi protein as being necessary for *D. melanogaster* germline development. The name Piwi itself is generated from this screen, "P-element induced wimpy testes." Piwi is clearly needed to silence the expression of the various TEs during germ cell specification [55,56]. As the full name of Piwi indicates, this factor is required for proper gametogenesis.

The mechanism of action appears similar to, but is distinct from the RNAi-mediated silencing described thus far. The TEs are initially transcribed into a long, single-stranded transcript. This RNA is exported to the cytoplasm and processed by an as of yet unknown enzyme to form primary piRNAs, which are loaded into a Piwi family member. One distinction that holds true for all piRNAs is that they are considerably longer than siRNAs (29–30 nucleotides versus 21–24). These primary piRNAs, loaded into a Piwi protein, are next used to target the cleavage of an antisense primary transcript of the TE. The product of this cleavage event is loaded into a different Piwi family member, which is imported back into the nucleus and signals for transcriptional gene silencing through

trans repression: PIWI/pRNAi control of Transposable Elements during gametogenesis

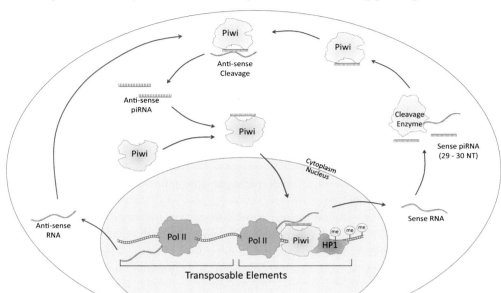

Figure 2 *The control of transposable elements by the Piwi proteins and piRNAs.* RNA polymerase II transcribes the repetitive transposable elements within developing germ cells. These RNAs are exported to the cytoplasm and degraded into 29–30 nucleotide RNAs by an unknown mechanism, and are bound by a member of the Piwi family. These small RNAs are thus called Piwi-interacting RNAs (piRNAs). The Piwi: piRNA complex is targeted to an antisense RNA, also transcribed from the transposable element. The specificity of this interaction is provided by Watson-Crick base pairing between the piRNA and the antisense transposable element transcript. Upon base pairing, the antisense transcript is cleaved by Piwi, and the resultant piRNA is loaded into a different member of the Piwi family. This protein: RNA complex is reimported into the nucleus and specifically targets the chromatin of the transposable element, causing transcriptional silencing.

heterochromatin formation (Figure 2). Thus, many of the concepts that were illustrated by the *S. pombe* silencing pathway are also utilized in piRNA.

3. LONG ncRNAs

3.1 Long intergenic noncoding RNAs (lincRNAs)

Although it was not known for over two decades, the discovery of Xist marked the first of a very large class of biomolecules, the long intergenic noncoding RNAs (lincRNAs) [5,6,57]. The discovery of this class of RNA was made possible by the development of two tangential fields and one revolutionary technological advancement. First, studies carried out on the *S. pombe* centromere made it clear that silent chromatin was not completely silent, and that ncRNAs could have important functional roles in the cell. Second, the advancement of the histone code hypothesis laid a foundation for the idea that a pattern of histone modifications can be both correlative and causative of the transcriptional status for a specific genomic location [58]. The existence of an actual histone code is greatly debated and may be a matter of semantics. However, that specific histone modification patterns are both a result of and an effector of specific biological processes is without doubt. Finally, the discovery of lincRNAs was accelerated by the ubiquity of high-throughput DNA sequencing technologies. These approaches allowed the identification of the genome-wide binding sites of numerous proteins and posttranslational histone modifications. They also facilitated the realization that a vast majority of mammalian genomes are transcribed, and that many previously undetected RNAs are present within the transcriptome [1,2].

Two of the most well-studied histone modification pattern functions can be assigned to H3K4me3 and H3K36me3. These two marks represent both transcriptional start sites and actively transcribed coding regions, respectively [59,60]. The juxtaposition of H3K4me3 peaks alongside long extended regions of H3K36me3 levels is an epigenetic mark characteristic on an actively transcribed gene. If one visualizes these histone modifications along with the annotated version of any mammalian genome, it is clear that these marks are most often associated with well-known genes. However, new genes can be discovered when a K4-K36 signature exists within a region of the genome that is devoid of any known genes. An analysis of RNA-seq data can illustrate that such regions of the genome are in fact expressed. This method of searching for novel K4-K36 signatures led to the discovery of thousands of genes that lack known open reading frames nor any ribosomal association [13]. These genes were termed long intergenic noncoding RNAs (lincRNAs).

An evolutionary analysis of lincRNAs across numerous species demonstrated that the sequences of their promoters were highly conserved whereas the coding regions were not [61]. However, the synteny of these lincRNAs was highly conserved. These lincRNAs, like Xist, are most often spliced and polyadenylated, and greater than 200 nucleotides (this is by definition). The precise numbers of annotated lincRNAs continue to grow, with current estimates indicating that there may be many more lincRNAs than protein-coding genes. It is difficult to estimate the total number of lincRNAs in a given species because the primary means by which they are identified is based on their K4-K36 chromatin signature, which requires them to be actively expressed, and their expression is highly cell-type specific.

The molecular function and physiological role of only a very small number of lincRNAs have been determined. This is in large part due to their only recently having been discovered. The successful knockdown of 147 lincRNAs in mouse embryonic stem cells (ESCs) demonstrated that 93% of these lincRNAs significantly altered gene expression levels to a similar extent as is seen in the knockdown of important pluripotent-specific protein-coding genes [62]. Furthermore, these gene expression effects were not limited to the vicinity of the lincRNAs, indicating that they are capable of functioning in *trans*. The knockdown of 26 lincRNAs caused an exit from the pluripotent state, as measured by changes in the gene expression levels of pluripotency markers. Indication that these lincRNAs were involved in the regulation of chromatin structure came from a large scale RNA-immunoprecipitation analysis of many known histone lysine methyltransferases. It was shown that 30 of the mouse ESC-specific lincRNAs interacted with either a writer, or reader, or eraser of H3K9, H3K4, or H3K27 methylation [63].

3.2 Repressive *cis*-acting long ncRNAs

3.2.1 *Xist and Rox1/2: dosage compensation and the original connection between ncRNAs and chromatin regulation*

A central problem for any diploid organism is the dosage compensation of the heterogametic sex. As is the case for mammals, the heterogametic sex (XY males) contain half as many copies of the genes that are found on the X chromosome than the females. The X-linked genes however, are not expressed at a twofold difference, but are mostly equally expressed within the two species. This fact necessitates that somehow the expression levels of each X-linked gene is normalized. There are two conceivable mechanisms by which this can occur: either the X-linked genes in the heterogametic sex (males in mammals) can be twofold upregulated, or the X-linked genes in the homogametic sex (females in mammals) can be twofold downregulated. Interestingly, different species have been identified that employ each of these approaches.

D. melanogaster handles the problems of dosage compensation by upregulating each of the genes on the male X chromosome twofold. This is also accomplished by virtue of an ncRNA, called Rox1 (RNA on X 1) [64]. Rox1 is a long ncRNA that localizes to chromatin in *cis*, but is capable of spreading laterally to ultimately coat the entire fly X chromosome. Rox1 recruits the Male Sex Lethal (MSL) complex to promote transcriptional activation [65]. Specifically, the recruitment of MSL increases the levels of acetylation of lysine 16 of histone H4 (H4K16ac), which allows the transcriptional activity of the paternal X to be equal to that of the two maternal X chromosomes.

Mammals achieve dosage compensation through the opposite method, which is to downregulate the expression of genes on the X-chromosomes in females (the homogametic sex). There are at least two conceivable approaches to downregulating the X-linked genes in female mammals. Both the paternal and maternal X-linked genes could be expressed at 50% of the levels observed in males. However, in actuality, evolution has favored a system whereby one of the two female X chromosomes are fully inactivated. The molecular decision to inactivate one of the two female X chromosomes occurs twice during mammalian embryogenesis. The first is an imprinted X-inactivation that occurs at the two- or four-cell stage. In this example, it is always the paternal X chromosome which becomes inactivated and the female remains active. The imprinted X-inactivation is removed within the inner cell mass, and both chromosomes once again become active. This sets the stage for the second wave of X-inactivation, which occurs randomly (with respect to the maternal and paternal chromosomes), and permanently. This inactivation can be considered a stable epigenetic mark, as each successive generation of cells derived from this early blastocyst stage will maintain the same pattern of X-inactivation.

The mechanism by which this random inactivation occurs also happens to be the oldest example of a long ncRNA controlling chromatin structure. In 1991, it was discovered that a single RNA was always expressed on the inactive X (Xi) [5,6,57]. This gene was not expressed on the active X (Xa). The gene product was termed Xist (X-inactivation specific transcript), and is found in a portion of the X chromosome that is called the X inactivation center (Xic) [66,67]. This RNA does not associate with ribosomes, has no detectable open reading frame, and is strictly nuclear localized. These observations led to the conclusion that it was functioning as an ncRNA.

The regulation of Xist expression is critical to mammalian dosage compensation. Interestingly, this regulation occurs in part by a second ncRNA, which is transcribed downstream of Xist and in the opposite orientation, resulting in a partial complementarity between the two transcripts. This ncRNA, termed Tsix, is always expressed in the opposite pattern as Xist [68]. Where Xist is expressed only on the Xi, Tsix is expressed only on the Xa. In fact, it is the expression of Xa which inhibits the expression of Xist and prevents the silencing of the Xa. The exact mechanism by which these two ncRNAs are alternately expressed on the Xa and the Xi remains poorly understood. A few plausible mechanisms involve the Tsix-mediated recruitment of enzymes that promote heterochromatin formation over the Xist promoter and prevent its transcription [69,70]. It is also possible that transcriptional interference of Xist is caused by the act of transcription of the Tsix gene. Another possible mechanism involves a transient pairing between the two X-inactivation centers. This pairing would bring all the known transcription factors involved in regulating the locus together in three-dimensional space. After the chromosomes come back apart, it is hypothesized that the activating factors needed for active Tsix asymmetrically distribute to only one chromosome. This chromosome hence expresses Tsix, represses Xist, and then becomes the active X. The other chromosome would then by default express Xist and become the inactive X.

Although details remain fuzzy regarding how the Xist and Tsix genes are regulated, more is known regarding the mechanism by which Xist physically causes the silencing of the Xi. In short, Xist binds to the PRC2 complex (Polycomb Repressive Complex 2), which promotes the formation of H3K27me3 in *cis*. The initial recruitment happens at a nucleation site within the first exons of the Xist and is done so by the 5′ region of Xist, called RepA [71]. The recruitment of PRC2 and subsequent increase in H3K27me3 levels are the primary cause for the transcriptional silencing throughout the entire chromosome. Xist initially recruits PRC2 to ~150 strong sites [72]. Once these strong sites have become enriched for H3K27me3, the PRC2 complex can cause H3K27me3 levels to spread laterally from these sites, which ultimately coats and silences the entire X chromosome (Figure 3).

One outstanding question, which applies for many long ncRNAs is the molecular mechanism by which they can bind to specific sites on the DNA. It is clear in the case of RNAi that specificity is generated from Watson-Crick base pairing interactions between the siRNAs and either nascent transcripts or one looped-out strand of the DNA double

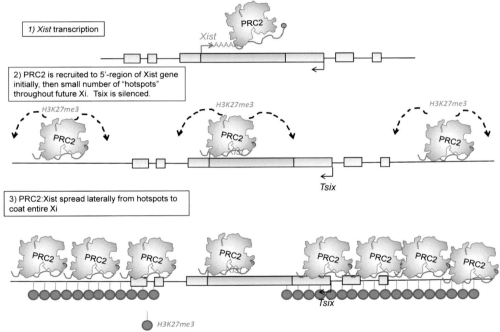

Figure 3 *The random inactivation of an X chromosome during mammalian development.* During female mammalian development, one of the two X chromosomes are epigenetically silenced within the inner cell mass. The process is random and is controlled by the ncRNA, Xist. The expression of Xist is regulated by the expression of its antisense counterpart Tsix; each X chromosome either expresses Xist or Tsix. Expression of Xist indicates the silencing of the entire chromosome whereas expression of Tsix allows the chromosome to remain actively transcribed. The heterochromatin-promoting complex, PRC2, binds directly to the 5′ region of the Xist RNA and is initially recruited to the region of the X chromosome that contains the Xist and Tsix genes, called the X chromosome inactivation center. The initial recruitment of PRC2 is followed by the binding of the Xist: PRC2 complex to ~150 hotspots throughout the future inactive X chromosome (Xi). Once loaded onto the chromosome at these hotspots, the PRC2 complex can spread laterally to coat the entire chromosome. The enzymatic activity of PRC2 is to promote H3K27 tri-methylation. This mark facilitates the transcriptional silencing of the entire X chromosome.

helix. However, because Xist is spread throughout the entire X chromosome it is impossible to invoke base pairing as a driving force for the specificity. Recent studies have shown the protein YY1 may function as an adapter protein, in that it can simultaneously bind to DNA and Xist [73]. It remains to be determined if RNA- and DNA-binding proteins can function as adapters to specify other ncRNAs to their chromatin targets.

There are two mechanisms by which a lincRNA can be considered to silence transcription in *cis*. The first mechanism is illustrated by Xist. In this case, the lincRNA itself has specific interaction with chromatin-modifying enzymes that are recruited at or very nearby the lincRNA gene via specific interactions. In the case of Xist, the spreading of the heterochromatin appears to be a function inherent to the proteins being recruited, namely the PRC2 complex. In this view, Xist is not an effector of chromatin modification but rather a scaffold to recruit proteins to the proper genomic locus. A second mechanism by which ncRNAs (or lincRNAs, the difference is within the way that lincRNAs are operationally defined) can repress gene expression in *cis* is termed transcriptional interference. In this mechanism, the RNA itself is unimportant. To achieve proper gene repression, the ncRNA must be continually transcribed, and it is this act of transcription that provides the signal. To prove the latter mechanism, one must show that destroying the RNA posttranscriptionally without affecting the act of transcription does not alleviate the suppressive effect. In this example, supplying the ncRNA in *trans* would not cause transcriptional silencing.

3.2.2 The repression of autosomally imprinted genes by ncRNAs

Two well-studied ncRNAs that repress transcription in *cis* are KCNQ1OT1 and AIR. The KCNQ1 domain spans about 1 Mb and contains about 10 genes, which are expressed exclusively from the maternal allele [74,75]. The KCNQ1OT1 ncRNA is transcribed from a promoter within intron 10 of the KCNQ1 gene in the antisense orientation [76]. This RNA is a 91 kb Pol II transcript, localized to chromatin, and not spliced. KCNQ1OT1 is transcribed exclusively on the paternal allele, and loss of KCNQ1OT1 causes biallelic expression of the entire KCNQ1 domain. Thus, this ncRNA is critical for (at least) the maintenance of genomic imprinting (the monoallelic expression of genes). The mechanism of action of KCNQ1OT1 is unclear. However, it appears likely that the RNA itself is important for maintaining the silent chromatin structure. KCNQ1OT1 has been shown to bind to the chromatin-modifying enzymes EED (embryonic ectoderm development) and G9A/EHMT2 (eukaryotic histone lysine n-methyltransferase 2), and the RNA itself has been shown to localize to the promoter regions of silent genes within the KCNQ1 imprinted domain [74]. This localization cannot be explained by transcriptional interference as the coding region of KCNQ1OT1 does not overlap with these promoters. Furthermore, these promoters were enriched for silent modifications that were likely deposited by either G9A or EED, and contained within a higher-order repressive chromatin environment along with KCNQ1OT1,

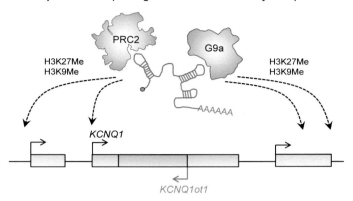

cis repression: Imprinting of the KCNQ1 locus by Kcnq1ot1

Figure 4 *Imprinting of the KCNQ1 locus by the KCNQ1OT1 transcript.* The allele-specific expression pattern (imprinting) of the KCNQ1 locus is mediated by the expression of the KCNQ1OT1 antisense transcript. The expression of KCNQ1OT1 is specific to the silent allele. The ncRNA is bound by both the PRC2 complex and G9A. The former contains the enzyme responsible for H3K27 trimethylation, and the latter is an enzyme responsible for H3K9 di- and trimethylation. Both of these histone modifications promote transcriptional gene silencing through heterochromatin formation. The mechanism by which KCNQ1OT1 promotes transcriptional repression is not limited to transcriptional interference; (1) the promoter of KCNQ1 does not overlap with the KCNQ1OT1 transcript; and (2) genes that are completely outside of the KCNQ1OT1 gene are also transcriptionally silenced by the KCNQ1OT1 RNA.

indicating that the physical interaction was involved in the transcriptional silencing [74,77,78]. The silencing effect mediated by KCNQ1OT1 requires a region within the RNA itself, termed the 5′ silencing domain (Figure 4).

The IGF2R domain is also imprinted, with expression exclusive to the maternal chromosome [79]. An antisense transcript is initiated at a promoter found within intron 2 of the IGF2R gene. The antisense transcript, AIR, interacts with G9A and promotes the formation of heterochromatin on the silent paternal chromosome [77,80]. Similar to KCNQ1OT1, the imprinting spreads beyond genes that have any overlap with AIR, and therefore, the RNA itself is likely important for the transcriptional silencing [81]. However, unlike KCNQ1OT1, the transcription of AIR does overlap the *igf2r* promoter (Figure 5). Although AIR is found enriched at some distal promoters, it is not found at the *igf2r* promoter. Thus, it may be possible that transcriptional interference facilitates the silencing of the IGF2R gene and that the other imprinted genes within the locus are silenced by a different mechanism.

There are many outstanding questions within all areas of ncRNA biology, and the three examples (Xist, KCNQ1OT1, and AIR) of *cis* silencing are no exceptions. In all three cases, there is a clear illustration of the ncRNA binding to specific genomic locations, and these locations lack any obvious ability to form canonical Watson–Crick base pairs. It remains unknown how a given RNA can form a specific interaction with DNA

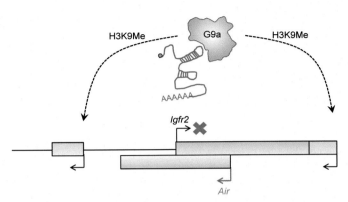

cis repression : Imprinting of the igf2r locus by Air.

- recruits G9a to repress neighboring genes
- Overlapping promoters prevents Igfr2 transcription

Figure 5 *Imprinting of the IGF2R locus by AIR.* The imprinting of the IGF2R gene locus is mediated by expression of the ncRNA, AIR. This ncRNA may function through two distinct mechanisms. It is likely that transcriptional interference may be responsible for the repression of the IGF2R gene itself, as its promoter overlaps with the AIR transcript. However, the numerous flanking genes, which do not overlap with the AIR transcript, are also imprinted. The promoters of these genes have been shown to be bound by AIR. Furthermore, AIR interacts with the H3K9me-specific enzyme G9A.

in the absence of base pairing. It may be that unique structures that are induced by protein binding allows for small regions of complementarity between RNA and DNA. These contacts may form noncanonical base pairs that are stabilized in the context of a specific chromatin environment. Furthermore, it is possible that the ncRNA forms a globular structure that can interact with the major groove of DNA in ways that have yet to be explored. A third, but not exclusive, possibility involves the contributions of both protein–DNA and protein–RNA contacts.

Each of these three ncRNAs is involved in setting up an imprinted genomic loci with monoallelic expression. The imprinted domain set up by Xist happens to span an entire X chromosome, whereas the patterns set up by the KCNQ1OT1 and AIR are much smaller. In each case, it is unclear how the initial asymmetry is established. There are plausible models to illustrate how the differential expression patterns of Xist and Tsix are initially established (see above). However, it remains unclear how this can happen for other autosomal imprinted loci.

3.3 Repressive *trans*-acting long ncRNAs

One of the newly identified lincRNAs that has been assigned a molecular function is termed HOTAIR, for Hox Antisense Intergenic RNA [82]. This 2.2 kb ncRNA is

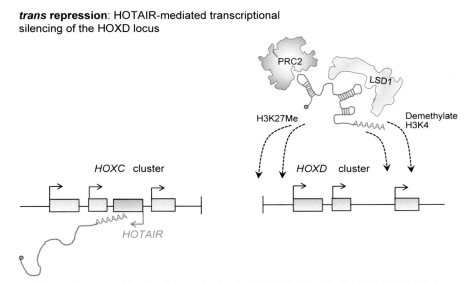

trans **repression**: HOTAIR-mediated transcriptional
silencing of the HOXD locus

Figure 6 *Trans-silencing of the* HoxD *locus by the lincRNA HOTAIR.* The lincRNA HOTAIR is transcribed within the *HOXC* cluster. This lincRNA is bound by both PRC2 and LSD1 protein complexes. The former facilitates H3K27me3 levels and the latter facilitates demethylation of H3K4me3. These two coordinated activities serve to silence active genes and maintain the silent state. The *HOXC* and *HoxD* loci exist on different chromosomes, and thus, HOTAIR represents an authentic *trans*-acting lincRNA. The molecular mechanism which provides HOTAIR specificity remains unclear.

transcribed within the *HoxC* locus by RNA polymerase II, and it is capped, spliced, and polyadenylated. Interestingly, knocking down the expression of HOTAIR does not affect the expression of genes within the *HoxC* locus, but instead it has a strong effect on the genes within the *HoxD* locus, which is located on a completely different chromosome. This indicates that HOTAIR must be functioning in *trans* (Figure 6).

Subsequent studies have indicated that HOTAIR binds to two different protein complexes simultaneously, the PRC2 and LSD1 (lysine-specific demethylase 1A) complexes [83]. In each case, HOTAIR functions as a molecular scaffold, which functions to target these molecules to their specific genomic targets. These two proteins function cooperatively to silence gene expression, as PRC2 promotes increases in H3K27 methylation and LSD1 catalyzes the demethylation of H3K4. Consistent with this activity, knocking down HOTAIR levels causes an increase in *HoxD* gene expression, a loss of K27 methylation and loss of SUZ12 binding across a >40 kb domain of the *HoxD* locus. The regulation of the *HoxD* locus by HOTAIR may be indicative of a more general mechanism. Using the novel RNA immunoprecipitation technique known as CHIRP (chromatin isolation by RNA purification), 832 additional HOTAIR–chromatin binding sites have been identified [84]. This technique allows the identification of chromatin regions that are contacted by a given

RNA by incubating a nuclear extract with biotinylated oligonucleotide probes that are complementary to the ncRNA of interest. The biotin can then be isolated through use of streptavidin-conjugated beads. Many reports have surfaced indicating that HOTAIR expression can be viewed as a prognostic indicator of various cancers [85].

3.4 Activating *cis*-acting long ncRNAs

3.4.1 *HOTTIP: a lincRNA example*

The widespread molecular mechanisms that are expected from the known lincRNAs can be illustrated by a contrast between HOTAIR and another lincRNA discovered within the *HoxA* locus, called HOTTIP [86]. The HOTTIP gene is found within the extreme 5′ region of the *HoxA* locus. The lincRNA binds to the adapter protein WDR5 (WD repeat containing protein 5) and the MLL (mixed lineage leukemia) complex. This interaction promotes gene expression throughout the *HoxA* locus by recruiting these complexes to their appropriate regulatory sequences across the locus. Unlike HOTAIR, this recruitment occurs in *cis* by facilitating DNA-looping between the location of the HOTAIR gene and the target regulatory sequence, a mechanism termed induced proximity (Figure 7).

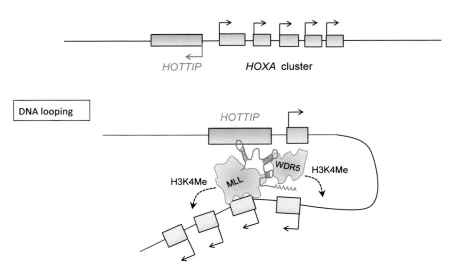

Figure 7 Cis-*activation of the* HOXA *cluster by the lincRNA HOTTIP.* The ncRNA HOTTIP functions locally to activate the gene expression of the *HOXA* gene cluster. HOTTIP binds to two transcriptional activators, the MLL complex and the WDR5 proteins. The lincRNA facilitates gene looping between its own site of transcription and the regulatory regions of the *HOXA* cluster. This novel mechanism was the first such connecting a specific lincRNA to long-range chromatin interactions and was termed "induced proximity."

3.4.2 eRNAs

A new class of ncRNA that has been termed the enhancer RNA (eRNA) will be classified as a *trans*-acting activating RNA. Whether these RNAs are *trans*-acting is debatable. Regardless of the precise mechanism by which they accomplish this feat, eRNAs appear important to the function of enhancers. Enhancers do not always act on the nearest gene and can sometimes jump over genes which lie in between them and their intended target. A given enhancer may not be capable of accessing the entire genome as a possible target, which could be considered a hallmark of a truly *trans*-acting factor. Rather, enhancers function locally without dissociating from chromatin (either looping to its promoter target or tracking). This necessitates an overall designation of *cis*-acting. However, some eRNAs may function in *trans*.

In a story reminiscent of the relationship between Xist and lincRNAs, the founding member of the class of ncRNA called eRNAs was discovered decades before the classification was even a consideration. This first eRNA was the locus control region (LCR) of the beta-globin locus. The transcription of the LCR, which is a functional enhancer, is needed to form the proper histone modification pattern throughout the entire domain [87,88]. The widespread identification of eRNAs was assisted greatly by the ability to identify enhancers based on a specific chromatin signature: H3K4me1 and H2K27ac [89]. In addition to containing this specific chromatin signature and being recognized by cell-type specific transcription factors, enhancers are often both transcribed and bound by RNA polymerase II [90–92]. The levels of eRNA expression correlate with the mRNA levels of the corresponding enhancers target gene. And, in cases where gene expression is activated by an experimentally controlled stimulation, an increase in the eRNA levels precedes the stimulus-induced increase in mRNA levels.

It can be challenging to distinguish the transcripts originating within enhancer regions from other ncRNAs. However, eRNAs are rarely spliced, not polyadenylated (although some are), and are often transcribed in both directions. Furthermore, the enhancers from which they originate are often not marked by H3K4me3, which epigenetically marks RNA polymerase II promoters. Similarly, within enhancer regions the CTD of the largest subunit of RNA polymerase II remains hypophosphorylated at serine 2, and and H3K36me3 is absent from within actively transcribed enhancer sequences. These are two marks that are strongly associated with actively transcribed protein-coding genes and lincRNAs.

The mechanisms by which eRNAs facilitate enhancer function remain unclear (Figure 8). In many cases it may be that the eRNA has no function and is merely a consequence of RNA polymerase II binding to open chromatin. As is always a possibility with ncRNAs, it may be that the act of transcription of the enhancer per se, and not the eRNA, is important for enhancer function. This exact mechanism was illustrated with the polycomb repressive element (PRE) in *D. melanogaster*. PREs promote gene silencing

cis activation: enhancer RNAs (eRNAs)

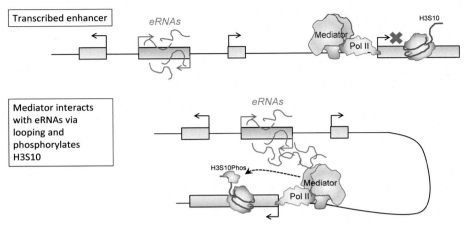

Figure 8 *Enhancer RNAs (eRNAs) promote enhancer-mediated transcriptional activation via a specific interaction with the mediator complex and promoting specific long range chromatin interactions.* Enhancer RNAs are ncRNAs that are often found associated with known enhancer sequences. Although many eRNAs have been discovered, the mechanism(s) by which they function as a part of their enhancers remain unclear. One example illustrated that eRNAs bind to the mediator complex, and promote its loading of Pol-II at the proper promoter region by facilitating long-range DNA looping.

through the recruitment of the PcG complexes. However, this gene repression can be alleviated and replaced by an activating signal transmitted through the TrxG (trithorax group) proteins. This switch from silent to active requires active transcription of the PRE [93]. However, the actual transcript is not relevant as it is equally capable of maintaining transcriptional activation in either the sense or antisense orientation [94].

There are currently two main approaches by which a function can be attributed to the eRNA per se: testing its activity within a heterologous reporter activity and/or knocking down the eRNA using an RNAi or like approach. In one study, ~3000 long ncRNAs were identified that are induced in primary keratinocytes in response to differentiation signals [95]. These eRNAs were confirmed to have enhancer function within a heterologous reporter assay and depletion of the ncRNA levels via RNAi reduced the enhancer activity. Similarly, a number of p53-bound enhancer regions (p53BERs), which are p53 binding sites distal from known p53-dependent genes, were shown to express eRNAs [96]. These eRNAs were also confirmed using a combination of a heterologous reporter assay and mutational analysis of the eRNA. One problem with using a standard reporter assay is that the sequence from which the eRNAs originate is also often the binding site for numerous known transcription factors. Thus, placing the DNA for a given eRNA upstream of a gene and demonstrating transcriptional activation does not prove eRNA function. This assay could be done in conjunction with RNAi knockdown of the eRNA, which would be consistent with eRNA function. Another approach, used in the p53BER study, is to force the eRNA to function in *trans* using a tethering system. To wit, a transgenic fusion eRNA is created that also contains 24 copies of the MS2 RNA stem-loop. This transgenic RNA can bind to an MS2 coat protein, which is fused to the GAL4 DNA-binding domain. If a third transgene is placed in these cells containing the GAL4 upstream activator sequence (UAS), the ability of the eRNA to promote transcription can be assayed. This assay can be done in combination with a mutational analysis of the eRNA, which can illuminate with precision the magnitude of the eRNA effect on transcription as well as which regions of the eRNA are critical.

The aforementioned two studies made great progress in proving that eRNAs themselves are functional as enhancers. However, they did not illustrate how these RNAs may function. To that end, the use of 3C (chromatin conformation capture)-type methods were employed to illustrate that estradiol-responsive eRNAs strengthen long-range looping interactions between the site of their transcription (the enhancer) and the target promoters [97]. It was also recently shown that eRNAs may facilitate loop formation through an interaction with the Mediator complex [98]. This interaction prevents Mediator-catalysis of H3S10-phosporylation levels, which is important for transcriptional activation. The interaction between Mediator and eRNAs also facilitates loop formation between enhancers and promoters.

4. NONCODING RNA-BASED GENE REGULATION BY ALLOSTERY

The previous examples illustrate simplistic cases where a given lincRNA binds to specific protein complexes, and these act in concert to alter the transcriptional output of a specific locus. Two examples were shown, one whereby the lincRNA and its associated proteins promoted gene activation and another promoting repression. A third mode of regulation can be considered whereby the binding of specific protein complex to one or more lincRNAs can be thought of as a molecular switch, toggling the cell between two alternate physiological states. One consideration regarding the many mechanisms of action for lincRNAs is the directness with which they regulate chromatin structure. The examples provided above of HOTAIR, HOTTIP, Xist, and Rox, can be considered as directly regulating chromatin structure. In these cases, the binding of a chromatin-modifying enzyme to the ncRNA is necessary for its binding to its target genomic location. Furthermore, the binding of the chromatin-modifying enzyme occurs to both chromatin and the ncRNA simultaneously.

There are many examples of lincRNAs that affect the regulation of gene expression. However, as there are many nonchromatin-altering mechanisms that could possibly explain this mode of gene regulation, it is better that these examples be pursued in one of the many available reviews that are specific to lincRNAs function. Nevertheless, there are examples of lincRNAs that definitely regulate gene expression through the alteration of chromatin structure, but perhaps the connection is not precisely a direct one (as defined for the previous examples). A catch-all phrase for these indirect regulators of chromatin structure can be said to be via allosteric gene regulation.

An example of such allosteric regulation was observed for the regulation of chromatin binding of Pc2 (Polycomb protein 2) [99]. Pc2 can be considered a member of the chromobox family of proteins, defined by the presence of chromodomain. As with the *S. pombe* Swi6 and Chp1, the presence of a chromodomain indicates an affinity for methylated lysine residues. Interestingly, the methylation/demethylation of lysine 191 within the Pc2 protein itself causes a switch in the protein's subnuclear localization. When Pc2 is methylated, it is located within the PcG bodies, and when it is demethylated, it is found within the interchromatin granules (ICGs). The significance of this localization is that transcriptionally silent genes are enriched for the PcG bodies and actively transcribed genes are found within the ICGs. The localization of Pc2 to the PcG bodies was shown to be dependent on interaction with the ncRNA TUG1 (taurine upregulated gene 1), and localization within the ICGs was dependent on interaction with the ncRNA MALAT/NEAT2 (metastasis associated lung adenocarcinoma transcript 1/noncoding nuclear-enriched abundant transcript 2). Although bacterially purified Pc2 proteins are a strong reader of H3K9me3, in the presence of TUG1, Pc2 switches its affinity to become a reader of H4R3me2 and H3K27me2. In the presence of MALAT/NEAT2, Pc2 becomes a reader of H2AK5ac and H2AK13ac. The former represent global markers of gene repression and the latter gene activation. Thus, the binding of Pc2 to either of two proteins represents an allosteric switch between the two completely different sets of histone modifications. This demonstrates a possible widespread mechanism by which a single set of proteins can be used to convert a set of genes from active to silent or vice versa and provides another fascinating framework to understand the mechanism by which lincRNAs regulate cellular physiology.

The allosteric regulation of protein binding via interactions with ncRNA has also been shown to contribute to X-inactivation (Figure 9). An ncRNA found roughly 10 kb upstream of the Xist gene, JPX, is capable of binding to CTCF (CCCTC binding factor) [100]. The interaction between CTCF and JPX prevents CTCF from binding to DNA. At this locus, the loss of CTCF binding causes an upregulation of Xist expression. Similarly, HP1 is capable of binding to RNA, but this binding prevents a simultaneous interaction with methylated lysines via its chromodomain. This allosteric regulation may be crucial for the RNAi-mediated regulation of chromatin structure.

allosteric regulation: MALAT-1/NEAT-2 & TUG1 regulation of Pc2 "reader" activity

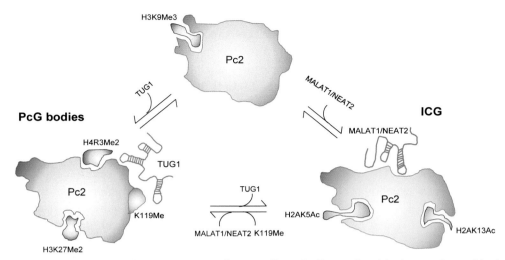

Figure 9 *The chromatin binding activity of Pc2 is allosterically regulated by interactions with the ncRNAs: MALAT-1/NEAT-2 or TUG1.* The protein Pc2 is a chromodomain protein that has been shown to bind H3K9me3 in its purified form. However, interaction with the ncRNA MALAT-1/NEAT-2 facilitates a switch in the histone modification reader activity of Pc2 to recognition of H2AK5ac/H2AK13ac. When bound to MALAT-1/NEAT-2, Pc2 is accordingly associated with active gene expression and is localized to the interchromatin granules. However, the methylation of Pc2 itself at lysine 119, facilitates an interaction with TUG1, which promotes localization to the PcG bodies. This corresponds with a switch in specificity into becoming a reader of H3K27me2 and H3R4me2 and promotion of transcriptional repression.

5. SUMMARY

5.1 *Cis* versus *trans* distinctions

The distinction between *cis*- and *trans*-acting has provided the basis for the hierarchical framework of this review, but it is not always the best method to categorize the function of ncRNAs. The functions of both the RITS complex and Xist are widely accepted to occur in *cis*, but they can be experimentally induced to function in *trans*. Furthermore, each of the long *cis*-acting repressive ncRNAs is able to spread laterally and silence genes at a distance from their site of synthesis. The piRNAs, which are considered *trans*-acting, may create even more confusion as they do not spread laterally and function to silence the expression of the genes from where they were generated. The distinction, which necessitates these factors being classified differentially, involves the biogenesis of the ncRNA and its mode of action, not the location of its genomic target. Xist assembles on the X chromosome and is not predicted to move into the nucleoplasm. Silencing is initiated at the Xist gene itself and spreads laterally, dependent upon direct contact between modified nucleosomes. Conversely, piRNAs silence only their own genes (or TEs), not neighboring genes, and thus could be considered to act in *cis* if this was all that was known. However, the mechanism of piRNA regulation involves movement into the cytoplasm and reimport to the nucleus. It is important to remember that these distinctions are only for the purposes of compartmentalizing our vast knowledge of the cell, and the cell does not always share the need to do this. One must consider each gene and each ncRNA individually, as each regulatory module will have unique properties.

5.2 Future directions

Much work remains in determining the molecular mechanisms of how ncRNAs affect chromatin structure. These examples should be thought of as a survey of some of the known mechanisms by which ncRNAs regulate chromatin structure. Surely, more progress in the field will discover new ncRNAs and with that, a more detailed mechanistic understanding. The following represent some of the most pressing questions that remain unanswered.

The role of the RNAi machinery in metazoans remains shrouded in mystery. This is in spite of the exquisite details that are understood within the *S. pombe* system. The conflicting reports from various systems, the expansion of the RNAi family members, and the complexity of metazoan gene regulation and genome organization all seem to collaborate to make understanding its exact role difficult. RNAi-mediated control of chromatin structure does provide a model for specificity, as the targeting of siRNA-bound protein complexes appears dependent on the specific Watson-Crick base pairing between the siRNA and nascent transcription. Conversely, in many cases, an ncRNA appears to provide specificity in targeting a protein complex to a given genomic locus without any obvious complementarity. The biochemical and molecular basis for such specificity remains unclear in many of these cases. Similarly, the biophysical basis for specificity between RNAs and proteins remains in its infancy. There is much to be learned regarding the function of eRNAs. However, the same can be said regarding the function of enhancers themselves. The future of the eRNA field can be thought of as having to proceed in parallel with the quest to understand how enhancers function, with eRNAs as a central part of that mechanism. This idea presents an ideal last thought to leave with the reader. The study of ncRNAs should not be considered as a novel niche field, but rather as something to be integrated into the whole of molecular biology.

REFERENCES

[1] Djebali S, Davis CA, Merkel A, Dobin A, Lassmann T, Mortazavi A, et al. Landscape of transcription in human cells. Nature September 6, 2012;489(7414):101–8. PubMed PMID: 22955620. Pubmed Central PMCID: 3684276.
[2] Consortium EP, Bernstein BE, Birney E, Dunham I, Green ED, Gunter C, et al. An integrated encyclopedia of DNA elements in the human genome. Nature September 6, 2012;489(7414):57–74. PubMed PMID: 22955616. Pubmed Central PMCID: 3439153.

[3] Brockdorff N, Kay G, Smith S, Keer JT, Hamvas RMJ, Brown SDM, et al. High-density molecular map of the central span of the mouse X-chromosome. Genomics May 1991;10(1):17–22. PubMed PMID: WOS: A1991FK40800003. English.

[4] Brockdorff N, Ashworth A, Kay GF, Cooper P, Smith S, Mccabe VM, et al. Conservation of position and exclusive expression of mouse xist from the inactive X-chromosome. Nature May 23, 1991;351(6324):329–31. PubMed PMID: WOS: A1991FM97600065. English.

[5] Brown SD. XIST and the mapping of the X chromosome inactivation centre. BioEssays November 1991;13(11):607–12. PubMed PMID: 1772416.

[6] Brockdorff N, Ashworth A, Kay GF, Cooper P, Smith S, McCabe VM, et al. Conservation of position and exclusive expression of mouse Xist from the inactive X chromosome. Nature May 23, 1991; 351(6324):329–31. PubMed PMID: 2034279.

[7] Timmons L, Fire A. Specific interference by ingested dsRNA. Nature October 29, 1998;395(6705):854. PubMed PMID: 9804418.

[8] Montgomery MK, Xu S, Fire A. RNA as a target of double-stranded RNA-mediated genetic interference in *Caenorhabditis elegans*. Proc Natl Acad Sci USA December 22, 1998;95(26):15502–7. PubMed PMID: 9860998. Pubmed Central PMCID: 28072.

[9] Montgomery MK, Fire A. Double-stranded RNA as a mediator in sequence-specific genetic silencing and co-suppression. Trends Genet July 1998;14(7):255–8. PubMed PMID: 9676523.

[10] Hall IM, Shankaranarayana GD, Noma K, Ayoub N, Cohen A, Grewal SI. Establishment and maintenance of a heterochromatin domain. Science September 27, 2002;297(5590):2232–7. PubMed PMID: 12215653.

[11] Volpe TA, Kidner C, Hall IM, Teng G, Grewal SI, Martienssen RA. Regulation of heterochromatic silencing and histone H3 lysine-9 methylation by RNAi. Science September 13, 2002;297(5588): 1833–7. PubMed PMID: 12193640.

[12] Grewal SI. RNAi-dependent formation of heterochromatin and its diverse functions. Curr Opin Genet Dev April 2010;20(2):134–41. PubMed PMID: 20207534. Pubmed Central PMCID: 3005588.

[13] Guttman M, Amit I, Garber M, French C, Lin MF, Feldser D, et al. Chromatin signature reveals over a thousand highly conserved large non-coding RNAs in mammals. Nature March 12, 2009;458(7235): 223–7. PubMed PMID: 19182780. Pubmed Central PMCID: 2754849.

[14] Rinn JL, Chang HY. Genome regulation by long noncoding RNAs. Annu Rev Biochem 2012;81:145–66. PubMed PMID: 22663078. Pubmed Central PMCID: 3858397.

[15] Martinez J, Patkaniowska A, Urlaub H, Luhrmann R, Tuschl T. Single-stranded antisense siRNAs guide target RNA cleavage in RNAi. Cell September 6, 2002;110(5):563–74. PubMed PMID: 12230974.

[16] Meister G, Landthaler M, Patkaniowska A, Dorsett Y, Teng G, Tuschl T. Human Argonaute2 mediates RNA cleavage targeted by miRNAs and siRNAs. Mol Cell July 23, 2004;15(2):185–97. PubMed PMID: 15260970.

[17] Hammond SM, Bernstein E, Beach D, Hannon GJ. An RNA-directed nuclease mediates post-transcriptional gene silencing in *Drosophila* cells. Nature March 16, 2000;404(6775):293–6. PubMed PMID: 10749213.

[18] Lee RC, Feinbaum RL, Ambros V. The *C. elegans* heterochronic gene lin-4 encodes small RNAs with antisense complementarity to lin-14. Cell December 3, 1993;75(5):843–54. PubMed PMID: 8252621.

[19] Lau NC, Lim LP, Weinstein EG, Bartel DP. An abundant class of tiny RNAs with probable regulatory roles in *Caenorhabditis elegans*. Science October 26, 2001;294(5543):858–62. PubMed PMID: 11679671.

[20] Yates LA, Norbury CJ, Gilbert RJ. The long and short of microRNA. Cell April 25, 2013;153(3): 516–9. PubMed PMID: 23622238.

[21] Sugiyama T, Cam H, Verdel A, Moazed D, Grewal SI. RNA-dependent RNA polymerase is an essential component of a self-enforcing loop coupling heterochromatin assembly to siRNA production. Proc Natl Acad Sci USA January 4, 2005;102(1):152–7. PubMed PMID: 15615848. Pubmed Central PMCID: 544066.

[22] Noma K, Sugiyama T, Cam H, Verdel A, Zofall M, Jia S, et al. RITS acts in *cis* to promote RNA interference-mediated transcriptional and post-transcriptional silencing. Nat Genet November 2004;36(11):1174–80. PubMed PMID: 15475954.

[23] Cam HP, Sugiyama T, Chen ES, Chen X, FitzGerald PC, Grewal SI. Comprehensive analysis of heterochromatin- and RNAi-mediated epigenetic control of the fission yeast genome. Nat Genet August 2005;37(8):809–19. PubMed PMID: 15976807.

[24] Campos EI, Reinberg D. Histones: annotating chromatin. Annu Rev Genet 2009;43:559–99. PubMed PMID: 19886812.

[25] Zhang K, Mosch K, Fischle W, Grewal SI. Roles of the Clr4 methyltransferase complex in nucleation, spreading and maintenance of heterochromatin. Nat Struct Mol Biol April 2008;15(4):381–8. PubMed PMID: 18345014.

[26] Iida T, Nakayama J, Moazed D. siRNA-mediated heterochromatin establishment requires HP1 and is associated with antisense transcription. Mol Cell July 25, 2008;31(2):178–89. PubMed PMID: 18657501. Pubmed Central PMCID: 2575423.

[27] Nakama M, Kawakami K, Kajitani T, Urano T, Murakami Y. DNA-RNA hybrid formation mediates RNAi-directed heterochromatin formation. Genes Cells March 2012;17(3):218–33. PubMed PMID: 22280061.

[28] Kloc A, Zaratiegui M, Nora E, Martienssen R. RNA interference guides histone modification during the S phase of chromosomal replication. Curr Biol April 8, 2008;18(7):490–5. PubMed PMID: 18394897. Pubmed Central PMCID: 2408823.

[29] Chen ES, Zhang K, Nicolas E, Cam HP, Zofall M, Grewal SI. Cell cycle control of centromeric repeat transcription and heterochromatin assembly. Nature February 7, 2008;451(7179):734–7. PubMed PMID: 18216783.

[30] Zaratiegui M, Castel SE, Irvine DV, Kloc A, Ren J, Li F, et al. RNAi promotes heterochromatic silencing through replication-coupled release of RNA Pol II. Nature November 3, 2011;479(7371): 135–8. PubMed PMID: 22002604. Pubmed Central PMCID: 3391703.

[31] Gullerova M, Moazed D, Proudfoot NJ. Autoregulation of convergent RNAi genes in fission yeast. Genes Dev March 15, 2011;25(6):556–68. PubMed PMID: 21357674. Pubmed Central PMCID: 3059830.

[32] Januszyk K, Lima CD. The eukaryotic RNA exosome. Curr Opin Struct Biol February 2014;24C: 132–40. PubMed PMID: 24525139. Pubmed Central PMCID: 3985421.

[33] LaCava J, Houseley J, Saveanu C, Petfalski E, Thompson E, Jacquier A, et al. RNA degradation by the exosome is promoted by a nuclear polyadenylation complex. Cell June 3, 2005;121(5):713–24. PubMed PMID: 15935758.

[34] Yamanaka S, Mehta S, Reyes-Turcu FE, Zhuang F, Fuchs RT, Rong Y, et al. RNAi triggered by specialized machinery silences developmental genes and retrotransposons. Nature January 24, 2013;493(7433):557–60. PubMed PMID: 23151475. Pubmed Central PMCID: 3554839.

[35] Buhler M, Spies N, Bartel DP, Moazed D. TRAMP-mediated RNA surveillance prevents spurious entry of RNAs into the *Schizosaccharomyces pombe* siRNA pathway. Nat Struct Mol Biol October 2008;15(10):1015–23. PubMed PMID: 18776903. Pubmed Central PMCID: 3240669.

[36] Reyes-Turcu FE, Zhang K, Zofall M, Chen E, Grewal SI. Defects in RNA quality control factors reveal RNAi-independent nucleation of heterochromatin. Nat Struct Mol Biol October 2011; 18(10):1132–8. PubMed PMID: 21892171. Pubmed Central PMCID: 3190054.

[37] Halic M, Moazed D. Dicer-independent primal RNAs trigger RNAi and heterochromatin formation. Cell February 19, 2010;140(4):504–16. PubMed PMID: 20178743. Pubmed Central PMCID: 3020400.

[38] Marasovic M, Zocco M, Halic M. Argonaute and Triman generate dicer-independent priRNAs and mature siRNAs to initiate heterochromatin formation. Mol Cell October 24, 2013;52(2):173–83. PubMed PMID: 24095277.

[39] Pal-Bhadra M, Leibovitch BA, Gandhi SG, Chikka MR, Bhadra U, Birchler JA, et al. Heterochromatic silencing and HP1 localization in *Drosophila* are dependent on the RNAi machinery. Science January 30, 2004;303(5658):669–72. PubMed PMID: 14752161.

[40] Deshpande G, Calhoun G, Schedl P. *Drosophila* argonaute-2 is required early in embryogenesis for the assembly of centric/centromeric heterochromatin, nuclear division, nuclear migration, and germ-cell formation. Genes Dev July 15, 2005;19(14):1680–5. PubMed PMID: 16024657. Pubmed Central PMCID: 1176005.

[41] Fagegaltier D, Bouge AL, Berry B, Poisot E, Sismeiro O, Coppee JY, et al. The endogenous siRNA pathway is involved in heterochromatin formation in *Drosophila*. Proc Natl Acad Sci USA December 15, 2009;106(50):21258–63. PubMed PMID: 19948966. Pubmed Central PMCID: 2795490.

[42] Cernilogar FM, Onorati MC, Kothe GO, Burroughs AM, Parsi KM, Breiling A, et al. Chromatin-associated RNA interference components contribute to transcriptional regulation in *Drosophila*. Nature December 15, 2011;480(7377):391–5. PubMed PMID: 22056986.

[43] Kavi HH, Birchler JA. Interaction of RNA polymerase II and the small RNA machinery affects heterochromatic silencing in *Drosophila*. Epigenet Chromatin 2009;2(1):15. PubMed PMID: 19917092. Pubmed Central PMCID: 2785806.

[44] Lei EP, Corces VG. RNA interference machinery influences the nuclear organization of a chromatin insulator. Nat Genet August 2006;38(8):936–41. PubMed PMID: 16862159.

[45] Moshkovich N, Nisha P, Boyle PJ, Thompson BA, Dale RK, Lei EP. RNAi-independent role for Argonaute2 in CTCF/CP190 chromatin insulator function. Genes Dev August 15, 2011;25(16):1686–701. PubMed PMID: 21852534. Pubmed Central PMCID: 3165934.

[46] Matzat LH, Dale RK, Lei EP. Messenger RNA is a functional component of a chromatin insulator complex. EMBO Rep October 2013;14(10):916–22. PubMed PMID: 23917615. Pubmed Central PMCID: 3807216.

[47] Fukagawa T, Nogami M, Yoshikawa M, Ikeno M, Okazaki T, Takami Y, et al. Dicer is essential for formation of the heterochromatin structure in vertebrate cells. Nat Cell Biol August 2004;6(8):784–91. PubMed PMID: 15247924.

[48] Kanellopoulou C, Muljo SA, Kung AL, Ganesan S, Drapkin R, Jenuwein T, et al. Dicer-deficient mouse embryonic stem cells are defective in differentiation and centromeric silencing. Genes Dev February 15, 2005;19(4):489–501. PubMed PMID: 15713842. Pubmed Central PMCID: 548949.

[49] Haussecker D, Proudfoot NJ. Dicer-dependent turnover of intergenic transcripts from the human beta-globin gene cluster. Mol Cell Biol November 2005;25(21):9724–33. PubMed PMID: 16227618. Pubmed Central PMCID: 1265824.

[50] Giles KE, Ghirlando R, Felsenfeld G. Maintenance of a constitutive heterochromatin domain in vertebrates by a Dicer-dependent mechanism. Nat Cell Biol January 2010;12(1)(94–9):1–6. PubMed PMID: 20010811. Pubmed Central PMCID: 3500915.

[51] Kim DH, Villeneuve LM, Morris KV, Rossi JJ. Argonaute-1 directs siRNA-mediated transcriptional gene silencing in human cells. Nat Struct Mol Biol September 2006;13(9):793–7. PubMed PMID: 16936726.

[52] Ameyar-Zazoua M, Rachez C, Souidi M, Robin P, Fritsch L, Young R, et al. Argonaute proteins couple chromatin silencing to alternative splicing. Nat Struct Mol Biol October 2012;19(10):998–1004. PubMed PMID: 22961379.

[53] Lee YS, Nakahara K, Pham JW, Kim K, He Z, Sontheimer EJ, et al. Distinct roles for *Drosophila* Dicer-1 and Dicer-2 in the siRNA/miRNA silencing pathways. Cell April 2, 2004;117(1):69–81. PubMed PMID: 15066283.

[54] Hock J, Meister G. The Argonaute protein family. Genome Biol 2008;9(2):210. PubMed PMID: 18304383. Pubmed Central PMCID: 2374724.

[55] Yin H, Lin H. An epigenetic activation role of Piwi and a Piwi-associated piRNA in *Drosophila melanogaster*. Nature November 8, 2007;450(7167):304–8. PubMed PMID: 17952056.

[56] Brower-Toland B, Findley SD, Jiang L, Liu L, Yin H, Dus M, et al. *Drosophila* PIWI associates with chromatin and interacts directly with HP1a. Genes Dev September 15, 2007;21(18):2300–11. PubMed PMID: 17875665. Pubmed Central PMCID: 1973144.

[57] Borsani G, Tonlorenzi R, Simmler MC, Dandolo L, Arnaud D, Capra V, et al. Characterization of a murine gene expressed from the inactive X chromosome. Nature May 23, 1991;351(6324):325–9. PubMed PMID: 2034278.

[58] Wang Y, Fischle W, Cheung W, Jacobs S, Khorasanizadeh S, Allis CD. Beyond the double helix: writing and reading the histone code. Novartis Found Symp 2004;259:3–17; discussion-21, 163-9. PubMed PMID: 15171244.

[59] Bell O, Wirbelauer C, Hild M, Scharf AN, Schwaiger M, MacAlpine DM, et al. Localized H3K36 methylation states define histone H4K16 acetylation during transcriptional elongation in *Drosophila*. EMBO J December 12, 2007;26(24):4974–84. PubMed PMID: 18007591. Pubmed Central PMCID: 2140113.

[60] Roh TY, Cuddapah S, Cui K, Zhao K. The genomic landscape of histone modifications in human T cells. Proc Natl Acad Sci USA October 24, 2006;103(43):15782–7. PubMed PMID: 17043231. Pubmed Central PMCID: 1613230.

[61] Cabili MN, Trapnell C, Goff L, Koziol M, Tazon-Vega B, Regev A, et al. Integrative annotation of human large intergenic noncoding RNAs reveals global properties and specific subclasses. Genes Dev September 15, 2011;25(18):1915–27. PubMed PMID: 21890647. Pubmed Central PMCID: 3185964.

[62] Guttman M, Donaghey J, Carey BW, Garber M, Grenier JK, Munson G, et al. lincRNAs act in the circuitry controlling pluripotency and differentiation. Nature September 15, 2011;477(7364): 295–300. PubMed PMID: 21874018. Pubmed Central PMCID: 3175327.

[63] Khalil AM, Guttman M, Huarte M, Garber M, Raj A, Rivea Morales D, et al. Many human large intergenic noncoding RNAs associate with chromatin-modifying complexes and affect gene expression. Proc Natl Acad Sci USA July 14, 2009;106(28):11667–72. PubMed PMID: 19571010. Pubmed Central PMCID: 2704857.

[64] Meller VH, Wu KH, Roman G, Kuroda MI, Davis RL. roX1 RNA paints the X chromosome of male *Drosophila* and is regulated by the dosage compensation system. Cell February 21, 1997;88(4):445–57. PubMed PMID: 9038336.

[65] Kelley RL, Lee OK, Shim YK. Transcription rate of noncoding roX1 RNA controls local spreading of the *Drosophila* MSL chromatin remodeling complex. Mech Dev 2008 Nov-Dec;125(11–12):1009–19. PubMed PMID: 18793722. Pubmed Central PMCID: 2659721.

[66] Tumer Z, Tommerup N, Tonnesen T, Kreuder J, Craig IW, Horn N. Mapping of the Menkes locus to Xq13.3 distal to the X-inactivation center by an intrachromosomal insertion of the segment Xq13.3-q21.2. Hum Genet March 1992;88(6):668–72. PubMed PMID: 1348049.

[67] Leppig KA, Brown CJ, Bressler SL, Gustashaw K, Pagon RA, Willard HF, et al. Mapping of the distal boundary of the X-inactivation center in a rearranged X chromosome from a female expressing XIST. Hum Mol Genet July 1993;2(7):883–7. PubMed PMID: 8364571.

[68] Lee JT, Davidow LS, Warshawsky D. Tsix, a gene antisense to Xist at the X-inactivation centre. Nat Genet April 1999;21(4):400–4. PubMed PMID: 10192391.

[69] Lee JT. Epigenetic regulation by long noncoding RNAs. Science December 14, 2012;338(6113):1435–9. PubMed PMID: 23229728.

[70] Froberg JE, Yang L, Lee JT. Guided by RNAs: X-inactivation as a model for lncRNA function. J Mol Biol October 9, 2013;425(19):3698–706. PubMed PMID: 23816838. Pubmed Central PMCID: 3771680.

[71] Zhao J, Sun BK, Erwin JA, Song JJ, Lee JT. Polycomb proteins targeted by a short repeat RNA to the mouse X chromosome. Science October 31, 2008;322(5902):750–6. PubMed PMID: 18974356. Pubmed Central PMCID: 2748911.

[72] Pinter SF, Sadreyev RI, Yildirim E, Jeon Y, Ohsumi TK, Borowsky M, et al. Spreading of X chromosome inactivation via a hierarchy of defined Polycomb stations. Genome Res October 2012;22(10):1864–76. PubMed PMID: 22948768. Pubmed Central PMCID: 3460182.

[73] Jeon Y, Lee JT. YY1 tethers Xist RNA to the inactive X nucleation center. Cell July 8, 2011;146(1): 119–33. PubMed PMID: 21729784. Pubmed Central PMCID: 3150513.

[74] Pandey RR, Mondal T, Mohammad F, Enroth S, Redrup L, Komorowski J, et al. Kcnq1ot1 antisense noncoding RNA mediates lineage-specific transcriptional silencing through chromatin-level regulation. Mol Cell October 24, 2008;32(2):232–46. PubMed PMID: 18951091.

[75] Smilinich NJ, Day CD, Fitzpatrick GV, Caldwell GM, Lossie AC, Cooper PR, et al. A maternally methylated CpG island in KvLQT1 is associated with an antisense paternal transcript and loss of imprinting in Beckwith-Wiedemann syndrome. Proc Natl Acad Sci USA July 6, 1999;96(14):8064–9. PubMed PMID: 10393948. Pubmed Central PMCID: 22188.

[76] Lee MP, DeBaun MR, Mitsuya K, Galonek HL, Brandenburg S, Oshimura M, et al. Loss of imprinting of a paternally expressed transcript, with antisense orientation to KVLQT1, occurs frequently in Beckwith-Wiedemann syndrome and is independent of insulin-like growth factor II imprinting. Proc Natl Acad Sci USA April 27, 1999;96(9):5203–8. PubMed PMID: 10220444. Pubmed Central PMCID: 21842.

[77] Wagschal A, Sutherland HG, Woodfine K, Henckel A, Chebli K, Schulz R, et al. G9a histone methyltransferase contributes to imprinting in the mouse placenta. Mol Cell Biol February 2008;28(3):1104–13. PubMed PMID: 18039842. Pubmed Central PMCID: 2223396.

[78] Terranova R, Yokobayashi S, Stadler MB, Otte AP, van Lohuizen M, Orkin SH, et al. Polycomb group proteins Ezh2 and Rnf2 direct genomic contraction and imprinted repression in early mouse embryos. Dev Cell November 2008;15(5):668–79. PubMed PMID: 18848501.

[79] Koerner MV, Pauler FM, Huang R, Barlow DP. The function of non-coding RNAs in genomic imprinting. Development June 2009;136(11):1771–83. PubMed PMID: 19429783. Pubmed Central PMCID: 2847617.

[80] Nagano T, Mitchell JA, Sanz LA, Pauler FM, Ferguson-Smith AC, Feil R, et al. The Air noncoding RNA epigenetically silences transcription by targeting G9a to chromatin. Science December 12, 2008;322(5908):1717–20. PubMed PMID: 18988810.

[81] Seidl CI, Stricker SH, Barlow DP. The imprinted Air ncRNA is an atypical RNAPII transcript that evades splicing and escapes nuclear export. EMBO J August 9, 2006;25(15):3565–75. PubMed PMID: 16874305. Pubmed Central PMCID: 1538572.

[82] Rinn JL, Kertesz M, Wang JK, Squazzo SL, Xu X, Brugmann SA, et al. Functional demarcation of active and silent chromatin domains in human HOX loci by noncoding RNAs. Cell June 29, 2007;129(7):1311–23. PubMed PMID: 17604720. Pubmed Central PMCID: 2084369.

[83] Tsai MC, Manor O, Wan Y, Mosammaparast N, Wang JK, Lan F, et al. Long noncoding RNA as modular scaffold of histone modification complexes. Science August 6, 2010;329(5992):689–93. PubMed PMID: 20616235. Pubmed Central PMCID: 2967777.

[84] Chu C, Quinn J, Chang HY. Chromatin isolation by RNA purification (ChIRP). J Vis Exp 2012;61. PubMed PMID: 22472705. Pubmed Central PMCID: 3460573.

[85] Gupta RA, Shah N, Wang KC, Kim J, Horlings HM, Wong DJ, et al. Long non-coding RNA HOTAIR reprograms chromatin state to promote cancer metastasis. Nature April 15, 2010;464(7291): 1071–6. PubMed PMID: 20393566. Pubmed Central PMCID: 3049919.

[86] Wang KC, Yang YW, Liu B, Sanyal A, Corces-Zimmerman R, Chen Y, et al. A long noncoding RNA maintains active chromatin to coordinate homeotic gene expression. Nature April 7, 2011;472(7341): 120–4. PubMed PMID: 21423168. Pubmed Central PMCID: 3670758.

[87] Gribnau J, Diderich K, Pruzina S, Calzolari R, Fraser P. Intergenic transcription and developmental remodeling of chromatin subdomains in the human beta-globin locus. Mol Cell February 2000;5(2): 377–86. PubMed PMID: 10882078.

[88] Ashe HL, Monks J, Wijgerde M, Fraser P, Proudfoot NJ. Intergenic transcription and transinduction of the human beta-globin locus. Genes Dev October 1, 1997;11(19):2494–509. PubMed PMID: 9334315 Pubmed Central PMCID: 316561.

[89] Heintzman ND, Stuart RK, Hon G, Fu Y, Ching CW, Hawkins RD, et al. Distinct and predictive chromatin signatures of transcriptional promoters and enhancers in the human genome. Nat Genet March 2007;39(3):311–8. PubMed PMID: 17277777.

[90] De Santa F, Barozzi I, Mietton F, Ghisletti S, Polletti S, Tusi BK, et al. A large fraction of extragenic RNA pol II transcription sites overlap enhancers. PLoS Biol May 2010;8(5):e1000384. PubMed PMID: 20485488. Pubmed Central PMCID: 2867938.

[91] Kim TK, Hemberg M, Gray JM, Costa AM, Bear DM, Wu J, et al. Widespread transcription at neuronal activity-regulated enhancers. Nature May 13, 2010;465(7295):182–7. PubMed PMID: 20393465. Pubmed Central PMCID: 3020079.

[92] Koch F, Andrau JC. Initiating RNA polymerase II and TIPs as hallmarks of enhancer activity and tissue-specificity. Transcription 2011 November–December;2(6):263–8. PubMed PMID: 22223044. Pubmed Central PMCID: 3265787.

[93] Rank G, Prestel M, Paro R. Transcription through intergenic chromosomal memory elements of the *Drosophila* bithorax complex correlates with an epigenetic switch. Mol Cell Biol November 2002;22(22):8026–34. PubMed PMID: 12391168. Pubmed Central PMCID: 134728.

[94] Schmitt S, Prestel M, Paro R. Intergenic transcription through a polycomb group response element counteracts silencing. Genes Dev March 15, 2005;19(6):697–708. PubMed PMID: 15741315. Pubmed Central PMCID: 1065723.

[95] Orom UA, Derrien T, Guigo R, Shiekhattar R. Long noncoding RNAs as enhancers of gene expression. Cold Spring Harbor Symp Quant Biol 2010;75:325–31. PubMed PMID: 21502407. Pubmed Central PMCID: 3779064.

[96] Melo CA, Drost J, Wijchers PJ, van de Werken H, de Wit E, Oude Vrielink JA, et al. eRNAs are required for p53-dependent enhancer activity and gene transcription. Mol Cell February 7, 2013;49(3):524–35. PubMed PMID: 23273978.

[97] Li W, Notani D, Ma Q, Tanasa B, Nunez E, Chen AY, et al. Functional roles of enhancer RNAs for oestrogen-dependent transcriptional activation. Nature June 27, 2013;498(7455):516–20. PubMed PMID: 23728302. Pubmed Central PMCID: 3718886.

[98] Lai F, Orom UA, Cesaroni M, Beringer M, Taatjes DJ, Blobel GA, et al. Activating RNAs associate with Mediator to enhance chromatin architecture and transcription. Nature February 28, 2013;494(7438): 497–501. PubMed PMID: 23417068.

[99] Yang L, Lin C, Liu W, Zhang J, Ohgi KA, Grinstein JD, et al. ncRNA- and Pc2 methylation-dependent gene relocation between nuclear structures mediates gene activation programs. Cell November 11, 2011;147(4):773–88. PubMed PMID: 22078878. Pubmed Central PMCID: 3297197.

[100] Sun S, Del Rosario BC, Szanto A, Ogawa Y, Jeon Y, Lee JT. Jpx RNA activates Xist by evicting CTCF. Cell June 20, 2013;153(7):1537–51. PubMed PMID: 23791181. Pubmed Central PMCID: 3777401.

CHAPTER 7

Epigenetic gene regulation and stem cell function

Aissa Benyoucef[1,2], Marjorie Brand[1,2]

[1]The Sprott Center for Stem Cell Research, Regenerative Medicine Program, Ottawa Hospital Research Institute, Ottawa, ON, Canada; [2]Department of Cellular and Molecular Medicine, University of Ottawa, Ottawa, ON, Canada

Contents

Epigenetic Gene Expression and Regulation
http://dx.doi.org/10.1016/B978-0-12-799958-6.00007-X
149

1. INTRODUCTION

1.1 Stem cells: what sets them apart?

The concept of stem cell was first introduced in the late nineteenth century as a reference to a proposed cell that would be present in an embryo to give rise to all cells of the body [1]. But it is not before the 1960s that the existence of stem cells, defined as cells that are capable of both self-renewal and differentiation, was demonstrated by Till and McCulloch, who showed that hematopoietic stem cells (HSCs) are capable of self-renewal and differentiation into all cell types present in the blood [2,3].

Since this seminal discovery, many different types of stem cells have been identified that fall into one of two categories:

1. Pluripotent stem cells (PSCs) that replicate indefinitely and have the capacity to differentiate into all derivatives of the three primary germ layers (ectoderm, endoderm, and mesoderm) (Figure 1).

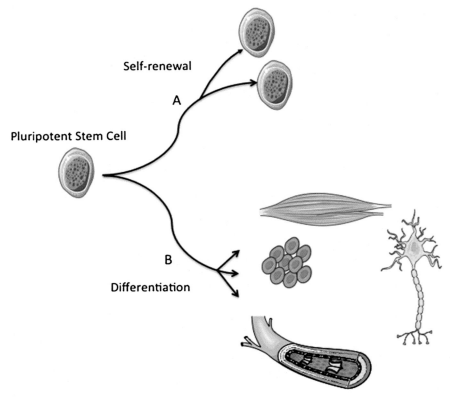

Figure 1 Pluripotent stem cells (PSCs) can either self-renew to maintain the pool of PSCs (A) or differentiate to give rise to all cells in the body, including muscle cells, red blood cells, neurons, and blood vessels (B).

2. Multipotent stem cells (MSCs) that have the capacity to self-renew for a limited number of cycles, and display a more restricted differentiation capacity, being limited to a specific lineage (e.g., blood (Figure 2), muscle, skin, neuron).

1.2 Pluripotent stem cells (PSCs)

In vivo, PSCs only exist as a transient population of cells that constitute the inner cell mass (ICM) of the pre- and postimplantation blastocyst. However, in cell culture, PSCs can be propagated indefinitely. Two states of PSCs have been described in both rodents and primates/humans: the early naïve PSCs (characterized by two active X chromosomes in females, and usage of the *OCT4* gene distal enhancer), and the late primed PSCs (characterized by X-inactivation in females and usage of the *OCT4* gene proximal enhancer) [4,5]. While it has been almost 35 years since naïve PSCs from rodents have been first cultured ex vivo [6,7], the existence of a cell culture-stable naïve human PSC state has only been demonstrated very recently [8,9]. And, similarly, primed rodent PSCs (also called epiblast stem cells or EpiSCs) were only recently derived [10,11]. This time delay partially explains the current nomenclature, where the commonly used term "embryonic stem cell (ESC)" refers to naïve PSCs in rodent and to primed PSCs in primate/human. To avoid confusion in this chapter, we will use the term PSCs rather than ESCs for both primed and naïve cell types. Currently, the exact relationship between

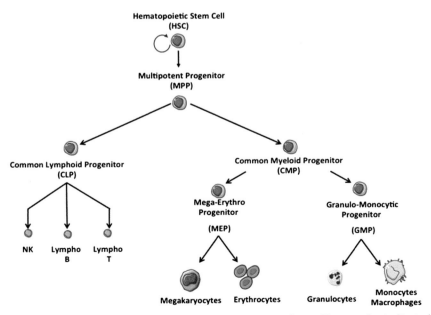

Figure 2 The multipotent hematopoietic stem cell (HSC) can either self-renew (as indicated by the round arrow) or differentiate to produce all cells present in the blood as indicated.

in vivo versus ex vivo, rodent versus human, and naïve versus primed PSCs remain to be clarified at the molecular, cellular, and developmental levels [4,5,12].

PSC lines can be obtained from several sources using different methods. First, they can be derived from the inner cell mass (ICM) of blastocysts that are obtained either through in vitro fertilization [13] or through somatic cell nuclear transfer [14,15]. Alternatively, PSCs can be derived by forced expression of transcription factors (e.g., OCT4, SOX2, KLF4, and c-MYC) in differentiated somatic cells, in which case they are termed "induced pluripotent stem cells" or iPSCs [16–20].

Because of their unique capacity to both self-renew and differentiate into all derivatives of the three primary germ layers, PSCs represent an unprecedented opportunity to devise entirely new ways of treating multiple human diseases that currently have no cure. Specifically, PSCs can potentially serve to model human diseases in the Petri dish, to screen for therapeutically efficient small molecule drugs, and ultimately to regenerate damaged organs or tissues ex vivo or in vivo [21,22]. However, the use of PSCs for therapeutic purposes remains entirely tied to our future capacity to control their differentiation, efficiently and exclusively toward the cell type(s) of interest. Furthermore, potential safety concerns (e.g., cancer development through partial reprogramming [23]) must be eliminated before PSCs may be considered for clinical use.

1.3 Multipotent stem cells (MSCs)

In contrast to PSCs, tissue-specific MSCs (also called adult stem cells) reside in almost all tissues of the body including bone marrow (HSCs [24] (Figure 2)), muscle (satellite stem cells [25]), skin (epidermal stem cells [26]), and brain (neural stem cells [27]). These cells exist mostly in a quiescent state and are essential for tissue homeostasis. Furthermore, they play critical roles in the response to tissue injury, stress, or aging through directing the production of new, functional differentiated cells that are capable of replenishing the stock of damaged or lost cells. The self-renewal capacity of adult stem cells is limited in time, however, and pools of MSCs may become exhausted in very old individuals (e.g., [28]) or in patients affected by degenerative diseases (e.g., Duchenne muscular dystrophy [29]). Despite these limitations, MSCs currently represent the most clinically advanced stem cell therapies as exemplified by HSC transplantations that have been successfully used for many years in the treatment of blood cancers [30] and thalassemias [31].

Interestingly, several recent reports suggest that old MSCs, including neuronal [32], muscle [33], and hematopoietic [34] stem cells can be rejuvenated through exposure to various agents—a finding that could potentially lead to new methods for overcoming the limited self-renewal property of these cells. Another advantage of MSCs is that these cells are prone to differentiate along their respective lineage(s), which greatly facilitates the design of efficient ex vivo differentiation protocols (e.g., erythropoiesis [35,36]) and may limit the risk of oncogenic transformation in vivo. However, it should be noted that

MSCs represent a rare population of cells within the adult tissue. Furthermore, they are difficult to isolate from their tissue-specific niches, and despite multiple attempts, it has been particularly difficult to expand them in culture without impairing their therapeutic potential (e.g., [37]). The above limitations are significant challenges that prevent many clinical applications since the small numbers of adult stem cells that can be isolated from tissues are often not sufficient for cell therapy. This is exemplified by HSCs isolated from cord blood, which provides enough cells for transplantation in children, but not adults. Therefore, similarly to the case of PSCs (see the above paragraph), only through a better understanding of the cellular and molecular mechanisms controlling cell fate determination (i.e., self-renewal versus differentiation) will we be able to expand MSCs to produce enough adult stem cells for therapies.

1.4 Epigenetic mechanisms at the heart of stem cell function and cell fate determination

From early experiments of cell fate conversion through forced expression of TFs (e.g., conversion of fibroblasts to myoblasts driven by a myogenic differentiation protein (MyoD) [38,39], or conversion of B cells to macrophages by GATA Binding Protein 1(GATA1)) [40] to the more recent description of somatic cells reprogramming into iPSCs by various combinations of TFs [16–20], it has always been clear that TFs are master regulators of cell fate determination (i.e., self-renewal versus differentiation into various lineages) in stem cells and other cells (also see a historical review by Thomas Graf [41]). Furthermore, it has been established for a long time that TFs work through the recruitment of cofactors (including RNA polymerase II) to specific regions of the genome to induce the activation or repression of transcriptional programs [42,43].

However, it is the discovery by David Allis in 1996 of a histone acetyltransferase enzymatic activity within the well-known transcriptional co-activator SAGA complex histone acetyltransferase catalytic subunit Gnc5 [44] that opened the way to the realization that TFs also work through the modification of chromatin structure. Pursuing this breakthrough finding, countless studies by a number of laboratories worldwide have led to the identification of additional histone-modifying enzymes (e.g., deacetylases, methyltransferases, and demethylases) [45], as well as enzymes that use the energy of ATP to remodel chromatin (e.g., SWI/SNF) [46]. Furthermore, these studies have confirmed that TFs and cofactors regulate gene expression at least partly through modification of chromatin structure, and vice versa chromatin structure regulates transcription by controlling the accessibility of TFs to their target sites in the genome, through creating sites that are refractory to binding or, on the contrary, by creating binding domains that attract and/or stabilize TFs and cofactors [47,48]. Therefore, chromatin-modifying enzymes are an integral part of the transcriptional regulatory network that defines each cell type and controls cell fate determination in stem cells and other cells.

Because chromatin structure regulates gene expression without altering DNA sequence, changes in histones posttranslational modifications (PTMs) and changes in the methylation status of DNA are typically referred to as epigenetic changes, even though not all chromatin modifications are maintained throughout cell divisions [49,50]. In this chapter, we use a broad definition of epigenetic regulation to include all types of chromatin-based changes in gene expression that do not alter DNA sequence, regardless of their transmittable status. Furthermore, we also consider changes in gene expression that correlate with the three-dimensional (3D) architecture of the nucleus to be epigenetically controlled, since they do not underlie alterations in the DNA sequence.

Given the importance of epigenetic mechanisms in controlling cell fate determination through gene regulatory networks, it may not be surprising that epigenetic factors are particularly important in stem cells. Indeed, knockout studies of epigenetic regulators, including polycomb and trithorax group proteins, histone modifiers and DNA-methylating enzymes have revealed key roles for these proteins in regulating pluripotency, multipotency, and differentiation [51–54]. Also interestingly, compared to differentiated cells, stem cells harbor a unique, more "open" and more dynamic chromatin structure characterized by pervasive transcription [55], higher global levels of histone acetylation [56], and higher histone mobility [57]. In addition, stem cells demonstrate a higher prevalence of "bivalent domains" that consist of regions where histone modifications with opposite functions (i.e., "active" H3K4me3 and "repressive" H3K27me3) overlap [58,59]. It is currently believed that this particular chromatin structure is essential for stem cells' pluripotency (or multipotency), since it allows rapid changes in gene expression in response to environmental cues.

Other evidence of the importance of epigenetic-based regulation in stem cells includes, for example, the observation that global epigenetic reorganization of chromatin (both H3K27 methylation and DNA methylation) occurs upon reprogramming to iPSCs [60], or the finding that chromatin modifiers play critical roles in iPSCs reprogramming by either increasing (e.g., enhancer of zeste 2, EZH2; lysine-specific demethylase 6A, UTX) or decreasing (e.g., DOT1-like histone H3K79 methyltransferase, DOT1L) its efficiency [51,61].

Finally, it is interesting to note that small molecule drugs are currently being developed that modulate the activity of epigenetic enzymes (e.g., [62,63]), providing an unprecedented opportunity to modulate stem cell function through rewiring the transcriptional regulatory network. For example, histone deacetylase (HDAC) inhibitors have been shown to facilitate the transition between primed and naïve human PSCs [9,64]. Most importantly, epigenetic drugs have the potential to greatly benefit stem cell-based therapies, both indirectly, through improving, for example, adult HSC expansion [65], or directly through enhancing the regenerative potential of endothelial stem cells in disease models [66].

In this chapter, we illustrate how epigenetic regulators contribute to the regulation of PSCs' and MSCs' function. We focus particularly on the epigenetic mechanisms that differentiate stem cells from other cells at the levels of 3D nuclear organization, DNA methylation, and histone modifications.

2. THE 3D NUCLEAR ARCHITECTURE OF STEM CELLS

2.1 The interphase nucleus is partitioned into different subcompartments that facilitate coordinated gene expression

The interphase nucleus is organized into substructures that play critical roles in facilitating coordinated gene expression (positively or negatively) for the propagation of cell fate memory and for the control of stem cell differentiation [67–69]. Repressive domains include the **nuclear lamina**, a meshwork of intermediate filaments (e.g., lamins) lining the inside of the nuclear envelope at the nuclear periphery, and **Polycomb (Pc) bodies** that correspond to discrete foci of silent genes bound by the polycomb group (PcG) proteins [70]. Interestingly, silent heterochromatin domains of 0.1–10 Mb in size have been identified that associate with the nuclear lamina (i.e., lamina-associated domains or LADs) and are enriched for dimethylated H3K9 (H3K9me2) and trimethylated H3K27 (H3K27me3) [71]. Interestingly, the LADs overlap, to a large extent, with G9a-dependent domains enriched for H3K9me2 that are termed LOCKS for large organized chromatin K9-modifications [72]. Furthermore, tethering experiments have shown that the nuclear lamina not only correlates with gene repression, but also has a causative role in decreasing gene expression [73,74]. While less is known about Pc bodies, it has been proposed that they may correspond to broad domains enriched for the repressive H3K27me3 mark [75,76]. Compartments that accumulate actively transcribed genes include **transcription factories** that specialize in: (1) RNA polymerase I-mediated transcription of ribosomal RNA in nucleoli [77]; (2) RNA polymerase III-mediated transcription of transfer RNA in the perinucleolar region [78]; and (3) RNA polymerase II (pol II)-mediated transcription of coregulated subsets of genes [79], as well as promyelocytic leukemia (**PML**) **bodies** that associate with transcriptionally active regions [80], **splicing speckles** [81], **Cajal bodies** [82] that are enriched for components of the splicing machinery, and finally **nuclear pore complexes** that export transcripts into the cytoplasm [83] (Figure 3).

2.1.1 *How do chromosomes organize themselves relative to these substructures?*

Cytological methods, such as fluorescence in situ hybridization and its derivative spectral karyotyping that allow visualization of entire chromosomes in interphase [84,85], combined with molecular approaches such as chromosome conformation capture (3C) and its derivatives 4C, 5C, or Hi-C that allow the identification of all pairwise chromatin interactions genome-wide (e.g., [86,87]), have revealed a previously unappreciated

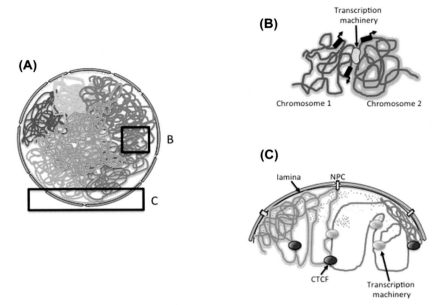

Figure 3 Model of the three-dimensional organization of the genome. (A) Global view of the genomic architecture in the interphase nucleus. Each chromosome (colored in blue, cyan, green, yellow, purple, and red) occupies a distinct territory. (B) Interchromosomal looping of active loci within a transcription factory. (C) The nuclear lamina is a globally repressive transcriptional compartment, except around nuclear pore complexes (NPCs) where active transcription occurs. CTCF plays an important role in mediating chromosomal looping.

hierarchical organization of the genome and provided us with important clues into the basic principles that link positioning in the nucleus to gene regulatory mechanisms. Altogether, these studies have shown that in the interphase nucleus, chromosomes occupy discrete, minimally overlapping domains (i.e., **chromosome territories** or CTs) that can intersect with one or more nuclear substructures [88,89]. On a global level, the positioning of CTs relative to one another and relative to the various nuclear subcompartments is not random, with small gene-rich chromosomes having a tendency to pair and to localize to the nuclear interior away, from the globally repressive nuclear periphery [86,90] (Figure 3(A)). In addition, these studies revealed an additional level of partitioning of the genome into two genome-wide compartments corresponding to open and closed chromatin. Finally, these studies revealed the existence of extensive, highly dynamic chromosomal interactions or loops (both intra- and interchromosomal), with particular loci actively moving in and out of active and repressive compartments depending on their expression status [91–94] (Figure 3(B)). Several types of loops have been identified (reviewed in [69]) including: (1) loops that join the 5′ end of genes to their transcription termination site to enhance transcriptional directionality [95]; (2) loops that bring distal enhancers in close proximity to gene promoters to activate their expression, either on the same chromosome (e.g., the β-globin locus control region [96,97]) or on different chromosomes (e.g., the regulatory H element [98]); (3) loops that bring several promoters together for coregulation [99]; and (4) loops that bring distal Pc response elements in proximity to gene promoters to repress their transcription [100]. While the positioning of specific loci within the 3D structure of the nucleus strongly correlates with their gene expression status, the extent to which this correlation reflects a cause or a consequence of transcriptional activity remains to be established [69].

2.2 The specificity of 3D nuclear organization in stem cells

As mentioned above, PSCs are characterized by a higher global level of transcriptional activity compared to differentiated somatic cells with both coding and noncoding DNA sequences, as well as DNA repeats, being transcribed at low levels [55]. It is generally believed that this state of pervasive transcription underlies pluripotency in stem cells, since it offers all genes the potential to be activated or repressed later during differentiation. Lineage-specific adult stem cells are usually considered as an intermediary state between PSCs and differentiated cells since they have begun turning off pluripotency genes (i.e., octamer-binding transcription factor 4, *OCT4*; the SRY-Box 2 element, *SOX2*, Nanog homeobox, *NANOG*) and genes characteristic of other lineages, but have kept many lineage-specific genes in a poised or active state.

2.2.1 Is the 3D organization of the nucleus important for maintaining the state of pervasive transcription that underlines pluripotency?

Several studies have suggested that the maintenance of pluripotency (and multipotency) in stem cells requires a particular topological organization of the genome characterized by a high degree of plasticity. Indeed, while the global chromosomal organization is similar in PSCs and differentiated cells with gene-poor chromosomes occupying positions closer to the repressive nuclear periphery, the nucleus of PSCs also possesses a number of unique features [101,102]. For instance, composition of the nuclear lamina is different in PSCs as these cells express lamin B1 and B2, but not lamin A/C. It has been suggested that the absence of lamin A/C in PSCs diminishes nuclear rigidity, which in turn increases chromatin mobility that is necessary to maintain genome plasticity [103,104]. Other nuclear substructures such as splicing speckles and Cajal bodies that are present in differentiated cells are more diffuse or absent in PSCs [105], whereas the opposite is true for Pc bodies that are present in PSCs but decrease upon differentiation [106].

PSCs are also characterized by increased levels of chromatin remodeling enzymes such as chromodomain helicase DNA-binding protein 1 (CHD1) [107] and Brahma-related gene 1 (*Brg1*) [108], increased levels of active chromatin modifications, and decreased levels of repressive histone marks [109,110]. Furthermore, partial decondensation of heterochromatic domains and centromeric foci have been observed in PSCs [55,109,111], which is consistent with the genome of these cells being in a more open configuration. Interestingly, it has been recently shown that transcriptionally inactive loci do not engage in as many specific long-range contacts in PSCs compared to differentiated cells (e.g., the taste-bud specific chemoreceptor *Tas2r110* gene engages in only three contacts in PSCs, but shows 34 contacts in astrocytes, even though it is repressed in both cell environments). This result is consistent with inactive chromatin being more randomly organized in the nucleus of PSCs than differentiated cells [112]. Finally, chromatin architectural proteins such as histones, heterochromatin proteins (HP1), and high mobility group (HMG) proteins are less stably attached to chromatin in PSCs [57,109]. Finally, it is important to note that pluripotency factors (e.g., OCT4, SOX2, and NANOG) have a tendency to cluster together in PSCs, and they play an essential role in establishing and maintaining the unique topological genomic organization of these cells [112].

2.2.2 What happens as stem cells differentiate?

As PSCs and MSCs differentiate, they rapidly lose their plasticity and progressively acquire new topological properties that are characteristic of differentiated cells. For instance, dramatic changes in the global genomic order have been observed during T-cell differentiation [113] and during differentiation of multipotent hematopoietic stem/progenitor cells (HSPCs) [84,85]. But the most prominent example of global chromosomal change during stem cell differentiation may be the process of X-chromosome inactivation (XCI) that occurs during the transition from naïve to primed female PSCs in order to equalize X-linked genes expression between sexes. In naïve female PSCs, both X chromosomes are transcriptionally active and as such, they both express the long noncoding RNA *Xist* (X-inactivation specific transcript). During a transient period prior to XCI, the two X chromosomes colocalize to allow pairing of the *Xist*-encoding inactivation centers, a process that is followed by upregulation of *Xist* exclusively on one of the X alleles (i.e., the one corresponding to the future inactive X chromosome) [114–116]. Even though the exact mechanism is unclear, these observations suggest that X chromosome pairing is essential for random allele inactivation. Interestingly, following pairing, the future inactive X chromosome relocates to the repressive perinucleolar region in a *Xist*-dependent manner [117]. The mechanism through which the noncoding RNA *Xist* triggers silencing involves coating the future inactive X chromosome to establish a repressive chromatin structure (i.e., enriched in repressive marks such as H3K27me3, DNA methylation, and macroH2A) [118] that excludes the RNA pol II transcriptional machinery [119]. Also interestingly, after differentiation, genes that escape XCI localize at the periphery of the inactive X chromosome territory [119] and colocalize with other active genes [120].

Altogether, these examples illustrate that global control of nuclear positioning is an important epigenetic mechanism that regulates stem cell self-renewal and differentiation. Furthermore, it has been proposed that the reciprocal influence of gene expression on genome topology (i.e., the function-structure-function model) constitutes an essential part of epigenetic regulation by allowing the formation of a self-enforcing, self-perpetuating system that is nevertheless able to respond to environmental changes by creating an alternate, self-enforcing and self-perpetuating environment [68,69].

3. DNA METHYLATION IN STEM CELLS

In mammalian cells, DNA methylation occurs on cytosine at carbon 5 (5-mC), predominantly in the context of CpG dinucleotides. DNA methylation promotes long-term gene silencing by affecting local histone modification patterns and by actively preventing the binding of factors involved in transcriptional activation. Furthermore, 5-mC is considered as a relatively stable epigenetic mark since once established, it can be transmitted to daughter cells through DNA replication [121]. Two classes of enzymes are able to mediate DNA methylation: (1) the maintenance DNA methyltransferase DNMT1 that recognizes hemi-methylated DNA after replication and copies the pattern of DNA methylation onto the nonmethylated daughter strand; and (2) the de novo DNA methyltransferases DNMT3a and 3b that use unmethylated DNA as a substrate, and are therefore involved in establishing novel methylation patterns during early development. While the distinct functions of DNA methyltransferases have been established for many years, recent studies have identified variations of this classical model with, for example, DNMT1 also being involved in de novo methylation [122]. Even though the DNA methyl mark is quite stable (relative to other epigenetic marks), it can be removed through two different mechanisms: (1) **passive demethylation**, whereby the 5-mC mark is not being maintained on the daughter strand during cell division and is therefore progressively lost after several replication cycles; and (2) **active demethylation**, whereby the ten eleven translocation (TET) enzymes catalyze the conversion of 5-mC to 5-hydroxymethyl cytosine (5-hmC), 5-formyl cytosine (5-fC), and 5-carboxyl cytosine (5-caC), which are then either lost through passive DNA-replication-dependent dilution, or actively replaced by an unmodified C through base excision repair [123]. Altogether, these studies have highlighted that the DNA methylation status is highly controlled through the combined action of DNA methylation, active and passive DNA demethylation, and the rate of replication [122,124].

DNA methylation plays a major role in epigenetic regulation, thanks to its ability to establish and maintain long-term silencing at specific loci during development [125]. Furthermore, DNA methylation is critically involved in mediating genomic imprinting, XCI, and silencing of transposable elements [126]. Since DNA methylation is able to transmit epigenetic memory through cell divisions, it is essential that this epigenetic mark be completely erased to reset the gene expression program in vivo in the totipotent zygote, as well as in cell culture during the process of cellular reprogramming to generate PSCs or iPSCs [124].

3.1 DNA demethylation during embryogenesis

During embryogenesis, there are two waves of massive DNA demethylation, the first one occurring shortly after fertilization and giving rise to the totipotent zygote, and the second one occurring later in primordial germ cells. Interestingly, it was shown that while the paternal pronucleus is subjected to active DNA demethylation through TET3-mediated oxidation [127–129], the maternal pronucleus is protected from conversion of 5-mC to 5-hmC by the maternal factor PGC7 [130], and therefore, DNA demethylation occurs more progressively [131]. Importantly, both processes are highly dependent on passive [127,132] and active DNA demethylation involving DNA repair [133–135]. This first wave of DNA demethylation is critical in establishing pluripotency in PSCs through erasing epigenetic memory.

3.2 DNA demethylation during cellular reprogramming in culture

Given that massive DNA demethylation is associated with the acquisition of totipotency and multipotency in vivo, it may not be surprising that DNA demethylation is critically important for the establishment and maintenance of PSCs in culture. Indeed, a number of studies have revealed the critical role of DNA demethylation in cellular reprogramming. First, PSCs grown in a cell culture medium containing two small-molecule kinase inhibitors (2i) that are known to promote a ground state of pluripotency [136] were recently shown to be hypomethylated compared to their PSCs counterparts grown in a conventional medium [137]. This result suggests that DNA hypomethylation stabilizes the state of pluripotency. Furthermore, when PSCs are transferred from the conventional medium to the 2i-containing medium, they rapidly lose DNA methylation at a number of sites including CpG islands, CG-rich and non-CG-rich promoters, gene bodies, bivalent promoters, and enhancers (see next section) as well as SINE and LINE1 elements, while differential DNA methylation at imprinted regions is preserved [137–139]. Interestingly, global DNA demethylation upon switching to the 2i-containing medium is associated to both a decrease in the expression of DNA methyltransferases (i.e., *Dnmt3a, 3b*) and to an increase in the activity of DNA demethylases TET1/2 leading to a rise in 5-hmC levels [137–139]. Taken together, these results strongly link DNA demethylation (both active and passive) to the establishment and maintenance of pluripotency in stem cells. Also consistent with this, a number of studies have shown that TET proteins play an important role in the process of iPSCs generation through forced expression of the pluripotency factors Oct4, Sox2, Klf4, and c-Myc (OSKM). For instance, forced expression of OSKM in mouse embryonic fibroblasts (MEFs) activates the expression of *Tet2* and induces an increase in 5-hmC levels [140]. Along the same line, TET1/2-depleted MEFs are unable to form iPSCs colonies while TET1/2 overexpression increases the efficiency of reprogramming [140,141]. Finally, TET1/2 can interact with NANOG, and TET1 was shown to be able to substitute for OCT4 during iPSCs generation [141,142]. Altogether, these studies emphasize the importance of DNA hypomethylation in pluripotency, and suggest that characterization of the DNA methylation status of PSCs is warranted prior to using these cells for therapeutic purposes.

4. HISTONE MODIFICATIONS IN STEM CELLS

4.1 The study of histone modifications in stem cells

Recent evidence has established the importance of histone proteins and their PTMs in regulating the function of PSCs and adult stem cells, including self-renewal and differentiation (e.g., [143]). The four core histones (H2A, H2B, H3, and H4) are subject to more than 20 covalent modifications, including acetylation, methylation, phosphorylation, and ubiquitination, many of which regulate gene expression through modifying chromatin structure and/or creating domains that stabilize or prevent the binding of TFs and cofactors [45,47,48]. Importantly, several histone PTMs have been shown to be heritably transmitted to daughter cells to ensure epigenetic memory [50].

Genome-wide maps of histone modifications have been reported in mammalian PSCs, adult stem cells, and differentiated cells, (e.g., [110,144–146]), providing us with a global picture of chromatin architecture and gene regulatory dynamics that occur during differentiation of stem cells. These critical insights have been made possible by major technological advances including (1) the generation of antibodies (Abs) that recognize specific combinations of histone PTMs; (2) the development of chromatin immunoprecipitation (ChIP) techniques that use Abs to enrich specific histone modifications, including native ChIP protocols that bypass the cross-link of chromatin to ensure a higher specificity [147]; and (3) the use of microarrays and next-generation sequencing that provide us with a genome-wide view of histone modifications [76]. While the availability of large numbers of PSCs (thanks to their efficient amplification in cell culture) has facilitated genome-scale analyses of histone modifications in these cells, the more elusive adult stem cells and differentiated cells have required the development of alternative protocols that use less cells such as nano-ChIP seq [148] or ChIP followed by linear amplification of DNA [149]. Finally, analyses of high-throughput datasets have required considerable computational efforts to develop new algorithms and pipelines that integrate different types of genome-wide measures, including histones PTMs, sequence-based genome annotations, nucleosome positioning, RNA expression, TFs binding, and others, to provide integrated data visualization and facilitate subsequent functional analyses [150] (e.g., SeqMINER [151], ChromHMM [152], and ChroModule [153]). It should be noted, however, that direct comparative analyses of ChIP-seq datasets remain challenging (mainly due to the difficulty of normalizing the noise between experiments) and may require tailored approaches (e.g., [154]). Going forward, it will be important to continue improving ChIP-seq methodologies toward better homogenization of protocols [155], as well as to develop novel methods allowing for example allele-specific ChIP (see below).

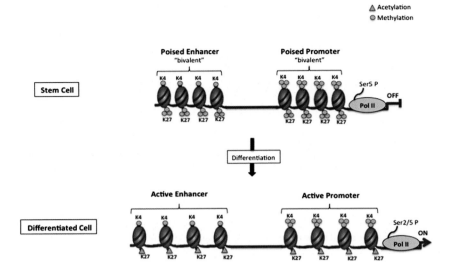

Figure 4 Model for the transition from a bivalent poised gene to an actively transcribed gene during stem cell differentiation. Stem cells are characterized by a high proportion of bivalent genes that are marked by opposing histone modifications (i.e., H3K27me3 and H3K4me3 marks for promoters, and H3K27me3 and H3K4me1 marks for enhancers). Bivalent genes promoters are also bound by the initiating form of RNA pol II (Ser5 P) and are expressed at very low levels. During stem cell differentiation, a number of bivalent genes transition to an active state characterized by the acetylation of H3K27. Active genes are also bound by the elongating form of RNA pol II (ser2 P).

Genome-wide mapping of histones PTMs in PSCs, adult stem cells, and differentiated cells revealed that specific combinations of histone modifications are present at particular genomic locations and strongly correlate with gene expression. For instance, monomethylated histone H3 Lys 4 (H3K4me1) is highly enriched within active enhancers while H3K4me3 and H3K27me3 modifications are present within active and repressed promoters, respectively [76,156] (Figure 4). Furthermore, histone PTMs are very dynamic across developmental stages and during lineage commitment with, for example, the promoter of the gene coding for the pluripotency factor Sox2 being marked with H3K4me3 when the gene is actively transcribed in PSCs, but being marked instead by the repressive histone mark H3K27me3 when the gene is turned off in differentiated fibroblasts [110].

4.2 The specifics of histone modifications in stem cells: high prevalence of bivalent loci

Comparative analyses of histone PTMs in murine and human PSCs revealed the surprising colocalization of active (i.e., H3K4me3) and repressive (i.e., H3K27me3) histone modifications on a large proportion of gene promoters [58,110,157,158]. Precisely, in PSCs, most promoters with a high CpG content (also called HCPs) are marked by

H3K4me3, and ~22% of those are simultaneously enriched for the repressive mark H3K27me3 [110,157]. Since H3K27me3 correlates with gene repression, while H3K4me3 is usually found on active genes, promoters that are simultaneously enriched for both marks have been called bivalent [58] (Figure 4). Interestingly, bivalent genes are expressed at extremely low levels, and it has been suggested that while the H3K27me3 mark is important to maintain these genes in a repressed state, the presence of H3K4me3 may poise them for rapid activation upon induction of differentiation by external signals. Consistent with the possibility that bivalent genes exist in a transcriptionally poised state, their promoters are bound by the initiating (i.e., phosphorylated on Ser5) but not the elongating (i.e., phosphorylated on Ser2) form of RNA Pol II and, as a consequence, they only express small transcripts that are not efficiently elongated [159]. Furthermore, the activation of bivalent genes upon PSCs differentiation is accompanied by removal of the repressive mark H3K27me3 (Figure 4) while bivalent genes that become silent retain H3K27me3 but lose H3K4me3 from their promoters [58,110,157,158]. Since bivalent genes are enriched for developmental regulators including TFs and signaling proteins, it has been proposed that the state of H3K4me3/H3K27me3 bivalency on promoters underlies pluripotency by preventing developmental genes from being expressed at high levels in PSCs while simultaneously priming these genes for rapid activation (or stable repression) upon differentiation.

Importantly, sequential ChIP experiments on polynucleosomes [58,157,160,161] or mononucleosomes [162,163] have confirmed that the active H3K4me3 and the repressive H3K27me3 histone marks are present together on neighboring nucleosomes or on the same nucleosome, respectively. These results demonstrate that, for the most part, bivalency does not result from heterogeneity within the cell population. Further supporting this conclusion, quantification of histone modifications by mass spectrometry of ChIP-ed mononucleosomes revealed that a significant proportion (i.e., 15%) of mononucleosomes marked by H3K4me3 are also enriched for H3K27me3 [163]. Together with a previous mass spectrometry study showing that H3K4me3 and H3K27me3 do not coexist on individual histones [164], this finding by Voigt et al. [163] also suggests that bivalent nucleosomes exist in an asymmetric conformation with each copy of the two histones H3 within a nucleosome, carrying either H3K4me3 or H3K27me3, but not both marks simultaneously [59]. Novel methods of allele-specific ChIPs are required, however, to determine the extent to which bivalent promoters may also reflect intracellular heterogeneity due to differential allelic enrichment.

In addition to being present in primed PSCs [58,110,157,158], bivalent promoters have also been identified in naïve PSCs (of both murine [165] and human [8] origin), in iPSCs, [166–168] and even in primordial germ cells [169]. Furthermore, they have been detected in vivo in the transient population of PSCs isolated from pre- and post-implantation blastocysts [170–172], which confirms that bivalent marks are not artifacts of cell culture. While bivalent promoters were first identified in PSCs, they are also present, albeit to a lesser extent, in adult MSCs, including neural progenitor cells [110] and HSCs [173], as well as differentiated cells such as embryonic fibroblasts [110] and CD4+ T-cells [144,160], the latter of which retain a restricted differentiation potential. Interestingly, bivalent promoters appear to be most prevalent in primed PSCs, since reduced numbers of bivalent H3K4me3/H3K27me3 marks have been reported both in naïve PSCs [8,165] and in cells with a more restricted potential (i.e., MSCs and differentiated cells). Finally, it should be noted that the bivalent state may also be gained during cell differentiation as was shown during differentiation of PSCs into glutamatergic pyramidal neurons [174]. While bivalent promoters have been proposed to be involved in the maintenance of plasticity in stem cells, their role in differentiated cells remains unclear.

More recently, histone modifications with seemingly opposing functions (i.e., active mark H3K4me1 and repressive mark H3K27me3) were also identified on distal enhancers [143,175]. Enhancers that harbor these two opposing marks simultaneously have also been called bivalent or poised and similarly to bivalent promoters, they are associated to developmental genes that are expressed at extremely low levels in PSCs. When those genes become activated upon differentiation, bivalent enhancers resolve to form active enhancers that are enriched in H3K4me1 and acetylated H3K27 (H3K27ac) (Figure 4). In contrast, enhancers that regulate pluripotency genes such as *Sox2* are in an active state (i.e., H3K4me1/H3K27ac) in PSCs, and upon differentiation they resolve into a poised state (i.e., H3K4me1/H3K27me3) in parallel with a decrease in expression of the associated gene [143,175].

In summary, signatures that are based on differential histone PTMs have been identified on gene promoters and enhancers in PSCs, adult stem cells, and differentiated cells. These signatures have been instrumental in distinguishing the transcriptional status of associated genes and allowed the identification of bivalent loci that are characterized by the co-occurrence of active and repressive histone modifications (Figure 4). Importantly, comparative analyses revealed that stem cells (particularly primed PSCs) display a characteristically high level of bivalent loci that decreases with differentiation. It is currently believed that this high degree of bivalency is critical for stem cells' pluripotency (and multipotency) by maintaining cell-specific genes in a repressed state that is stable, yet easily reversible in response to developmental cues [59].

4.3 Mechanisms/enzymes involved in establishing bivalent marks in stem cells

4.3.1 The importance of CpG islands for generation of bivalent domains in PSCs

A number of studies have established the critical role of CpG islands in the formation of bivalent domains. Firstly, it was recognized early on that bivalent promoters strongly correlate with CpG islands, particularly in PSCs where virtually all H3K4me3/H3K27me3 bivalent promoters map to CpG-rich domains [58,110,157], such domains being also devoid of DNA methylation [176,177]. Importantly, the artificial introduction of

CpG-rich DNA clusters into PSCs is sufficient to trigger the formation of bivalent domains as long as the DNA sequence does not contain strong binding sites for TFs [178–180]. These findings demonstrate that CpG islands are both necessary and sufficient to mediate the formation of bivalent domains and that this process can be modulated by TFs.

4.3.2 Establishment of the H3K4me3 mark by trithorax group (TrxG) proteins

Methylation of H3K4 is mediated by the SET domain-containing family of enzymes that include SET1A/B, and the TrxG proteins MLL1/2 (mixed lineage leukemia, homologs of *Drosophila* Trx) and MLL3/4 (homologs of *Drosophila* Trr, trithorax-related). Each of these H3K4 methyltransferases associates with the WRAD complex (containing WDR5, RbBP5, Ash2L, and Dpy30 subunits) that is required to promote their stability and catalytic activity [181,182]. While SET1A/B-containing complexes are likely responsible for the bulk of H3K4me3 [183], MLL2 has been recently identified as the enzyme that is primarily responsible for trimethylation of H3K4 on bivalent promoters in PSCs [184,185]. Furthermore, MLL1 also appears to be able to perform this function in the absence of MLL2 while the SET1 complex seems to be mostly excluded from bivalent promoters [185]. Finally, it is interesting to note that depletion of each individual subunit of the WRAD complex results in a global decrease of H3K4me3 level [186–189], a result that further emphasizes the critical role of this subcomplex in promoting the methyltransferase activity of SET1-like enzymes in vivo.

The importance of MLL1/2 complexes for PSCs function is globally supported by the phenotypes observed upon depletion of the different complexes' subunits. For instance, deletion of either MLL1 or MLL2 leads to early embryonic lethality [190,191], suggesting that both enzymes have important, nonoverlapping roles in development. Furthermore, it was demonstrated that even though MLL2 deficiency does not compromise pluripotency per se, it does lead to a significant decrease in PSCs proliferation, survival, and differentiation, and also compromises the timing and coordination of lineage commitment [192]. Finally, depletion of either the WDR5 or the Ash2L subunit of the WRAD complex in PSCs leads to severe pluripotency defects [189,193] while knockdown of the Dpy30 subunit leads to defects in PSCs cell-fate specification, but does not appear to affect self renewal [188]. While these results globally support a significant role for the SET1 family of H3K4 methyltransferases in regulating stem cell functions, it is important to emphasize that the observed phenotypes may also reflect the role of these protein complexes on nonbivalent genes, as well as their potential methyltransferase-independent functions.

4.3.3 How are the MLL1/2 complexes recruited to chromatin?

Recruitment of MLL1/2-containing complexes to CpG islands likely encompasses several mechanisms that may act separately or in combination, depending on gene loci and the cellular environment. First, MLL1 and MLL2 each possess a CxxC DNA-binding domain that recognizes specifically unmethylated CpG domains [194,195]. In addition to promoting the binding of MLLs to nonmethylated CpG islands, the CxxC domain may also be involved in excluding MLL complexes from repressed genes that are marked by DNA methylation, thereby preventing spurious activation. MLL1/2 complexes can also be recruited to chromatin through interactions with sequence-specific DNA-binding TFs (e.g., OCT4 [193], estrogen receptor [196], NFE2 [197]), histone variants (e.g., H2A.Z [198]), long noncoding RNA (e.g., HOTTIP [199]), as well as TET/OGT proteins through their interaction with the HCF1 subunit of MLL complexes [200]. Interestingly, TET/OGT proteins are also enriched on CpG islands and have been proposed to be involved in active DNA demethylation, thereby functionally linking H3K4 trimethylation with DNA demethylation [201,202]. Finally, interaction of MLL1/2 complexes with the initiating RNA pol II through its Ser5 phosphorylated domain [203] may be involved in stabilizing MLL complexes binding to active and bivalent promoters (Figure 5).

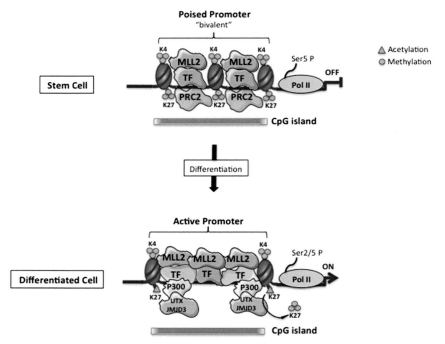

Figure 5 Epigenetic enzymes are involved in transitioning from a bivalent poised gene to an actively transcribed gene during stem cell differentiation. Bivalent marks on CpG-rich promoters are established by the combined action of the MLL2 complex that introduces the H3K4me3 mark, and the PRC2 complex that introduces the H3K27me3 mark. Recruitment of MLL2 and PRC2 complexes to a gene promoter entails several mechanisms including CpG-island binding and transcription factor (TF)-mediated binding. The transition from a poised to an active transcriptional state is facilitated by TF-mediated recruitment of a H3K27 demethylase (e.g., UTX or JMJD3) that removes the repressive H3K27me3 mark, followed by acetylation of this residue by a histone acetyltransferase (e.g., p300).

4.3.4 Establishment of the H3K27me3 mark by polycomb group (PcG) proteins

Polycomb proteins (PcG) are an evolutionarily conserved family of chromatin regulators best known for their function in establishing and maintaining epigenetic memory during development [204]. Methylation of H3K27 is mediated by the polycomb repressive complex 2 (PRC2) that is made of four core subunits: the SET domain–containing H3K27 methyltransferase *enhancer of zeste 2* Ezh2 (or its homolog Ezh1), the *embryonic ectoderm development* protein Eed, the *suppressor of zeste 12* protein Suz12, and the histone chaperones RbAp46/48 (also known as RBBP7/4) [205].

Even though Ezh1 and Ezh2 are both capable of methylating H3K27 [206,207], these two enzymes differ in several key aspects. Firstly, the PRC2-Ezh1 complex displays a considerably weaker methyltransferase activity than PRC2-Ezh2 [206,208]. Furthermore, PRC2-Ezh1 possesses additional intrinsic activities such as nucleosome binding [208] and chromatin compaction [206] that are not found in PRC2-Ezh2. Also importantly, the expression profile of Ezh1 significantly differs from that of Ezh2 with both enzymes being expressed in actively dividing cells (e.g., muscle progenitors), while Ezh2 levels decrease during terminal differentiation such that Ezh1 is the predominant form of H3K27 methyltransferase in nondividing differentiated cells [206,209–211]. Also interestingly, the expression of Ezh1 and Ezh2 in somatic stem cells can differ depending on the developmental stage with Ezh1 being highly expressed in HSPCs from the bone marrow, compared to HSPCs from fetal liver whereas Ezh2 is expressed at similar levels in both cell environments [212]. Altogether, these results suggest that PRC2-Ezh1 and PRC2-Ezh2 complexes display redundant but also specific functions. Consistent with the two protein complexes having redundant roles are the findings that Ezh1 and Ezh2 bind to a partially overlapping subset of genes in murine PSCs, and that Ezh1 has a compensatory role in H3K27 methylation and transcriptional silencing in the absence of Ezh2 [206,207]. However, PRC2-Ezh1 cannot fully rescue defects in mesoendodermal development during $Ezh2^{-/-}$ cell differentiation, suggesting nonredundant functions of the two PRC2 complexes in lineage commitment [207]. Also consistent with a differential function of Ezh1 versus Ezh2, mice with an Ezh2 knockout die around the postimplantation period of embryogenesis due to gastrulation defects [213], whereas mice lacking Ezh1 are viable [214]. Finally, a recent study showed that the conditional knockout of Ezh1 in the hematopoietic lineage severely affects the self-renewal capacity of adult HSCs, thereby decreasing the size of the quiescent population of HSCs [215], while deletion of Ezh2 does not appear to affect adult HSCs [212]. Taken together, these results emphasize the differential functions of PRC2-Ezh1 and PRC2-Ezh2 in stem cell maintenance and differentiation.

Similarly to H3K4 methyltransferases (see above), the H3K27 methyltransferases Ezh1 and Ezh2 require other subunits of the PRC2 core complex for their stability and/or enzymatic activity. Specifically, Eed leads to allosteric activation of the methyltransferase activity of PRC2 through binding to its product H3K27me3 [216], and Suz12 is important for Ezh2 stability and catalytic activity [217]. Also consistent with the importance of Ezh1/2-associated proteins for PRC2 activity in vivo, depletion of Eed and Suz12 in PSCs leads to a global decrease of H3K27me2/3 and to the derepression of lineage-specific genes, including bivalent genes [218–220]. In addition, it is interesting to note that even though individual knockouts of PRC2 core complex members (including Ezh2, Eed, and Suz12, but not Ezh1) are lethal in mouse postimplantation due to gastrulation defects [213,217,221], PSCs depleted for subunits of the PRC2 complex are still viable and able to self-renew [207,220,222,223]. Nevertheless, these cells are defective in their capacity to differentiate [207,220,222,223], which confirms the importance of PRC2 for stem cell differentiation.

4.3.5 How are the PRC2 complexes recruited to chromatin?

Similarly to the case of MLL complexes, CpG islands appear to play a critical role in recruiting PRC2 complexes to gene promoters. Indeed, in PSCs, most PRC2-bound promoters contain CpG islands [224,225]. Furthermore, introduction of ectopic GC-rich sequences is sufficient to recruit the PRC2 complex [178,180]. Interestingly, it was noted that the small number of CpG islands that are not bound by Ezh2 in PSCs contain binding sites for highly expressed transcriptional activators, suggesting that high-CpG promoters adopt a bivalent conformation as a default state unless activating TFs are present [225]. Whereas high-CpG domains are undoubtedly important for the binding of PRC2 to gene promoters, the recruitment mechanism remains unclear, especially that none of the PRC2 core complex members possess the nonmethylated DNA-binding domain CxxC. Several nonmutually exclusive recruitment mechanisms have been proposed based on different types of proteins that interact with PRC2.

1. Interaction with sequence-specific DNA-binding TFs
    Several transcriptional repressors that interact with PRC2 (e.g., YY1, REST, and SNAIL) have been proposed to mediate its recruitment to a subset of CpG island-containing promoters [226–228].

2. Interaction with long and short noncoding RNAs
    The long noncoding RNA *Xist*, known for its role in XCI, interacts with Ezh2 and is thought to play a role in recruiting PRC2 to the inactive X chromosome [229]. Other long noncoding RNAs have been proposed to recruit PRC2 to particular loci, including HOTAIR [230] and Kcnqot1 [231]. In addition, small transcripts (50–200 nucleotides) spanning CpG-rich promoters have been shown to interact with subunits of PRC2 [232,233]. However, so far it is unclear whether noncoding RNAs are implicated in the recruitment of PRC2 to bivalent promoters [234].

3. Interaction with ancillary proteins that bind to histone modifications
    In addition to the proteins noted above, PRC2 interacts with the three mammalian homologs of *drosophila* Polycomb-like (PCL): PCL1, PCL2, and PCL3 (also called

PHF1, MTF2, and PHF19, respectively). These proteins contain two PHD fingers and a Tudor domain, the latter of which mediates binding to the histone marks H3K36me2 and H3K36me3 that are usually enriched in the coding region of active genes, but are also present on some bivalent promoters [235–238]. In addition, it has been shown that depletion of PCL2 in murine PSCs leads to upregulation of pluri-potency factors and failure to properly differentiate [239], suggesting a model in which PCL proteins could be involved in recruiting PRC2 to silence a subset of active genes during cell fate transitions. Along the same line, the knockdown of PCL3 in PSCs decreases Suz12 binding to bivalent promoters and leads to self-renewal defects [240]. Finally, it should be noted that PRC2 binds to its own product H3K27me3, which may help its stabilization at promoter sites and/or its propagation to large stretches of chromatin [216,241].

4. Interaction with ancillary proteins that bind to GC-rich sequences
The proteins AEBP and Jarid2 that copurify with PRC2 and display unspecific DNA binding with increased affinity for GC-rich sequences have been proposed to medi-ate PRC2-binding to CpG-containing bivalent promoters [223,242–246]. Further-more, AEBP and Jarid2 are also able to enhance PRC2 catalytic activity. Interestingly, a recent study showed that Jarid2 is required to mediate the binding of PRC2-Ezh2 but not that of PRC2-Ezh1 to nucleosomes [208]. This result, together with the finding that Ezh2 and Jarid2 levels decrease during cell differentiation, prompted Son et al. [208] to propose a model whereby PRC2-Ezh1 and PRC2-Ezh2 have distinct roles in PSCs vs. adult stem cells vs. differentiated cells. Specifically, they proposed that in actively dividing PSCs, Jarid2 recruits PRC2-Ezh2 to chromatin to introduce the H3K27me3 mark, thanks to its robust methyltransferase activity, while PRC2-Ezh1 is mostly involved in chromatin compaction. In adult stem cells that display reduced levels of Jarid2, PRC2-Ezh2 can still be recruited to chromatin to establish new H3K27me3 marks through dimerization with PRC2-Ezh1. In contrast, in ter-minally differentiated cells that express low levels of both Jarid2 and Ezh2, the weak methyltransferase activity of PRC2-Ezh1 is sufficient for the maintenance (but not the establishment) of H3K27me3 [208]. In summary, multiple mechanisms may be involved in mediating the binding of PRC2 to CpG islands containing bivalent promoters.

4.4 Other histone modifications that regulate stem cell functions

While we have focused mainly on protein complexes that underlie the formation of bivalent marks on gene promoters, additional histone-modifying enzymes have been identified that are important for stem cell function. For instance, the methyltransferase G9a (and its homolog and heterodimerization partner GLP) mediates the formation of large domains of repressive H3K9me2 mark (called the LOCKS) in PSCs [72]. Further-more, G9a has been involved in silencing of a number of genes in PSCs, MSCs, and differentiated cells [247]. Also interestingly, G9a mediates dimethylation of H3K9 but also H3K27 (both in vitro and in vivo) [248–250], suggesting that G9a could potentially be involved in regulating the function of some bivalent promoters. While it would be tempting to suggest that G9a/GLP collaborates with PRC2 to mediate gene repression, this hypothesis is unlikely, given that (1) the bulk of G9a/GLP complex does not associate with PRC2, either in PSCs or in progenitors [207,251]; (2) no binding of G9a has been detected on bivalent promoters in PSCs [207]; and (3) the Jarid1a H3K4me3 demethylase that stably (and directly) interacts with G9a in PSCs [251] binds to genomic regions that do not overlap with PRC2 binding [243].

Another class of protein complexes that regulate the function of bivalent genes is called PRC1 for polycomb repressor complex 1 (reviewed in [252,253]). While several types of PRC1 complexes have been identified, they all contain the RING1A/B ubiquitin ligase that catalyzes monoubiquitination of H2A on K119. Furthermore, PRC1 complexes mediate chromatin compaction and are considered effectors of PcG-mediated long-term epigenetic memory [204]. Interestingly, RING1B was shown to occupy approximately half of all bivalent domains in PSCs, and the genes bound by PRC1 are also more likely to remain repressed upon PSCs differentiation [225]. In summary, PRC1 is a major player in gene expression in PSCs.

5. CONCLUSIVE REMARKS ON THE IMPORTANCE OF EPIGENETIC-BASED METHODS TO OPTIMIZE THE USE OF STEM AND PROGENITOR CELLS FOR REGENERATIVE MEDICINE

In this chapter, we have used specific examples to illustrate how epigenetic mechanisms (i.e., DNA methylation, histones PTMs, and the 3D organization of the nucleus) are critically involved in controlling stem cell functions, including self-renewal, cell fate determination, and differentiation. Going forward, an important challenge will be to combine our knowledge of the different levels of epigenetic regulation to provide a unified view of the mechanisms that control cell fate memory, stem cell proliferation, self-renewal, and differentiation. While the task is immense, the efficient and safe use of stem cells for clinical applications depends entirely on our ability to decipher the molecular mechanisms governing stem cell function, such that we can expand and efficiently differentiate them toward the cell type(s) of interest. On a positive note, two recent studies have provided proof-of-principle that small molecule drugs that modulate the activities of epigenetic enzymes (i.e., the so-called epigenetic drugs) can be utilized to enhance stem cell functions [66,254]. Specifically, one study showed that the HDAC inhibitor valproic acid significantly increases ex vivo expansion of HSPCs that subsequently engraft in a recipient bone marrow [254]. This finding could allow for the expansion of HSPCs in cell culture for use in transplantation, and therefore has important implications for all diseases that require HSC transplantation, such as leukemia, thalassemia, and sickle cell anemia. The other study used another HDAC inhibitor (i.e., trichostatin A or TSA) to enhance the vascular repair property of cord blood derived endothelial stem/progenitor cells. Specifically, Palii et al. [66] showed that ex vivo treatment with TSA increases the kinetic of revascularization by transplanted endothelial progenitors in a preclinical model of limb ischemia. In addition, this finding has potential implications for all diseases that require rapid revascularization, such as heart attacks and strokes. While these two studies have opened the way to using epigenetic drugs to boost stem cell functions, much remains to be done before we can fully control the function of stem cells through epigenetics-based methods.

ACKNOWLEDGMENTS

The authors would like to thank F.J. Dilworth (OHRI) for critically reading the manuscript. M.B. holds the Canadian Research Chair in the regulation of gene expression. Funding in the Brand lab is provided by the Canadian Institutes of Health Research.

REFERENCES

[1] Ramalho-Santos M, Willenbring H. On the origin of the term "stem cell". Cell Stem Cell 2007;1(1):35–8.
[2] Till JE, Mc CE. A direct measurement of the radiation sensitivity of normal mouse bone marrow cells. Radiat Res 1961;14:213–22.
[3] Till JE, McCulloch EA, Siminovitch L. A stochastic model of stem cell proliferation, based on the growth of spleen colony-forming cells. Proc Natl Acad Sci USA 1964;51:29–36.
[4] De Los Angeles A, Loh YH, Tesar PJ, Daley GQ. Accessing naive human pluripotency. Curr Opin Genet Dev 2012;22(3):272–82.
[5] Mascetti VL, Pedersen RA. Naivete of the human pluripotent stem cell. Nat Biotechnol 2014;32(1): 68–70.
[6] Evans MJ, Kaufman MH. Establishment in culture of pluripotential cells from mouse embryos. Nature 1981;292(5819):154–6.
[7] Martin GR. Isolation of a pluripotent cell line from early mouse embryos cultured in medium conditioned by teratocarcinoma stem cells. Proc Natl Acad Sci USA 1981;78(12):7634–8.
[8] Gafni O, Weinberger L, Mansour AA, et al. Derivation of novel human ground state naive pluripotent stem cells. Nature 2013;504(7479):282–6.
[9] Ware CB, Nelson AM, Mecham B, et al. Derivation of naive human embryonic stem cells. Proc Natl Acad Sci USA 2014;111(12):4484–9.
[10] Tesar PJ, Chenoweth JG, Brook FA, et al. New cell lines from mouse epiblast share defining features with human embryonic stem cells. Nature 2007;448(7150):196–9.
[11] Brons IG, Smithers LE, Trotter MW, et al. Derivation of pluripotent epiblast stem cells from mammalian embryos. Nature 2007;448(7150):191–5.
[12] Rossant J. Stem cells and early lineage development. Cell 2008;132(4):527–31.
[13] Thomson JA, Itskovitz-Eldor J, Shapiro SS, et al. Embryonic stem cell lines derived from human blastocysts. Science 1998;282(5391):1145–7.
[14] Tachibana M, Amato P, Sparman M, et al. Human embryonic stem cells derived by somatic cell nuclear transfer. Cell 2013;153(6):1228–38.
[15] Chung YG, Eum JH, Lee JE, et al. Human somatic cell nuclear transfer using adult cells. Cell Stem Cell 2014.

[16] Okita K, Ichisaka T, Yamanaka S. Generation of germline-competent induced pluripotent stem cells. Nature 2007;448(7151):313–7.

[17] Takahashi K, Tanabe K, Ohnuki M, et al. Induction of pluripotent stem cells from adult human fibroblasts by defined factors. Cell 2007;131(5):861–72.

[18] Yu J, Vodyanik MA, Smuga-Otto K, et al. Induced pluripotent stem cell lines derived from human somatic cells. Science 2007;318(5858):1917–20.

[19] Wernig M, Meissner A, Foreman R, et al. In vitro reprogramming of fibroblasts into a pluripotent ES-cell-like state. Nature 2007;448(7151):318–24.

[20] Yamanaka S. Induced pluripotent stem cells: past, present, and future. Cell Stem Cell 2012;10(6): 678–84.

[21] Robinton DA, Daley GQ. The promise of induced pluripotent stem cells in research and therapy. Nature 2012;481(7381):295–305.

[22] Inoue H, Nagata N, Kurokawa H, Yamanaka S. iPS cells: a game changer for future medicine. EMBO J 2014;33(5):409–17.

[23] Ohnishi K, Semi K, Yamamoto T, et al. Premature termination of reprogramming in vivo leads to cancer development through altered epigenetic regulation. Cell 2014;156(4):663–77.

[24] Orkin SH, Zon LI. Hematopoiesis: an evolving paradigm for stem cell biology. Cell 2008;132(4): 631–44.

[25] Aziz A, Sebastian S, Dilworth FJ. The origin and fate of muscle satellite cells. Stem Cell Rev 2012;8(2):609–22.

[26] Blanpain C, Fuchs E. Epidermal stem cells of the skin. Annu Rev Cell Dev Biol 2006;22:339–73.

[27] Butti E, Cusimano M, Bacigaluppi M, Martino G. Neurogenic and non-neurogenic functions of endogenous neural stem cells. Front Neurosci 2014;8:92.

[28] Holstege H, Pfeiffer W, Sie D, et al. Somatic mutations found in the healthy blood compartment of a 115-yr-old woman demonstrate oligoclonal hematopoiesis. Genome Res 2014;24(5):733–42.

[29] Sacco A, Mourkioti F, Tran R, et al. Short telomeres and stem cell exhaustion model Duchenne muscular dystrophy in mdx/mTR mice. Cell 2010;143(7):1059–71.

[30] Hamilton BK, Copelan EA. Concise review: the role of hematopoietic stem cell transplantation in the treatment of acute myeloid leukemia. Stem Cells 2012;30(8):1581–6.

[31] Lucarelli G, Isgro A, Sodani P, Gaziev J. Hematopoietic stem cell transplantation in thalassemia and sickle cell anemia. Cold Spring Harb Perspect Med 2012;2(5):a011825.

[32] Katsimpardi L, Litterman NK, Schein PA, et al. Vascular and neurogenic rejuvenation of the aging mouse brain by young systemic factors. Science 2014;344(6184):630–4.

[33] Bernet JD, Doles JD, Hall JK, Kelly Tanaka K, Carter TA, Olwin BB. p38 MAPK signaling underlies a cell-autonomous loss of stem cell self-renewal in skeletal muscle of aged mice. Nat Med 2014;20(3):265–71.

[34] Florian MC, Nattamai KJ, Dorr K, et al. A canonical to non-canonical Wnt signalling switch in haematopoietic stem-cell ageing. Nature 2013;503(7476):392–6.

[35] Giarratana MC, Kobari L, Lapillonne H, et al. Ex vivo generation of fully mature human red blood cells from hematopoietic stem cells. Nat Biotechnol 2005;23(1):69–74.

[36] Palii CG, Pasha R, Brand M. Lentiviral-mediated knockdown during ex vivo erythropoiesis of human hematopoietic stem cells. J Vis Exp 2011;53.

[37] Walasek MA, van Os R, de Haan G. Hematopoietic stem cell expansion: challenges and opportunities. Ann NY Acad Sci 2012;1266:138–50.

[38] Lassar AB, Paterson BM, Weintraub H. Transfection of a DNA locus that mediates the conversion of 10T1/2 fibroblasts to myoblasts. Cell 1986;47(5):649–56.

[39] Davis RL, Weintraub H, Lassar AB. Expression of a single transfected cDNA converts fibroblasts to myoblasts. Cell 1987;51(6):987–1000.

[40] Xie H, Ye M, Feng R, Graf T. Stepwise reprogramming of B cells into macrophages. Cell 2004;117(5):663–76.

[41] Graf T. Historical origins of transdifferentiation and reprogramming. Cell Stem Cell 2011;9(6): 504–16.

[42] Dilworth FJ, Chambon P. Nuclear receptors coordinate the activities of chromatin remodeling complexes and coactivators to facilitate initiation of transcription. Oncogene 2001;20(24):3047–54.

[43] Roeder RG. Transcriptional regulation and the role of diverse coactivators in animal cells. FEBS Lett 2005;579(4):909–15.

[44] Brownell JE, Zhou J, Ranalli T, et al. Tetrahymena histone acetyltransferase A: a homolog to yeast Gcn5p linking histone acetylation to gene activation. Cell 1996;84(6):843–51.

[45] Bannister AJ, Kouzarides T. Regulation of chromatin by histone modifications. Cell Res 2011;21(3): 381–95.

[46] Ho L, Crabtree GR. Chromatin remodelling during development. Nature 2010;463(7280):474–84.

[47] Taverna SD, Li H, Ruthenburg AJ, Allis CD, Patel DJ. How chromatin-binding modules interpret histone modifications: lessons from professional pocket pickers. Nat Struct Mol Biol 2007;14(11): 1025–40.

[48] Hosey AM, Chaturvedi CP, Brand M. Crosstalk between histone modifications maintains the developmental pattern of gene expression on a tissue-specific locus. Epigenetics 2010;5(4):273–81.

[49] Probst AV, Dunleavy E, Almouzni G. Epigenetic inheritance during the cell cycle. Nat Rev Mol Cell Biol 2009;10(3):192–206.

[50] Margueron R, Reinberg D. Chromatin structure and the inheritance of epigenetic information. Nat Rev Genet 2010;11(4):285–96.

[51] Mansour AA, Gafni O, Weinberger L, et al. The H3K27 demethylase Utx regulates somatic and germ cell epigenetic reprogramming. Nature 2012;488(7411):409–13.

[52] Langlois T, da Costa Reis Monte Mor B, Lenglet G, et al. TET2 deficiency inhibits mesoderm and hematopoietic differentiation in human embryonic stem cells. Stem Cells 2014.

[53] Pasini D, Bracken AP, Agger K, et al. Regulation of stem cell differentiation by histone methyltransferases and demethylases. Cold Spring Harb Symp Quant Biol 2008;73:253–63.

[54] Laugesen A, Helin K. Chromatin repressive complexes in stem cells, development, and cancer. Cell Stem Cell 2014;14(6):735–51.

[55] Efroni S, Duttagupta R, Cheng J, et al. Global transcription in pluripotent embryonic stem cells. Cell Stem Cell 2008;2(5):437–47.

[56] Meshorer E, Misteli T. Chromatin in pluripotent embryonic stem cells and differentiation. Nat Rev Mol Cell Biol 2006;7(7):540–6.

[57] Boskovic A, Eid A, Pontabry J, et al. Higher chromatin mobility supports totipotency and precedes pluripotency in vivo. Genes Dev 2014;28(10):1042–7.

[58] Bernstein BE, Mikkelsen TS, Xie X, et al. A bivalent chromatin structure marks key developmental genes in embryonic stem cells. Cell 2006;125(2):315–26.

[59] Voigt P, Tee WW, Reinberg D. A double take on bivalent promoters. Genes Dev 2013;27(12): 1318–38.

[60] Polo JM, Anderssen E, Walsh RM, et al. A molecular roadmap of reprogramming somatic cells into iPS cells. Cell 2012;151(7):1617–32.

[61] Onder TT, Kara N, Cherry A, et al. Chromatin-modifying enzymes as modulators of reprogramming. Nature 2012;483(7391):598–602.

[62] Kubicek S, O'Sullivan RJ, August EM, et al. Reversal of H3K9me2 by a small-molecule inhibitor for the G9a histone methyltransferase. Mol Cell 2007;25(3):473–81.

[63] Kruidenier L, Chung CW, Cheng Z, et al. A selective jumonji H3K27 demethylase inhibitor modulates the proinflammatory macrophage response. Nature 2012;488(7411):404–8.

[64] Bao S, Tang F, Li X, et al. Epigenetic reversion of post-implantation epiblast to pluripotent embryonic stem cells. Nature 2009;461(7268):1292–5.

[65] Broxmeyer HE. Inhibiting HDAC for human hematopoietic stem cell expansion. J Clin Invest 2014;124(6):2365–8.

[66] Palii CG, Vulesevic B, Fraineau S, et al. Trichostatin a enhances vascular repair by injected human endothelial progenitors through increasing the expression of TAL1-dependent genes. Cell Stem Cell 2014;14(5):644–57.

[67] Sexton T, Schober H, Fraser P, Gasser SM. Gene regulation through nuclear organization. Nat Struct Mol Biol 2007;14(11):1049–55.

[68] Rajapakse I, Groudine M. On emerging nuclear order. J Cell Biol 2011;192(5):711–21.

[69] Cavalli G, Misteli T. Functional implications of genome topology. Nat Struct Mol Biol 2013; 20(3):290–9.

[70] Saurin AJ, Shiels C, Williamson J, et al. The human polycomb group complex associates with pericentromeric heterochromatin to form a novel nuclear domain. J Cell Biol 1998;142(4):887–98.

[71] Guelen L, Pagie L, Brasset E, et al. Domain organization of human chromosomes revealed by mapping of nuclear lamina interactions. Nature 2008;453(7197):948–51.

[72] Wen B, Wu H, Shinkai Y, Irizarry RA, Feinberg AP. Large histone H3 lysine 9 dimethylated chromatin blocks distinguish differentiated from embryonic stem cells. Nat Genet 2009;41(2):246–50.

[73] Finlan LE, Sproul D, Thomson I, et al. Recruitment to the nuclear periphery can alter expression of genes in human cells. PLoS Genet 2008;4(3):e1000039.

[74] Reddy KL, Zullo JM, Bertolino E, Singh H. Transcriptional repression mediated by repositioning of genes to the nuclear lamina. Nature 2008;452(7184):243–7.

[75] Pauler FM, Sloane MA, Huang R, et al. H3K27me3 forms BLOCs over silent genes and intergenic regions and specifies a histone banding pattern on a mouse autosomal chromosome. Genome Res 2009;19(2):221–33.

[76] Zhou VW, Goren A, Bernstein BE. Charting histone modifications and the functional organization of mammalian genomes. Nat Rev Genet 2011;12(1):7–18.

[77] Boisvert FM, van Koningsbruggen S, Navascues J, Lamond AI. The multifunctional nucleolus. Nat Rev Mol Cell Biol 2007;8(7):574–85.

[78] Pombo A, Jackson DA, Hollinshead M, Wang Z, Roeder RG, Cook PR. Regional specialization in human nuclei: visualization of discrete sites of transcription by RNA polymerase III. EMBO J 1999;18(8):2241–53.

[79] Schoenfelder S, Sexton T, Chakalova L, et al. Preferential associations between co-regulated genes reveal a transcriptional interactome in erythroid cells. Nat Genet 2010;42(1):53–61.

[80] Wang J, Shiels C, Sasieni P, et al. Promyelocytic leukemia nuclear bodies associate with transcriptionally active genomic regions. J Cell Biol 2004;164(4):515–26.

[81] Hall LL, Smith KP, Byron M, Lawrence JB. Molecular anatomy of a speckle. Anat Rec A Discov Mol Cell Evol Biol 2006;288(7):664–75.

[82] Morris GE. The cajal body. Biochim Biophys Acta 2008;1783(11):2108–15.

[83] Strambio-De-Castillia C, Niepel M, Rout MP. The nuclear pore complex: bridging nuclear transport and gene regulation. Nat Rev Mol Cell Biol 2010;11(7):490–501.

[84] Kosak ST, Scalzo D, Alworth SV, et al. Coordinate gene regulation during hematopoiesis is related to genomic organization. PLoS Biol 2007;5(11):e309.

[85] Rajapakse I, Perlman MD, Scalzo D, Kooperberg C, Groudine M, Kosak ST. The emergence of lineage-specific chromosomal topologies from coordinate gene regulation. Proc Natl Acad Sci USA 2009;106(16):6679–84.

[86] Lieberman-Aiden E, van Berkum NL, Williams L, et al. Comprehensive mapping of long-range interactions reveals folding principles of the human genome. Science 2009;326(5950):289–93.

[87] Sexton T, Yaffe E, Kenigsberg E, et al. Three-dimensional folding and functional organization principles of the Drosophila genome. Cell 2012;148(3):458–72.

[88] Branco MR, Pombo A. Intermingling of chromosome territories in interphase suggests role in translocations and transcription-dependent associations. PLoS Biol 2006;4(5):e138.

[89] Cremer T, Cremer M. Chromosome territories. Cold Spring Harb Perspect Biol 2010;2(3):a003889.

[90] Boyle S, Gilchrist S, Bridger JM, Mahy NL, Ellis JA, Bickmore WA. The spatial organization of human chromosomes within the nuclei of normal and emerin-mutant cells. Hum Mol Genet 2001;10(3):211–9.

[91] Brown KE, Guest SS, Smale ST, Hahm K, Merkenschlager M, Fisher AG. Association of transcriptionally silent genes with Ikaros complexes at centromeric heterochromatin. Cell 1997;91(6):845–54.

[92] Skok JA, Brown KE, Azuara V, et al. Nonequivalent nuclear location of immunoglobulin alleles in B lymphocytes. Nat Immunol 2001;2(9):848–54.

[93] Kosak ST, Skok JA, Medina KL, et al. Subnuclear compartmentalization of immunoglobulin loci during lymphocyte development. Science 2002;296(5565):158–62.

[94] Ragoczy T, Bender MA, Telling A, Byron R, Groudine M. The locus control region is required for association of the murine beta-globin locus with engaged transcription factories during erythroid maturation. Genes Dev 2006;20(11):1447–57.

[95] Tan-Wong SM, Zaugg JB, Camblong J, et al. Gene loops enhance transcriptional directionality. Science 2012;338(6107):671–5.

[96] Tolhuis B, Palstra RJ, Splinter E, Grosveld F, de Laat W. Looping and interaction between hypersensitive sites in the active beta-globin locus. Mol Cell 2002;10(6):1453–65.

[97] Deng W, Lee J, Wang H, et al. Controlling long-range genomic interactions at a native locus by targeted tethering of a looping factor. Cell 2012;149(6):1233–44.

[98] Lomvardas S, Barnea G, Pisapia DJ, Mendelsohn M, Kirkland J, Axel R. Interchromosomal interactions and olfactory receptor choice. Cell 2006;126(2):403–13.

[99] Li G, Ruan X, Auerbach RK, et al. Extensive promoter-centered chromatin interactions provide a topological basis for transcription regulation. Cell 2012;148(1–2):84–98.

[100] Tiwari VK, Cope L, McGarvey KM, Ohm JE, Baylin SB. A novel 6C assay uncovers polycomb-mediated higher order chromatin conformations. Genome Res 2008;18(7):1171–9.

[101] Wiblin AE, Cui W, Clark AJ, Bickmore WA. Distinctive nuclear organisations of centromeres and regions involved in pluripotency in human embryonic stem cells. J Cell Sci 2005;118(Pt 17):3861–8.

[102] Bartova E, Galiova G, Krejci J, Harnicarova A, Strasak L, Kozubek S. Epigenome and chromatin structure in human embryonic stem cells undergoing differentiation. Dev Dyn 2008;237(12):3690–702.

[103] Constantinescu D, Gray HL, Sammak PJ, Schatten GP, Csoka AB. Lamin A/C expression is a marker of mouse and human embryonic stem cell differentiation. Stem Cells 2006;24(1):177–85.

[104] Pajerowski JD, Dahl KN, Zhong FL, Sammak PJ, Discher DE. Physical plasticity of the nucleus in stem cell differentiation. Proc Natl Acad Sci USA 2007;104(40):15619–24.

[105] Butler JT, Hall LL, Smith KP, Lawrence JB. Changing nuclear landscape and unique PML structures during early epigenetic transitions of human embryonic stem cells. J Cell Biochem 2009;107(4):609–21.

[106] Ren X, Vincenz C, Kerppola TK. Changes in the distributions and dynamics of polycomb repressive complexes during embryonic stem cell differentiation. Mol Cell Biol 2008;28(9):2884–95.

[107] Gaspar-Maia A, Alajem A, Polesso F, et al. Chd1 regulates open chromatin and pluripotency of embryonic stem cells. Nature 2009;460(7257):863–8.

[108] Ho L, Miller EL, Ronan JL, Ho WQ, Jothi R, Crabtree GR. esBAF facilitates pluripotency by conditioning the genome for LIF/STAT3 signalling and by regulating polycomb function. Nat Cell Biol 2011;13(8):903–13.

[109] Meshorer E, Yellajoshula D, George E, Scambler PJ, Brown DT, Misteli T. Hyperdynamic plasticity of chromatin proteins in pluripotent embryonic stem cells. Dev Cell 2006;10(1):105–16.

[110] Mikkelsen TS, Ku M, Jaffe DB, et al. Genome-wide maps of chromatin state in pluripotent and lineage-committed cells. Nature 2007;448(7153):553–60.

[111] Fussner E, Djuric U, Strauss M, et al. Constitutive heterochromatin reorganization during somatic cell reprogramming. EMBO J 2011;30(9):1778–89.

[112] de Wit E, Bouwman BA, Zhu Y, et al. The pluripotent genome in three dimensions is shaped around pluripotency factors. Nature 2013;501(7466):227–31.

[113] Kim SH, McQueen PG, Lichtman MK, Shevach EM, Parada LA, Misteli T. Spatial genome organization during T-cell differentiation. Cytogenet Genome Res 2004;105(2–4):292–301.

[114] Xu N, Tsai CL, Lee JT. Transient homologous chromosome pairing marks the onset of X inactivation. Science 2006;311(5764):1149–52.

[115] Bacher CP, Guggiari M, Brors B, et al. Transient colocalization of X-inactivation centres accompanies the initiation of X inactivation. Nat Cell Biol 2006;8(3):293–9.

[116] Masui O, Bonnet I, Le Baccon P, et al. Live-cell chromosome dynamics and outcome of X chromosome pairing events during ES cell differentiation. Cell 2011;145(3):447–58.

[117] Zhang LF, Huynh KD, Lee JT. Perinucleolar targeting of the inactive X during S phase: evidence for a role in the maintenance of silencing. Cell 2007;129(4):693–706.

[118] Minkovsky A, Patel S, Plath K. Concise review: pluripotency and the transcriptional inactivation of the female mammalian X chromosome. Stem Cells 2012;30(1):48–54.

[119] Chaumeil J, Le Baccon P, Wutz A, Heard E. A novel role for Xist RNA in the formation of a repressive nuclear compartment into which genes are recruited when silenced. Genes Dev 2006;20(16):2223–37.

[120] Splinter E, de Wit E, Nora EP, et al. The inactive X chromosome adopts a unique three-dimensional conformation that is dependent on Xist RNA. Genes Dev 2011;25(13):1371–83.

[121] Lande-Diner L, Cedar H. Silence of the genes–mechanisms of long-term repression. Nat Rev Genet 2005;6(8):648–54.

[122] Jeltsch A, Jurkowska RZ. New concepts in DNA methylation. Trends Biochem Sci 2014;39(7): 310–8.

[123] Kohli RM, Zhang Y. TET enzymes, TDG and the dynamics of DNA demethylation. Nature 2013;502(7472):472–9.

[124] Bagci H, Fisher AG. DNA demethylation in pluripotency and reprogramming: the role of tet proteins and cell division. Cell Stem Cell 2013;13(3):265–9.

[125] Klose RJ, Bird AP. Genomic DNA methylation: the mark and its mediators. Trends Biochem Sci 2006;31(2):89–97.

[126] Bestor TH, Bourc'his D. Transposon silencing and imprint establishment in mammalian germ cells. Cold Spring Harb Symp Quant Biol 2004;69:381–7.

[127] Gu TP, Guo F, Yang H, et al. The role of Tet3 DNA dioxygenase in epigenetic reprogramming by oocytes. Nature 2011;477(7366):606–10.

[128] Iqbal K, Jin SG, Pfeifer GP, Szabo PE. Reprogramming of the paternal genome upon fertilization involves genome-wide oxidation of 5-methylcytosine. Proc Natl Acad Sci USA 2011;108(9):3642–7.

[129] Wossidlo M, Nakamura T, Lepikhov K, et al. 5-Hydroxymethylcytosine in the mammalian zygote is linked with epigenetic reprogramming. Nat Commun 2011;2:241.

[130] Nakamura T, Liu YJ, Nakashima H, et al. PGC7 binds histone H3K9me2 to protect against conversion of 5mC to 5hmC in early embryos. Nature 2012;486(7403):415–9.

[131] Santos F, Hendrich B, Reik W, Dean W. Dynamic reprogramming of DNA methylation in the early mouse embryo. Dev Biol 2002;241(1):172–82.

[132] Rougier N, Bourc'his D, Gomes DM, et al. Chromosome methylation patterns during mammalian preimplantation development. Genes Dev 1998;12:2108–13.

[133] Inoue A, Shen L, Dai Q, He C, Zhang Y. Generation and replication-dependent dilution of 5fC and 5caC during mouse preimplantation development. Cell Res 2011;21(12):1670–6.

[134] Hajkova P, Jeffries SJ, Lee C, Miller N, Jackson SP, Surani MA. Genome-wide reprogramming in the mouse germ line entails the base excision repair pathway. Science 2010;329(5987):78–82.

[135] Wossidlo M, Arand J, Sebastiano V, et al. Dynamic link of DNA demethylation, DNA strand breaks and repair in mouse zygotes. EMBO J 2010;29(11):1877–88.

[136] Ying QL, Wray J, Nichols J, et al. The ground state of embryonic stem cell self-renewal. Nature 2008;453(7194):519–23.

[137] Leitch HG, McEwen KR, Turp A, et al. Naive pluripotency is associated with global DNA hypomethylation. Nat Struct Mol Biol 2013;20(3):311–6.

[138] Ficz G, Hore TA, Santos F, et al. FGF signaling inhibition in ESCs drives rapid genome-wide demethylation to the epigenetic ground state of pluripotency. Cell Stem Cell 2013;13(3):351–9.

[139] Habibi E, Brinkman AB, Arand J, et al. Whole-genome bisulfite sequencing of two distinct interconvertible DNA methylomes of mouse embryonic stem cells. Cell Stem Cell 2013;13(3):360–9.

[140] Doege CA, Inoue K, Yamashita T, et al. Early-stage epigenetic modification during somatic cell reprogramming by Parp1 and Tet2. Nature 2012;488(7413):652–5.

[141] Costa Y, Ding J, Theunissen TW, et al. NANOG-dependent function of TET1 and TET2 in establishment of pluripotency. Nature 2013;495(7441):370–4.

[142] Gao Y, Chen J, Li K, et al. Replacement of Oct4 by Tet1 during iPSC induction reveals an important role of DNA methylation and hydroxymethylation in reprogramming. Cell Stem Cell 2013;12(4): 453–69.

[143] Gifford CA, Ziller MJ, Gu H, et al. Transcriptional and epigenetic dynamics during specification of human embryonic stem cells. Cell 2013;153(5):1149–63.

[144] Barski A, Cuddapah S, Cui K, et al. High-resolution profiling of histone methylations in the human genome. Cell 2007;129(4):823–37.

[145] Heintzman ND, Stuart RK, Hon G, et al. Distinct and predictive chromatin signatures of transcriptional promoters and enhancers in the human genome. Nat Genet 2007;39(3):311–8.

[146] Wang Z, Zang C, Rosenfeld JA, et al. Combinatorial patterns of histone acetylations and methylations in the human genome. Nat Genet 2008;40(7):897–903.

[147] Brand M, Rampalli S, Chaturvedi CP, Dilworth FJ. Analysis of epigenetic modifications of chromatin at specific gene loci by native chromatin immunoprecipitation of nucleosomes isolated using hydroxyapatite chromatography. Nat Protoc 2008;3(3):398–409.

[148] Adli M, Bernstein BE. Whole-genome chromatin profiling from limited numbers of cells using nano-ChIP-seq. Nat Protoc 2011;6(10):1656–68.

[149] Shankaranarayanan P, Mendoza-Parra MA, van Gool W, Trindade LM, Gronemeyer H. Single-tube linear DNA amplification for genome-wide studies using a few thousand cells. Nat Protoc 2012;7(2):328–38.

[150] Hawkins RD, Hon GC, Ren B. Next-generation genomics: an integrative approach. Nat Rev Genet 2010;11(7):476–86.

[151] Ye T, Krebs AR, Choukrallah MA, et al. seqMINER: an integrated ChIP-seq data interpretation platform. Nucleic Acids Res 2011;39(6):e35.

[152] Ernst J, Kellis M. ChromHMM: automating chromatin-state discovery and characterization. Nat Methods 2012;9(3):215–6.

[153] Won KJ, Zhang X, Wang T, et al. Comparative annotation of functional regions in the human genome using epigenomic data. Nucleic Acids Res 2013;41(8):4423–32.

[154] Bardet AF, He Q, Zeitlinger J, Stark A. A computational pipeline for comparative ChIP-seq analyses. Nat Protoc 2012;7(1):45–61.

[155] Felsani A, Gudmundsson B, Nanni S, et al. Impact of different ChIP-Seq protocols on DNA integrity and quality of bioinformatics analysis results. Briefings Funct Genomics 2014.

[156] Calo E, Wysocka J. Modification of enhancer chromatin: what, how, and why? Mol Cell 2013;49(5):825–37.

[157] Pan G, Tian S, Nie J, et al. Whole-genome analysis of histone H3 lysine 4 and lysine 27 methylation in human embryonic stem cells. Cell Stem Cell 2007;1(3):299–312.

[158] Zhao XD, Han X, Chew JL, et al. Whole-genome mapping of histone H3 Lys4 and 27 trimethylations reveals distinct genomic compartments in human embryonic stem cells. Cell Stem Cell 2007;1(3):286–98.

[159] Brookes E, de Santiago I, Hebenstreit D, et al. Polycomb associates genome-wide with a specific RNA polymerase II variant, and regulates metabolic genes in ESCs. Cell Stem Cell 2012;10(2):157–70.

[160] Roh TY, Cuddapah S, Cui K, Zhao K. The genomic landscape of histone modifications in human T cells. Proc Natl Acad Sci USA 2006;103(43):15782–7.

[161] De Gobbi M, Garrick D, Lynch M, et al. Generation of bivalent chromatin domains during cell fate decisions. Epigenet Chromatin 2011;4(1):9.

[162] Seenundun S, Rampalli S, Liu QC, et al. UTX mediates demethylation of H3K27me3 at muscle-specific genes during myogenesis. EMBO J 2010;29(8):1401–11.

[163] Voigt P, LeRoy G, Drury 3rd WJ, et al. Asymmetrically modified nucleosomes. Cell 2012;151(1):181–93.

[164] Young NL, DiMaggio PA, Plazas-Mayorca MD, Baliban RC, Floudas CA, Garcia BA. High throughput characterization of combinatorial histone codes. Mol Cell Proteomics 2009;8(10):2266–84.

[165] Marks H, Kalkan T, Menafra R, et al. The transcriptional and epigenomic foundations of ground state pluripotency. Cell 2012;149(3):590–604.

[166] Maherali N, Sridharan R, Xie W, et al. Directly reprogrammed fibroblasts show global epigenetic remodeling and widespread tissue contribution. Cell Stem Cell 2007;1(1):55–70.

[167] Mikkelsen TS, Hanna J, Zhang X, et al. Dissecting direct reprogramming through integrative genomic analysis. Nature 2008;454(7200):49–55.

[168] Guenther MG, Frampton GM, Soldner F, et al. Chromatin structure and gene expression programs of human embryonic and induced pluripotent stem cells. Cell Stem Cell 2010;7(2):249–57.

[169] Sachs M, Onodera C, Blaschke K, Ebata KT, Song JS, Ramalho-Santos M. Bivalent chromatin marks developmental regulatory genes in the mouse embryonic germline in vivo. Cell Rep 2013;3(6):1777–84.

[170] Alder O, Lavial F, Helness A, et al. Ring1B and Suv39h1 delineate distinct chromatin states at bivalent genes during early mouse lineage commitment. Development 2010;137(15):2483–92.

[171] Dahl JA, Reiner AH, Klungland A, Wakayama T, Collas P. Histone H3 lysine 27 methylation asymmetry on developmentally-regulated promoters distinguish the first two lineages in mouse preimplantation embryos. PLoS One 2010;5(2):e9150.

[172] Rugg-Gunn PJ, Cox BJ, Ralston A, Rossant J. Distinct histone modifications in stem cell lines and tissue lineages from the early mouse embryo. Proc Natl Acad Sci USA 2010;107(24):10783–90.

[173] Cui K, Zang C, Roh TY, et al. Chromatin signatures in multipotent human hematopoietic stem cells indicate the fate of bivalent genes during differentiation. Cell Stem Cell 2009;4(1):80–93.

[174] Mohn F, Weber M, Rebhan M, et al. Lineage-specific polycomb targets and de novo DNA methylation define restriction and potential of neuronal progenitors. Mol Cell 2008;30(6):755–66.

[175] Rada-Iglesias A, Bajpai R, Swigut T, Brugmann SA, Flynn RA, Wysocka J. A unique chromatin signature uncovers early developmental enhancers in humans. Nature 2011;470(7333):279–83.

[176] Weber M, Hellmann I, Stadler MB, et al. Distribution, silencing potential and evolutionary impact of promoter DNA methylation in the human genome. Nat Genet 2007;39(4):457–66.

[177] Meissner A, Mikkelsen TS, Gu H, et al. Genome-scale DNA methylation maps of pluripotent and differentiated cells. Nature 2008;454(7205):766–70.

[178] Mendenhall EM, Koche RP, Truong T, et al. GC-rich sequence elements recruit PRC2 in mammalian ES cells. PLoS Genet 2010;6(12):e1001244.

[179] Thomson JP, Skene PJ, Selfridge J, et al. CpG islands influence chromatin structure via the CpG-binding protein Cfp1. Nature 2010;464(7291):1082–6.

[180] Lynch MD, Smith AJ, De Gobbi M, et al. An interspecies analysis reveals a key role for unmethylated CpG dinucleotides in vertebrate polycomb complex recruitment. EMBO J 2012;31(2):317–29.

[181] Ernst P, Vakoc CR. WRAD: enabler of the SET1-family of H3K4 methyltransferases. Briefings Funct Genomics 2012;11(3):217–26.

[182] Couture JF, Skiniotis G. Assembling a COMPASS. Epigenetics 2013;8(4):349–54.

[183] Shilatifard A. The COMPASS family of histone H3K4 methylases: mechanisms of regulation in development and disease pathogenesis. Annu Rev Biochem 2012;81:65–95.

[184] Hu D, Garruss AS, Gao X, et al. The Mll2 branch of the COMPASS family regulates bivalent promoters in mouse embryonic stem cells. Nat Struct Mol Biol 2013;20(9):1093–7.

[185] Denissov S, Hofemeister H, Marks H, et al. Mll2 is required for H3K4 trimethylation on bivalent promoters in embryonic stem cells, whereas Mll1 is redundant. Development 2014;141(3):526–37.

[186] Wysocka J, Swigut T, Milne TA, et al. WDR5 associates with histone H3 methylated at K4 and is essential for H3 K4 methylation and vertebrate development. Cell 2005;121(6):859–72.

[187] Dou Y, Milne TA, Ruthenburg AJ, et al. Regulation of MLL1 H3K4 methyltransferase activity by its core components. Nat Struct Mol Biol 2006;13(8):713–9.

[188] Jiang H, Shukla A, Wang X, Chen WY, Bernstein BE, Roeder RG. Role for Dpy-30 in ES cell-fate specification by regulation of H3K4 methylation within bivalent domains. Cell 2011;144(4):513–25.

[189] Wan M, Liang J, Xiong Y, et al. The trithorax group protein Ash2l is essential for pluripotency and maintaining open chromatin in embryonic stem cells. J Biol Chem 2013;288(7):5039–48.

[190] Yu BD, Hess JL, Horning SE, Brown GA, Korsmeyer SJ. Altered Hox expression and segmental identity in Mll-mutant mice. Nature 1995;378(6556):505–8.

[191] Glaser S, Schaft J, Lubitz S, et al. Multiple epigenetic maintenance factors implicated by the loss of Mll2 in mouse development. Development 2006;133(8):1423–32.

[192] Lubitz S, Glaser S, Schaft J, Stewart AF, Anastassiadis K. Increased apoptosis and skewed differentiation in mouse embryonic stem cells lacking the histone methyltransferase Mll2. Mol Biol Cell 2007;18(6):2356–66.

[193] Ang YS, Tsai SY, Lee DF, et al. Wdr5 mediates self-renewal and reprogramming via the embryonic stem cell core transcriptional network. Cell 2011;145(2):183–97.

[194] Birke M, Schreiner S, Garcia-Cuellar MP, Mahr K, Titgemeyer F, Slany RK. The MT domain of the proto-oncoprotein MLL binds to CpG-containing DNA and discriminates against methylation. Nucleic Acids Res 2002;30(4):958–65.

[195] Bach C, Mueller D, Buhl S, Garcia-Cuellar MP, Slany RK. Alterations of the CxxC domain preclude oncogenic activation of mixed-lineage leukemia 2. Oncogene 2009;28(6):815–23.

[196] Mo R, Rao SM, Zhu YJ. Identification of the MLL2 complex as a coactivator for estrogen receptor alpha. J Biol Chem 2006;281(23):15714–20.

[197] Demers C, Chaturvedi CP, Ranish JA, et al. Activator-mediated recruitment of the MLL2 methyltransferase complex to the beta-globin locus. Mol Cell 2007;27(4):573–84.

[198] Hu G, Cui K, Northrup D, et al. H2A.Z facilitates access of active and repressive complexes to chromatin in embryonic stem cell self-renewal and differentiation. Cell Stem Cell 2013;12(2):180–92.

[199] Wang KC, Chang HY. Molecular mechanisms of long noncoding RNAs. Mol Cell 2011;43(6): 904–14.

[200] Capotosti F, Guernier S, Lammers F, et al. O-GlcNAc transferase catalyzes site-specific proteolysis of HCF-1. Cell 2011;144(3):376–88.

[201] Deplus R, Delatte B, Schwinn MK, et al. TET2 and TET3 regulate GlcNAcylation and H3K4 methylation through OGT and SET1/COMPASS. EMBO J 2013;32(5):645–55.

[202] Delatte B, Fuks F. TET proteins: on the frenetic hunt for new cytosine modifications. Briefings Funct Genomics 2013;12(3):191–204.

[203] Milne TA, Dou Y, Martin ME, Brock HW, Roeder RG, Hess JL. MLL associates specifically with a subset of transcriptionally active target genes. Proc Natl Acad Sci USA 2005;102(41):14765–70.

[204] Schuettengruber B, Chourrout D, Vervoort M, Leblanc B, Cavalli G. Genome regulation by polycomb and trithorax proteins. Cell 2007;128(4):735–45.

[205] Margueron R, Reinberg D. The Polycomb complex PRC2 and its mark in life. Nature 2011;469(7330):343–9.

[206] Margueron R, Li G, Sarma K, et al. Ezh1 and Ezh2 maintain repressive chromatin through different mechanisms. Mol Cell 2008;32(4):503–18.

[207] Shen X, Liu Y, Hsu YJ, et al. EZH1 mediates methylation on histone H3 lysine 27 and complements EZH2 in maintaining stem cell identity and executing pluripotency. Mol Cell 2008;32(4):491–502.

[208] Son J, Shen SS, Margueron R, Reinberg D. Nucleosome-binding activities within JARID2 and EZH1 regulate the function of PRC2 on chromatin. Genes Dev 2013;27(24):2663–77.

[209] Visser HP, Gunster MJ, Kluin-Nelemans HC, et al. The Polycomb group protein EZH2 is upregulated in proliferating, cultured human mantle cell lymphoma. Br J Haematol 2001;112(4):950–8.

[210] Stojic L, Jasencakova Z, Prezioso C, et al. Chromatin regulated interchange between polycomb repressive complex 2 (PRC2)-Ezh2 and PRC2-Ezh1 complexes controls myogenin activation in skeletal muscle cells. Epigenet Chromatin 2011;4:16.

[211] Mousavi K, Zare H, Wang AH, Sartorelli V. Polycomb protein Ezh1 promotes RNA polymerase II elongation. Mol Cell 2012;45(2):255–62.

[212] Mochizuki-Kashio M, Mishima Y, Miyagi S, et al. Dependency on the polycomb gene Ezh2 distinguishes fetal from adult hematopoietic stem cells. Blood 2011;118(25):6553–61.

[213] O'Carroll D, Erhardt S, Pagani M, Barton SC, Surani MA, Jenuwein T. The polycomb-group gene Ezh2 is required for early mouse development. Mol Cell Biol 2001;21(13):4330–6.

[214] Ezhkova E, Lien WH, Stokes N, Pasolli HA, Silva JM, Fuchs E. EZH1 and EZH2 cogovern histone H3K27 trimethylation and are essential for hair follicle homeostasis and wound repair. Genes Dev 2011;25(5):485–98.

[215] Hidalgo I, Herrera-Merchan A, Ligos JM, et al. Ezh1 is required for hematopoietic stem cell maintenance and prevents senescence-like cell cycle arrest. Cell Stem Cell 2012;11(5):649–62.

[216] Margueron R, Justin N, Ohno K, et al. Role of the polycomb protein EED in the propagation of repressive histone marks. Nature 2009;461(7265):762–7.

[217] Pasini D, Bracken AP, Jensen MR, Lazzerini Denchi E, Helin K. Suz12 is essential for mouse development and for EZH2 histone methyltransferase activity. EMBO J 2004;23(20):4061–71.

[218] Azuara V, Perry P, Sauer S, et al. Chromatin signatures of pluripotent cell lines. Nat Cell Biol 2006;8(5):532–8.

[219] Boyer LA, Plath K, Zeitlinger J, et al. Polycomb complexes repress developmental regulators in murine embryonic stem cells. Nature 2006;441(7091):349–53.

[220] Pasini D, Bracken AP, Hansen JB, Capillo M, Helin K. The polycomb group protein Suz12 is required for embryonic stem cell differentiation. Mol Cell Biol 2007;27(10):3769–79.

[221] Faust C, Schumacher A, Holdener B, Magnuson T. The eed mutation disrupts anterior mesoderm production in mice. Development 1995;121(2):273–85.

[222] Chamberlain SJ, Yee D, Magnuson T. Polycomb repressive complex 2 is dispensable for maintenance of embryonic stem cell pluripotency. Stem Cells 2008;26(6):1496–505.

[223] Leeb M, Pasini D, Novatchkova M, Jaritz M, Helin K, Wutz A. Polycomb complexes act redundantly to repress genomic repeats and genes. Genes Dev 2010;24(3):265–76.

[224] Tanay A, O'Donnell AH, Damelin M, Bestor TH. Hyperconserved CpG domains underlie Polycomb-binding sites. Proc Natl Acad Sci USA 2007;104(13):5521–6.

[225] Ku M, Koche RP, Rheinbay E, et al. Genomewide analysis of PRC1 and PRC2 occupancy identifies two classes of bivalent domains. PLoS Genet 2008;4(10):e1000242.

[226] Herranz N, Pasini D, Diaz VM, et al. Polycomb complex 2 is required for E-cadherin repression by the Snail1 transcription factor. Mol Cell Biol 2008;28(15):4772–81.

[227] Dietrich N, Lerdrup M, Landt E, et al. REST-mediated recruitment of polycomb repressor complexes in mammalian cells. PLoS Genet 2012;8(3):e1002494.

[228] Arnold P, Scholer A, Pachkov M, et al. Modeling of epigenome dynamics identifies transcription factors that mediate Polycomb targeting. Genome Res 2013;23(1):60–73.

[229] Plath K, Fang J, Mlynarczyk-Evans SK, et al. Role of histone H3 lysine 27 methylation in X inactivation. Science 2003;300(5616):131–5.

[230] Rinn JL, Kertesz M, Wang JK, et al. Functional demarcation of active and silent chromatin domains in human HOX loci by noncoding RNAs. Cell 2007;129(7):1311–23.

[231] Pandey RR, Mondal T, Mohammad F, et al. Kcnq1ot1 antisense noncoding RNA mediates lineage-specific transcriptional silencing through chromatin-level regulation. Mol Cell 2008;32(2):232–46.

[232] Kanhere A, Viiri K, Araujo CC, et al. Short RNAs are transcribed from repressed polycomb target genes and interact with polycomb repressive complex-2. Mol Cell 2010;38(5):675–88.

[233] Zhao J, Ohsumi TK, Kung JT, et al. Genome-wide identification of polycomb-associated RNAs by RIP-seq. Mol Cell 2010;40(6):939–53.

[234] Brockdorff N. Noncoding RNA and Polycomb recruitment. RNA 2013;19(4):429–42.

[235] Ballare C, Lange M, Lapinaite A, et al. Phf19 links methylated Lys36 of histone H3 to regulation of Polycomb activity. Nat Struct Mol Biol 2012;19(12):1257–65.

[236] Brien GL, Gambero G, O'Connell DJ, et al. Polycomb PHF19 binds H3K36me3 and recruits PRC2 and demethylase NO66 to embryonic stem cell genes during differentiation. Nat Struct Mol Biol 2012;19(12):1273–81.

[237] Musselman CA, Avvakumov N, Watanabe R, et al. Molecular basis for H3K36me3 recognition by the Tudor domain of PHF1. Nat Struct Mol Biol 2012;19(12):1266–72.

[238] Cai L, Rothbart SB, Lu R, et al. An H3K36 methylation-engaging Tudor motif of polycomb-like proteins mediates PRC2 complex targeting. Mol Cell 2013;49(3):571–82.

[239] Walker E, Chang WY, Hunkapiller J, et al. Polycomb-like 2 associates with PRC2 and regulates transcriptional networks during mouse embryonic stem cell self-renewal and differentiation. Cell Stem Cell 2010;6(2):153–66.

[240] Hunkapiller J, Shen Y, Diaz A, et al. Polycomb-like 3 promotes polycomb repressive complex 2 binding to CpG islands and embryonic stem cell self-renewal. PLoS Genet 2012;8(3):e1002576.

[241] Hansen KH, Bracken AP, Pasini D, et al. A model for transmission of the H3K27me3 epigenetic mark. Nat Cell Biol 2008;10(11):1291–300.

[242] Kim H, Kang K, Kim J. AEBP2 as a potential targeting protein for Polycomb Repression Complex PRC2. Nucleic Acids Res 2009;37(9):2940–50.

[243] Peng JC, Valouev A, Swigut T, et al. Jarid2/Jumonji coordinates control of PRC2 enzymatic activity and target gene occupancy in pluripotent cells. Cell 2009;139(7):1290–302.

[244] Shen X, Kim W, Fujiwara Y, et al. Jumonji modulates polycomb activity and self-renewal versus differentiation of stem cells. Cell 2009;139(7):1303–14.

[245] Li G, Margueron R, Ku M, Chambon P, Bernstein BE, Reinberg D. Jarid2 and PRC2, partners in regulating gene expression. Genes Dev 2010;24(4):368–80.

[246] Landeira D, Sauer S, Poot R, et al. Jarid2 is a PRC2 component in embryonic stem cells required for multi-lineage differentiation and recruitment of PRC1 and RNA Polymerase II to developmental regulators. Nat Cell Biol 2010;12(6):618–24.

[247] Collins R, Cheng X. A case study in cross-talk: the histone lysine methyltransferases G9a and GLP. Nucleic Acids Res 2010;38(11):3503–11.

[248] Tachibana M, Sugimoto K, Fukushima T, Shinkai Y. Set domain-containing protein, G9a, is a novel lysine-preferring mammalian histone methyltransferase with hyperactivity and specific selectivity to lysines 9 and 27 of histone H3. J Biol Chem 2001;276(27):25309–17.

[249] Chaturvedi CP, Hosey AM, Palii C, et al. Dual role for the methyltransferase G9a in the maintenance of {beta}-globin gene transcription in adult erythroid cells. Proc Natl Acad Sci USA 2009.

[250] Wu H, Chen X, Xiong J, et al. Histone methyltransferase G9a contributes to H3K27 methylation in vivo. Cell Res 2011;21(2):365–7.

[251] Chaturvedi CP, Somasundaram B, Singh K, et al. Maintenance of gene silencing by the coordinate action of the H3K9 methyltransferase G9a/KMT1C and the H3K4 demethylase Jarid1a/KDM5A. Proc Natl Acad Sci USA 2012;109(46):18845–50.

[252] Simon JA, Kingston RE. Occupying chromatin: polycomb mechanisms for getting to genomic targets, stopping transcriptional traffic, and staying put. Mol Cell 2013;49(5):808–24.

[253] Di Croce L, Helin K. Transcriptional regulation by Polycomb group proteins. Nat Struct Mol Biol 2013;20(10):1147–55.

[254] Chaurasia P, Gajzer DC, Schaniel C, D'Souza S, Hoffman R. Epigenetic reprogramming induces the expansion of cord blood stem cells. J Clin Invest 2014;124(6):2378–95.

CHAPTER 8

Epigenetic inheritance

Benjamin B. Mills, Christine M. McBride, Nicole C. Riddle
Department of Biology, The University of Alabama at Birmingham, Birmingham, AL, USA

Contents

1. INTRODUCTION

Ever since the discovery that all cells within an organism have the same DNA content despite their vastly different phenotypes, scientists have been trying to understand how this array of phenotypes arises. This question is at the core of the field of epigenetics, which is concerned with the study of mitotically and/or meiotically heritable changes in phenotypes that occur in the absence of DNA sequence changes [131]. Decades of research into epigenetic phenomena such as X-chromosome inactivation, parental imprinting, posttranscriptional gene silencing, RNA interference (RNAi), and others, revealed many of the molecular pathways responsible. These pathways include DNA methylation, histone modifications, chromatin structure, and an ever expanding array of noncoding RNAs. Their study over the last half century has revealed how multiple cellular phenotypes arise from a single genotype.

Epigenetic Gene Expression and Regulation
http://dx.doi.org/10.1016/B978-0-12-799958-6.00008-1

As evident from the definition given above, inheritance is required for a phenomenon or mechanism to be considered within the realm of epigenetics. Two types of epigenetic inheritance can be distinguished: inheritance from cell to cell through mitosis and inheritance from parent to offspring through meiosis. Cell-to-cell epigenetic inheritance is typically concerned with maintaining a specific collection of epigenetic marks, i.e., an epigenetic state [1]. The epigenetic state determines a cell's identity; while all cells of an individual possess the same genetic information, they differ in their epigenetic marks [2]. Epigenetic marks are thus cell type-specific, and passing on the epigenetic state of the parent cell to daughter cells means that the daughter cells will maintain the same cellular identity as the parent cell. While epigenetic states mostly are maintained during mitosis, epigenetic information does change as cell differentiation occurs and can change in response to environmental factors [2,3]. However, maintaining an epigenetic state through mitotic cell divisions is essential to propagate specific cell lineages, and the faithful inheritance of epigenetic information from cell to cell will be one focus of our review.

The second focus will be the inheritance of epigenetic information from parent to offspring, which is quite different from cell-to-cell inheritance. The transition from one generation to the next typically requires erasing most parental epigenetic marks [4]. The act of erasing the parental epigenetic state allows for the formation of a zygote that is undifferentiated and totipotent and thus can give rise to the many different cell types needed for proper development [5]. However, this resetting is not perfect, and thus, not all epigenetic traces are removed. Some epigenetic marks or signals can be passed down from parent to offspring and can influence gene expression and phenotypes in the offspring [6]. Although opposite in their requirements, both cell-to-cell and parent-to-offspring epigenetic inheritance are types of inheritance that are vital for an organism's health and development. When considering epigenetic inheritance, it is essential to specify the inheritance type, as they differ considerably in the conditions necessary for inheritance to occur.

Many epigenetic systems have been introduced in detail previously in this book. In this chapter, we will focus on the major molecular mechanisms mediating the inheritance of epigenetic information (Table 1). Specifically, we will discuss the inheritance of DNA modifications, which are covalent alterations on the DNA bases themselves, and the inheritance of covalent modifications of histones. We will also discuss RNA-based epigenetic information systems and their inheritance, as well as present some examples of epigenetic inheritance where the underlying molecular mechanisms are currently unknown. In addition, we will discuss similarities and differences between two types of epigenetic inheritance systems—cell to cell and parent to offspring. For each epigenetic system, we will illustrate them through the use of examples. This broad overview will demonstrate common themes governing epigenetic inheritance and highlight differences that are specific to individual epigenetic information systems.

Table 1 Molecules mediating epigenetic inheritance. Molecules involved in epigenetic inheritance and our current understanding of their transmission

Molecules	DNA modification	Histone modification	miRNA	siRNA	piRNA	lncRNA	Prions
Cell to cell?	Yes	Yes	Yes	Yes	Yes	Yes	Yes
Parent to offspring (PO)?	Yes	Yes	Yes	Yes	Yes	?	?
	• Typically reset in germ cells • PO inheritance might be exceptional cases	• Mechanism for copying histone modification patterns is unclear • Fate of histone modifications transmitted to the zygote is unclear	• PO transmission not well understood	• PO transmission not well understood	• Function in somatic tissues not well understood • Fate of piRNAs transmitted to the zygote is unclear	• PO transmission is likely but not well studied	• Horizontal transmission between individuals • Horizontal transmission between species • Unclear if PO transmission occurs

2. DYNAMICS OF EPIGENETIC INFORMATION THROUGHOUT THE LIFE CYCLE

As noted above, inheritance of information, genetic or epigenetic, occurs by two different ways during the life cycle of most species, either through mitosis (cell to cell) or through meiosis and zygote formation (parent to-offspring) (Figure 1). While the fate of genetic information encoded in the DNA during mitosis and meiosis is well understood, what happens to epigenetic information is often less clear, and highly variable depending on the type of epigenetic information in question and sometimes also depending on the species under study. Despite this variability, some general patterns emerge for how epigenetic information is maintained and/or altered during the life cycle of multicellular organisms.

Typically, pluripotent cells in the early developing embryo have a characteristic set of epigenetic marks [7,8]. These marks change as the cells differentiate, eventually leading to the establishment of a cell type-specific epigenetic profile characteristic for a particular cell

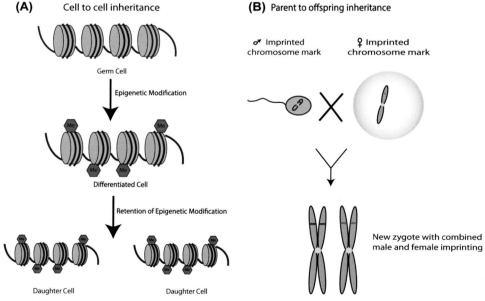

Figure 1 Types of epigenetic inheritance. (A) Inheritance from cell to cell. Early in development, pluripotent germ cells lack most epigenetic marks (top). As the cells differentiate their epigenetic profile, they acquire marks characteristic of their cell type (middle; DNA methylation indicated in red). When the differentiated cell divides mitotically, the daughter cells will retain the epigenetic profile of the parental cells and thus maintain the same differentiated cell identity (bottom). (B) Inheritance from parent to offspring. This type of inheritance is illustrated for the case of imprinting. The two germ cells, sperm and egg, carry different epigenetic marks at an imprinted locus (pink versus blue). The imprint from both parents is passed to the zygote, thus inherited from parent to offspring.

type [2]. This tissue specificity of epigenetic profiles is illustrated well by examining the variability of DNA methylation patterns within and between individuals. DNA methylation patterns are more similar among groups of samples from identical tissues from different individuals than among groups of different tissues from the same individual (for example, see Ref. [9]). Unless perturbed by external factors, the cell type-specific epigenetic profiles are maintained through mitosis in a given tissue, and daughter cells will exhibit the same epigenetic marks and phenotype as their parent cells.

While in somatic tissues, epigenetic profiles are mostly maintained and inherited unchanged, and epigenetic profiles have to be reset to allow for the return to the pluripotent state in the embryo when transitioning from one generation to the next [10]. In many species, this resetting—also called reprogramming—begins with the formation of the germ cells. Often, erasure of epigenetic marks is part of the developmental program leading up to the formation of mature germ cells, sperm and egg [10]. Because these processes are best understood for DNA methylation, we will focus our discussion of reprogramming on this epigenetic system.

In mammals, DNA methylation patterns are erased in the primordial germ cells (PGCs) (recently reviewed in Ref. [11]). Prior to the final migration of germ cells in the mouse, the PGCs exhibit DNA methylation patterns reflecting their somatic past. As the PGCs mature, their DNA methylation profile is reset to a more undifferentiated state, and much of the DNA methylation is removed, including parental imprints. In addition, the inactive X chromosome is reactivated in females. Following this erasure of epigenetic marks, a germ cell-specific epigenetic profile is established that differs between the sexes. These germline-specific epigenetic profiles include the parental imprints that are present at imprinted regions of the genome [11]. Thus, germ cell development includes the erasure of somatic parental epigenetic information and the establishment of germ cell-specific epigenetic profiles.

In the zygote, a second phase of epigenetic reprogramming occurs. The germ cell-specific epigenetic patterns are replaced by embryonic epigenetic information that returns the cell to a pluripotent state [11]. In mouse, this process involves a second erasure of epigenetic information, in addition to the erasure that occurred during germ cell development [11]. For example, DNA methylation is removed from the paternal and maternal genomes with the exception of imprinted regions, which maintain the DNA methylation profile characteristic of their parental origin [11]. This second phase of reprogramming is necessary to achieve pluripotency and allows for proper embryonic development. After this process, embryonic cells have the epigenetic profile characteristic of pluripotent cells, which will gradually change into cell type-specific profiles through differentiation as described in the previous section.

Reprogramming of epigenetic profiles prior to the parent–offspring transition is a conserved feature of many eukaryotes. However, there are significant differences in the specific sequence of events employed to achieve this resetting in the various species.

While mammals use a two-phasic resetting—first in the germline and then in the early embryo—methylome data from developing zebrafish indicate that sperm DNA methylation patterns are inherited in the early embryo and that oocyte DNA methylation patterns are altered to resemble those found in the sperm nucleus [12,13]. Thus, genomic regions that are unmethylated in the oocyte and methylated in the sperm will be methylated during this reprogramming step in the maternally derived genome. In contrast, genomic regions that are methylated in the oocyte and unmethylated in the sperm will be demethylated in the maternally derived genome [12,13]. Reprogramming in zebrafish thus involves both methylation and demethylation and is decidedly different from what is observed in mammals, where both the sperm- and oocyte-derived genomes are affected by demethylation in the developing embryo.

Other eukaryotes, such as plants, also undergo epigenetic reprogramming when transitioning between generations. This reprogramming is similar to what is observed in mammals, but as described for zebrafish, differences occur due to the specifics of sexual reproduction in plants. DNA demethylation in *Arabadopsis*, for example, is observed during both male and female gametogenesis [14]. However, this demethylation differs between male and female gametogenesis both in timing and in the sequence contexts affected (CG and CHG, versus CHH; where CG = CpG; CHG = CpHpG with H = A, C, or T; CHH methylation is primarily associated with transposable elements (TEs)) [15]. In the early embryo, DNA methylation levels increase quickly to the typical somatic level, and a second erasure of DNA methylation as seen in mammals does not appear to occur [16]. These examples illustrate for DNA methylation how epigenetic information is reset at the parent-to-offspring transition in many eukaryotes. While the above examples focus on DNA methylation, where the reprogramming pathways are best understood, it is likely that other epigenetic information is reset as well. If inheritance of epigenetic information is to be achieved, it thus has to be able to resist this reprogramming.

3. EPIGENETIC INHERITANCE IN CILIATES

Ciliates are a group of protozoans characterized by the presence of multiple cilia [17]. These organisms, specifically the protozoan *Tetrahymena thermophila*, have been influential in the epigenetics field, leading to discoveries such as the identification of histone variants, the characterization of the first histone acetyltransferase, and the detection of histone lysine methylation and histone phosphorylation [18]. These discoveries were possible due to ciliates such as *Tetrahymena* having two distinct types of nuclei. *T. thermophila*, for example, has a macronucleus that carries out somatic functions and a micronucleus that is mostly inactive and represents the germline [19]. Because of its limited activity outside of sexual reproduction, the epigenotype of the micronucleus is not altered by life events affecting the somatic macronucleus and its epigenome [20]. This division of labor between the macro- and micronucleus allows ciliates to bypass reprogramming before passing on epigenetic and genetic information to an offspring. Because the micronucleus' epigenome is not altered by life events, the genome and epigenome can be replicated without the need to erase epigenetic marks. The macronucleus, on the other hand, undergoes extensive alterations throughout the ciliate's life and passes down epigenetic information via the mechanisms discussed below [20]. Nuclear dimorphism makes ciliates a powerful system to study epigenetic differences between somatic and germline and provides an elegant solution to the problem of how to pass on a consistent and unaltered epigenotype along with the genome in parent-to-offspring inheritance.

Macronucleus development illustrates an interesting special case of epigenetic inheritance, as it is regenerated from a copy of the micronucleus after each sexual reproduction. During macronuclear development, the genome is highly modified and rearranged, a process that is controlled by scanRNAs that move between the old macronucleus and the new developing macronucleus, delivering information about sequence content and epigenetic modifications [20]. This information transfer depends on sequence homology between the sequences in the old and new macronuclei. To be included in the new developing macronucleus, a sequence must be present in the parental macronucleus; sequences that are only present in the micronucleus are eliminated [21]. The conservation of sequences from old to new macronuclei was discovered because microinjection of specific DNA sequences into the parental macronucleus prevents injected sequences from being eliminated from the progeny's somatic genome (macronucleus) [22]. Thus, epigenetic elements arising from the parental macronucleus determine the fate of the offspring's macronucleus. This direct relationship between parental and offspring genomes, coupled with the dichotomy of the nuclei, further emphasizes the convenience of using ciliates for epigenetic inheritance research. Ciliates represent a special case of epigenetic inheritance that does not perfectly fit into either category of epigenetic inheritance, cell to cell or parent to offspring. However, through their use of scanRNAs during macronuclear development, they provide support for RNA-based epigenetic inheritance, which is discussed in more detail in Section 6.

4. EPIGENETIC INHERITANCE VIA DNA MODIFICATIONS

Among epigenetic information systems, DNA modifications are unique in that the mode of their inheritance is relatively straightforward. DNA modifications are covalent modifications of the DNA bases, with the most commonly studied eukaryotic DNA modification being methylation at the five position of cytosine, leading to the formation of 5-methyl-cytosine (5mC; for a recent review of the functions of 5mC in mammals see Ref. [23]). This epigenetic mark is often referred to simply as DNA methylation. In addition to 5mC, several other modifications occur at cytosine residues as well. These include 5-hydroxymethylcytosine, 5-formylcytosine, and 5-carboxylcytosine, all of which are oxidative derivatives of 5mC [24]. While 5mC has been the subject of study since the 1970s, the importance of the various 5mC derivatives has been recognized only recently, and they are thus not as well understood.

Given that the DNA modifications are covalently attached to the DNA, they will be inherited along with the DNA during cell divisions, either mitosis or meiosis. While the inheritance of individual modified bases in the DNA is relatively straightforward, the inheritance of specific modification patterns over the course of several cell division cycles is a more complicated problem. During DNA replication, the template strand will retain the modified base; however, the newly synthesized strand will not contain any modified bases, creating hemimethylated DNA. If no new corresponding DNA modifications are added to the newly synthesized strand, during the next division cycle, a double-stranded DNA molecule will be created that contains no DNA modifications, in addition to a molecule with one modified strand and one unmodified strand. Thus, to reliably inherit DNA modification patterns, mechanisms to modify the newly synthesized DNA strand at the appropriate bases have to be in place.

How DNA modification patterns are maintained through cell division is best understood for 5mC (for a review see Ref. [25]). In species such as plants and mammals, much of the 5mC occurs in a CpG (or in case of plants also CpNpG) sequence context. These sites are unusual because by default, they have a cytosine that can be methylated close-by on the opposing strand (symmetrical sites). Specialized DNA methyltransferase enzymes can recognize hemimethylated DNA templates, such as a double-stranded DNA molecule immediately after synthesis of the new strand. These "maintenance methyltransferases" then copy the DNA methylation onto the new strand, ensuring that 5mC patterns can be inherited reliably over several cell division cycles [25]. Thus, inheritance of DNA methylation patterns through mitosis and meiosis is possible for these symmetrical sites, as long as the maintenance methyltransferase enzyme is present.

Inheritance of DNA methylation in other sequence contexts is more difficult to achieve, as there are not necessarily suitable cytosines close-by on the newly synthesized strand. 5mC in these nonsymmetrical sequence contexts has to be reestablished after every cell division by the action of de novo methyltransferases that can produce 5mC in any sequence context [26]. In plants like *Arabidopsis thaliana*, the de novo DNA (cytosine-5) methyltransferase enzyme (DRM2) can be targeted to specific sites in the genome by small interfering RNAs in a process called RNA-directed DNA methylation (RdDM) (reviewed by refs [27,28]). The small RNAs direct the DRM2 enzyme to specific regions of the genome, providing specificity through sequence complementarity [27]. This process allows for the replication of 5mC patterns in any sequence context, but it does require the small RNA triggers to be passed on to daughter cells (cell-to-cell inheritance) or to the germ cells (parent-to-offspring inheritance) to facilitate the inheritance of the DNA methylation patterns.

How nonsymmetrical DNA methylation patterns are inherited in mammals is less clear, as they lack the plant specific RdDM pathway [26]. As nonsymmetrical DNA methylation is mostly limited to embryonic cells in mammals [29], it has not been studied as well as in plants where it is pervasive. Currently available data point to two pathways that might contribute to the inheritance of nonsymmetrical DNA methylation in mammals. One pathway is via interaction with histone modifications, as the interaction of DNA methyltransferases with target sites can be inhibited by the presence of H3K4 methylation [30]. This pathway would require the faithful inheritance of histone modifications (discussed in Section 5). The second pathway that has been suggested involves the Piwi-interacting RNA (piRNA) machinery (where the term Piwi is derived from P-element induced wimpy testis in *Drosophila melanogaster*), and similarly to what is well-established in plants, postulates a role for piRNAs in the targeting of de novo DNA methyltransferase enzymes [31,32]. If inheritance of nonsymmetrical DNA methylation patterns relies on this pathway, it would require simply the inheritance of the piRNA trigger to propagate any DNA methylation pattern. Evidence for the inheritance of noncoding RNAs such as short interfering RNAs (siRNAs) and piRNAs is discussed in Section 6 in detail, and suggests that they are a likely mechanism for mediating the inheritance of nonsymmetrical DNA methylation patterns.

5. EPIGENETIC INHERITANCE VIA HISTONE MODIFICATIONS

Histone modifications are another type of epigenetic mark that can potentially be transmitted from cell to cell and from parent to offspring [6,33]. Various posttranslational modifications can occur at histones, and the most common modifications include phosphorylation, acetylation, methylation, and ubiquitination, but others occur as well [34]. Many of the sites of modification are in the N-terminal histone tails that protrude from the nucleosome core particle and govern the interactions between the nucleosome and other chromosomal proteins [35]. Modifications within the histone core, on the other hand, often affect the interactions between the nucleosome and the DNA [36]. Together, both histone core modification and histone tail modifications greatly impact chromatin structure and modulate access to the DNA. Compared to DNA modifications, our understanding of how histone modifications are transmitted from cell to cell or parent to offspring is limited.

The inheritance of histone modification patterns across the genome is complicated by the fact that histones are evicted from the DNA during DNA replication [37]. Chromatin is disrupted in front of the replication fork, and nucleosomes are disassembled. Behind the replication fork, nucleosomes have to be reformed on the two DNA molecules. Available data suggest that the nucleosome core particle disassociates into an $(H3-H4)_2$ tetramer and two H2A-H2B dimers upon eviction from the DNA [36]. Experiments focusing on H3 and H4 indicate that the parental tetramers stay in close proximity and are randomly distributed to the daughter strands in a location that is close to the location they were evicted from [38,39]. The parental histones are supplemented by newly synthesized histones that carry a characteristic set of histone modifications [40]. Thus, after replication, the newly synthesized DNA strands carry a mixture of histone modifications—some of which reflect the parental histone modification (recycled histones) and some that lack those modifications (newly synthesized histones).

To achieve long-term inheritance of histone modification patterns across the genome, newly synthesized histones have to be modified in such a way to match the parental histones. As specific enzymes produce histone modifications, the problem becomes one of recruiting the appropriate enzyme to a specific sequence, a problem that is very similar to the problem of the inheritance of DNA methylation at nonsymmetrical sites discussed in the previous section. For some histone modifications, we know how the recruitment of the histone-modifying enzymes is achieved. For example, methylation of histone 3 at lysine 9 (H3K9) in humans is produced by suppressor of variegation 3-9 homolog 1 (SUV39H1) H3K9 methyltransferase that binds to H3K9me2/me3 via its chromodomain [41]. Thus, SUV39H1 can be recruited to specific genomic regions if H3K9me2/me3 is present on the recycled parental histones and propagate the H3K9 methylation mark to adjacent nucleosomes. A similar mechanism exists for H3K27 methylation, where the polycomb repressive complex 2 (PRC2) includes subunits that both recognize the H3K27 methylation mark and produce it [42]. These examples illustrate how histone modification patterns can be inherited from cell to cell.

Similar mechanisms are potentially at work in the transition between generations through the female germline. However, in a significant number of species, sperm chromatin differs from chromatin in somatic cells. In sperm, the majority of histones are replaced by protamines [43]. In human sperm, only approximately 10% of the DNA remains associated with histones, and in mouse, it is only 1% [43]. Thus, through the sperm, only a small percentage of histone modifications can potentially be inherited from parent to offspring, and new histones will be incorporated into the paternal genome in the zygote. How histone modifications are targeted to these newly incorporated histones in the paternal genome is not well understood. Because histone-modifying enzymes can interact with each other (for example, see Ref. [44]), with DNA-modifying enzymes (for example, see Ref. [45]), and also with proteins in the various small RNA pathways (for example, see Ref. [46]), it is possible that these processes contribute to the inheritance of histone modifications between generations.

A 2014 study from *Caenorhabditis elegans* illustrates some of the processes involved in the inheritance of histone modifications. Gaydos and colleagues investigated the inheritance of H3K27 methylation during development and across generations by studying the offspring of crosses between wild type *C. elegans* and *C. elegans* lacking H3K27me3 [47]. They were able to detect H3K27me3 and H3K9me2 in mature sperm as well as its transmission to the zygote. In the absence of the PRC2 H3K27 methyltransferase complex, the paternally derived H3K27 methylation was maintained on a subset of chromosomes—it did not spread to the maternally derived chromosomes—up to the 24-cell stage through several rounds of cell division. These findings demonstrate that modified histones can stay associated with specific DNA sequences through multiple rounds of cell division. Additional experiments showed that H3K27me3 did not spread to the paternal chromosome set during embryo development if the sperm chromatin was devoid of H3K27me3, even in the presence of the PRC2 complex. This result is specific to H3K27me3, as in analogous experiments with H3K9me3, the modification quickly spreads to all chromosomes [47]. Together, these experiments demonstrate the epigenetic inheritance of a histone modification, H3K27me3, in *C. elegans* from cell to cell and across generations, and suggest that inheritance methods might to some degree depend on the specific histone modification.

6. EPIGENETIC INHERITANCE THROUGH NONCODING RNA

Starting in the 1990s, noncoding RNAs became known as major players in gene regulation, and it was soon discovered that they were able to contribute to epigenetic inheritance as well [48]. These noncoding RNAs include a number of small RNA (sRNA) classes, which affect multiple epigenetic pathways. The main classes of sRNAs are microRNAs (miRNAs), siRNAs, and piRNAs. In addition to these sRNAs, long noncoding RNAs (lncRNAs) also appear to play a role in epigenetic pathways. While much remains to be discovered about noncoding RNAs, it is apparent that they have a major impact on cellular identity, cellular memory, and epigenetic inheritance.

6.1 sRNA

sRNAs were discovered in the 1990s due to their role in gene silencing [49]. sRNAs typically are 20–25 nucleotides in length and represent a subclass of noncoding RNAs [50]. Most sRNAs are produced endogenously—although some are derived from viruses—and many function in gene regulation [51]. Apart from their function in gene-specific transcriptional silencing, sRNAs contribute to the maintenance of heterochromatin and participate in antiviral defense [48,52]. Because of their role in gene regulation, sRNAs have become the focus of drug discovery research in the hope that they can be used to silence harmful gene expression [53]. Despite their fairly recent discovery, sRNAs have greatly impacted our understanding of gene regulation, specifically gene silencing, and have improved our perception of epigenetic inheritance. The following sections will detail different types of sRNAs and how they might function in epigenetic inheritance.

6.1.1 MicroRNA

miRNAs function in the regulation of gene expression, and they are derived from endogenous hairpin transcripts to yield ~25 nucleotide RNA species [54]. miRNAs tend to originate from intragenic sequences, often oriented antisense to neighboring genes, suggesting that they are transcribed as individual, defined entities, with their own regulatory sequences [55]. Some miRNAs are transcribed coordinately with their host gene, allowing for coupled regulation of the protein-coding gene and the miRNA [56]. miRNAs are estimated to regulate ~60% of all protein-coding genes in the human genome [57]. This regulatory function is accomplished by the miRNA's association with the RNA-induced silencing complex (RISC), forming miRISC that is guided to the 3′ UTR of a sequence-matched mRNA. Binding of an miRNA to an mRNA inhibits protein synthesis, through both destabilization and degradation of the mRNA message or through translational inhibition [58,59]. Thus, the presence of miRNAs impacts a cell's expression profile and shapes its protein pool.

Through their impact on gene expression levels, miRNAs play a vital role in the establishment of cellular identity. miRNAs are important in determining what genes a cell expresses or silences, and inheritance of miRNAs could provide a form of epigenetic memory that is passed on from cell to cell or from parent to offspring [60]. Depending on their stability, inherited miRNAs could contribute to the establishment of expression profiles similar to those of the parent cell in the daughter cells. miRNA-mediated epigenetic inheritance has been described in *D. melanogaster* [61], zebrafish [61], mice [62], and other organisms.

An example from human cancer cells illustrates the importance of miRNAs for the maintenance of expression states [63]. Octamer-binding transcription factor 4 (OCT4) is a transcription factor essential for pluripotent embryonic stem cells. It is part of a negative feedback loop with miRNA-145; OCT4 suppresses expression of miRNA-145, and OCT4 expression is in turn suppressed by miRNA-145. High levels of OCT4 cause low expression levels of miRNA-145 in stem cells, causing them to remain undifferentiated. This pattern is reversed when cells undergo differentiation; cells switch from a high OCT4 and low miRNA-145 state in stem cells to a low OCT 4 and high miRNA-145 state as the cells differentiate [64]. This type of feedback loop can be very powerful to strengthen and reinforce the cell's current expression state, as shown in general terms in Figure 2. Once established, a state is maintained through the feedback loop. As long as the mRNAs or miRNAs involved in the feedback loop are equally divided to daughter cells, the parental expression state will be inherited from cell to cell. Such a feedback system can be considered an example of epigenetic inheritance because the new cell remembers the expression state of the parent cell (see Ref. [65] for an alternative interpretation). Thus, miRNAs can contribute to cellular memory and help ensure that expression states are maintained within specific cell lineages.

In addition to being inherited from cell to cell through mitosis, miRNAs can also be passed on from parent to offspring. miRNAs, for example, have been found in human spermatozoa, suggesting that they may play a role in early development by regulating expression of various genes [66]. The importance of inherited miRNAs for embryonic development is illustrated by the fact that the sperm-borne miRNA-34c is essential for

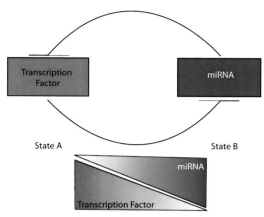

Figure 2 Model illustrating a potential regulatory loop between a transcription factor and an miRNA. High levels of the transcription factor are essential for maintaining state A, whereas high levels of the miRNA are required for state B. The mutual repression of the transcription factor and the miRNA ensures that one specific state is passed on to daughter cells. *Adapted from Ref. [60].*

the first cleavage division in the mouse embryo [67]. This example illustrates that miRNAs are an important part of sperm- and oocyte-cytoplasmic transfer, allowing information from the parent to affect gene expression in the offspring, a case of epigenetic inheritance.

6.1.2 Small interfering RNA

siRNAs are another sRNA species involved in the regulation of gene expression, often specifically targeting TEs via posttranscriptional gene silencing [68]. siRNAs are derived from long double-stranded RNA molecules through the action of the endoribonuclease Dicer, which cuts the long double-stranded RNAs into small siRNAs [69]. siRNAs recognize matching mRNAs by sequence complementarity and initiate degradation of the targeted mRNA. To achieve this degradation, the siRNA becomes part of an effector complex, similar to the miRISC called siRISC [70]. This siRISC contains an Argonaute protein, which has RNase activity, to cleave the targeted mRNA [71]. Due to their ability to degrade target mRNAs, both the miRISC and the siRISC effector complexes are a potent defense against RNA viruses. siRNAs also protect the genome against TEs in the germline, where other gene silencing mechanisms tend to be inactive due to the epigenetic reprogramming discussed in Section 2 [71]. The presence of siRNAs in germline cells is well-documented, for example in *C. elegans* [72], *D. melanogaster* [73], and mice [74], illustrating the importance of this function of siRNAs.

A case from the plant kingdom exemplifies how siRNAs can mediate transgenerational epigenetic inheritance. In *A. thaliana*, transposon activity is usually repressed by DNA methylation via the function of DNA methyltransferase 1 (MET1) [75]. Loss-of-function mutants of *met1* show increased TE activity, but not complete loss of TE silencing [75]. The remaining silencing activity is due to siRNAs, which are able to regulate TEs independent of DNA methylation through a posttranscriptional mechanism [76]. Repression by siRNAs prevents the amplification and reinsertion of TEs that would otherwise be passed down to daughter cells [76]. This function is essential in germ cells, where MET1 is absent and any transposition would be highly deleterious, much more so than in somatic cells. In plants, these TE-repressing siRNAs are inherited from parent to offspring through the gametes, and through the surrounding gametophytic and sporophytic cells [77]. Thus, upon fertilization, the zygote receives not only genetic information from both parents, but also a supply of siRNAs corresponding to the TEs present in both parental genomes [48]. This example illustrates how siRNAs can be inherited from parent to offspring after fulfilling an essential function in the germline, a case of epigenetic inheritance.

In addition to the germline TE-suppression function of siRNAs in plants, they have another function in transgenerational inheritance of DNA methylation via the RdDM pathway introduced in Section 3. This additional function is shown elegantly in an experiment using genetically wild-type *Arabidopsis* strains with altered epigenomes. These lines, called epigenetic recombinant inbred lines, had DNA methylation loss restricted to specific regions of the genome [78]. DNA methylation could not be restored to regions normally methylated if it had been lost for one or more generations [79]. However, the loss of DNA methylation could be progressively restored across several generations if siRNAs matching the locus were still produced by the organism [80]. These findings demonstrate that siRNAs are important for maintaining DNA methylation patterns across generations, facilitating the inheritance of epigenetic information from parent to offspring. Thus, siRNAs provide information regarding the parental DNA methylation pattern and direct the DNA methylation machinery to replicate this epigenetic state in the offspring.

6.1.3 Piwi-interacting RNA

The third main category of small noncoding RNAs is piRNA. This class of RNA was named for its associations with Piwi family proteins [60]. Members of the Piwi protein family include Piwi, Aubergine (AUB), and Argonaute 3 (AGO3) in *Drosophila* and PIWIL1, PIWIL2, and PIWIL4 in mice. Several different research groups working on mice, rats, and *Drosophila* spermatogenic cells discovered that piRNAs are connected to posttranscriptional gene silencing of TEs and the control of gene expression in germ cells [52]. Source regions for piRNAs are found in piRNA clusters across the genome [81]. These clusters contain varying numbers of piRNAs, ranging from dozens to thousands, and they can be anywhere from 1 to 100 kb in length [82]. piRNAs have been found in both the male and female germ cells of many different organisms [83]. piRNAs are present in the brain as well, and Piwi knockdown studies demonstrated that the piRNA pathway leads to increased DNA methylation of Creb2 (cyclic AMP response element binding protein B), reducing its expression in the brain and promoting memory [84]. While piRNAs may have important epigenetic effects in organs such as the brain, their established function is in the epigenetic regulation of gene expression in germ cells.

piRNAs are produced differently from siRNA and miRNA (Figure 3). Unlike the other two sRNA generating pathways, no Dicer nuclease is involved in the production of piRNAs [60]. Two possible pathways contribute to the production of a mature piRNAs [85]. The primary pathway is thought to initiate piRNA biogenesis by using maternally deposited piRNA as a template for more piRNA creation [86]. Current models from *Drosophila* suggest that the nuclease Zucchini (ZUC) cleaves the primary transcript derived from the piRNA clusters; most likely it also produces the 5′ end of the primary piRNAs [52]. The Argonaute proteins Piwi or AUB bind the cleaved RNA with assistance of Shutdown (SHU) and heat shock protein 83 (Hsp83) [87]. Next, a currently unidentified enzyme trims from the 3′ end of the RNA until it reaches the footprint protected by the Piwi protein [88]. Then HEN1 (small RNA 2′-O-methyltransferase, *Arabidopsis*) methylates the 3′ end of the piRNA [89]. The Piwi protein is then guided by the mature piRNA to complementary nascent transcripts and, by recruitment of histone methyltransferases, can lead to heterochromatin formation at the target locus.

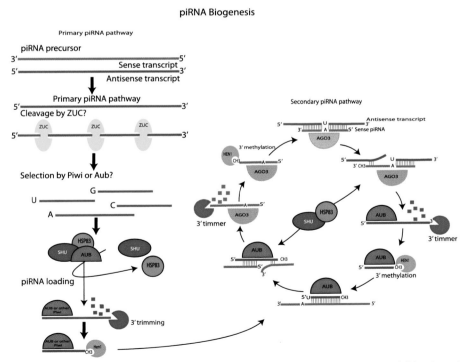

Figure 3 Pathways leading to piRNA biogenesis. The diagram shows the current model for the primary and secondary piRNA biogenesis pathway. For details, see text. *Adapted from Ref. [98].*

Therefore, Piwi proteins are able to recruit epigenetic modifiers to piRNA clusters, and these modifiers can upregulate production of complementary piRNA transcripts [90]. By this process, maternally deposited piRNA can lead to the generation of more piRNA in the developing embryo. This relationship between maternally deposited piRNA and piRNA generated in the embryo is akin to a positive feedback loop. Initially, piRNA is needed to kick-start this biogenesis, and could explain why maternal deposits of piRNA are required for proper cluster expression in offspring [60].

The second piRNA biogenesis mechanisms that might explain how maternal piRNA deposits initiate the creation of cognate piRNA is the "Ping-Pong" pathway [81]. While the existence of this second piRNA biogenesis pathway is well supported by experimental evidence, it is unclear if individual piRNAs go through one or both of the two pathways. In the ping-pong pathway, AUB is loaded with a piRNA, allowing it to recognize complementary transcripts and cleave them to produce the 5′ end of a new piRNA [91]. The 3′ end of the freshly cleaved piRNA precursor is then bound by AGO3 with the help of SHU, and it is cleaved again to generate a mature piRNA identical to the piRNA initially bound to AUB [87]. While the primary piRNA biogenesis pathway is present in somatic and germline tissues, the ping-pong pathway only exists in germline tissues [60]. Both the primary and the secondary piRNA biogenesis pathway rely on maternally deposited piRNA triggers that were passed from parent to offspring.

piRNAs have been shown to affect epigenetic inheritance in germ cells in silkworm, *Drosophila*, mice, and zebrafish [92]. In addition to their germline function, piRNAs can mediate parent-to-offspring inheritance of epigenetic information, which can be seen in *C. elegans*. Here, the Piwi pathway affects the chromatin structure at target loci through the creation of repressive histone marks [93]. Interestingly, the *C. elegans* Piwi protein PRG-1 (Piwi (fruitfly) related gene) does not actually enter the nucleus to change chromatin structure [94]. Instead, another class of small RNAs, called 22G RNA, is produced using the PRG-1 targeted RNA as a template for transcription [95]. This transcription takes place in the cytoplasm of the cell and is carried out by RNA-dependent RNA polymerase [95]. Next, the 22G RNA is bound by the *C. elegans* Argonaute protein WAGO-9, and this complex shuttles the RNA into the nucleus [96]. Once in the nucleus, WAGO-9 bound 22G RNAs initiate targeted transcriptional gene silencing, eventually altering chromatin structure with the production of H3K9me3 [93]. The heterochromatin protein 1 (HP1) like proteins HPL-1 and HPL-2 are involved in this silencing effects as well [93]. Thus, piRNAs in *C. elegans* can precipitate chromatin structure changes and target the formation of repressive chromatin.

After the initial chromatin structure change, this process can become independent of PRG-1 through an unknown mechanism. By this mechanism, the transcriptional silencing caused by the piRNA pathway is converted to a form of silencing that is stably inherited across multiple generations [97]. This phenomenon is called RNA-induced epigenetic silencing (RNAe) [98]. A chromatin and expression change initiated through this mechanism will remain for several parent-to-offspring generations without continuous upkeep. This shift does not happen by default because genes, targeted by PRG-1, can remain dependent on PRG-1 for silencing in wild type animals [97]. How the shift from PRG-1 dependent to PRG-1 independent silencing happens is currently unknown, but is truly a novel example of epigenetic inheritance and demonstrates the impact of piRNAs on transgenerational epigenetic inheritance.

Transmission of the silenced chromatin state initiated by the piRNA pathway through meiosis is dependent on WAGO-9 and other RNAi factors [93]. In WAGO-9 mutants, germ cells lose characteristic immortality, resulting in sterility after several generations. These mutants also lose the H3K9me3 marks [99]. This finding implies that PRG-1 and the WAGO-9 proteins are required for maintaining proper gene expression in germ cells over multiple generations. Most likely, in the absence of PRG-1 and WAGO-9, the sporadic ectopic activation of germline-incompatible genes normally repressed by PRG-1 and WAGO-9 leads to sterility. If the piRNA pathway remains intact, the epigenetic changes initiated by piRNAs remain and are inherited across generations.

Another example of the inheritance of piRNAs comes from *Drosophila*. Here, mutations in the piRNA pathway lead to the derepression of TEs in the germ cells [32]. Increased expression of TEs is associated with sterility, possibly caused by DNA double strand breaks created by transposition events [100]. piRNAs thus act as an immune system, protecting the genome against parasitic sequences; they serve as a memory system to remember all previous interaction with various types of TEs [60]. Because the piRNAs have to recognize and protect the genome against multiple types of TEs, piRNA sequences are extremely diverse. The pool of piRNAs present within a particular strain of an organism correlates well with the TEs present in the genome of that organism [101]. It appears that any sequence that resides within a piRNA cluster region can be transcribed into a piRNA [102]. When an exogenous reporter gene is inserted into a piRNA cluster experimentally, it is processed into piRNAs that match the reporter gene and can repress additional reporter genes matching its sequence [102]. This experiment reveals how piRNAs can provide epigenetic memory and defend the genome against any particular TE type. Any novel TE will eventually insert into a piRNA cluster by random transposition, which allows the piRNA pathways to create new piRNAs matching the novel TE, resulting in its repression at all locations in the genome [103]. This epigenetic mechanism of TE suppression is extremely important for germ cells because of the absence of repressive chromatin marks due to the reprogramming events. Lack of piRNAs in the germline leads to large-scale TE activation and large numbers of transpositions, with the potential to disrupt vital gene functions. After fulfilling their function in the germline, piRNAs are inherited to the next generation, creating the maternal deposits necessary to set up piRNA pools through the process described above. These cases illustrate the link between the germline functions of piRNAs and their functions in the embryo, and support a role for piRNAs in epigenetic inheritance.

6.2 Long noncoding RNA

Another type of noncoding RNA is lncRNA, non-protein-coding transcripts greater than 200 nucleotides in length [104]. lncRNA transcripts are produced from intergenic regions of the genome, and next generation sequencing experiments examining overall RNA pools have shown that lncRNAs number in the tens of thousands in mammals [105]. The majority of these transcripts appear to have effects within a cell, but only a small amount of them have been fully characterized [106]. While much of the function of lncRNAs has yet to be discovered, some of the best characterized lncRNAs are known to be important for imprinting and might mediate epigenetic inheritance.

The first mammalian lncRNA found to have gene silencing effects was the *Xist* lncRNA, a noncoding X-inactivation specific transcript derived from the inactive X chromosome. *Xist* is essential for the inactivation of one X chromosome in female mammals, which results in Barr body formation [107]. The *Xist* lncRNA coats the inactive X chromosome, attracting repressive DNA and histone modifications and resulting in permanent X chromosome inactivation [108]. Once inactivated by the action of *Xist*, the inactive X chromosome is maintained in this inactive state through mitosis [109]. This X-inactivation is reversed only during reprogramming of the epigenetic profile during germ cell development (see Section 2). Thus, X-inactivation is a classic example of epigenetic inheritance that involves an lncRNA.

Imprinting, the selective silencing of a specific locus from either the maternally or paternally derived chromosomal copy, has some similarities to X-inactivation. Because both imprinting and X-inactivation involve silencing of just one of two copies of a specific part of the genome, it has been suggested that imprinting may use lncRNAs in a mechanism similar to X-inactivation [110]. In mammals, imprinting occurs at a small number of loci throughout the genome (<1% of loci), but it is required for proper development [111]. Imprinted genes tend to be clustered, implying that domains rather than individual genes are imprinted [112]. However, while X-inactivation acts on a whole chromosome, imprinting usually affects cluster of only three to 15 genes, and any lncRNA-mediated silencing mechanism would have to function on a much smaller scale than *Xist*.

lncRNAs associated with imprinting clusters are important for the regulation of gene expression from the imprinting cluster [113,114]. For example, lncRNAs can repress adjacent genes by transcriptional interference, as is the case for *Airn* (macro ncRNA, mouse), which represses *Igf2r* (insulin-like growth factor 2 receptor, mouse) [115]. In addition, lncRNAs can carry out regulatory functions in *trans*, as they can serve as guides for chromatin-modifying enzymes [113,114]. lncRNAs with roles in genomic imprinting inherited from parent to offspring is currently unclear. However, like other RNAs in the cell, lncRNAs are distributed to daughter cells during cell division, where they can contribute to the maintenance of specific epigenetic states at imprinted gene clusters.

A more general example of cell-to-cell inheritance of an lncRNA is provided by studies of the lncRNA AF118081 in human lung cancer cells [116]. lncRNA AF118081 is overexpressed in malignant transformed 16HBE cells compared to normal cells. siRNA knockdown of AF118081 demonstrated a role for the lncRNA AF118081 in cancer; its knockdown inhibited cell growth and tumor invasion, implicating the lncRNA in cancer growth. High levels of lncRNA AF118081 are maintained in the cultured cells, suggesting that the lncRNA and the expression state of the lncRNA locus are passed down to future cell generations. These examples illustrate that similar to sRNAs, lncRNAs have the potential to contribute to epigenetic inheritance, as they can be readily passed down from cell to cell and from parent to offspring as part of the parentally supplied RNA pool.

7. PRIONS

Prions provide a unique example of epigenetic inheritance that is often overlooked. Prions are self-propagating, misfolded proteins that can be inherited via a conformation templating mechanism [117]. Prion proteins exist in two different structural conformations. In the normal, functional conformation, the protein carries out its cellular functions and does not differ in its behavior from any other protein. In the aberrant conformation, the protein can convert any prion protein in the normal conformation to the aberrant conformation [117]. For example, the mammalian prion protein PrP occurs in the nonpathogenic form PrP^C, and as an infectious pathogenic form PrP^{Sc} [117]. If a PrP protein in the PrP^C conformation encounters a PrP protein in the PrP^{Sc} conformation, the PrP^C protein will be changed into PrP^{Sc} [117]. Thus, the protein conformation of PrP^{Sc} is self-propagating, and the inheritance mediated by the infectious PrP^{Sc} protein can be considered a form of epigenetic inheritance.

In mammals, studies of prion inheritance are rare; prions cause a variety of neurodegenerative disorders, including Creutzfeldt-Jakob disease (CJD), kuru, and variant CJD in humans, scrapie in sheep, and bovine spongiform encephalopathy in cows [118], and studies have been most concerned with horizontal transmission of these diseases between individuals. However, prions also occur in fungi, and in the yeast *Saccharomyces cerevisiae* several prion proteins exist that can be either beneficial or detrimental [119]. The inheritance of prions, from cell to cell and also through mating, has been studied in detail in *S. cerevisiae* [120] because of the tools available in yeast and because all prions are not detrimental in yeast [119,120]. Cell-to-cell transmission of prions has also been seen in human tissue culture cells [119]. These data illustrate that prions can mediate epigenetic inheritance in several species.

8. UNKNOWN MECHANISM STUDIES

In addition to the cases of epigenetic inheritance illustrated above for the main epigenetic information systems, there are a number of cases of documented epigenetic inheritance where the underlying molecular mechanisms are unknown.

8.1 Dutch Hunger Winter

A famous human example of epigenetic inheritance of unknown mechanisms comes from studies of the Dutch Hunger Winter of 1944–1945 [121]. During this time, famine affected the western Netherlands, caused by an embargo on transport of food supplies instituted by the occupying German forces. The famine lasted for approximately 5 months and ended abruptly when allied forces liberated the area [121]. This unfortunate event has given researchers an unprecedented opportunity to look at the effects of poor maternal nutrition on the health and well-being of progeny in utero and at the epigenetic effects on offspring after two generations.

As other types of studies have demonstrated, in the Dutch Hunger Winter studies, strong correlations exist between in utero malnutrition and a host of health problems, some that begin at an early age and others that appear later in life. In addition to the effects seen in the individuals that experienced the famine in utero (F1), effects of famine also appeared to manifest themselves in the F2 generation through unknown mechanisms of epigenetic inheritance. In a cohort study done on women exposed to famine during the Dutch Hunger Winter in utero (F1) and their offspring (F2), it was found that not all effects of famine, like lower birth weight, were passed on, but some effects, including birth length and poor health later in life, were passed down and observed in the F2 as well as the F1 [122]. How these epigenetic modifications are passed down multiple generations is still not understood. Likely, several epigenetic mechanisms contribute to the impact on the phenotype for offspring exposed to in utero malnutrition. In addition, multiple studies suggest that there are critical windows in embryonic development where the epigenome is susceptible to the introduction of epimutations through exposure to environmental factors such as toxicants or diet [123]. These findings indicate that additional studies of historical datasets, such as the Dutch Hunger Winter, could yield valuable insights into how epigenetic changes are precipitated by large-scale environmental impacts and passed down in humans. These studies are of great interest, as they suggest that the environments previous generations were exposed to can still impact individuals today. Currently, it is unknown how often epigenetic information about the parental environment is transmitted to offspring over multiple generations and how large the phenotypic effects are; these questions need to be resolved by further studies.

8.2 Paramutation

Another epigenetic phenomenon is paramutation, which despite a long history of study, is still poorly understood on the molecular level. Paramutation refers to an interaction between two alleles at a single locus that occurs in the heterozygote; one allele induces a heritable, phenotypic change in the other allele. The alleles interact in *trans* to establish meiotically heritable expression states [124]. Different from *cis* epigenetic modifications (modifications that change chromatin structure through DNA or histone modifications [125]), *trans* epigenetic signals are maintained by soluble molecules like transcription factors or sRNAs acting in feedback loops of self-regulation of expression (see Figure 2) [126]. The first description of paramutation was in the 1950s for the maize *r1* (*colored 1*) locus, which encodes a transcription factor in the anthocyanin biosynthetic pathway [127]. How paramutation occurs is still not fully understood, but is an important epigenetic phenomenon because it so clearly impacts the phenotype of future generations. Studies in maize have revealed some of the molecular mechanisms involved in paramutation. Current models suggest that a pathway similar to the RdDM pathways might be involved, because paramutation requires the function of *mop1* (*mediator of paramutation 1*), an RNA-dependent RNA polymerase gene [128]. In addition, studies from the *b1* (*colored plant 1*) locus indicate that siRNAs and a repeat locus upstream of *b1* are important for paramutation and that DNA methylation plays a role as well [129]. These data suggest that several epigenetic systems are involved in this unique case of epigenetic inheritance.

An example of paramutation in mammals occurs at the *Kit* locus in mice. Homozygous deletions of the *Kit* gene result in lethality; however, heterozygous mice with one allele inactivated by a *LacZ* insertion within the *Kit* gene show a white tail phenotype [130]. This phenotype is passed down to future generations at a variable frequency to offspring in the absence of the *LacZ* insertion, indicating that paramutation has occurred and the wild type allele has been somehow changed in the heterozygous state. Interestingly, this effect could be replicated by injecting RNA from the heterozygotes into wild-type fertilized eggs; the RNA injection could also induce the white tail phenotype [130], indicating that the paramutation event at the *Kit* locus relies on an RNA-mediated mechanism. These examples illustrate that paramutation clearly involves parent-to-offspring epigenetic inheritance of a yet-to-be fully characterized molecular mechanism.

9. CONCLUSIONS

The field of epigenetics covers a wide variety of phenomena and includes a diverse array of molecular mechanisms such as DNA modifications, histone modifications, and a variety of noncoding RNAs. While the available data suggest that the majority of these epigenetic information systems are capable of transmitting information from cell to cell and from parent to offspring, the sequence of events controlling this inheritance is often not fully understood. As the examples in this chapter demonstrate, evidence for cell-to-cell inheritance of the epigenetic information is strong in most cases, while additional data are necessary to properly evaluate the prevalence and importance of epigenetic inheritance from parent to offspring. Genetic predisposition, environmental factors, and epigenetic mechanisms all contribute together to determine the phenotype of an individual. While the importance of genetic and environmental factors is well-established, only recently has the role of epigenetic mechanisms begun to be appreciated. Of vital importance are studies that (1) determine how frequently epigenetic information is passed on from one generation to the next and escapes the resetting that is thought to normally occur; and (2) estimate the effects on the offspring phenotypes. If such epigenetic inheritance occurs frequently and/or has large effects on phenotypes, a reevaluation of our models for the inheritance of phenotypes might be required.

ACKNOWLEDGMENTS

We would like to thank the members of the Riddle lab for their helpful comments on the manuscript. In addition, we would like to apologize to all our colleagues whose work we were unable to cite due to space limitations.

REFERENCES

[1] Ng RK, Gurdon JB. Epigenetic inheritance of cell differentiation status. Cell Cycle (Georgetown, Tex) 2008;7(9):1173–7.
[2] Chen T, Dent SY. Chromatin modifiers and remodellers: regulators of cellular differentiation. Nat Rev Genet 2014;15(2):93–106.
[3] Becker C, Weigel D. Epigenetic variation: origin and transgenerational inheritance. Curr Opin Plant Biol 2012;15(5):562–7.
[4] Bond DM, Finnegan EJ. Passing the message on: inheritance of epigenetic traits. Trends Plant Sci 2007;12(5):211–6.
[5] Mitalipov S, Wolf D. Totipotency, pluripotency and nuclear reprogramming. Adv Biochem Eng Biotechnol 2009;114:185–99.
[6] Campos EI, Stafford JM, Reinberg D. Epigenetic inheritance: histone bookmarks across generations. Trends Cell Biol 2014;24(11):664–74.
[7] Buszczak M, Spradling AC. Searching chromatin for stem cell identity. Cell 2006;125(2):233–6.
[8] Leeb M, Wutz A. Establishment of epigenetic patterns in development. Chromosoma 2012;121(3):251–62.
[9] Fernandez AF, Assenov Y, Martin-Subero JI, et al. A DNA methylation fingerprint of 1628 human samples. Genome Res 2012;22(2):407–19.
[10] Feng S, Jacobsen SE, Reik W. Epigenetic reprogramming in plant and animal development. Science (New York, NY) 2010;330(6004):622–7.
[11] Messerschmidt DM, Knowles BB, Solter D. DNA methylation dynamics during epigenetic reprogramming in the germline and preimplantation embryos. Genes Dev 2014;28(8):812–28.
[12] Jiang L, Zhang J, Wang JJ, et al. Sperm, but not oocyte, DNA methylome is inherited by zebrafish early embryos. Cell 2013;153(4):773–84.
[13] Potok ME, Nix DA, Parnell TJ, Cairns BR. Reprogramming the maternal zebrafish genome after fertilization to match the paternal methylation pattern. Cell 2013;153(4):759–72.
[14] Yang H, Lu P, Wang Y, Ma H. The transcriptome landscape of *Arabidopsis* male meiocytes from high-throughput sequencing: the complexity and evolution of the meiotic process. Plant J 2011;65(4): 503–16.
[15] Calarco JP, Borges F, Donoghue MT, et al. Reprogramming of DNA methylation in pollen guides epigenetic inheritance via small RNA. Cell 2012;151(1):194–205.
[16] Kawashima T, Berger F. Epigenetic reprogramming in plant sexual reproduction. Nat Rev Genet 2014;15(9):613–24.
[17] Yi Z, Song W, Clamp JC, Chen Z, Gao S, Zhang Q. Reconsideration of systematic relationships within the order Euplotida (Protista, Ciliophora) using new sequences of the gene coding for small-subunit rRNA and testing the use of combined data sets to construct phylogenies of the Diophrys-complex. Mol Phylogenet Evol 2009;50(3):599–607.
[18] Nanney DL, Simon EM. Laboratory and evolutionary history of *Tetrahymena thermophila*. Methods Cell Biol 2000;62:3–25.
[19] Chalker DL, Meyer E, Mochizuki K. Epigenetics of ciliates. Cold Spring Harb Perspect Biol 2013;5(12):a017764.
[20] Nowacki M, Landweber LF. Epigenetic inheritance in ciliates. Curr Opin Microbiol 2009;12(6):638–43.

[21] Jessop-Murray H, Martin LD, Gilley D, Preer Jr JR, Polisky B. Permanent rescue of a non-Mendelian mutation of *Paramecium* by microinjection of specific DNA sequences. Genetics 1991;129(3):727–34.

[22] Duharcourt S, Butler A, Meyer E. Epigenetic self-regulation of developmental excision of an internal eliminated sequence on *Paramecium tetraurelia*. Genes Dev 1995;9(16):2065–77.

[23] Guibert S, Weber M. Functions of DNA methylation and hydroxymethylation in mammalian development. Curr Top Dev Biol 2013;104:47–83.

[24] Kohli RM, Zhang Y. TET enzymes, TDG and the dynamics of DNA demethylation. Nature 2013;502(7472):472–9.

[25] Jeltsch A, Jurkowska RZ. New concepts in DNA methylation. Trends Biochem Sci 2014;39(7):310–8.

[26] He X-J, Chen T, Zhu J-K. Regulation and function of DNA methylation in plants and animals. Cell Res 2011;21(3):442–65.

[27] Matzke MA, Mosher RA. RNA-directed DNA methylation: an epigenetic pathway of increasing complexity. Nat Rev Genet 2014;15(6):394–408.

[28] Gao Z, Liu HL, Daxinger L, et al. An RNA polymerase II- and AGO4-associated protein acts in RNA-directed DNA methylation. Nature 2010;465(7294):106–9.

[29] Ramsahoye BH, Biniszkiewicz D, Lyko F, Clark V, Bird AP, Jaenisch R. Non-CpG methylation is prevalent in embryonic stem cells and may be mediated by DNA methyltransferase 3a. Proc Natl Acad Sci USA 2000;97(10):5237–42.

[30] Ooi SK, Qiu C, Bernstein E, et al. DNMT3L connects unmethylated lysine 4 of histone H3 to de novo methylation of DNA. Nature 2007;448(7154):714–7.

[31] Kuramochi-Miyagawa S, Kimura T, Ijiri TW, et al. Mili, a mammalian member of piwi family gene, is essential for spermatogenesis. Development (Cambridge, England) 2004;131(4):839–49.

[32] Aravin AA, Sachidanandam R, Girard A, Fejes-Toth K, Hannon GJ. Developmentally regulated piRNA clusters implicate MILI in transposon control. Science (New York, NY) 2007;316(5825):744–7.

[33] Rivera C, Gurard-Levin ZA, Almouzni G, Loyola A. Histone lysine methylation and chromatin replication. Biochim Biophys Acta 2014;1839(12):1433–9.

[34] Rothbart SB, Strahl BD. Interpreting the language of histone and DNA modifications. Biochim Biophys Acta 2014;1839(8):627–43.

[35] Lee JS, Smith E, Shilatifard A. The language of histone crosstalk. Cell 2010;142(5):682–5.

[36] Campos EI, Reinberg D. Histones: annotating chromatin. Annu Rev Genet 2009;43:559–99.

[37] MacAlpine DM, Almouzni G. Chromatin and DNA replication. Cold Spring Harb Perspect Biol 2013;5(8):a010207.

[38] Annunziato AT. Split decision: what happens to nucleosomes during DNA replication? J Biol Chem 2005;280(13):12065–8.

[39] Annunziato AT. Assembling chromatin: the long and winding road. Biochim Biophys Acta 2013;1819(3–4):196–210.

[40] De Koning L, Corpet A, Haber JE, Almouzni G. Histone chaperones: an escort network regulating histone traffic. Nat Struct Mol Biol 2007;14(11):997–1007.

[41] Wang T, Xu C, Liu Y, et al. Crystal structure of the human SUV39H1 chromodomain and its recognition of histone H3K9me2/3. PloS One 2012;7(12):e52977.

[42] Margueron R, Justin N, Ohno K, et al. Role of the polycomb protein EED in the propagation of repressive histone marks. Nature 2009;461(7265):762–7.

[43] Rathke C, Baarends WM, Awe S, Renkawitz-Pohl R. Chromatin dynamics during spermiogenesis. Biochim Biophys Acta 2014;1839(3):155–68.

[44] Fritsch L, Robin P, Mathieu JR, et al. A subset of the histone H3 lysine 9 methyltransferases Suv39h1, G9a, GLP, and SETDB1 participate in a multimeric complex. Mol Cell 2010;37(1):46–56.

[45] Rountree MR, Bachman KE, Baylin SB. DNMT1 binds HDAC2 and a new co-repressor, DMAP1, to form a complex at replication foci. Nat Genet 2000;25(3):269–77.

[46] Cho S, Park JS, Kang YK. AGO2 and SETDB1 cooperate in promoter-targeted transcriptional silencing of the androgen receptor gene. Nucleic Acids Res 2014;42(22):13545–56.

[47] Gaydos LJ, Wang W, Strome S. Gene repression. H3K27me and PRC2 transmit a memory of repression across generations and during development. Science (New York, NY) 2014;345(6203):1515–8.

[48] Mirouze M. The small RNA-based odyssey of epigenetic information in plants: from cells to species. DNA Cell Biol 2012;31(12):1650–6.

[49] Zamore PD, Haley B. Ribo-gnome: the big world of small RNAs. Science (New York, NY) 2005;309(5740):1519–24.

[50] Guo W, Chen W, Yu W, Huang W, Deng W. Small interfering RNA-based molecular therapy of cancers. Chin J Cancer 2013;32(9):488–93.

[51] Elbashir SM, Harborth J, Lendeckel W, Yalcin A, Weber K, Tuschl T. Duplexes of 21-nucleotide RNAs mediate RNA interference in cultured mammalian cells. Nature 2001;411(6836):494–8.

[52] Siomi MC, Sato K, Pezic D, Aravin AA. PIWI-interacting small RNAs: the vanguard of genome defence. Nat Rev Mol Cell Biol 2011;12(4):246–58.

[53] Mraheil MA, Billion A, Kuenne C, et al. Comparative genome-wide analysis of small RNAs of major Gram-positive pathogens: from identification to application. Microb Biotechnol 2010;3(6):658–76.

[54] Miska EA. How microRNAs control cell division, differentiation and death. Curr Opin Genet Dev 2005;15(5):563–8.

[55] Lagos-Quintana M, Rauhut R, Lendeckel W, Tuschl T. Identification of novel genes coding for small expressed RNAs. Science (New York, NY) 2001;294(5543):853–8.

[56] Mraz M, Dolezalova D, Plevova K, et al. MicroRNA-650 expression is influenced by immunoglobulin gene rearrangement and affects the biology of chronic lymphocytic leukemia. Blood 2012;119(9):2110–3.

[57] Friedman RC, Farh KK, Burge CB, Bartel DP. Most mammalian mRNAs are conserved targets of microRNAs. Genome Res 2009;19(1):92–105.

[58] Stroynowska-Czerwinska A, Fiszer A, Krzyzosiak WJ. The panorama of miRNA-mediated mechanisms in mammalian cells. Cell Mol Life Sci 2014;71(12):2253–70.

[59] Bazzini AA, Lee MT, Giraldez AJ. Ribosome profiling shows that miR-430 reduces translation before causing mRNA decay in zebrafish. Science (New York, NY) 2012;336(6078):233–7.

[60] Stuwe E, Toth KF, Aravin AA. Small but sturdy: small RNAs in cellular memory and epigenetics. Genes Dev 2014;28(5):423–31.

[61] Soni K, Choudhary A, Patowary A, et al. miR-34 is maternally inherited in *Drosophila melanogaster* and *Danio rerio*. Nucleic Acids Res 2013;41(8):4470–80.

[62] Gapp K, Jawaid A, Sarkies P, et al. Implication of sperm RNAs in transgenerational inheritance of the effects of early trauma in mice. Nat Neurosci 2014;17(5):667–9.

[63] Wu Y, Liu S, Xin H, et al. Up-regulation of microRNA-145 promotes differentiation by repressing OCT4 in human endometrial adenocarcinoma cells. Cancer 2011;117(17):3989–98.

[64] Xu N, Papagiannakopoulos T, Pan G, Thomson JA, Kosik KS. MicroRNA-145 regulates OCT4, SOX2, and KLF4 and represses pluripotency in human embryonic stem cells. Cell 2009;137(4):647–58.

[65] Ptashne M. Epigenetics: core misconcept. Proc Natl Acad Sci USA 2013;110(18):7101–3.

[66] Ostermeier GC, Goodrich RJ, Moldenhauer JS, Diamond MP, Krawetz SA. A suite of novel human spermatozoal RNAs. J Androl 2005;26(1):70–4.

[67] Liu W-M, Pang RTK, Chiu PCN, et al. Sperm-borne microRNA-34c is required for the first cleavage division in mouse. Proc Natl Acad Sci USA 2012;109(2):490–4.

[68] Cogoni C, Irelan JT, Schumacher M, Schmidhauser TJ, Selker EU, Macino G. Transgene silencing of the al-1 gene in vegetative cells of *Neurospora* is mediated by a cytoplasmic effector and does not depend on DNA-DNA interactions or DNA methylation. Embo J 1996;15(12):3153–63.

[69] Ketting RF, Fischer SE, Bernstein E, Sijen T, Hannon GJ, Plasterk RH. Dicer functions in RNA interference and in synthesis of small RNA involved in developmental timing in *C. elegans*. Genes Dev 2001;15(20):2654–9.

[70] Agrawal N, Dasaradhi PV, Mohmmed A, Malhotra P, Bhatnagar RK, Mukherjee SK. RNA interference: biology, mechanism, and applications. Microbiol Mol Biol Rev 2003;67(4):657–85.

[71] Tomari Y, Zamore PD. Perspective: machines for RNAi. Genes Dev 2005;19(5):517–29.

[72] Tijsterman M, May RC, Simmer F, Okihara KL, Plasterk RH. Genes required for systemic RNA interference in *Caenorhabditis elegans*. Curr Biol 2004;14(2):111–6.

[73] Klenov MS, Lavrov SA, Stolyarenko AD, et al. Repeat-associated siRNAs cause chromatin silencing of retrotransposons in the *Drosophila melanogaster* germline. Nucleic Acids Res 2007;35(16):5430–8.

[74] Watanabe T, Takeda A, Tsukiyama T, et al. Identification and characterization of two novel classes of small RNAs in the mouse germline: retrotransposon-derived siRNAs in oocytes and germline small RNAs in testes. Genes Dev 2006;20(13):1732–43.

[75] Kato M, Miura A, Bender J, Jacobsen SE, Kakutani T. Role of CG and non-CG methylation in immobilization of transposons in *Arabidopsis*. Curr Biol 2003;13(5):421–6.

[76] Mirouze M, Reinders J, Bucher E, et al. Selective epigenetic control of retrotransposition in *Arabidopsis*. Nature 2009;461(7262):427–30.
[77] Autran D, Baroux C, Raissig MT, et al. Maternal epigenetic pathways control parental contributions to *Arabidopsis* early embryogenesis. Cell 2011;145(5):707–19.
[78] Johannes F, Porcher E, Teixeira FK, et al. Assessing the impact of transgenerational epigenetic variation on complex traits. PLoS Genetics 2009;5(6):e1000530.
[79] Reinders J, Wulff BB, Mirouze M, et al. Compromised stability of DNA methylation and transposon immobilization in mosaic *Arabidopsis* epigenomes. Genes Dev 2009;23(8):939–50.
[80] Teixeira FK, Heredia F, Sarazin A, et al. A role for RNAi in the selective correction of DNA methylation defects. Science (New York, NY) 2009;323(5921):1600–4.
[81] Brennecke J, Aravin AA, Stark A, et al. Discrete small RNA-generating loci as master regulators of transposon activity in *Drosophila*. Cell 2007;128(6):1089–103.
[82] O'Donnell KA, Boeke JD. Mighty Piwis defend the germline against genome intruders. Cell 2007;129(1):37–44.
[83] Houwing S, Kamminga LM, Berezikov E, et al. A role for Piwi and piRNAs in germ cell maintenance and transposon silencing in Zebrafish. Cell 2007;129(1):69–82.
[84] Iyengar BR, Choudhary A, Sarangdhar MA, Venkatesh KV, Gadgil CJ, Pillai B. Non-coding RNA interact to regulate neuronal development and function. Front Cell Neurosci 2014;8:47.
[85] Ghildiyal M, Zamore PD. Small silencing RNAs: an expanding universe. Nat Rev Genet 2009;10(2):94–108.
[86] Brennecke J, Malone CD, Aravin AA, Sachidanandam R, Stark A, Hannon GJ. An epigenetic role for maternally inherited piRNAs in transposon silencing. Science (New York, NY) 2008;322(5906): 1387–92.
[87] Olivieri D, Senti KA, Subramanian S, Sachidanandam R, Brennecke J. The cochaperone shutdown defines a group of biogenesis factors essential for all piRNA populations in *Drosophila*. Mol Cell 2012;47(6):954–69.
[88] Kawaoka S, Izumi N, Katsuma S, Tomari Y. 3′ end formation of Piwi-interacting RNAs in vitro. Mol Cell 2011;43(6):1015–22.
[89] Saito K, Sakaguchi Y, Suzuki T, Suzuki T, Siomi H, Siomi MC. Pimet, the *Drosophila* homolog of HEN1, mediates 2′-O-methylation of Piwi- interacting RNAs at their 3′ ends. Genes Dev 2007;21(13):1603–8.
[90] Gu T, Elgin SC. Maternal depletion of Piwi, a component of the RNAi system, impacts heterochromatin formation in *Drosophila*. PLoS Genetics 2013;9(9):e1003780.
[91] Gunawardane LS, Saito K, Nishida KM, et al. A slicer-mediated mechanism for repeat-associated siRNA 5′ end formation in *Drosophila*. Science (New York, NY) 2007;315(5818):1587–90.
[92] Brasset E, Chambeyron S. Epigenetics and transgenerational inheritance. Genome Biol 2013;14(5):306.
[93] Luteijn MJ, van Bergeijk P, Kaaij LJ, et al. Extremely stable Piwi-induced gene silencing in *Caenorhabditis elegans*. Embo J 2012;31(16):3422–30.
[94] Das PP, Bagijn MP, Goldstein LD, et al. Piwi and piRNAs act upstream of an endogenous siRNA pathway to suppress Tc3 transposon mobility in the *Caenorhabditis elegans* germline. Mol Cell 2008;31(1):79–90.
[95] Batista PJ, Ruby JG, Claycomb JM, et al. PRG-1 and 21U-RNAs interact to form the piRNA complex required for fertility in *C. elegans*. Mol Cell 2008;31(1):67–78.
[96] Ashe A, Sapetschnig A, Weick EM, et al. piRNAs can trigger a multigenerational epigenetic memory in the germline of *C. elegans*. Cell 2012;150(1):88–99.
[97] Bagijn MP, Goldstein LD, Sapetschnig A, et al. Function, targets, and evolution of *Caenorhabditis elegans* piRNAs. Science (New York, NY) 2012;337(6094):574–8.
[98] Luteijn MJ, Ketting RF. Piwi-interacting RNAs: from generation to transgenerational epigenetics. Nat Rev Genet 2013;14(8):523–34.
[99] Buckley BA, Burkhart KB, Gu SG, et al. A nuclear *Argonaute* promotes multigenerational epigenetic inheritance and germline immortality. Nature 2012;489(7416):447–51.
[100] Klattenhoff C, Xi H, Li C, et al. The *Drosophila* HP1 homolog Rhino is required for transposon silencing and piRNA production by dual-strand clusters. Cell 2009;138(6):1137–49.
[101] Rozhkov NV, Hammell M, Hannon GJ. Multiple roles for Piwi in silencing *Drosophila* transposons. Genes & Dev 2013;27(4):400–12.
[102] Muerdter F, Olovnikov I, Molaro A, et al. Production of artificial piRNAs in flies and mice. RNA (New York, NY) 2012;18(1):42–52.

[103] Khurana JS, Wang J, Xu J, et al. Adaptation to P element transposon invasion in *Drosophila melanogaster*. Cell 2011;147(7):1551–63.

[104] Perkel JM. Visiting "noncodarnia". BioTechniques 2013;54(6):301, 3–4.

[105] Carninci P, Kasukawa T, Katayama S, et al. The transcriptional landscape of the mammalian genome. Science (New York, NY) 2005;309(5740):1559–63.

[106] Mercer TR, Dinger ME, Mattick JS. Long non-coding RNAs: insights into functions. Nat Rev Genet 2009;10(3):155–9.

[107] McCarrey JR, Dilworth DD. Expression of Xist in mouse germ cells correlates with X-chromosome inactivation. Nat Genet 1992;2(3):200–3.

[108] Thorvaldsen JL, Verona RI, Bartolomei MS. X-tra! X-tra! News from the mouse X chromosome. Dev Biol 2006;298(2):344–53.

[109] Jonkers I, Monkhorst K, Rentmeester E, Grootegoed JA, Grosveld F, Gribnau J. Xist RNA is confined to the nuclear territory of the silenced X chromosome throughout the cell cycle. Mol Cell Biol 2008;28(18):5583–94.

[110] Lyon MF. Epigenetic inheritance in mammals. Trends Genet 1993;9(4):123–8.

[111] Wilkinson LS, Davies W, Isles AR. Genomic imprinting effects on brain development and function. Nat Rev Neurosci 2007;8(11):832–43.

[112] Zwart R, Sleutels F, Wutz A, Schinkel AH, Barlow DP. Bidirectional action of the Igf2r imprint control element on upstream and downstream imprinted genes. Genes Dev 2001;15(18):2361–6.

[113] Adalsteinsson BT, Ferguson-Smith AC. Epigenetic control of the genome-lessons from genomic imprinting. Genes 2014;5(3):635–55.

[114] Barlow DP, Bartolomei MS. Genomic imprinting in mammals. Cold Spring Harb Perspect Biol 2014;6(2).

[115] Latos PA, Pauler FM, Koerner MV, et al. Airn transcriptional overlap, but not its lncRNA products, induces imprinted Igf2r silencing. Science (New York, NY) 2012;338(6113):1469–72.

[116] Yang Q, Zhang S, Liu H, et al. Oncogenic role of long noncoding RNA AF118081 in anti-benzo[a]pyrene-trans-7,8-dihydrodiol-9,10-epoxide-transformed 16HBE cells. Toxicol Lett 2014;229(3):430–9.

[117] Colby DW, Prusiner SB. Prions. Cold Spring Harb Perspect Biol 2011;3(1):a006833.

[118] Prusiner SB. Biology and genetics of prions causing neurodegeneration. Annu Rev Genet 2013;47:601–23.

[119] Hofmann J, Vorberg I. Life cycle of cytosolic prions. Prion 2013;7(5):369–77.

[120] Uptain SM, Lindquist S. Prions as protein-based genetic elements. Annu Rev Microbiol 2002;56:703–41.

[121] Stein AD, Lumey LH. The relationship between maternal and offspring birth weights after maternal prenatal famine exposure: the Dutch Famine Birth Cohort Study. Hum Biol 2000;72(4):641–54.

[122] Painter RC, Osmond C, Gluckman P, Hanson M, Phillips DI, Roseboom TJ. Transgenerational effects of prenatal exposure to the Dutch famine on neonatal adiposity and health in later life. BJOG 2008;115(10):1243–9.

[123] Jirtle RL, Skinner MK. Environmental epigenomics and disease susceptibility. Nat Rev Genet 2007;8(4):253–62.

[124] Chandler V, Alleman M. Paramutation: epigenetic instructions passed across generations. Genetics 2008;178(4):1839–44.

[125] Talbert PB, Henikoff S. Histone variants–ancient wrap artists of the epigenome. Nat Rev Mol Cell Biol 2010;11(4):264–75.

[126] Bonasio R, Tu S, Reinberg D. Molecular signals of epigenetic states. Science (New York, NY) 2010;330(6004):612–6.

[127] Pilu R. Paramutation: just a curiosity or fine tuning of gene expression in the next generation? Curr Genomics 2011;12(4):298–306.

[128] Alleman M, Sidorenko L, McGinnis K, et al. An RNA-dependent RNA polymerase is required for paramutation in maize. Nature 2006;442(7100):295–8.

[129] Belele CL, Sidorenko L, Stam M, Bader R, Arteaga-Vazquez MA, Chandler VL. Specific tandem repeats are sufficient for paramutation-induced trans-generational silencing. PLoS Genetics 2013;9(10):e1003773.

[130] Rassoulzadegan M, Grandjean V, Gounon P, Vincent S, Gillot I, Cuzin F. RNA-mediated non-mendelian inheritance of an epigenetic change in the mouse. Nature 2006;441(7092):469–74.

[131] Riggs AD. Epigenetic mechanisms of gene regulation, vol. 32. Cold Spring Harbor; 1996.

CHAPTER 9

Transgenerational epigenetic regulation by environmental factors in human diseases

Yuanyuan Li

Department of Medicine, Division of Hematology and Oncology, University of Alabama at Birmingham, Birmingham, AL, USA; Comprehensive Cancer Center, University of Alabama at Birmingham, Birmingham, AL, USA; Nutrition Obesity Research Center, University of Alabama at Birmingham, Birmingham, AL, USA

Contents

1. INTRODUCTION

Although the precise mechanisms for disease etiology are still under intense investigation, scientists have reached a consensus that multiple factors contribute to human diseases that include genetic mutation, polymorphisms, or chromosomal abnormality, as well as many other mechanisms. Because the genome is evolutionarily and chemically stable, the role of the environment is considered crucial in disease etiology [1]. Environmental factors generally regulate genome activity, independent of DNA sequence manipulation (e.g., epigenetics) [2]. An additional consideration for environmental influences on disease etiology is the importance of the role of the environment in developmental processes and its importance in shaping phenotypes during embryogenesis [3–6]. The mammalian embryogenesis process involves complicated regulations that are mainly controlled by genetic and epigenetic mechanisms [7,8]. Early embryonic developmental processes are controlled by highly conserved and accurate developmental genetic/epigenetic programs, leading to precisely time-controlled tissue-specific gene expression, global gene silencing, and subsequent diverse phenotypes between cells and organs, as well as individuals. Recent studies have shown that epigenetic regulations such as DNA

Epigenetic Gene Expression and Regulation
http://dx.doi.org/10.1016/B978-0-12-799958-6.00009-3

methylation and histone modification play crucial roles in the reprogramming process during early development including gametogenesis, embryogenesis, and fetal development [7,9–11]. Exposure to environmental compounds during a crucial time of development induces the establishment of specific epigenetic patterns that influence phenotypic variation, which, in some cases, lead to disease states at the later adult stage of development [3–6].

Epidemiological studies have suggested significant environmental impacts on disease etiology that could not be explained solely by genetic features. The most important examples of such observations are studies in monozygotic twins (same genetic composition), which reveal striking discordances in the prevalence of many diseases, suggesting epigenetic mechanisms play a major role in promoting these diseases [12]. This chapter focuses on how epigenetic mechanisms regulate adult-onset disease transgenerationally and interact with environmental factors. Although studies on epigenetic mechanisms in environment-associated transgenerational human disease are just emerging, understanding the process of epigenetic reprogramming during embryogenesis and how environmental factors affect this process will lead to beneficial health outcomes in the next generation.

2. EPIGENETIC REGULATIONS DURING EARLY DEVELOPMENT

The newest definition for epigenetics is "molecular factors/processes around DNA that regulates genome activity independent of DNA sequencing, and that are mitotically stable" [3]. Epigenetic mechanisms control gene expression via changing the accessibility of chromatin to transcriptional regulation locally and globally through modifications of the DNA and modification or rearrangement of nucleosomes [13–15]. The most common types of molecular processes involving epigenetic regulation are DNA methylation, histone modifications, and noncoding RNA (ncRNA). Although the history and definition of epigenetics has evolved, the majority of the molecular elements of the epigenetic regulatory processes have only been recently elucidated, especially with regards to studies on a subconcept of epigenetics known as transgenerational epigenetics. Transgenerational epigenetics have recently received extensive attention due to its ability to prewrite individual epigenetic profiles that lead to consequences on the adult phenotype formation and disease onset, which is called the developmental origin of diseases [3–6,16]. Transgenerational epigenetics is defined as "the germline (egg or sperm) transmission of epigenetic information between generations in the absence of any environmental exposure" [17]. Direct environmental exposure does not involve a generational phenotype. Thus, studies extending to generations with no direct environmental exposure can be considered transgenerational because the germline is the only cell type to transmit the epigenetic information generationally (Figure 1). Therefore, an exposure to a certain environmental component at any time during the life course of an individual (F0) results in the exposure to the germline (sperm or egg) that will generate the next generation (F1 generation) (Figure 1, left panel). Thus, the F2 generation will be the first generation influenced by this exposure due to transgenerational effects. However, if the exposure occurs during gestation, it will affect not only the F0 generation female and the F1 generation fetus, but also the F1 germline that will generate the F2 generation (Figure 1, right panel). In this case, transgenerational effects will firstly exhibit in F3 generation via germline transmission without direct environmental exposure. Studies that involve direct exposure to a generation, such as maternal intervention, should refer to multigenerational exposure effects rather than transgenerational effects. The altered epigenetic information will be transmitted through the germline, and this allows environmental factors to influence long-term regulation of gene expression later in life that underlies the processes of phenotype formation. Environmentally induced epigenetic changes may also result in a wide range of phenotypic consequences, such as different disease conditions (e.g., cancer, reproductive defects, and obesity) [18–26].

The best characterized epigenetic factor to be involved in germline transmission of epigenetic information is DNA methylation. DNA methylation plays a critical role during early embryonic development [27]. One of the most important periods during

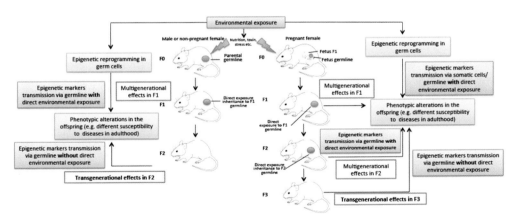

Figure 1 Environmental factors influence epigenetic transgenerational inheritance. An environmental factor acts on the F0 generation either in male and nonpregnant female or gestating female to influence epigenetic reprogramming in germline cells. For the exposed male and nonpregnant female parents, the altered epigenetic markers are transferred through the germline to subsequent F1 generations via multigenerational inheritance (direct environmental exposure) and F2 generation via transgenerational inheritance (indirect environmental exposure). For the exposed pregnant female parents, the altered epigenetic markers are transferred through the germline to subsequent F1–F2 generations via multigenerational inheritance (direct environmental exposure) and F3 generation via transgenerational inheritance (indirect environmental exposure). The embryo generated from this germline starts with an altered epigenome that affects developing somatic cells and tissues to have an altered transcriptome. This altered somatic cell transcriptome can then promote adult-onset disease associated with the transgenerational phenotype.

embryogenesis is fetal gonadal sex determination; when the germline is undergoing epigenetic programming and DNA remethylation. This genome-wide DNA methylation reprogramming process is overwhelming, as a new methylation profile will be established, leading to phenotypic variation in progeny. In particular, both maternal and paternal chromosomes undergo progressive demethylation by a passive mechanism, which erases most of the epigenetic marks in the zygote [28,29]. After implantation, global embryonic de novo methylation patterns are reestablished, which are then maintained throughout life in the somatic cells [30]. This dynamic methylation reprogramming, including global demethylation and remethylation processes, is essential for fetal development during early embryogenesis. The genome-wide demethylation processes might lead to chromatin decondensation, contributing to the transcriptional activation in the zygotic genes that are essential for early development. Subsequent de novo methylation processes might facilitate development of gene-specific methylation patterns, which determine tissue-specific transcription through a global silencing state.

Although most genomic DNA undergoes genome-wide demethylation and de novo methylation processes during early embryogenesis, the methylation marks on imprinted genes escape from this prevailing reprogramming, and thus, are preserved as parental imprints, leading to the differential expression profiles of several imprinted genes in the paternal and maternal alleles during development [31]. Therefore, incorrect development of DNA methylation patterns during this critical period may lead to embryonic lethality [32], developmental malformations [33], and increased risk of certain diseases [11]. Although DNA methylation plays a critical role in fetal germline development and early embryonic development, all other epigenetic processes (such as histone modifications and ncRNAs) are also involved and have unique functions in regulating development [34,35]. Further studies regarding the role of all epigenetic processes in regulating development are required.

3. ENVIRONMENT AND TRANSGENERATIONAL EPIGENETIC INHERITANCE

Epigenetic variability appears to be susceptible to modulation by environmental changes. Certain exposures during crucial periods of early development can shift the normal differentiation process and lead to varied phonotypes through a modification of the epigenome. This heritable transmission of environmentally induced phenotypes is referred to as transgenerational inheritance [36]. An altered epigenome induced by an environmental factor impacts the phenotype, which can create altered epigenetic hallmarks in somatic and germline cells that persist throughout the life of the individual. The phenotypic alterations will appear long after the exposure to environmental compounds, when these epigenetic markers are transmitted to the next generation through germline transmission. Germline epigenomic changes can be induced by environmental factors when the window of epigenetic reprogramming process is targeted during early development. In the event the germline epigenome is modified, impacts on the generation derived from that germline can occur and produce a transgenerational effect [3–6]. This would explain how early life events have consequences on the adult phenotype formation and disease onset, which is called the developmental origin of diseases [16].

A number of studies have shown the effects of environmental factors on influencing epigenetics and phenotypes. These environmental compounds include nutritional factors [37] such as methyl donors, e.g., folate [38]; and dietary compounds with properties in regulating epigenetic processes, e.g., green tea polyphenols (GTPs), soybean isoflavone, and broccoli sprouts sulforaphane (SFN) [39–41]; inorganic contaminants, e.g., arsenic [42,43]; drugs, e.g., cocaine [44]; habits such as drinking and smoking; endocrine disruptors, e.g., plastic component, bisphenol A (BPA) [45–47]; and the fungicide such as Vinclozolin [19,48,49] (Table 1). Some studies have also demonstrated maternal behavioral effects, including nursing behavior [50], stress, and depression [51], on epigenetics and disease outcome in the offspring. Nevertheless, nutritional factors are considered the most prevalent environmental factors that have ever been studied to influence epigenetic hallmarks, leading to transgenerational effects on human diseases. Not only the type of nutritional compounds, but also the quality and quantity of nutritive play a role on epigenetic profiles and subsequent phenotypes in the offspring [52–55] (Table 1).

In the following section, I will focus on nutritional factors and their effects on epigenetic inheritance and transgenerational impact on human diseases. Although the exact mechanisms that explain how nutritional factors regulate epigenetic reprogramming during early development is not fully understood, a better understanding of these processes will explain how an early life exposure can impact adult-onset disease, and further help to develop novel early-life prevention approaches to protect our offspring from various diseases in their adult life.

4. EPIGENETIC DIET AND HUMAN DISEASES

Epigenetic events frequently occur during early embryonic development, and aberrant epigenetic programming leads to various congenital disorders in the offspring [9]. Nevertheless, this vulnerability to environmental exposure during embryogenesis provides an excellent opportunity to reprogram epigenetic profiles, leading to beneficial outcomes such as cancer prevention in the offspring. One of the periods most sensitive to environmental exposures is when the germline is undergoing epigenetic reprogramming and DNA remethylation [28,29]. When an environmental factor promotes a permanent change of the gamete epigenome during this critical period, this altered epigenome and associated phenotypes can be transgenerationally transmitted to subsequent generations and progeny [3]. As a major component of environmental factors, nutritional factors such as dietary components with properties influencing epigenetic processes, so-called

Table 1 Environmentally induced epigenetic inheritance

Exposure	Resource	Pathology	Transgenerational effects	References
Nutrition				
Folate	Leafy vegetables	Congenital malformations	Folate insufficiency increased the risk for neural tube defects in the fetus and genome-wide DNA hypomethylation and genomic instability in *agouti* mice	[60–64]
Genistein	Soybean products	Reproductive effects, prostate, breast cancer protection	Maternal genistein may cause diverse effects of breast cancer development in later life; maternal genistein leads to genome-wide DNA hypermethylation in *agouti* mice	[65–70]
Sulforaphane	Cruciferous vegetables	Prostate, breast cancer protection	Early-life consumption of cruciferous vegetables exhibits more effective prevention effects against cancers	[37,40,71,73,74]
EGCG	Green tea	Prevention of various cancer, cardiovascular disease	Maternal ingestion of green tea may provide transplacental protection against carcinogenesis and chronic diseases	[37,72,75]
Caloric restriction	n/a	Impaired function in various organs	Maternal caloric restriction promotes metabolic disease in the F3 generation	[52,77]
High fat diet	Various	Growth and insulin sensitivity	Metabolic disease and obesity	[53,54]

Table 1 Environmentally induced epigenetic inheritance—cont'd

Exposure	Resource	Pathology	Transgenerational effects	References
Over nutrition	Various	Growth and insulin sensitivity	Neonatal over-growth induces metabolic syn-drome including obesity, insulin resistance, and glucose intolerance (F0–F2).	[55]
Others				
BPA	Endocrine disruptors (toxicant)	Prostate cancer, impaired reproduction system	BPA-induced transgenerational testicular abnor-mality (F1–F3)	[19]
Vinclozolin	Endocrine disruptors (toxicant)	Impaired male fertility	Embryonic rats exposed to maternal vinclozo-lin caused adult-onset disease including testicular, prostate, and renal abnormalities, and increased the incidence of tumors due to male germline changes in DNA methyla-tion.	[19,48,49]

EGCG, (−)-epigallocatechin-3-gallate; BPA, bisphenol A.

"epigenetic diets," are believed to have effects on multigenerational/transgenerational inheritance of adult-onset disease phenotypes [37,56,57].

Numerous studies have examined the potential for epigenetic inheritance of nutritional metabolic risk in human and mouse populations. A classic model for studying nutritional epigenetic effects has been the Agouti mouse, which consisted of detecting changes in methylation of the A^{vy} allele [58,59]. The coat color of mice carrying the *agouti* viable yellow gene varies from yellow, to mottled, to the wild type black/brown Agouti mouse phenotype, where each phenotype was dependent on the methylation status of the transgene. Therefore, the coat color and other aspects of the Agouti mouse phenotypes provided a direct readout of the methylation status of the allele. The agouti mouse model system has been successfully used to detect the methylation status in mammals when administrating folate-deficient dietary supplementation [60].

Folate is a water-soluble B vitamin from dietary sources or supplements, and believed to act as a methyl donor diet for the synthesis of S–adenosylmethionine (SAM), the universal methyl donor of biological methylation [61]. Folate deficiency leads to a decrease in SAM and is associated with genome-wide hypomethylation in both humans and animal models [62,63]. Evidence has indicated that folate insufficiency during pregnancy increases the risk for development of neural tube defects in the fetus and may also influence the risk for other human diseases in later life [64]. Studies have shown that black Agouti dams fed with methyl donor supplement (folate, methionine, choline, and vitamin B12) during pregnancy increased DNA methylation in the A^{vy} allele by hypermethylation [60]. The percentage of phenotypes with a more mottled "agouti coat" (combined black and yellow pigment in the hair) was higher when increasing levels of folate supplements were added to the diet. This shift in the distribution of mouse coat colors was correlated with DNA methylation of the A^{vy} allele, suggesting that maternal folate diet can have a significant effect on DNA methylation of the offspring.

The Agouti model was also employed to investigate the effects of a plant nutritive on the fetal epigenome. Genistein is a botanical isoflavone enriched in soybean products and has been of particular interest because of its bioactive roles inhibiting various human cancers and other chronic noncommunicable diseases [65–68]. Genistein represents a class of phytoestrogens that are active in multiple biological systems, including estrogen and non-estrogen receptor-mediated signaling pathways [69]. In the Agouti model, early exposure to genistein *in utero* alters multiple phenotypes of the offspring of the Agouti mice, such as reduction of prevalence of obesity and brown coat due to hypermethylation of the *agouti* gene [70]. The altered DNA methylation is found in three germ layers (brain, kidney, and liver), indicating that genistein may influence DNA methylation during early embryonic development. The observed effects of genistein on the epigenome serve as a plausible explanation for the lowered incidence of breast cancers in Asians, where the genistein-rich soy products have been in their traditional daily diet through generations, as well as the increased breast cancer incidence in Asians who immigrated to the United States [65]. These results suggest that folate and soybean genistein can act as epigenetic modulators to interfere with early epigenetic reprogramming leading to altered phenotypes, such as different susceptibility to diseases in the offspring. Our pilot study also shows that maternal (*in utero*) exposure to soybean genistein can reduce breast cancer incidence of the offspring in adult life. By comparison, the combined use of these two maternal nutritional supplementations (folate and genistein) has been found to counteract BPA, an endocrine-active compound known to cause adverse effects on the reproduction and the development of animals, induce DNA hypomethylation, and shift coat color distribution to black in the Agouti mouse system [46]. These findings demonstrate that appropriate maternal dietary regimen can not only protect against the deleterious effects of environmental toxicants on the fetal epigenome, but also prevent future diseases in the offspring.

Other epigenetic diets, such as SFN, an inhibitor of histone deacetylases (HDACs) [71] and GTPs, an inhibitor of DNA methyltransferases (DNMTs), [72] have been studied for their transgenerational effects [37]. Epidemiological studies have also suggested that early-life consumption of SFN through dietary cruciferous vegetables is more effective than later-life consumption [40,73]. Asian populations who consume more cruciferous vegetables at earlier ages than Caucasians have less incidence of certain diseases such as breast cancer, suggesting that maternal consumption of cruciferous vegetables during pregnancy may reduce the subsequent risk of cancer in their children [74]. In addition, consumption of green tea in a small amount (<2 cups per day) is safe during pregnancy; and maternal ingestion of green tea during pregnancy and nursing may provide transplacental protection against carcinogenesis [75].

Not only the type of dietary components, but also the quality and quantity of diets will affect the fetal epigenome, leading to varied phenotypes in the offspring. It has been proposed that unhealthy maternal diets (high-fat or low-protein diets) or a prior history of *in utero* exposure to maternal caloric restriction can result in increased metabolic risk in offspring [76]. This impact can be transmitted to several generations, leading to epigenetic transgenerational inheritance. Examples include maternal caloric restriction, which promotes metabolic disease phenotypes observed in the F3 generation [52,77], and a high-fat diet resulting in transgenerational adult-onset metabolic disease and obesity [54,55]. For example, individuals exposed to famine in WWII during gestation had a poorer glucose tolerance than those born the year before the famine [78]. Studies have found increased neonatal adiposity among the grandchildren of women who had been undernourished during pregnancy [79]. Furthermore, offspring of prenatally undernourished fathers, but not mothers, were heavier and more obese than the offspring of fathers and mothers who had not been undernourished prenatally. Thus, under certain nutritional conditions (poor, rich, or unbalanced maternal diets), the fetal environment could modify the development of the embryo to prepare the resultant offspring for a future environment during adult life.

5. EPIGENETIC INHERITANCE OF HUMAN DISEASES

Early evidence has revealed that developmental plasticity is affected, at least in part, by epigenetic changes that are established in early life and modulate gene expression during embryogenesis. Several diseases and syndromes have abnormal epigenetic regulations, leading to various pathologies. For example, abnormal DNA methylation on imprinting genes, *H19* and *IGF2* (insulin-like growth factor 2 receptor), can cause several congenital disorders, such as Silver–Russell syndrome [80]. Another epigenetic disease caused by abnormal DNA methylation of the X-chromosome is Fragile X syndrome [81]. Cancer is also considered a disease with an epigenetic origin. Epidemiological studies have provided first-hand evidence indicating a tight correlation of epigenetic diet consumption with cancer prevention in the progeny. For example, maternal consumptions of folate (a methyl-donor), epigenetic diet, or cruciferous vegetables that are enriched with SFN (an HDAC inhibitor), have been shown to reduce the risk of common acute lymphoblastic leukemia or breast cancer in later life, respectively [73,82]. In addition, our pilot studies have shown that maternal exposure to soybean genistein (an HDAC/DNMT inhibitor) can prevent onset of breast tumorigenesis in a spontaneous breast cancer mouse model.

Dietary enrichment of methyl donors (folic acid, selenium choline, vitamin B12, diallyl sulfide) has been shown to prevent transgenerational obesity [25]. Maternal folate deficiency leads to genome-wide changes in the fetal epigenome, and these resultant offsprings are more prone to develop diabetes and high blood pressure by adult age [83], indicating a potential role of maternal folate on prevention of diabetes and cardiovascular diseases in adult life. Dietary inhibitors of histone acetylases and deacetylases (HATs and HDACs) may influence chromatin states as well by affecting genes of the adipogenic pathways [84,85]. Studies on the agouti mice exposed to soybean genistein *in utero* show a change of obesity status in the offspring, suggesting that early exposure to soybean products may influence the adipogenesis processes in later life, and thus influence the susceptibility to certain human diseases that are closely related to obesity, such as type 2 diabetes [70]. A growing list of diseases with an epigenetic component suggests that epigenetics will play a crucial role in disease etiology for many of the disease states. However, environmental factors, such as nutrition, show great potential in the prevention of diseases at very early life stages of development and throughout multiple generations via epigenetic control.

6. CONCLUSIONS

Transgenerational epigenetic inheritance has recently received considerable attention due to its important implication in disease etiology. The ability of environmental exposures to influence transgenerational epigenetics significantly impacts our understanding of the basic regulation of biology. Nutritional factors with properties that influence epigenetic processes are believed to have effects on multigenerational/transgenerational inheritance of adult-onset disease phenotypes. Therefore, early exposures to certain nutritional factors are critical in disease etiology, due to the sensitivity of epigenetic systems during early stages of development. Better understanding of the transgenerational effects of bioactive dietary compounds, as well as the precise mechanisms during these processes, will facilitate the discovery of novel approaches linking early dietary or pharmaceutical interventions to human disease prevention.

ACKNOWLEDGMENTS AND FUNDING

This work was supported by grants from the National Cancer Institute (R01 CA178441; R03 CA176766) and the American Institute for Cancer Research.

REFERENCES

[1] Jirtle RL, Skinner MK. Environmental epigenomics and disease susceptibility. Nat Rev Genet April 2007;8(4):253–62.
[2] Jaenisch R, Bird A. Epigenetic regulation of gene expression: how the genome integrates intrinsic and environmental signals. Nat Genet March 2003;33(Suppl.):245–54.
[3] Skinner MK, Manikkam M, Guerrero-Bosagna C. Epigenetic transgenerational actions of environmental factors in disease etiology. Trends Endocrinol Metab 2010;21:214–22.
[4] Skinner MK. Environmental stress and epigenetic transgenerational inheritance. BMC Med September 5, 2014;12:153.
[5] Guerrero-Bosagna C, Skinner MK. Environmentally induced epigenetic transgenerational inheritance of phenotype and disease. Mol Cell Endocrinol May 6, 2012;354(1–2):3–8.
[6] Heard E, Martienssen RA. Transgenerational epigenetic inheritance: myths and mechanisms. Cell March 27, 2014;157(1):95–109.
[7] Li E. Chromatin modification and epigenetic reprogramming in mammalian development. Nat Rev Genet September 2002;3(9):662–73.
[8] Byrne C, Tainsky M, Fuchs E. Programming gene expression in developing epidermis. Development September 1994;120(9):2369–83.
[9] Cantone I, Fisher AG. Epigenetic programming and reprogramming during development. Nat Struct Mol Biol 2013;20(3):282–9.
[10] Ho L, Crabtree GR. Chromatin remodelling during development. Nature 2010;463(7280):474–84.
[11] Gopalakrishnan S, Van Emburgh BO, Robertson KD. DNA methylation in development and human disease. Mutat Res 2008;647(1–2):30–8.
[12] Bell JT, Spector TD. A twin approach to unraveling epigenetics. Trends Genet March 2011;27(3):116–25.
[13] Baylin SB, Ohm JE. Epigenetic gene silencing in cancer – a mechanism for early oncogenic pathway addiction? Nat Rev Cancer 2006;6:107–16.
[14] Kouzarides T. Chromatin modifications and their function. Cell 2007;128:693–705.
[15] Herman JG, Baylin SB. Gene silencing in cancer in association with promoter hypermethylation. N Engl J Med 2003;349:2042–54.
[16] Hanson MA, Gluckman PD. Developmental origins of health and disease: new insights. Basic Clin Pharmacol Toxicol February 2008;102(2):90–3.
[17] Skinner MK. Environmental epigenetic transgenerational inheritance and somatic epigenetic mitotic stability. Epigenetics July 2011;6(7):838–42.
[18] Anway MD, Skinner MK. Transgenerational effects of the endocrine disruptor vinclozolin on the prostate transcriptome and adult onset disease. Prostate April 1, 2008;68(5):517–29.
[19] Anway MD, Cupp AS, Uzumcu M, Skinner MK. Epigenetic transgenerational actions of endocrine disruptors and male fertility. Science June 3, 2005;308(5727):1466–9.
[20] Cheng RY, Hockman T, Crawford E, Anderson LM, Shiao YH. Epigenetic and gene expression changes related to transgenerational carcinogenesis. Mol Carcinog May 2004;40(1):1–11.
[21] Guerrero-Bosagna CM, Sabat P, Valdovinos FS, Valladares LE, Clark SJ. Epigenetic and phenotypic changes result from a continuous pre and post natal dietary exposure to phytoestrogens in an experimental population of mice. BMC Physiol September 15, 2008;8:17.
[22] Howdeshell KL, Hotchkiss AK, Thayer KA, Vandenbergh JG, vom Saal FS. Exposure to bisphenol A advances puberty. Nature October 21, 1999;401(6755):763–4.
[23] Newbold RR, Padilla-Banks E, Jefferson WN. Adverse effects of the model environmental estrogen diethylstilbestrol are transmitted to subsequent generations. Endocrinology June 2006;147(6 Suppl.):S11–7.

[24] Newbold RR, Padilla-Banks E, Jefferson WN, Heindel JJ. Effects of endocrine disruptors on obesity. Int J Androl April 2008;31(2):201–8.

[25] Waterland RA, Travisano M, Tahiliani KG, Rached MT, Mirza S. Methyl donor supplementation prevents transgenerational amplification of obesity. Int J Obes (Lond) September 2008;32(9):1373–9.

[26] Yamasaki H, Loktionov A, Tomatis L. Perinatal and multigenerational effect of carcinogens: possible contribution to determination of cancer susceptibility. Environ Health Perspect November 1992;98: 39–43.

[27] Razin A, Shemer R. DNA methylation in early development. Hum Mol Genet 1995;4(Spec No):1751–5.

[28] Mayer W, Niveleau A, Walter J, Fundele R, Haaf T. Demethylation of the zygotic paternal genome. Nature 2000;403(6769):501–2.

[29] Kafri T, Ariel M, Brandeis M, Shemer R, Urven L, McCarrey J, et al. Developmental pattern of gene-specific DNA methylation in the mouse embryo and germ line. Genes Dev 1992;6(5):705–14.

[30] Santos F, Hendrich B, Reik W, Dean W. Dynamic reprogramming of DNA methylation in the early mouse embryo. Dev Biol 2002;241(1):172–82.

[31] Jaenisch R. DNA methylation and imprinting: why bother? Trends Genet 1997;13(8):323–9.

[32] Li E, Bestor TH, Jaenisch R. Targeted mutation of the DNA methyltransferase gene results in embryonic lethality. Cell 1992;69(6):915–26.

[33] Matsuda M, Yasutomi M. Inhibition of cephalic neural tube closure by 5-azacytidine in neurulating rat embryos in vitro. Anat Embryol Berl 1992;185(3):217–23.

[34] Rugg-Gunn PJ, Cox BJ, Ralston A, Rossant J. Distinct histone modifications in stem cell lines and tissue lineages from the early mouse embryo. Proc Natl Acad Sci USA 2010;107(24):10783–90.

[35] Amaral PP, Mattick JS. Noncoding RNA in development. Mamm Genome August 2008;19(7–8): 454–92.

[36] Whitelaw NC, Whitelaw E. Transgenerational epigenetic inheritance in health and disease. Curr Opin Genet Dev June 2008;18(3):273–9.

[37] Li Y, Saldanha SN, Tollefsbol TO. Impact of epigenetic dietary compounds on transgenerational prevention of human diseases. AAPS J January 2014;16(1):27–36.

[38] Cooney CA, Dave AA, Wolff GL. Maternal methyl supplements in mice affect epigenetic variation and DNA methylation of offspring. J Nutr August 2002;132(8 Suppl.):2393S–400S.

[39] Vanhees K, Coort S, Ruijters EJ, Godschalk RW, van Schooten FJ, Barjesteh van Waalwijk van Doorn-Khosrovani S. Epigenetics: prenatal exposure to genistein leaves a permanent signature on the hematopoietic lineage. FASEB J 2011;25(2):797–807.

[40] Nelson NJ. Migrant studies aid the search for factors linked to breast cancer risk. J Natl Cancer Inst 2006;98(7):436–8.

[41] Yang P, Li H. Epigallocatechin-3-gallate ameliorates hyperglycemia-induced embryonic vasculopathy and malformation by inhibition of Foxo3a activation. Am J Obstet Gynecol 2010;203(1):75. e1–6.

[42] Singh KP, DuMond Jr JW. Genetic and epigenetic changes induced by chronic low dose exposure to arsenic of mouse testicular Leydig cells. Int J Oncol January 2007;30(1):253–60.

[43] Waalkes MP, Liu J, Chen H, Xie Y, Achanzar WE, Zhou YS, et al. Estrogen signaling in livers of male mice with hepatocellular carcinoma induced by exposure to arsenic in utero. J Natl Cancer Inst March 17, 2004;96(6):466–74.

[44] Novikova SI, He F, Bai J, Cutrufello NJ, Lidow MS, Undieh AS. Maternal cocaine administration in mice alters DNA methylation and gene expression in hippocampal neurons of neonatal and prepubertal offspring. PLoS One April 2, 2008;3(4):e1919.

[45] Ho SM, Tang WY, Belmonte de Frausto J, Prins GS. Developmental exposure to estradiol and bisphenol A increases susceptibility to prostate carcinogenesis and epigenetically regulates phosphodiesterase type 4 variant 4. Cancer Res June 1, 2006;66(11):5624–32.

[46] Dolinoy DC, Huang D, Jirtle RL. Maternal nutrient supplementation counteracts bisphenol A-induced DNA hypomethylation in early development. Proc Natl Acad Sci USA August 7, 2007;104(32): 13056–61.

[47] Yaoi T, Itoh K, Nakamura K, Ogi H, Fujiwara Y, Fushiki S. Genome-wide analysis of epigenomic alterations in fetal mouse forebrain after exposure to low doses of bisphenol A. Biochem Biophys Res Commun November 21, 2008;376(3):563–7.

[48] Anway MD, Memon MA, Uzumcu M, Skinner MK. Transgenerational effect of the endocrine disruptor vinclozolin on male spermatogenesis. J Androl November–December 2006;27(6):868–79. Epub July 12, 2006.

[49] Anway MD, Leathers C, Skinner MK. Endocrine disruptor vinclozolin induced epigenetic transgenerational adult-onset disease. Endocrinology December 2006;147(12):5515–23.

[50] Champagne FA, Weaver IC, Diorio J, Dymov S, Szyf M, Meaney MJ. Maternal care associated with methylation of the estrogen receptor-alpha1b promoter and estrogen receptor-alpha expression in the medial preoptic area of female offspring. Endocrinology June 2006;147(6):2909–15.

[51] Oberlander TF, Weinberg J, Papsdorf M, Grunau R, Misri S, Devlin AM. Prenatal exposure to maternal depression, neonatal methylation of human glucocorticoid receptor gene (NR3C1) and infant cortisol stress responses. Epigenetics March–April 2008;3(2):97–106.

[52] Le Clair C, Abbi T, Sandhu H, Tappia PS. Impact of maternal undernutrition on diabetes and cardiovascular disease risk in adult offspring. Can J Physiol Pharmacol March 2009;87(3):161–79. http://dx.doi.org/10.1139/y09-006.

[53] Wu Q, Suzuki M. Parental obesity and overweight affect the body-fat accumulation in the offspring: the possible effect of a high-fat diet through epigenetic inheritance. Obes Rev May 2006;7(2):201–8.

[54] Dunn GA, Bale TL. Maternal high-fat diet effects on third-generation female body size via the paternal lineage. Endocrinology June 2011;152(6):2228–36.

[55] Pentinat T, Ramon-Krauel M, Cebria J, Diaz R, Jimenez-Chillaron JC. Transgenerational inheritance of glucose intolerance in a mouse model of neonatal overnutrition. Endocrinology December 2010;151(12):5617–23.

[56] Meeran SM, Ahmed A, Tollefsbol TO. Epigenetic targets of bioactive dietary components for cancer prevention and therapy. Clin Epigenet 2010;1(3–4):101–16.

[57] Hardy TM, Tollefsbol TO. Epigenetic diet: impact on the epigenome and cancer. Epigenomics 2011;3(4):503–18.

[58] Yen TT, Gill AM, Frigeri LG, Barsh GS, Wolff GL. Obesity, diabetes, and neoplasia in yellow A(vy)/- mice: ectopic expression of the agouti gene. FASEB J 1994;8(8):479–88.

[59] Michaud EJ, van Vugt MJ, Bultman SJ, Sweet HO, Davisson MT, Woychik RP. Differential expression of a new dominant agouti allele (Aiapy) is correlated with methylation state and is influenced by parental lineage. Genes Dev 1994;8(12):1463–72.

[60] Wolff GL, Kodell RL, Moore SR, Cooney CA. Maternal epigenetics and methyl supplements affect agouti gene expression in Avy/a mice. FASEB J 1998;12(11):949–57.

[61] Hoffman DR, Cornatzer WE, Duerre JA. Relationship between tissue levels of S-adenosylmethionine, S-adenylhomocysteine, and transmethylation reactions. Can J Biochem 1979;57(1):56–65.

[62] Li Y, Tollefsbol TO. Impact on DNA methylation in cancer prevention and therapy by bioactive dietary components. Curr Med Chem 2010;17(20):2141–51.

[63] McKay JA, Williams EA, Mathers JC. Folate and DNA methylation during in utero development and aging. Biochem Soc Trans 2004;32(Pt 6):1006–7.

[64] Pitkin RM. Folate and neural tube defects. Am J Clin Nutr 2007;85(1):285S–8S.

[65] Fang C, Tseng M, Daly M. Correlates of soy food consumption in women at increased risk for breast cancer. J Am Diet Assoc 2005;105(10):1552–8.

[66] Barnes S. Effect of genistein on in vitro and in vivo models of cancer. J Nutr 1995;125(3 Suppl.): 777S–83S.

[67] Li Y, Chen H, Hardy TM, Tollefsbol TO. Epigenetic regulation of multiple tumor-related genes leads to suppression of breast tumorigenesis by dietary genistein. PLoS One 2013;8(1):e54369.

[68] Li Y, Meeran SM, Patel SN, Chen H, Hardy TM, Tollefsbol TO. Epigenetic reactivation of estrogen receptor-α (ERα) by genistein enhances hormonal therapy sensitivity in ERα-negative breast cancer. Mol Cancer 2013;12:9.

[69] Wang TT, Sathyamoorthy N, Phang JM. Molecular effects of genistein on estrogen receptor mediated pathways. Carcinogenesis 1996;17(2):271–5.

[70] Dolinoy DC, Weidman JR, Waterland RA, Jirtle RL. Maternal genistein alters coat color and protects Avy mouse offspring from obesity by modifying the fetal epigenome. Environ Health Perspect 2006;114(4):567–72.

[71] Myzak MC, Karplus PA, Chung FL, Dashwood RH. A novel mechanism of chemoprotection by sulforaphane: inhibition of histone deacetylase. Cancer Res 2004;64(16):5767–74.

[72] Fang MZ, Wang Y, Ai N, Hou Z, Sun Y, Lu H, et al. Tea polyphenol (-)-epigallocatechin-3-gallate inhibits DNA methyltransferase and reactivates methylation-silenced genes in cancer cell lines. Cancer Res 2003;63(22):7563–70.

[73] Ziegler RG, Hoover RN, Pike MC, Hildesheim A, Nomura AM, West DW, et al. Migration patterns and breast cancer risk in Asian-American women. J Natl Cancer Inst 1993;85(22):1819–27.

[74] Lee HP, Gourley L, Duffy SW, Estéve J, Lee J, Day NE. Dietary effects on breast-cancer risk in Singapore. Lancet 1991;337(8751):1197–200.

[75] Castro DJ, Yu Z, Löhr CV, Pereira CB, Giovanini JN, Fischer KA, et al. Chemoprevention of dibenzo[a,l]pyrene transplacental carcinogenesis in mice born to mothers administered green tea: primary role of caffeine. Carcinogenesis 2008;29(8):1581–6.

[76] Hales CN, Barker DJ. Type 2 (non-insulin-dependent) diabetes mellitus: the thrifty phenotype hypothesis. 1992. Int J Epidemiol October 2013;42(5):1215–22.

[77] Waterland RA. Epigenetic mechanisms affecting regulation of energy balance: many questions, few answers. Annu Rev Nutr 2014;34:337–55.

[78] Painter RC, Osmond C, Gluckman P, Hanson M, Phillips DI, Roseboom TJ. Transgenerational effects of prenatal exposure to the Dutch famine on neonatal adiposity and health in later life. BJOG September 2008;115(10):1243–9.

[79] Veenendaal MV, Painter RC, de Rooij SR, Bossuyt PM, van der Post JA, Gluckman PD, et al. Transgenerational effects of prenatal exposure to the 1944–45 Dutch famine. BJOG April 2013;120(5):548–53.

[80] Yamazawa K, Kagami M, Nagai T, Kondoh T, Onigata K, Maeyama K, et al. Molecular and clinical findings and their correlations in Silver-Russell syndrome: implications for a positive role of IGF2 in growth determination and differential imprinting regulation of the IGF2-H19 domain in bodies and placentas. J Mol Med Berl October 2008;86(10):1171–81.

[81] Walter E, Mazaika PK, Reiss AL. Insights into brain development from neurogenetic syndromes: evidence from fragile X syndrome, Williams syndrome, Turner syndrome and velocardiofacial syndrome. Neuroscience November 24, 2009;164(1):257–71.

[82] Thompson JR, Gerald PF, Willoughby ML, Armstrong BK. Maternal folate supplementation in pregnancy and protection against acute lymphoblastic leukaemia in childhood: a case-control study. Lancet 2001;358(9297):1935–40.

[83] Sinclair KD, Allegrucci C, Singh R, Gardner DS, Sebastian S, Bispham J, et al. DNA methylation, insulin resistance, and blood pressure in offspring determined by maternal periconceptional B vitamin and methionine status. Proc Natl Acad Sci USA 2007;104(49):19351–6.

[84] Ejaz A, Wu D, Kwan P, Meydani M. Curcumin inhibits adipogenesis in 3T3-L1 adipocytes and angiogenesis and obesity in C57/BL mice. J Nutr 2009;139(5):919–25.

[85] Campión J, Milagro FI, Martínez JA. Individuality and epigenetics in obesity. Obes Rev 2009;10(4):383–92.

CHAPTER 10

Identification of intergenic long noncoding RNA by deep sequencing

Gangqing Hu, Keji Zhao
Systems Biology Center, National Heart, Lung, and Blood Institute, NIH, Bethesda, MD, USA

Contents

1. INTRODUCTION

The structure of the human genome was once conceived as tens of thousands of protein-coding genes separated by noncoding sequences that may harbor regulatory DNA elements but are transcriptionally silent. Recent genome-wide transcriptomic studies challenged this view by revealing that the majority of human genome is transcribed though in a context-dependent manner [1,2]. A large portion of the transcripts are longer than 200 base pairs and do not appear to code functional proteins, and thus are termed as long noncoding RNAs (lncRNA) [3]. LncRNAs are similar to protein-coding genes in that they are mostly transcribed by RNA polymerase II, capped, spliced, and polyadenylated [4]. However, they are often expressed at a level one magnitude lower than protein-coding genes in a tissue- and lineage-specific manner, and exhibit marginal conservation in both primary sequence and secondary structure [5–9]. One group of lncRNAs receiving particular interest (known as lincRNAs, or long intergenic noncoding RNAs) are those not overlapping with any protein-coding gene and non-lincRNA gene. The study on their regulation, expression, and function appears to be easier due to the lack of complications from the overlapping with protein-coding genes [10].

Epigenetic Gene Expression and Regulation
http://dx.doi.org/10.1016/B978-0-12-799958-6.00010-X

LncRNAs contribute to almost every aspect of cellular and molecular biology, including cell cycle, genomic imprinting, cell identity, apoptosis, epigenetic regulation, posttranscriptional regulation of gene expression, etc. [9,11]. An appealing model is that many lncRNAs play roles in epigenetic regulation by directing chromatin-modifying enzymes to specific target sites [11]. In fact, several large-scale studies revealed that 10–30% of lincRNAs are associated with chromatin regulators, though some of the interactions may be indirect [12–14]. As specific examples, lincRNA Xist recruits Polycomb repressive complex 2 (PRC2) to the X-inactivation center during X chromosome inactivation [15], and lincRNA HOTAIR also interacts with PRC2 to establish H3K27me3-mediated repression at the *HOXD* locus [16]. Another group of lncRNAs are competitive endogenous RNA (ceRNA), which contains DNA elements to base pair with microRNAs and thereby derepresses their target genes [17]. In addition to their function in epigenetic regulation and as microRNA decoys, lincRNAs originating from active intergenic enhancers are more prevalent than previously thought [18]. While the evolution, tissue specificity, and coexpression with nearby genes of these enhancer-associated lincRNAs are distinct from those of promoter-associated ones [19], it is unclear whether the action of transcription matters or the product matters in regulating proximal targets [20]. Despite the above possibilities, from a perspective of evolutionary biology, most of the lncRNAs are assumed to be nonfunctional, because of their limited conservations and because of a minimal fitness cost being imposed to maintain spurious transcripts [10]. Nevertheless, a combination of knockout and "rescue" experiments is useful in understanding the function of a specific lincRNA.

While lncRNAs potentially outnumber protein-coding genes in mammalian genomes, only a few have been functionally characterized. Precise prediction of lncRNA is an essential step to explore the full spectrum of lncRNA expression and function in a life system. Because of the lack of sequence conservation, identification of lincRNA based solely on sequence features has been proven difficult, though remarkable attempts have been made recently [21,22]. On the other hand, a large number of lncRNAs have been identified through analyzing transcriptomic data generated by next generation sequencing [5–9]. Meanwhile, numerous in silico methods have been developed to facilitate the study of regulation, expression, and function of lncRNAs. In this book chapter, we start by describing several common technical considerations and computational procedures for lncRNA identification based on RNA-seq data, followed by summarizing current views on the regulation of lncRNA expression based on the integration with ChIP-Seq data generated for the genome-wide occupancy of transcription factors (TFs) and histone modifications. We chose to focus on intergenic lncRNAs, but expect the discussions in many cases to also apply to other groups of lncRNAs.

2. AN OVERVIEW OF METHODS FOR lncRNA DETECTION

Developments in genomic technologies combined with computational tools have paved the way for scientists to identify lncRNAs and study their structure and function in unprecedented detail. Early work on genome-wide discovery of lncRNAs involved the isolation of full-length cDNA clones followed by expensive Sanger sequencing [23]. However, due to the low sequencing throughput, the method fell short in estimating their expression levels. Traditional microarray technique employed hybridization of DNA sequences to oligonucleotides targeting specific genomic regions, providing a qualitative view for gene expression, but it relied on prior knowledge of gene structures [24]. DNA tiling arrays consisted of oligonucleotides spanning an entire genomic region with the advantage to discover new genes [25]. LncRNAs can also be detected by considering the epigenetic features of active genomic region. For instance, H3K4me3 is associated with promoters, and H3K36me3 is associated with transcribed regions of underlying lincRNA genes [26]. However, the strategy is unable to assign exon–intron structure and is constrained to intergenic regions.

A direct method for lncRNA detection is to profile transcriptomes by RNA-seq, which has the advantage of detection of a greater dynamic range in expression, the discovery of new transcripts, and the quantification of splice variants [26]. For the application of this strategy, several factors should be considered before library preparation and sequencing. The first is to decide whether to use polyadenylated RNA or total RNA. While the majority of protein-coding transcripts carry poly-A tails, non-polyadenylated transcripts are more prevalent for lncRNA genes than for protein-coding genes, [8] and thus starting with total RNA works better than polyA RNA in detecting as many lincRNAs as possible. Because over 90% of the total RNAs are ribosomal RNAs, they are depleted before library construction. However, due to the enrichment of unspliced RNAs in total RNAs, splice sites are more difficult to detect with sequencing reads from total RNAs than with sequencing reads from Poly-A selected RNAs. The second is to use single-end or pair-end sequencing. While both strategies are capable of revealing new transcript loci, pair-end sequencing is preferred. Because the genomic regions of lncRNA are often enriched with repetitive sequences [27], and that the transcripts of lincRNAs are far less abundant than protein-coding genes [8], paired-end sequencing can enhance the accurate assignment of sequencing reads to the reference genome. The third factor to consider is the read length and sequencing depth. The read length is more often preset by sequencing instruments, for example 50–200 bps for a HiSeq 2000 machine; other instruments such as 454 and PacBio produce longer reads of hundreds or thousands of base pairs at the cost of a low sequencing throughout. For sequencing depth, the ENCODE consortium recommends 20–25 million mappable reads (>30 bps, PE) for calling differentially expressed genes, and 100–200 million reads (76 bps, PE) for defining novel transcripts for polyA$^+$ samples. However, given that lncRNAs are much less abundant than protein-coding genes, a deeper sequence depth is crucial for accurate noncoding RNA analysis. In practice, we adhere to the regular throughput to screen lineage/tissue-specific lncRNAs, and then choose those which are moderately expressed for follow-up experiments, based on the rationale that more abundant transcripts are more likely to have biological functions; accurate quantification of the lncRNA expression can be done by conventional quantitative RT-PCR. Biological repeats should always be obtained to minimize the number of false positives. Other factors to consider are whether to use strand-specific RNA-seq [28] or regular RNA-seq of fractionated nucleic or cytoplasmic RNAs [29]. Strand-specific RNA-seq has the advantage to reveal noncoding transcript antisense to existing genes and to distinguish two lincRNAs divergently transcribed from a shared promoter. The profile of transcripts from different cellular compartments offer an opportunity to identify lncRNAs implicated in transcriptional regulation or translational regulation.

The different methods for lncRNA detection are not mutually exclusive. Instead, a combination of different methods could overcome the shortcomings of each single method. For instance, about 10 billion sequencing reads may be required to reveal the full spectrum of transcription complexity of lincRNAs for the human genome if solely based on the use of deep sequencing [30], which is infeasible. An alternative way is to first use conventional RNA-seq to define genomic regions of interest, followed by using tiling oligonucleotides arrays to specifically enrich RNAs from these genomic regions, and then further sequencing of the enriched species. John Rinn's group applied this strategy to human foot fibroblast cell lines and obtained an ultra fine resolution on the transcription complexity of tens of protein-coding regions and intergenic regions. Remarkably, the method robustly detects the transcripts occurring six times per 10,000 cells in human fibroblasts [30].

3. LincRNA IDENTIFICATION BASED ON RNA-SEQ

3.1 Alignment of sequencing reads and transcript assembly

One important step in deep sequencing data analysis is to align the short reads to a reference genome. The alignment of RNA-seq reads is more complicated than the mapping of ChIP-Seq reads due to the non contiguous nature of transcripts as a result of splicing. Several splice-sensitive aligners were developed, for example MapSplice [31], STAR [32], and TopHat [33]. Methods such as MapSplice [31] and TopHat [33] involve a two-step approach in which reads or parts of the reads are first aligned to the genome to define expressed regions and create an exon-junction library, to which the remaining unaligned reads are then mapped; genome annotation and splicing regulatory elements such as the AG-GT signal are often incorporated to define the junctions [31,33]. These methods can be considered as an extension of a short reads aligner such as Bowtie [34] and BWA [35]. Other methods handling alignment of spliced sequencing reads in one-pass were also developed such as GSNAP [36] and STAR [32]. Despite these progresses, transcript read alignment is not a fully resolved problem, with future challenges including correctly placing of multi-mapped reads (note, lncRNAs are enriched in repetitive sequences), reducing false positive exon junctions, and handling longer reads with higher error rates [37].

The second step in the process is to reconstruct transcript from the RNA-seq reads. One strategy uses reads aligned to the genome and exon–intron junctions from a splice-sensitive aligner. Examples are Cufflinks [38] and Scripture [9]. Cufflinks implements a bipartite-graph-based approach to construct a parsimonious set of transcripts that explain the read coverage from an experiment [38], while Scripture uses a statistic segmentation approach to infer paths from a transcript graph, where each path corresponds to a gene isoform [9]. As expected, both approaches require a reference from the whole genome sequence. However, although genome annotation is not necessary, it will generally improve the performance [39]. Another strategy borrows the idea of de novo genome assembly by using short reads; examples are Trinity [40] and Velvet [41]. A comprehensive assessment of 14 transcript reconstruction methods for RNA-seq data reveals that the rate of exon/isoform recovery is particularly lower in lncRNAs, largely due to their low expression levels, a lack of translational feature, and a loosening requirement on splicing signals [39]. In addition, many transcripts reconstructed for lncRNAs may represent partial isoforms, providing that the current RNA-seq data is frequently generated from a sequence depth recommended for protein-coding genes [39].

Transcribed genomic regions can also be identified from RNA-seq data using methods designed for peak calling from ChIP-Seq data. For example, in a recent attempt to identify lincRNAs during T cell development and differentiation, we used SICER to identify genomic islands enriched in RNA-seq reads in intergenic regions [42]; SICER utilizes the read enrichment information from neighboring bins to identify clusters of signals unlikely to appear by chance [43]. Islands from the same intergenic region are then grouped into different clusters based on their coexpression across different experimental conditions [42]. One cluster may contain several lincRNAs with similar expression patterns. Because it is also possible that several clusters within one locus may represent different parts of a lincRNA gene due to insufficient sequencing depth, we focused on clusters that show a signature of a promoter such as the appearance of H3K4me3 peaks for follow-up data analysis [42].

3.2 Definition of intergenic regions

Interpretation of the results from lncRNA knockdown and/or knockout experiments could be complicated if the lncRNA overlaps with a protein-coding gene or other non-lncRNA elements. Therefore, functional elucidation of lncRNA genes located in intergenic regions is preferred by many researchers. Transcripts reconstructed from RNA-seq are first examined based on their positions relative to protein-coding genes. Gene annotations from public databases such as RefSeq [43], Ensembl [44], and GENCODE [45] are frequently used to define coding regions. Because of current limitation in sequencing depth, some of reconstructed transcripts, although showing no overlap with any protein-coding genes, could be unannotated extensions of neighboring protein-coding genes. One way to minimize this possibility is to focus on transcripts that are located certain distance away from protein-coding genes. An alternative way is to supply the genome annotation together with the RNA-seq reads to a transcript reconstruction program to reannotate the genes about their 5′ and 3′ ends; the refined annotations are then used to define the boundaries of protein-coding genes [42]. Due to the imperfection in gene annotation, some noncoding genes may be misannotated as protein-coding. It is therefore attractive to define coding genes from the whole genome transcripts through coding potential assessment (see next section) and then use them to redefine intergenic regions.

3.3 Computational methods for assessment of coding potentials

Prediction of protein-coding genes from a DNA sequence is a long-standing issue in the field of bioinformatics. The task is relatively easier for RNA transcript than for genomic DNA sequence without the need to delineate splicing events, which is similar for predicting genes from microbial genomes [46]. Several strategies have been described in literature. First, a coding transcript can be roughly separated from a noncoding transcript based on the size of the open reading frame (ORF): ORFs of over 100 amino acids are not likely to have occurred by chance and thus, indicate a potential in coding capacity [22]. However, many of the well-studied lincRNAs are exceptions to this rule, including *Xist* and *HOTAIR* [47], both of which have potential ORFs >300 bps. The second strategy is to conceptually translate the transcript from six potential ORFs and search the corresponding amino acid sequences against public protein databases (e.g., SWISPROT and NCBI nr) or protein–domain database (e.g., CDD). Coding Potential Calculator (CPC) serves as a representative method that combines ORF prediction and similarity hits to known proteins for coding potential assessment [66]. It may fail to predict lncRNAs originated from pseudo genes because of their similarity to parent genes. Since transposal elements are prevalent within lncRNAs, additional care should be taken to handle similarity hits to transposons associated protein-coding genes. In addition, many of the NCBI nr protein annotations are from in silico prediction such that some lncRNAs may be misannotated as protein-coding genes. Therefore, we recommend to search against manually curated protein databases [42].

The availability of genome sequences for different organisms facilitates another way of coding potential assessment based on comparative sequence analysis. The sequence contents of codon positions of protein-coding genes are presumed to evolve under purifying selection such that synonymous substitution is preferred over nonsynonymous substitution. The difference in substitution between coding and noncoding regions as calculated from multiple species alignment is characterized by a codon substitution frequency (CSF) metric [48]. However, it does not utilize all the information from multiple alignments, and the output score is from empirical distributions in known coding and noncoding regions. A latter development reformulates the CSF metric into a statistic model comparison problem, in which the best explanation of a sequence alignment is chosen between two phylogenetic models, one estimated from known coding regions and the other from noncoding regions [49]. One limitation of the alignment-based methods is the availability of genomes (sequence and annotation) at reasonable phylogenetic distances to the research subject of interest.

Methods for ab initio gene prediction based on intrinsic DNA sequence have been recently proposed for lncRNA prediction. CPAT [50] and CNCI [51] are two examples. While both methods consider sequence features such as ORF size, nucleotide composition, and codon usage bias, they utilize different statistic models for discrimination analysis: logistic regression by CPAT [50] and support vector machine by CNCI [51]. As supervised machine learning methods, both CPAT and CNCI requires training data from protein-coding genes and nonprotein-coding genes. Interestingly, CNCI trained on humans performs well on other vertebrates, but less favorable on invertebrates and plants [51].

4. REGULATION OF lncRNA TRANSCRIPTION

The regulation of transcription of lncRNA genes is presumed to be similar to that of protein-coding genes, given that many of the lncRNAs are transcribed by RNA Pol II and are 5′ capped and 3′ polyadenylated [8]. Epigenetic mechanisms are well known for regulating Pol II mediated transcription in the context of chromatin, consisting of DNA and core histones in the eukaryotic nuclei. DNA sequence can be methylated, which is linked to transcriptional repression. Histones are subject to various posttranslational modifications, which are associated with different transcriptional activities [52–54]. For instance, H3K4me3 and H3K36me3 are enriched at promoter and gene body regions, respectively, and are positively associated with transcriptional activity, while trimethylation of H3K27 is enriched at promoter and/or gene body regions of silent genes and is associated with gene repression [55]. A comparison of H3K4me3, H3K36me3, and H3K27me3 between lncRNA and protein-coding genes at promoters suggests that epigenetic regulation of lncRNA genes shares common features with protein-coding genes [56]. Nevertheless, the same study suggested that DNA methylation and H3K9me3 do not seem to regulate lncRNA expression [56]. Another group reported that H3R2me1 is differentially distributed between lncRNA and mRNA genes [21]. An integration of histone marks and genomic features into a logistic regulation model indicated that H3K9me3, H3K27ac, and H3K4me1 are among many of the factors that mostly contribute to the discrimination between lncRNA and mRNA genes [21]. Therefore, the epigenetic regulation of at least a subgroup of lncRNAs appears to be different from mRNA genes.

Transcriptional regulation of lncRNA by TFs is a topic of several recent publications. Yang et al. integrated hundreds of ChIP-Seq data from six organisms into a database and implemented two genome browsers to facilitate the examination of TF binding at lncRNA genes of interest [57]. Given that most lncRNAs are lineage-specifically expressed, it is expected that the expression may be controlled by lineage-specific TFs. Indeed, during the differentiation from naïve CD4$^+$ T cell into distinctive T helper cells, many lineage-specific lincRNA genes are bound and regulated by master TFs such as T-bet and STAT4 in T$_H$1 cells and GATA-3 and STAT6 in T$_H$2 cells [42]. Regulatory regions of lncRNAs expressed in embryonic stem cells are preferentially bound by TFs associated with pluripotency, *AK028326* (OCT4-activated) and *AK141205* (Nanog-repressed) for examples [58]. Interestingly, from an in silico binding-site analysis, Necsulea et al. found that homeobox TFs (OCT4, for example), which function in embryonic development, are more often associated with lncRNA promoters than with protein-coding promoters [59], suggesting that lncRNAs are important for embryonic development, pluripotency, and differentiation.

5. FUTURE DIRECTIONS

The number of reported lincRNAs increases rapidly due to recent applications of RNA-seq techniques to different organisms, lineages/tissues, and under different experimental conditions. An accurate delineation of the full-length structure of lincRNA from RNA-seq data is still challenging. In many cases, the transcripts reconstructed from RNA-seq data are parts of a lincRNA. However, this can be improved by either increasing sequencing depth or by other sequencing technique such as targeted sequencing [30]. LincRNAs from genomic regions with highly repetitive sequence could be more difficult to discover than those from other regions. A combinational use of short and long sequencing reads in transcript reconstruction and new single molecule sequencing techniques would help to address this issue [30]. Coding potential assessments of lincRNA have relied on similarity search and/or comparative sequence analysis in most applications, while tools for ab initio prediction of coding sequence from gene transcripts have recently been made available. However, these methods are supervised and require well-curated training sets [50,51], and therefore, it remains promising to develop an unsupervised method that can be universally adapted to any organism in this regard.

While many features of lincRNAs are clear, such as their low expression level, lack of sequence conservation, lineage/tissue-specific expression, and similar mechanisms of transcription with protein-coding genes, the function of the majority of lincRNAs remains unclear. Computational methods based on coexpression with protein-coding genes have led to the inference of function for several lincRNAs followed by experimental validation [26,60–62]. The survey of lincRNA expression and the inference of their functions at single-cell level have been less exploited, largely because the current technique for single-cell RNA-seq cannot reliably and accurately measure the expression dynamics of lowly expressed species, which includes lincRNAs [63]. Genome editing techniques, such as CRISPR/Cas [64,65], offer a new opportunity to the study of lincRNA function in vivo. Because of the lack of ORF, targeting DNA sequences can be alternatively chosen at key regulatory elements such as promoter and enhancers; the development of computational tools to optimize the target selection of CRISPR for lincRNA is still at its infancy but is expected to emerge in the near future.

Summary of tools and resource for lincRNA analysis		Notes
ENCODE recommendation for RNA-seq		
20–25 mil reads (mappable) 100–200 mil reads	>30 bps 76 bps, PE	Differentially expressed genes PE helps to better define novel transcripts, especially for polyA$^+$ samples
Equipment		
HiSeq 454 PacBio	50–200 bps 100s bps 1000s bps	
Databases		
RefSeq	Public gene annotation database	The online USCS genome browser Web sites provide user-friendly downloads of annotation for the human genome, the mouse genome, and others
Ensembl	Public gene annotation database	
GENCODE	Public gene annotation database	
Transcript inference related tools		
Bowtie & BWA	Backward search with Burrows–Wheeler transform	Short read alignment to a reference genome
MapSplice & TopHat	Two-step method (read alignment followed by splice inference)	Splice junction discovery
GSNAP & STAR	One-step method (read alignment and splice junction called at the same time)	
Cufflinks	Bipartite-graph-based method	Reference-based transcriptome assembly
Scripture	A statistical segmentation method and spliced/unspliced reads within a connectivity graph to infer transcripts	
Trinity & Velvet	Partition the genome into de Bruijn graphs from raw reads and process the graphs to infer transcripts	Transcriptome assembly without reference genome

Continued

—cont'd

Summary of tools and resource for lincRNA analysis		Notes
Transcript inference related tools		
SICER	Utilizes read enrichment from neighbor bins to call genomic regions enriched with NGS short reads	A method originally designed to call peaks from ChIP-Seq data, but can also be applied to call peaks (genomic regions with active transcription) from RNA-Seq data. Recommended to process total RNA-Seq or to call genomic region with very low read coverage (e.g., regions with long noncoding RNA).
Tools for coding potential assessment		
CNCI	SVM classification based on sequence intrinsic composition	http://www.bioinfo.org/software/cnci/
CPAT	A logistic regression model built with four sequence features	http://lilab.research.bcm.edu/cpat/
CPC	SVM classification based on six sequence features	http://cpc.cbi.pku.edu.cn/
CSF & PhyloCSF	A metric based on the difference in codon substitution frequencies between coding and noncoding sequences	http://compbio.mit.edu/PhyloCSF

ACKNOWLEDGMENTS

The research in the authors' laboratory was supported by Division of Intramural Research, National Heart, Lung and Blood Institute, NIH.

REFERENCES

[1] Bernstein BE, et al. An integrated encyclopedia of DNA elements in the human genome. Nature 2012;489(7414):57–74.
[2] Djebali S, et al. Landscape of transcription in human cells. Nature 2012;489(7414):101–8.
[3] Mercer TR, Dinger ME, Mattick JS. Long non-coding RNAs: insights into functions. Nat Rev Genet 2009;10(3):155–9.
[4] Gibb EA, Brown CJ, Lam WL. The functional role of long non-coding RNA in human carcinomas. Mol Cancer 2011;10.
[5] Cabili MN, et al. Integrative annotation of human large intergenic noncoding RNAs reveals global properties and specific subclasses. Genes Dev 2011;25(18):1915–27.

[6] Sigova AA, et al. Divergent transcription of long noncoding RNA/mRNA gene pairs in embryonic stem cells. Proc Natl Acad Sci USA 2013;110(8):2876–81.

[7] Pauli A, et al. Systematic identification of long noncoding RNAs expressed during zebrafish embryogenesis. Genome Res 2012;22(3):577–91.

[8] Derrien T, et al. The GENCODE v7 catalog of human long noncoding RNAs: analysis of their gene structure, evolution, and expression. Genome Res 2012;22(9):1775–89.

[9] Guttman M, et al. Ab initio reconstruction of cell type-specific transcriptomes in mouse reveals the conserved multi-exonic structure of lincRNAs. Nat Biotechnol 2010;28(5):503–10.

[10] Ulitsky I, Bartel DP. lincRNAs: genomics, evolution, and mechanisms. Cell 2013;154(1):26–46.

[11] Rinn JL, Chang HY. Genome regulation by long noncoding RNAs. Annu Rev Biochem 2012;81:145–66.

[12] Khalil AM, et al. Many human large intergenic noncoding RNAs associate with chromatin-modifying complexes and affect gene expression. Proc Natl Acad Sci USA 2009;106(28):11667–72.

[13] Zhao J, et al. Genome-wide identification of polycomb-associated RNAs by RIP-seq. Mol Cell 2010;40(6):939–53.

[14] Guttman M, et al. lincRNAs act in the circuitry controlling pluripotency and differentiation. Nature 2011;477(7364):295–300.

[15] Simon MD, et al. High-resolution Xist binding maps reveal two-step spreading during X-chromosome inactivation. Nature 2013;504(7480):465–9.

[16] Rinn JL, et al. Functional demarcation of active and silent chromatin domains in human HOX loci by noncoding RNAs. Cell 2007;129(7):1311–23.

[17] Salmena L, et al. A ceRNA hypothesis: the Rosetta Stone of a hidden RNA language? Cell 2011;146(3):353–8.

[18] Orom UA, Shiekhattar R. Long noncoding RNAs usher in a new era in the biology of enhancers. Cell 2013;154(6):1190–3.

[19] Marques AC, et al. Chromatin signatures at transcriptional start sites separate two equally populated yet distinct classes of intergenic long noncoding RNAs. Genome Biol 2013;14(11):R131.

[20] Natoli G, Andrau JC. Noncoding transcription at enhancers: general principles and functional models. Annu Rev Genet 2012;46:1–19.

[21] Lv J, et al. Long non-coding RNA identification over mouse brain development by integrative modeling of chromatin and genomic features. Nucleic Acids Res 2013;41(22):10044–61.

[22] Wang Y, et al. Computational identification of human long intergenic non-coding RNAs using a GA-SVM algorithm. Gene 2014;533(1):94–9.

[23] Ilott NE, Ponting CP. Predicting long non-coding RNAs using RNA sequencing. Methods 2013;63(1):50–9.

[24] Schena M, et al. Quantitative monitoring of gene expression patterns with a complementary DNA microarray. Science 1995;270(5235):467–70.

[25] Mockler TC, et al. Applications of DNA tiling arrays for whole-genome analysis. Genomics 2005;85(1):1–15.

[26] Guttman M, et al. Chromatin signature reveals over a thousand highly conserved large non-coding RNAs in mammals. Nature 2009;458(7235):223–7.

[27] Kelley D, Rinn J. Transposable elements reveal a stem cell-specific class of long noncoding RNAs. Genome Biol 2012;13(11):R107.

[28] Borodina T, Adjaye J, Sultan M. A strand-specific library preparation protocol for RNA sequencing. Methods Enzymol 2011;500:79–98.

[29] Clark MB, et al. Genome-wide analysis of long noncoding RNA stability. Genome Res 2012;22(5):885–98.

[30] Mercer TR, et al. Targeted RNA sequencing reveals the deep complexity of the human transcriptome. Nat Biotechnol 2012;30(1):99–104.

[31] Wang K, et al. MapSplice: accurate mapping of RNA-seq reads for splice junction discovery. Nucleic Acids Res 2010;38(18):e178.

[32] Dobin A, et al. STAR: ultrafast universal RNA-seq aligner. Bioinformatics 2013;29(1):15–21.

[33] Trapnell C, Pachter L, Salzberg SL. TopHat: discovering splice junctions with RNA-Seq. Bioinformatics 2009;25(9):1105–11.

[34] Langmead B, et al. Ultrafast and memory-efficient alignment of short DNA sequences to the human genome. Genome Biol 2009;10(3):R25.

[35] Li H, Durbin R. Fast and accurate short read alignment with Burrows-Wheeler transform. Bioinformatics 2009;25(14):1754–60.

[36] Wu TD, Nacu S. Fast and SNP-tolerant detection of complex variants and splicing in short reads. Bioinformatics 2010;26(7):873–81.

[37] Engstrom PG, et al. Systematic evaluation of spliced alignment programs for RNA-seq data. Nat Methods 2013;10(12):1185–91.

[38] Trapnell C, et al. Transcript assembly and quantification by RNA-Seq reveals unannotated transcripts and isoform switching during cell differentiation. Nat Biotechnol 2010;28(5):511–5.

[39] Steijger T, et al. Assessment of transcript reconstruction methods for RNA-seq. Nat Methods 2013;10(12):1177–84.

[40] Grabherr MG, et al. Full-length transcriptome assembly from RNA-Seq data without a reference genome. Nat Biotechnol 2011;29(7):644–52.

[41] Zerbino DR, Birney E. Velvet: algorithms for de novo short read assembly using de Bruijn graphs. Genome Res 2008;18(5):821–9.

[42] Hu G, et al. Expression and regulation of intergenic long noncoding RNAs during T cell development and differentiation. Nat Immunol 2013;14(11):1190–8.

[43] Zang CZ, et al. A clustering approach for identification of enriched domains from histone modification ChIP-Seq data. Bioinformatics 2009;25(15):1952–8.

[44] Flicek P, et al. Ensembl 2013. Nucleic Acids Res 2013;41(Database issue):D48–55.

[45] Harrow J, et al. GENCODE: the reference human genome annotation for the ENCODE project. Genome Res 2012;22(9):1760–74.

[46] Zhu H, et al. MED: a new non-supervised gene prediction algorithm for bacterial and archaeal genomes. BMC Bioinf 2007;8:97.

[47] Hangauer MJ, Vaughn IW, McManus MT. Pervasive transcription of the human genome produces thousands of previously unidentified long intergenic noncoding RNAs. PLoS Genet 2013;9(6): e1003569.

[48] Lin MF, et al. Revisiting the protein-coding gene catalog of *Drosophila melanogaster* using 12 fly genomes. Genome Res 2007;17(12):1823–36.

[49] Lin MF, Jungreis I, Kellis M. PhyloCSF: a comparative genomics method to distinguish protein-coding and non-coding regions. Bioinformatics 2011;27(13):i275–82.

[50] Wang L, et al. CPAT: Coding-Potential Assessment Tool using an alignment-free logistic regression model. Nucleic Acids Res 2013;41(6):e74.

[51] Sun L, et al. Utilizing sequence intrinsic composition to classify protein-coding and long non-coding transcripts. Nucleic Acids Res 2013;41(17):e166.

[52] Schones DE, Zhao K. Genome-wide approaches to studying chromatin modifications. Nat Rev Genet 2008;9(3):179–91.

[53] Barski A, et al. High-resolution profiling of histone methylations in the human genome. Cell 2007;129(4):823–37.

[54] Wang Z, et al. Combinatorial patterns of histone acetylations and methylations in the human genome. Nat Genet 2008;40(7):897–903.

[55] Wang Z, Schones DE, Zhao K. Characterization of human epigenomes. Curr Opin Genet Dev 2009;19(2):127–34.

[56] Sati S, et al. Genome-wide analysis reveals distinct patterns of epigenetic features in long non-coding RNA loci. Nucleic Acids Res 2012;40(20):10018–31.

[57] Yang JH, et al. ChIPBase: a database for decoding the transcriptional regulation of long non-coding RNA and microRNA genes from ChIP-Seq data. Nucleic Acids Res 2013;41(Database issue):D177–87.

[58] Sheik Mohamed J, et al. Conserved long noncoding RNAs transcriptionally regulated by Oct4 and Nanog modulate pluripotency in mouse embryonic stem cells. RNA 2010;16(2):324–37.

[59] Necsulea A, et al. The evolution of lncRNA repertoires and expression patterns in tetrapods. Nature 2014;505(7485):635–40.

[60] Guo X, et al. Long non-coding RNAs function annotation: a global prediction method based on bi-colored networks. Nucleic Acids Res 2013;41(2):e35.

[61] Huarte M, et al. A large intergenic noncoding RNA induced by p53 mediates global gene repression in the p53 response. Cell 2010;142(3):409–19.

[62] Loewer S, et al. Large intergenic non-coding RNA-RoR modulates reprogramming of human induced pluripotent stem cells. Nat Genet 2010;42(12):1113–7.

[63] Wu AR, et al. Quantitative assessment of single-cell RNA-sequencing methods. Nat Methods 2014;11(1):41–6.

[64] Cong L, et al. Multiplex genome engineering using CRISPR/Cas systems. Science 2013;339(6121): 819–23.

[65] Mali P, et al. RNA-guided human genome engineering via Cas9. Science 2013;339(6121):823–6.

[66] Kong L, et al. CPC: assess the protein-coding potential of transcripts using sequence features and support vector machine. Nucleic Acids Res 2007;36:W345–9.

CHAPTER 11

Regulation of erythroid cell differentiation by transcription factors, chromatin structure alterations, and noncoding RNA

Alex Xiucheng Fan, Mir A. Hossain, Jared Stees, Ekaterina Gavrilova, Jörg Bungert
Department of Biochemistry and Molecular Biology, Center for Epigenetics, Genetics Institute, Powell Gene Therapy Center, College of Medicine, University of Florida, Gainesville, Florida, USA

Contents

1. INTRODUCTION

Erythroid cells differentiate from hematopoietic stem cells (HSCs) that reside within specific niches in the adult bone marrow [1,2]. HSCs are slowly dividing cells that exhibit long–term repopulating activity (LT-HSCs) when transferred to the bone marrow.

Epigenetic Gene Expression and Regulation
http://dx.doi.org/10.1016/B978-0-12-799958-6.00011-1

Within the bone marrow, the HSCs differentiate along various hematopoietic lineages, a process that is regulated mainly by the activity of cell-type-specific transcription factors. The process of erythroid differentiation occurs over multiple defined intermediary cell types that exhibit a decreasing ability to differentiate into other cell types (Figure 1). The process is intrinsically stochastic and determined by the balance of transcription factors that mediate differentiation along one hematopoietic lineage and restrict differentiation along another hematopoietic lineage. This is illustrated by the transcription factors PU.1 and GATA-1/2 (GATA-binding factor) [3]. PU.1 mediates differentiation of HSCs along the lymphoid lineage and inhibits differentiation of myeloid/erythroid cells. GATA family members GATA-1 and GATA-2 have the opposite activity and promote differentiation along the erythroid lineage. In the following, the role of transcription factors, chromatin structure alterations, and noncoding RNA during the process of erythroid differentiation will be described.

2. ERYTHROID CELL DIFFERENTIATION

Upon leaving the niche in the bone marrow, LT-HSCs differentiate into cells that only exhibit short-term repopulating activity (ST-HSCs) [1,2]. ST-HSCs differentiate into either common lymphoid progenitors or common myeloid progenitors (CMPs). The CMPs further differentiate into granulocyte/monocyte precursors or megakaryocyte/erythroid cell precursors (MEPs). MEPs differentiate into blast-forming unit-erythroid cells (BFU-Es) followed by colony-forming unit-erythroid cells (CFU-Es), which then mature into orthochromatic normoblasts and reticulocytes (Figure 1(A)). Enucleation occurs primarily at the endothelial cell stage and leads to the release of mature erythrocytes, or red blood cells (RBCs), into the bloodstream [4]. The differentiation and rapid cycling of erythroid progenitors is regulated by erythropoietin (EPO), which binds to the EPO receptor and initiates a cascade of events that stimulate proliferation and differentiation of erythroid cells [5]. Due to the action of EPO, a single CFU-E yields about 30–50 enucleated erythroid cells within a timeframe of approximately three days [1,2]. The lifetime of RBCs in circulation is about 120 days, after which they are removed by macrophages. Adult humans contain between 10 and 20 trillion circulating RBCs, which represent about one quarter of all eukaryotic cells of the body. Each second, 2.5 million new erythrocytes are released into the bloodstream.

Differentiating erythroid cells become specialized for the transport of O_2 from the lungs to the tissues. The main characteristics of mature erythrocytes are thus a high content of hemoglobin and a flexible cell wall structure that allows smooth rides through tiny blood vessels (Figure 1(B)). Genes encoding for these components are increasingly transcribed during the differentiation of erythroid cells.

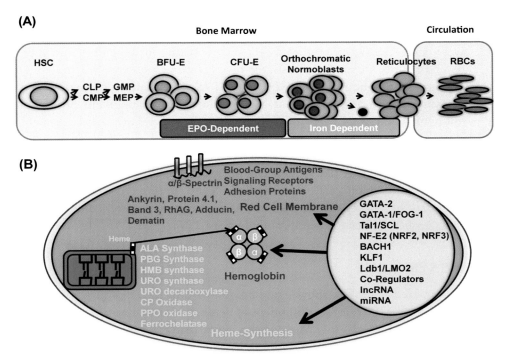

Figure 1 Differentiation of erythroid cells. (A) Erythroid cells differentiate from hematopoietic stem cells (HSC) in the bone marrow. The long-term HSCs successively differentiate into the multipotent progenitors CLP (common lymphoid progenitors) and CMPs (common myeloid progenitors). The CMPs differentiate to GMPs (granulocyte monocyte precursors), or MEPs (megakaryocyte/erythroid precursors). MEPs differentiate into committed BFU-Es (blast-forming unit-erythroid cells), which then give rise to CFU-Es (colony-forming unit-erythroid cells). The CFU-Es mature along various intermediate stages (referred to here as orthochromatic normoblasts). The final maturation stage is the generation of reticulocytes, which then enucleate and are released into the blood stream as red blood cells (RBCs). (B) Main components of the differentiating erythroid cell. Erythroid cells become specialized for the gas transport and express specific cell membrane components that render the cells flexible and provide resistance to shear stress, enzymes for the synthesis of heme (see text for abbreviations), and hemoglobin subunits. The erythroid cell-specific activities are encoded by genes that are regulated by erythroid-specific and ubiquitously expressed transcription factors, chromatin regulators, and noncoding RNA.

3. HEMOGLOBIN

Hemoglobin is a tetramer composed of two α-type and two β-type subunits. The α- and β-type subunits are encoded from clustered genes located on different chromosomes, Chromosome 11 for the β-type *globin* genes and Chromosome 16 for the α-type *globin* genes [6]. Hemoglobin also contains four heme groups harboring a central Fe^{2+}, which interacts via histidine residues within the globin chains. The composition of hemoglobin changes during development [7]. In the early human embryo, until about 6 weeks after conception, the embryonic ε- and ζ-*globin* genes (from the β- and α-*globin* gene locus, respectively) are expressed in erythroid cells differentiating in the embryonic yolk sac. After the end of yolk sac erythropoiesis, the α-*globin* genes are expressed and continue to be expressed throughout adulthood. During fetal liver hematopoiesis, the γ-*globin* genes are expressed and combine with α-chains to form fetal hemoglobin. Around birth, the site of hematopoiesis switches to the bone marrow, which correlates with silencing of the fetal γ-*globin* genes and activation of the adult δ- and β-*globin* genes. The δ-*globin* gene is expressed at levels less than 5% of that of the main β-*globin* gene, due to a point mutation of the TATA-like sequence in the promoter. During adult hematopoiesis, a small fraction of red blood cells, the so-called F cells, continue to express the γ-*globin* genes. In certain hereditary conditions known as hereditary persistence of fetal hemoglobin, the number of F cells is significantly increased [8].

4. HEME SYNTHESIS

Heme is composed of protoporphyrin and iron and synthesized in a variety of different cells types [9]. Heme is a critical component of hemoglobin and reversibly binds oxygen via the positively charged Fe^{2+}. Most of the genes that express enzymes involved in heme biosynthesis are under control of housekeeping as well as erythroid-specific regulatory DNA elements [10]. These regulatory DNA elements recruit transcription factors and associated coregulators that introduce epigenetic changes that are maintained during subsequent cell divisions. In addition, many of the RNAs that are translated into Heme synthesizing enzymes are subject to erythroid-specific alternative splicing.

Heme synthesis begins with the synthesis of δ-aminolevulinic acid (ALA) by the ALA synthase (ALAS) in the inner mitochondrial matrix. In mammalian species, there are two genes encoding for ALAS, ALAS-1 and ALAS-2. ALAS-1 is ubiquitously expressed, while ALAS2 is erythroid-specific [11]. Both ALAS isoforms contain mitochondrial targeting sequences and heme regulatory motifs (HRMs). Heme binds to the HRMs and inhibits mitochondrial import of ALAS. ALA is transported to the outer mitochondria membrane and used for the formation of porphobilinogen (PBG), which is catalyzed by the porphobilinogen synthase (PBGS). PBG is then converted to hydroxymethylbilane (HMB) by the HMB synthase. Uroporphyrinogen (URO) synthase (UROS) and URO decarboxylase (UROD) mediate the formation of coproporphyrinogen (CP) which is transported into the intermembrane space. Two oxidation steps catalyzed by CP oxidase and protoporphyrinogen oxidase (PPOX) lead to the formation of protoporphyrin IX in the inner mitochondria membrane facing the matrix. Ferrochelatase (FECH) carries out the final reaction, which is the insertion of ferrous iron into protoporphyrin, a step shown to be regulated by the mitochondrial ATPase inhibitory factor 1 (Atpif1).

5. RED CELL MEMBRANE

The plasma membrane of RBCs is unique in that it provides specific mechanical features that allow blood cells to flow through tiny blood vessels [12,13]. RBCs have a characteristic discoid shape and excellent elastic properties exhibiting high resistance to fluid stresses. The unique characteristics of RBC membranes are governed by a special structure that is composed of the lipid bilayer and a complex network of associated proteins that are anchored to the membrane by transmembrane proteins. The lipid bilayer is asymmetric with phosphatidylserine and phosphatidylethanolamine, restricted to the inner monolayer, and phosphatidylcholine and sphingomyelin restricted to the outer monolayer. There are more than 50 proteins associated with the RBC membrane. These proteins define the blood group antigens and/or provide transport function, serve as adhesion or signaling receptor proteins, or play structural roles. The membrane skeletal proteins form a complex spectrin–dominated network that is anchored to the lipid bilayer by two macromolecular membrane complexes containing either ankyrin or protein 4.1R [12,13]. The membrane proteins are linked to the bilayer by band 3 (EB3) and RhAG (Rhesus-associated glycoprotein), which interact with ankyrin, or by glycophorin C, Rh (Rhesus protein), XK (X-linked Kx blood group), or Duffy (Duffy blood group factor, ACKR1) through interactions with protein 4.1R. Additional linker proteins are adducin and dematin. The skeletal proteins are composed of α- and β-spectrin, actin, tropomyosin, and tropomodulin, as well as dematin, adducin, and protein 4.1R. The interactions of the spectrin dimer and of spectrin/actin with the protein 4.1R provide crucial mechanical stability and contribute to resistance to shear stress-induced fragmentation of the membrane during circulation [13].

6. REGULATION OF GENE EXPRESSION DURING DIFFERENTIATION OF ERYTHROID CELLS

There are many inherited diseases that are characterized by malfunctions or altered regulation of critical components of the erythroid cell. This is best illustrated by mutations affecting expression or function of hemoglobin, for example, sickle cell anemia and thalassemia [8]. As increased fetal globin gene expression represents a potential therapy of β-hemoglobinopathies, understanding how the globin genes are regulated during development and differentiation has been a major effort in the field over the last four decades. Many of these studies led to fundamental knowledge about basic principles involved in transcription, gene regulation, and the role of chromatin structure [14–18]. In the following, we will review the major players involved in the regulation of transcriptional programs during differentiation of erythroid cells and will provide a more detailed discussion on the current state of globin gene regulation by transcription factors, epigenetic mechanisms, noncoding RNA, and large-scale alterations of chromatin conformations.

6.1 Erythroid transcription factors

There are many transcription factors involved in the regulation of erythroid-specific genes [19]. In general, ubiquitous transcription factors collaborate with tissue-restricted transcription factors to activate and repress transcription of genes during erythroid differentiation. In addition, transcription factors act in conjunction with cofactors that modulate chromatin structure, recruit transcription complexes, or regulate the transcription elongation rate of genes. The main erythroid DNA-binding transcriptional regulators are GATA-1, GATA-2, KLF1 (EKLF, Erythroid Krüpple Like Factor), NF-E2 (Nuclear Factor, Erythroid 2), and related proteins, as well as Tal1/SCL (T-cell acute lymphocytic leukemia/stem cell leukemia). There are many additional ubiquitously expressed transcription factors and coregulators that collaborate with the tissue-specific activities to regulate erythroid transcription programs during development and differentiation.

6.1.1 GATA-1 and GATA-2

GATA factors are zinc finger proteins that play crucial roles during development and differentiation of many cell types and bind to the consensus DNA sequence (A/T) GATA(A/G) (WGATAR) [20]. The only two members of the GATA-family of transcription factors expressed in erythroid cells are GATA-1 and GATA-2. GATA factors are characterized by the presence of two zinc fingers. While each of the zinc fingers can bind the GATA consensus sequence, often only one of the zinc fingers, the C-terminal one, is engaged in DNA binding. However, a recent study demonstrates that the N- and C-terminal zinc fingers of GATA-3 can both bind to DNA [21]. The zinc finger domains interact with the major groove of the DNA, and additional amino acids contact the minor groove to increase the binding affinity. The N-terminal zinc finger of GATA-1 interacts with the coregulator friend of GATA-1 (FOG-1) [22]. Many erythroid genes are regulated by GATA-1 in a FOG-1 dependent manner, including the adult α- and β-globin genes, but there are also a number of loci at which GATA-1 functions in a FOG-1 independent manner, for example, the dematin gene, and at these loci, GATA-1 may engage in DNA interactions utilizing both zinc fingers [23]. How GATA factors select target sites *in vivo* is not completely understood. It is estimated that GATA-1 only occupies 0.07% of all WGATAR motifs in erythroid cells. The accessibility of a target site likely plays a major role in target site selection, but the interaction with other transcription factors or coregulators also modulates GATA-1 binding occupancy. This has been shown for FOG-1 which, even though not a DNA binding protein, appears to assist GATA-1 in selecting specific WGATAR target sites [24]. Furthermore, it was shown that the zinc finger DNA binding domain of KLF3 alone reveals promiscuous binding in the genome [25]. Inclusion of a protein interaction domain that binds to the histone acetyltransferase CBP (CREB-binding protein) restricts binding of KLF3 to specific target sites. Many coregulator proteins have so-called reader domains that recognize specific modifications in histone tails [26].

Thus, interactions with coregulators and other transcription factors may provide the context for gene-specific interactions of GATA factors.

GATA-1 is regarded as a master regulator of erythroid-specific genes, and many erythroid promoters are characterized by the presence of GATA-binding sites, including those encoding transcription factors KLF1, Tal1, GATA-1, and GATA-2. GATA-2 is expressed at high levels in HSCs, CMPs, and MEPs, and its expression gradually declines after the onset of erythroid commitment [19,20,23]. In contrast, GATA-1 is expressed at low levels in stem and progenitor cells, and its expression increases during differentiation, but declines at the late basophilic/orthoblastic stages of erythropoiesis. Many studies have shown that GATA-2 positively regulates both its own gene and that of GATA-1 in progenitor cells. In committed erythroblasts, GATA-1 displaces GATA-2 from the two promoters and suppresses expression of GATA-2, while auto-activating its own gene [27,28]. The displacement of GATA-2 by GATA-1 with a concomitant reversal of GATA factor function (activation versus repression) is known as the GATA-switch and has been documented to operate at several erythroid-specific genes, including *GATA-2*.

GATA factors modulate gene expression, at least in part, by shaping the epigenetic landscape at erythroid-specific gene loci, thereby altering chromatin accessibility. GATA-1 associates with a large number of proteins that exhibit diverse activities [29,30]. These proteins include transcription factors, coregulators, nucleosome remodeling factors, histone modifying activities, and proteins that alter chromatin configurations. Earlier studies have shown that GATA-1 interacts with the histone acetyltransferase CBP/p300 [31]. The acetylation of specific histone tail residues correlates with chromatin accessibility and activity of specific gene loci [26]. Recent genome-wide analysis comparing GATA-1 occupancy with the presence of specific histone modifications revealed that GATA-1 binding sites are associated with increased levels of H4K16ac and H3K27ac [32]. Acetylation of these residues is mediated by CBP/p300 and is often associated with active gene loci. The H3K27ac mark is found more often in distal regulatory elements, and this is consistent with recent observations showing that GATA-1 also binds more often in regulatory elements located far away from the genes it regulates. Dot1L is a DOT1-like histone methyltransferase specific for H3K79, which lies in the core of histone H3 [33]. Previous studies established connections between GATA-1 and Dot1L and, indeed, GATA-1-bound loci reveal increased levels of H3K79me2 [32]. The H3K79me2 modification is associated with transcription elongation [26]. Mutations in Dot1L are associated with defects in yolk sac hematopoiesis [34]. This is in part due to the misregulation of two transcription factors regulating lineage commitment of hematopoietic progenitor cells, GATA-2 and PU.1. Normally, the PU.1 levels are low in erythroid progenitors while GATA-2 levels are high. In Dot1L mutant mice, erythroid progenitors in the yolk sac reveal an increase in the level of PU.1 and a decrease in the levels of GATA-2 [34].

Chromatin accessibility is also regulated by nucleosome remodeling complexes, which shift nucleosomes in an ATP consuming process [35]. Brg1 (Brahma Related Gene 1) contains ATPase activity and is part of an SWI/SNF (switch/sucrose nonfermenting) chromatin-remodeling complex in mammalian cells. A hypomorphic mutation in the active site of Brg1 causes defects in erythropoiesis in mice [36]. Recent studies have shown that GATA-1 is involved in the recruitment of Brg1-containing SWI/SNF complexes to erythroid regulatory DNA elements [37]. This likely leads to the repositioning of nucleosomes at GATA-sites containing promoter and enhancer elements. Enhancers and promoters are often found to be free of nucleosomes (NRF: nucleosome free region) or to be occupied by alternative, unstable nucleosomes containing the histone variants H3.3 and H2A.Z [38].

GATA-1 forms stable protein complexes with the transcription factor Tal1, a helix-loop-helix protein that interacts with the Ebox (CANNTG) and is required for erythroid differentiation [39]. Complex formation is mediated by additional proteins, including LMO2 (lim only 2) and Ldb1. The GATA-1/Tal1 complex binds to composite sites containing both GATA and Ebox motifs separated from each other by 9–11 bp. The composite GATA/Ebox motif is present in many erythroid specific gene-encoding components of hemoglobin or the red blood cell membrane [19,20,23]. The Ldb1 subunit contains a homodimerization domain, and it was shown that Ldb1-containing protein complexes can mediate the interaction between enhancers and promoters [40]. This has been elegantly demonstrated by artificially tethering Ldb1 to the β-*globin* gene promoter using an artificial DNA binding domain [41]. The presence of Ldb1 at the promoter was sufficient for the establishment of interactions with the locus control region, located far upstream of the globin genes, and for activation of β-*globin* gene expression. Ldb1 has been shown to bind a large number of erythroid regulatory elements and is part of protein complexes that bring enhancers and promoters in close proximity [40]. The data suggest that Ldb1 is an architectural protein that contributes to the establishment of chromatin configurations required for erythroid-specific gene expression patterns.

GATA-1 has been shown to interact with the mediator complex [42]. The mediator is a large protein complex involved in transcription activation [43]. The individual subunits interact with different transcription factors or with RNA polymerase II (RNA Pol II), and thus the mediator forms a bridge between transcription activators and the basal transcription machinery. Like Ldb1, many chromatin interaction events are characterized by the presence of the mediator, suggesting that this co-activator may also be involved in mediating interactions between promoters and long-distance control elements. Mutations in mediator subunits cause defects in erythropoiesis and anemia [44].

In addition to interacting with coregulators involved in activation of transcription, GATA-1 also interacts with transcription repressors, including the NuRD (nucleosome remodeling and deacetylase) complex [29,30]. The genome-wide analysis of GATA-1 occupancy revealed three clusters of genes differentially regulated by GATA-1 [32]. The first class contained genes that are expressed in other hematopoietic cell lineages but are repressed in erythroid cells. These repressed genes are also associated with high levels of H3K27me3, a histone mark introduced by the Polycomb repressor complex (PRC). The second cluster contained genes that are expressed at increasing levels during differentiation of erythroid cells and encode erythroid-specific proteins involved in heme synthesis or components of hemoglobin, and the red blood cell membrane associated protein skeleton. The third cluster contained gene-encoding components of housekeeping functions. Expression of these genes is downregulated during erythroid differentiation. This demonstrates that GATA-1 provides important functions, not only for the activation of erythroid-specific genes, but also for the repression or complete silencing of genes not required for the function of the specialized red blood cell (Figure 2).

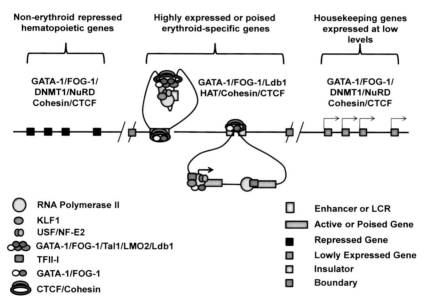

Figure 2 Regulation of erythroid gene expression patterns by tissue-specific and ubiquitously expressed transcription factors. This is a simplified model of gene regulation in erythroid cells and focuses mainly on GATA-1-mediated events. GATA-1/FOG-1 as well as the CTCF and cohesin protein complexes establish and separate from each other accessible and inaccessible domains in the nucleus during erythroid differentiation. In the accessible domains, tissue-specific transcription factors like KLF1, NF-E2, and Tal1 work in conjunction with ubiquitously expressed proteins like USF (upstream stimulatory factor) to regulate accessibility of regulatory DNA elements and to mediate recruitment and activity of transcription complexes. The GATA-1/Tal1/LMO2/Ldb1 complex mediates proximity between enhancer and promoter elements. Stress responsive erythroid genes (e.g., *HIRA*) are occupied by RNA Pol II and TFII-I (transcription factor of RNA Pol II-I). Housekeeping genes that are expressed at low levels in erythroid cells are also occupied by repressive GATA-1 protein complexes.

6.1.2 KLF1

Krüpple–like factor 1 (KLF1, formerly known as EKLF) is a critical regulator of erythroid specific gene transcription [45]. It contains three zinc fingers and binds to the CACCC box. ChIP-sequencing analysis in erythroid cells revealed that KLF1 occupies a large number of genes expressed in erythroid cells. Often, KLF1 binds in close proximity to GATA-1 and Tal1, suggesting that these proteins together orchestrate erythroid transcription programs. The expression and activity of KLF1 is tightly regulated during erythroid differentiation [45]. Sumoylation of KLF1 at lysine 74 allows it to interact with the Mi-2b/HDAC/NuRD (chromodomain helicase DNA binding protein 4/histone deacetylase/nucleosome remodeling and deacetylation complex) repressor complex. Likewise, acetylation of KLF1 at lysine 302 promotes interactions with the Sin3A/HDAC1 (SIN3 transcription regulator family member A/histone deacetylase 1) repressor complex. On the other hand, acetylation of KLF1 at lysine 288 recruits the Brg1/SWI/SNF complex, which at least in part mediates the transcription activation function of KLF1. The phosphorylation of KLF1 by casein kinase II augments the transcriptional activation function.

The interaction with nucleosome remodeling complexes is consistent with the observation that KLF1-deficient mice reveal defects in DNase I hypersensitive site (HS) formation in the β-*globin* gene locus [46]. KLF1 binds to locus control region DNase I hypersensitive sites (LCR HSs) as well as to the adult β-*globin* gene promoter. These sites are no longer accessible in KLF1-deficient mice. These data show that KLF1 plays an important role in establishing epigenetic signatures at erythroid-specific gene loci. KLF1 also interacts with components of the basal transcription machinery including TAF9 (TBP-associated factor 9), a component of the transcription factor TFII-D complex, and the transcription factor TFII-H [46,47]. TFII-D is the primary promoter-binding complex for RNA Pol II [48]. The TFII-D subunit TATA-binding-protein interacts with the TATA box, often located between 25 and 30 bp upstream of the transcription start site of RNA Pol II transcribed genes. The TAF9 subunit has been shown to contact the downstream promoter element (DPE). Importantly, KLF1-mediated transcription of the DPE-containing adult β-*globin* gene is TAF9 dependent, while that of the DPE-deficient *AHSP* (alpha hemoglobin stabilizing protein) gene occurs independent from TAF9. TFII-H contains helicase, ATPase, and kinase activities [49]. The helicase activity is important for the unwinding of the DNA double-strand at the promoter, and the kinase activity mediates phosphorylation of serine 5 of the RNA Pol II C-terminal domain (CTD). Promoter opening and CTD serine 5 phosphorylation are critical for transcription initiation. The interactions with these basal transcription factors illustrates that KLF1 may play an important role in the recruitment and activity of transcription initiation complexes.

KLF1 is critical for the expression of a large number of erythroid genes. It was shown that KLF1 regulated genes often colocalize within the nucleus and that KLF1 is required for the association of coregulated genes [45,50]. It appears that at least some genes specifically expressed in erythroid cells are transcribed within special nuclear domains enriched for transcription complexes. These nuclear domains are called transcription factories [51]. KLF1 may be involved in the establishment of erythroid-specific transcription factories or may recruit genes to these domains.

The switch in expression from the fetal to the adult β-*globin* gene is regulated by complex mechanisms that involve a large number of proteins and coregulators. KLF1 plays a prominent role in this process. KLF1 interacts with the adult β-*globin* gene promoter and recruits chromatin remodeling complexes and components of the basal transcription apparatus. Furthermore, KLF1 also interacts with and activates the promoter of the gene encoding for the repressor protein BCL11A (B-cell CLL/lymphoma 11A) [52,53]. BLC11A is directly or indirectly involved in the silencing of the fetal γ-*globin* genes [6]. Thus, KLF1 directly activates β-*globin* gene expression and indirectly represses expression of the fetal globin genes.

Several KLF1-related proteins have been identified to play a role in globin gene regulation and erythroid differentiation. KLF2 binds to the embryonic globin gene promoters and activates transcription. In contrast, KLF3 and KLF8 function together to repress γ-*globin* gene expression [54].

6.1.3 *Tal1*

Tal1, also known as SCL factor, is a helix-loop-helix protein that binds to DNA as a heterodimer with members of the E-family of transcription factor proteins (E2A, HEB, and E2-2) [55,56]. Deficiency of Tal1 causes early embryonic lethality in mice at 9.5 days postcoitum (dpc) due to impaired yolk sac hematopoiesis [57]. Tal1 is required for early commitment of hematopoietic progenitors and also plays an important role during differentiation of erythroid cells and megakaryocytes [55,56]. Tal1, like many other transcription factors, functions as both repressor and activator of transcription. The opposite activities are mediated by Tal1-interacting coregulators. Tal1 interacts with the histone deacetylases 1 and 2 (HDAC1 and HDAC2), with the histone demethylase LSD1 (lysine (K)-specific demethylase 1A), which removes methyl groups from H3K4, as well as with ETO-2 and related proteins [55,56]. ETO-2 interacts with HDACs and the transcriptional mSin3A corepressor and is expressed at high levels in undifferentiated erythroid cells [58]. Activating Tal1 associates with p300/CBP, P/CAF (p300/CBP-associated chromatin assembly factor), and Brg1 containing chromatin remodeling complexes.

Genome-wide ChIP-Seq studies demonstrated that Tal1 interacts with many regulatory DNA elements in undifferentiated erythroid cells, including erythroid-specific promoters and enhancers that are not yet active [59]. It thus seems that Tal1 binds to poised regulatory elements in erythroid progenitor cells together with corepressing activities. At later differentiation stages, Tal1 binds to the same regulatory elements but functions as an activator. The switch in Tal1 activity during erythroid differentiation is likely driven by P/CAF1-mediated acetylation of Tal1, which causes dissociation of corepressor complexes, and by decreased expression of ETO-2 during differentiation [55,56,58]. Furthermore, during differentiation, Tal1 becomes part of a large DNA protein complex containing GATA-1 and the cofactors Ldb1 and LMO2, as mentioned before [39].

Tal1-deficient mice rescued with a DNA-binding defective Tal1 mutant survived beyond 9.5 dpc, showing that Tal1 functions, at least at a subset gene, in a DNA-binding independent manner [60]. For example, Tal1 associates with regulatory elements driving expression of transcription factors in erythroid cells, for example, GATA-1, LMO2, Ldb1, E2A, Bcl11A, NF-E2, BACH1 (BTB and CNC homolog 1), and KLF1. Consistently, the genome-wide binding studies revealed that Tal1 peaks are not only enriched for the preferred CAGCTG Ebox but are also frequently associated with CACCC and CTGCCA/TGNNG sequences, potential binding sites for KLF1 and GATA-1, respectively [55,56,59].

The data demonstrate that Tal1 is a critical regulator of erythropoiesis and functions in the context of other transcription factors, particularly GATA-1 and KLF1, to regulate expression of genes. Its association with the histone acetyltransferases p300/CBP and P/CAF1, as well as with HDACs, suggests that one of the main functions of Tal1 is to regulate histone modifications at erythroid-specific enhancer and promoter elements. In addition, through the recruitment of Ldb1, together with GATA-1, Tal1 is involved in bringing enhancers and promoters in close proximity and in stimulating transcription elongation.

6.1.4 NF-E2

Transcription factor NF-E2 is a heterodimer consisting of the erythroid-specific subunit p45 and broadly expressed small Maf proteins (p18, MafF, MafG, and MafK) [61]. The p45 subunit is expressed during erythropoiesis and megakaryopoiesis. There are two NF-E2-related factors (NRF1 and NRF2) expressed in erythroid cells. All three proteins belong to the cap'n'collar b-zip family of transcription factors, which contain a leucine zipper flanked by a basic DNA binding domain. The small Maf proteins also interact with the BTB (broad complex tramtrack bric a brac)-domain containing proteins BACH1 and BACH2 [61]. NF-E2 and related proteins, as well as BACH family proteins, bind to a subset of Maf recognition elements (MARE) with the sequence 5'-TGACTCA-3' [61].

The small Maf (p18) subunits only harbor DNA binding and protein:protein interaction motifs but no effector domains [61]. Small Maf homodimers can thus function as repressors of transcription by competing with the activating heterodimers. NF-E2 and related factors contain activation domains with which these proteins interact with coregulators or components of the basal transcription apparatus [61]. For example, p45 and MafK have been shown to interact with TAF-4 (formerly known as TAF(II)-130) and RNA Pol II, respectively [62,63]. It appears that just like other transcription factors, MafK and interacting partners can function as activators or repressors of transcription.

A proteomic analysis of MafK during erythroid cell differentiation revealed that it associates with Bach1 and corepressor complexes, including the NuRD/Mi2 complex and Brg1-containing chromatin remodeling complexes in undifferentiated cells [63]. In differentiated erythroid cells, MafK associates with p45 or related factors and interacts with coactivating factors like CBP and the methyltransferase MLL (mixed lineage leukemia). Thus, the same MARE sequences can recruit repressing or activating protein complexes during differentiation of erythroid cells. This has been elegantly demonstrated in the β-*globin* gene locus. In undifferentiated erythroid cells, BACH1 binds a tandem MARE sequence in locus control region HS2 [64]. This likely contributes to keeping the LCR in an accessible but inactive (poised) configuration in progenitor cells. During differentiation, heme dissociates the BACH1 complex from the DNA, thus allowing MafK/NF-E2 heterodimers to interact and recruit coactivating activities.

At least for the β-*globin* locus, it has been shown that NF-E2 is involved in the transfer of RNA Pol II from the LCR to the adult β-*globin* gene promoter [65]. Furthermore, data show that NF-E2 interacts with transcription factor USF2 (upstream stimulatory factor 2) and that both proteins are required for the recruitment of transcription complexes to the β-*globin* gene locus [66]. The genome-wide analysis of USF2 and NF-E2 DNase I footprints in erythroid cells suggest that NF-E2 directly binds to distal regulatory elements and indirectly to proximal promoter regions, likely mediated by USF2 [67]. In contrast, USF2 interacts directly with promoter proximal sequences and indirectly with distal regulatory elements, which appears to be mediated by NF-E2. The data suggest that USF2 and NF-E2 cooperate in the recruitment of transcription complexes to the β-*globin* gene locus and perhaps to other erythroid specific genes.

Mice deficient for p45 do not reveal significant defects in erythropoiesis but exhibit a dramatic reduction in platelets, showing that p45 expression is more critical during megakaryopoiesis [68]. The absence of a strong erythroid phenotype suggests that NF-E2-related proteins compensate for the lack of p45 in erythroid cells [61]. This is supported by observations showing that in the absence of p45, there is an increased association of NRF2 with the β-*globin* gene locus [69].

6.1.5 BCL11A

The transcription factor BCL11A (B-cell lymphoma/leukemia 11A) recently emerged as an important regulator of globin gene expression [70]. BCL11A is a zinc finger protein and critically involved in the generation of B-cell progenitor cells. Genome-wide association studies aimed at identifying modifiers of β-type *globin* gene regulation identified single nucleotide polymorphisms (SNPs) within or proximal to the *BCL11A* gene [71]. These SNPS are associated with elevated expression of the γ-*globin* genes, suggesting that BCL11A is a repressor of the fetal globin genes. Indeed, subsequent studies in cell culture and transgenic mice demonstrated that BCL11A is involved in the silencing of the fetal *globin* genes [72]. So far, it is not clear if BCL11A directly binds to sequences in the *β-globin* gene locus, if it is recruited to the globin locus indirectly via other DNA-binding transcription factors, or if it acts indirectly via regulation of a modulator of fetal *globin* gene expression. Some data suggest that BCL11A is recruited to an intergenic region between the Aγ- and δ-*globin* genes as part of a large protein complex that contains GATA-1, Sox2 (sex determining region Y-Box 2), and corepressors [70].

6.1.6 Other transcription factors

There are a number of other transcription factors that are implicated in the process of erythropoiesis. Transcription factor c-Myc (cellular myelocytomatosis) is a helix-loop-helix (HLH) protein that stimulates proliferation by activating a large number of genes, including those transcribed by RNA Pol I, II, and III [73]. In embryonic stem cells, c-Myc regulates transcription, mostly at the level of elongation, by recruiting the transcription elongation factor pTEFB, which phosphorylates serine 2 of the RNA Pol II CTD [74]. c-Myc contains a leucine-zipper and a helix-loop-helix domain and is structurally related to USF, a dimer composed of USF1 and USF2. c-Myc normally associates with another HLH protein called Max and activates transcription by binding to Ebox elements [73]. c-Myc stimulates the proliferation of erythroid progenitor cells and inhibits the differentiation of these cells [56]. c-Myc negatively regulates expression of transcription factor genes, including *GATA-1, NF-E2,* and *LMO2*, and represses expression of p27, an inhibitor of cyclin-dependent kinases. Expression of c-Myc declines during erythroid differentiation [56].

USF is a heterodimer composed of USF1 and USF2 [56]. As mentioned before, previous studies have shown that USF interacts with NF-E2 and regulates recruitment of transcription complexes in the *β-globin* gene locus [66]. However, USF appears to be required at an early step during erythroid cell specification. Mice expressing a dominant negative USF protein exhibit severe anemia and reduced expression of all key erythroid transcription factors [56]. Furthermore, it was shown that USF recruits Set1A (SET domain containing 1A, where SET is an acronym for a domain containing Su(var)3–9, Enhancer-of-zeste (EHZ), and trithorax (Trx) or its homologs), an H3K4-specific methyltransferase to the *Hoxb4* (*Homeobox B4*) gene [75]. USF-mediated activation of HoxB4 is required for erythroid cell differentiation. USF, which is ubiquitously expressed, operates in conjunction with lineage-restricted activities to regulate erythroid differentiation.

Transcription factor c-Myb belongs to the myeloblastosis family of transcription factors [76]. It contains DNA binding as well as activation and repression domains. c-Myb is an important regulator of hematopoiesis and plays a positive role in regulating the commitment to the erythroid lineage, possibly through activating expression of KLF1 and LMO2 [77]. c-Myb has also been implicated in the repression of the *γ-globin* genes in adult erythroid cells. Genome-wide association studies identified mutations in an intergenic region between the *c-Myb* and the *HBS1L* (HBS1-like translational GTPase) gene that are linked to increased expression of the fetal globin genes [6].

Additional proteins are involved in mediating the switch from fetal to adult *globin* gene expression during development. The DRED (direct repeat erythroid definitive) protein complex consists of the orphan nuclear receptors TR2 (nuclear receptor subfamily 2, group C, member 1) and TR4 (nuclear receptor subfamily 2, group C, member 2) DNA-binding subunits and corepressors [78]. This complex is recruited to direct repeat elements located in the embryonic and fetal β-type *globin* genes and represses expression of these genes in adult definitive cells. In contrast, transcription factor IKAROS (IKAROS family zinc finger 1) operates in conjunction with GATA-1 to mediate recruitment of Brg1-containing nucleosome remodeling complexes and transcription elongation factors to the γ-*globin* gene promoters [79]. IKAROS has previously been identified as a component of the Pyr-complex, a large protein complex that includes a long polypyrimidine-rich region recognition binding domain and HDAC, and has chromatin remodeling activity. IKAROS and the Pyr-complex have been implicated in the silencing of γ-*globin* transcription and are thus shown to contribute to the switch in globin gene expression during development [80]. These data thus show that during development, IKAROS functions as an epigenetic regulator and that it switches from activator to repressor with respect to γ-*globin* gene expression. It is unknown what mediates this switch in activity.

6.2 Noncoding RNA
6.2.1 MicroRNAs (miRNAs)
Noncoding RNAs play important roles during cell differentiation and the response to extracellular stimuli. miRNAs are relatively short RNAs that inhibit gene expression by decreasing the stability of or translation from target RNAs [81]. They interact with Argonaute and are incorporated into the RNA-induced silencing complex, which they recruit to target RNAs depending on complementarity to the miRNA sequence. miR-NAS are usually transcribed by RNA Pol II as longer precursors (pre-miRNA), often containing multiple miRNAs, and are processed by RNAse type III endonucleases Drosha or Dicer.

Several erythroid-specific miRNA genes are regulated by GATA-1, and they target mRNAs that encode important activities for erythropoiesis [82,83]. miRNAs-144/451 are transcribed from the same locus and transcription is regulated by GATA-1. miRNA 451 deficiency in zebrafish causes defects in erythropoiesis. A potential target of miRNA 451 is the GATA-2 mRNA and indeed, knockdown of GATA-2 expression rescued the defect in erythropoiesis caused by lack of miRNA 451 [84]. Loss of the miRNA 144/451 gene locus in mice caused mild anemia and increased sensitivity to oxidative stress [85]. This is mediated by increased expression of 14-3-3ζ, which inhibits nuclear localization of FOXO3 (Forkhead Box O3). FOXO3 is an activator of genes expressing important antioxidant proteins [82,83].

Some erythroid-specific miRNAs target mRNAs encoding erythroid cell surface receptors, like miRNAs *221/222* targeting c-kit and miRNA *24* targeting the *activin type I receptor* mRNA. The activin type I receptor positively regulates erythroid differentiation through phosphorylation of the coregulator SMAD2 (SMAD family member 2) and downstream events, leading to production of hemoglobin. Thus, miRNA *24* is a negative regulator of erythropoiesis. Many of the miRNAs regulating erythropoiesis act on progenitor cells and are downregulated during erythroid differentiation [82,83].

Expression of transcription factor c-Myb is regulated by miRNA 15a, which targets the c-Myb 3′UTR [82,83]. c-Myb binds to the promoter of the *miRNA 15a* gene and positively activates transcription, suggesting an autoregulatory loop. The miRNA *15* and *miRNA 16* genes are located on chromosome 13 and upregulated in human trisomy 13, which is associated with reduced expression of c-Myb and elevated levels of *γ-globin* gene expression further supporting an important role of c-Myb in fetal globin gene expression. miRNA *150* also targets c-Myb and is involved in regulating the megakaryocyte versus erythrocyte lineage differentiation. miRNA *223* targets LMO2, which is a component of the GATA-1/Tal1/Ldb1-containing protein complex that positively regulates a large number of erythroid-specific genes [82,83].

Terminal erythroid differentiation is accompanied by a dramatic condensation of chromatin and subsequent enucleation [4]. This process requires the downregulation of the histone acetyl-transferase GCN5 (SAGA complex histone acetyltransferase catalytic subunit Gnc5) that is activated by c-Myc. Mxi, a c-Myc antagonist, is required for the downregulation of GCN5. miRNA *191* inhibits nuclear condensation by preventing the downregulation of GCN5. Mxi1, as well as Riok3 (RIO kinase 3), another protein involved in nuclear condensation, has been identified as targets of miRNA 191 [82,83]. Finally, miRNA 96 was shown to directly bind to the γ-*globin* mRNA and to decrease its expression. Curiously, while most miRNA binding sites are in the 3′UTR or in other noncoding parts of the RNA, miRNA *96* interacts with the open reading frame (ORF) of the γ-*globin* mRNA [82,83].

6.2.2 Long noncoding RNA (lncRNA)

Mammalian genomes are extensively transcribed and generate a large number of noncoding RNAs lacking ORFs [86]. Noncoding RNAs larger than 200 nucleotides and either containing or lacking poly-A tails are referred to as long noncoding RNA, or lncRNA. The lncRNAs are a heterogeneous group of RNAs that exert different functions and include antisense RNA (asRNA), enhancer RNA (eRNA), or long intergenic noncoding RNA (lincRNA). Antisense transcription affects expression of the sense RNA in different ways. The process of antisense transcription can lead to transcription interference and can block either transcription initiation or elongation of the sense strand. Alternatively, the antisense transcript can interact with the sense transcript and decrease its stability.

It has been known for a long time that specific long distance regulatory elements produce lnc- or lincRNAs. One of the first examples was the *β-globin* gene locus. The Tuan laboratory found long transcripts originating from within LCR HS2 [87]. Subsequent studies confirmed that HS2 and other LCR HSs harbor promoter activity [88]. Generation of LCR transcripts appeared to precede activation of the globin genes during erythroid differentiation. It is currently not known if the LCR associated transcripts play a role in globin gene activation or repression. It is possible that it is the act of transcription and not the transcripts that contributes to activation of the globin gene locus. Either transcription of HS sites and flanking DNA maintains a dynamic open chromatin structure over the LCR or the transcription toward the genes and promotes interactions between the LCR and the globin genes, a process termed facilitated tracking [87].

In other gene loci, eRNAs have been shown to regulate genes by mediating chromatin accessibility at promoter regions and subsequent recruitment of mediator and transcription complexes [86]. Recent genome-wide analysis of RNA sequencing data demonstrated that a large fraction of enhancer elements are transcribed bidirectionally, leading to formation of eRNAs [89]. The eRNAs are produced as efficiently as mRNAs but exhibit different processing features that lead to low abundance, due to instability or technical limitations in detecting eRNAs [86]. Enhancers are characterized by distinct histone posttranslational modifications and proteins that bind to them [90]. Enhancers are typically hypomethylated and contain flanking nucleosomes that are acetylated at H3K27 and monomethylated at H3K4. They are also marked by the presence of the coregulator p300, which is known to acetylate H3K27. However, in addition to these distinct epigenetic marks, the most predictive indicator for active enhancers is the presence of eRNA [86].

lncRNAs can act locally in close proximity to the site of synthesis or distally in *cis* or *trans* [86]. Not much is known about how lncRNAs are targeted to genomic loci. Using novel techniques to map the genome-wide interaction of lncRNAs, e.g., ChIRP (chromatin isolation by RNA purification), it was shown that some lncRNAs bind to hundreds or thousands of genomic loci [86]. While not all of the binding events may be functional, it appears that individual lncRNAs regulate expression of multiple genes. This is illustrated by *PCGEM1* (prostate-specific transcript 1), which interacts with the androgen receptor (AR) and occupies AR-responsive enhancers [91]. More than 2000 binding sites were identified, and it was shown that PCGEM1 stimulates AR-mediated transcription activation. Some lncRNAs may be targeted to DNA by association with DNA binding transcription factors, as may be the case for *PCGEM1*. Other lncRNAs bind directly to DNA, either in the context of a triple helix, or immediately at the site of synthesis with the template DNA strand, as has been shown for the lncRNA *Mistral* [86]. It is also possible that lncRNAs engage in RNA/RNA interactions, and this process could aide in bringing enhancer and promoter elements in close proximity [86]. There is evidence suggesting that lncRNAs localize to specific foci in the nucleus, and the folding of the chromatin and colocalization of active gene loci may facilitate the transfer of lncRNAs from one locus to another over relative short distances within these foci.

lncRNAs exhibit diverse functions and can both activate and repress transcription. *HOTAIR* (HOX antisense intergenic RNA) is transcribed from within the *HoxC* gene locus but represses transcription of genes in the *HoxD* gene locus [92]. It associates with PRC2 and binds to extended chromosomal domains. PRC2 is a co-repressor that contains an H3K27-specific histone methyltransferase [26]. H3K27 methylation is associated with repressed chromatin domains [26]. The phosphorylation of a PRC2 subunit facilitates interactions with *HOTAIR* [93]. *HOTAIR* also interacts with other proteins involved in silencing, suggesting that it may serve as a scaffold that assembles repressive protein complexes and targets them to specific gene loci [86,93]. *Mistral* recruits the MLL1 (lysine-specific methyltransferase 2A) histone methyltransferase complex to specific sites in the genome [86]. As MLL1 specifically methylates H3K4, *Mistral* acts as a positive regulator of transcription. The methylation of H3K4 is recognized by a subunit of TFIID, suggesting that *Mistral* is indirectly involved in the recruitment of transcription complexes [86]. It should also be mentioned that lncRNAs not always have a functional role. In the human growth hormone (hGH) gene locus, a lncRNA generated from a distal regulatory element was shown not to be involved in the regulation of transcription in this gene locus in transgenic mice [94]. Rather, the process of transcription through the locus altered the chromatin structure and activated transcription. As mentioned before, a similar mechanism of transcription-driven enhancement by a distal regulatory element has been proposed for the human β-*globin* gene locus [87].

Several lincRNAs have been shown to be involved in the regulation of erythropoiesis. The β-*globin* locus associated lincRNAs have already been discussed. It is currently unknown if any of the globin associated lncRNAs play a functional role in *globin* gene regulation or if the process of transcription that generates these RNAs is involved in modulating expression patterns in the β-*globin* gene locus. The Lodish laboratory recently identified more than 600 lncRNAs in murine fetal liver erythroid cells using RNA sequencing [95]. More than 100 of these lncRNAs have previously not been identified and appear to be erythroid-specific. Indeed, among those RNAs were multiple that were shown to be regulated by erythroid-specific transcription factors, including GATA-1, Tal1, and/or KLF1. One such lncRNA, *alncRNA-EC7*, was generated from a locus occupied by GATA-1, Tal1, KLF1, RNA Pol II, and enriched for H3K4 monomethylation and H3K27 acetylation, modifications typically found in the enhancer [95]. Transcription within the enhancer occurred bidirectionally, and the RNA was found to be spliced and polyadenylated. Knockdown experiments showed that *alncRNA-EC7* was required for high-level expression of the *BAND 3* gene, which is located 10 kb upstream of the lncRNA locus.

6.3 Chromatin structure alteration

During the differentiation of erythroid cells, the chromatin becomes gradually more condensed. The chromatin condensation represses all genes not required for the function of erythroid cells [20,23,32]. It appears that the transcription factor GATA-1, through association with chromatin modifying activities, contributes to the repression of large chromatin domains. At the same time, GATA-1 also maintains and increases expression of erythroid-specific genes together with erythroid transcription factors, including Tal1, KLF1, and NF-E2 [20,23]. Active and inactive chromatin are separated from each other by boundary elements, which prevent the spread of heterochromatin formation into active domains [96]. Active chromatin domains are often flanked by sites that recruit cohesin and other boundary proteins. Within active domains insulators restrict the actions of enhancers on particular target genes [97]. While boundaries and insulators have been well characterized in *Drosophila*, our knowledge about the proteins that mediate these activities in mammalian cells is limited. The most prominent insulator binding protein studied in mammalian cells is CTCF, a CCCTC-DNA binding protein containing 11 zinc fingers [98]. CTCF often acts in conjunction with cohesin to mediate large chromatin loops as determined by the capture chromatin conformation (3C) method or related assays. It appears that within these larger looped domains, enhancers can operate on target genes by coming in close proximity, which may be mediated by proteins binding to enhancers and promoters.

Interactions between enhancers or LCRs and target genes have been documented for a number of gene loci in erythroid cells. For example, in the *c-kit* gene locus GATA-factors bind to regulatory elements and mediate interactions between distal regulatory elements and the promoter [99]. *c-kit* encodes a receptor tyrosine kinase required for hematopoietic differentiation. It is expressed in erythroid progenitor cells and down-regulated during differentiation of erythroid cells. In progenitor cells, GATA-2 binds to a distal element located −114 kb upstream of the transcription start site and mediates interactions between this enhancer and the promoter. Increased expression of GATA-1, which inhibits expression of GATA-2 as mentioned earlier, disrupts enhancer promoter interactions. Instead, loops are formed between the promoter and downstream GATA-1 binding sites, which correlates with repression of *c-kit* gene expression [99].

Interactions between long-distance regulatory elements and gene promoters have also been found in the *α-* and *β-globin* gene loci [100]. The LCR comes in close proximity to the active genes during development and differentiation. Furthermore, other erythroid-specific genes also associate with the LCR in differentiated cells. This suggests that the LCR may form an erythroid-specific transcription factory into which erythroid genes are recruited to, or it colocalizes with other genes to these nuclear domains. Several transcription factors have been shown to be required for the establishment of proximity between the LCR and the *globin* genes, including GATA-1, KLF1, Tal1, FOG-1, and Brg1 [100]. In erythroid progenitor cells, the LCR interacts with other distant HSs, flanking the globin gene locus to form a poised chromatin configuration that has been referred to as a chromatin hub [101]. In differentiated cells, the active genes interact with the LCR complex to form an active chromatin hub. The insulator protein and cohesin are part of the loops that form in the poised and active configurations [100].

The chromatin loops in the β-*globin* locus are erythroid-specific, suggesting that they are mediated by activities only expressed in erythroid cells. The role of CTCF and cohesin is likely to strengthen and maintain interactions between regulatory elements in the globin gene locus.

The role of DNA methylation in the regulation of erythroid gene expression has not been extensively investigated. Recent studies demonstrated that, surprisingly, the overall DNA methylation levels decrease during erythroid differentiation [102]. Furthermore, DNA hydroxymethylation patterns are dynamically regulated during erythroid differentiation, suggesting an important role for DNA demethylation during erythroid differentiation. Recent studies suggest that TET (ten-eleven translocation) methylcytosine deoxygenase 2 (TET2) is important for erythroid differentiation [103]. Mutations in the *TET2* gene are associated with chronic myelomonocytic leukemia and defects in erythroid differentiation in humans and lead to anemic phenotypes in mice [104]. Hydroxymethylation was shown to increase or to be maintained at regulatory DNA elements bound by erythroid-specific transcription factors [104]. This suggests that the hydroxymethylation and subsequent demethylation of DNA in erythroid-specific regulatory elements, or those elements that remain active during erythroid differentiation, is an important step in establishing erythroid-specific expression patterns.

Chromatin structure alterations in erythroid cells play important roles in establishing gene expression patterns. In addition, during terminal erythroid differentiation, the nuclear chromatin becomes extremely condensed, which is required for the process of enucleation [4]. Among the factors that mediate the overall condensation of erythroid nuclear chromatin are HDACs and inhibitors of HATs (histone acetyltransferases).

6.4 Gene regulation during erythropoiesis

The gene regulatory mechanisms operating during differentiation of erythroid cells are not unique and likely resemble scenarios occurring during differentiation of other cell types. One notable exception is the extreme condensation of chromatin during terminal erythroid maturation [4]. The differentiation of erythroid cells from multipotent progenitor cells is regulated by transcription factors, noncoding RNA, and post-transcriptional mechanisms. This appears to be an intrinsically stochastic process, which is mediated by the balance of transcription factors that drive lineage determination as illustrated by the contrasting and cross-regulatory activities of PU.1 and GATA transcription factors [3]. Many recent reports highlight the importance of pioneer transcription factors during cellular differentiation [105]. These are DNA-binding proteins that occupy regulatory DNA elements before they are engaged in activating transcription. Pioneer transcription factors associate with chromatin modifying activities and bind regulatory DNA elements to establish or maintain accessibility during cellular differentiation. GATA-2 and Tal1 are among the earliest erythroid transcription factors expressed during differentiation [23,56]. These proteins bind to many regulatory elements in the genome and may mark regulatory elements that become activated during subsequent differentiation stages. At later stages, GATA-1 is increasingly expressed and displaces GATA-2 from many genomic locations [20,23]. It appears that one function of GATA-1 is to establish topological chromatin domains in which repressed chromatin is separated from active chromatin [32]. CTCF and cohesins likely assist GATA-1 and other proteins in establishing and separating from each other active chromatin domains. Erythroid-specific active chromatin domains appear to colocalize in the nucleus, which may be mediated by KLF1 and other proteins and/or noncoding RNA [50]. During the differentiation of erythroid cells, genes encoding important functional components of the red blood cell are expressed at increased levels, likely mediated by erythroid-specific transcription factors and enhancer elements. Expression of housekeeping genes is maintained but reduced. All genes that encode functions not required for the red blood cells are turned off and packed into inaccessible heterochromatin. Multipotent progenitors have a more plastic chromatin structure that allows activation of genes involved in different cell lineages [106]. Once the cell commits to a particular lineage gene, encoding functions for other cell lineages are silenced.

The *β-globin* gene locus has been the subject of intense studies for more than 30 years [107]. Fundamental principles involved in the regulation of cell- and developmental-stage-specific gene expression patterns have first been discovered studying globin genes. The discovery of altered chromatin configurations in the globin loci in embryonic versus adult erythroid cells by Groudine and Weintraub was a hallmark in our current understanding of gene regulation and the impact of chromatin structure [108]. LCR HSs are accessible and bound by proteins in erythroid progenitor cells that do not express the globin genes (Figure 3) [109,110]. This suggests that the LCR is the primary site in the globin gene locus from which activation occurs during erythroid differentiation. The LCR contains a large number of binding sites for ubiquitously expressed and erythroid-specific transcription factors. It is possible that erythroid-specific transcription factors, like GATA-2, together with ubiquitous DNA binding proteins, like USF, keep the LCR HSs in a dynamic semi-open configuration. The recruitment of RNA Pol II and nongenic transcription likely contributes to maintaining dynamic accessibility during several, differentiation-driving cell division cycles. The increased expression of erythroid transcription factors like GATA-1, KLF1, and NF-E2 (or related proteins) would lead to the formation of stable enhancer complexes at the LCR. The increased binding of multiple proteins with different functions leads to the enrichment of activities involved in transcription activation (chromatin remodeling and modifying complexes, mediator, basal transcription machinery, elongation factors) and perhaps the transfer of these activities to the promoters once proximity is established (Figure 3) [111]. Alternatively, proximity between the LCR and the promoters drives efficient recruitment of elongation competent transcription complexes. Promoters and enhancers are distinguished by different histone modifications. Enhancer-flanked nucleosomes are enriched for H3K27ac and H3K4me, while promoter proximal

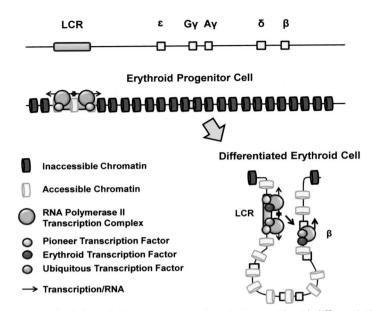

Figure 3 Regulation of adult β-*globin* gene expression during erythroid differentiation. Simplified model of LCR-mediated regulation of globin gene expression. In erythroid progenitor cells, only the LCR HSs are in an accessible configuration. This may be mediated by pioneer transcription factors that act in conjunction with ubiquitously expressed proteins. Transcription of the LCR HS sites maintains an open structure. During differentiation, increased expression of erythroid transcription factors leads to formation of stable LCR HS protein complexes. Interaction of these complexes with the adult *globin* gene promoter leads to efficient recruitment and elongation of transcription complexes.

nucleosomes are enriched for H3K4me3. Different histone modifying enzymes are involved in establishing these epigenetic marks. MLL2 (lysine-specific methyltransferase 2D) trimethylates H3K4 and interacts with NF-E2 [112]. It is thus possible that NF-E2 transfers MLL2 to the promoter, which then modifies the histones. H3K4me3-modified histones interact with the TFIID complex and are thus implicated in the recruitment of transcription complexes [26].

Several stress response genes that encode proteins playing important roles during erythropoiesis, like the heme-regulated eIF2α kinase (HRI) and ATF3 (activating transcription factor 3), harbor paused RNA Pol II at the promoters [113]. Transcription factor TFII-I was found to occupy sites immediately downstream of paused RNA Pol II at these genes and inhibition of TFII-I causes a reduction in the recruitment of elongin A to the ATF3 gene and a failure to activate the gene in response to cellular stress [113]. This suggests that TFII-I plays an important role in regulating the release of paused RNA Pol II from stress-inducible genes in erythroid cells (Figure 2).

The developmental regulation of the *globin* genes is regulated by stage-specific activities that mediate activation and repression of transcription at the embryonic,

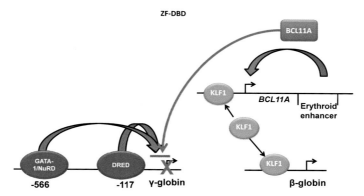

Figure 4 Repression of γ-*globin* gene transcription in adult erythroid cells. Shown is a simplified model of γ-*globin* gene regulation. The γ-*globin* gene promoters contain multiple binding sites for repressor protein complexes. Only two are shown in the graph, a binding site for GATA-1, located 522 upstream of the transcription start site, and a binding site for the DRED complex, located 117 bp from the start site of transcription. In addition, the β-*globin* genes are repressed by BCL11A ,which is positively regulated by KLF1, an activator of adult β-*globin* gene expression, and an intronic erythroid-specific enhancer.

fetal, and adult stages (Figure 4). Much attention has been focused on the regulation of the fetal γ-*globin* genes, as higher-level expression of these genes ameliorates severe phenotypes associated with sickle cell disease and thalassemias [8]. As mentioned before, KLF1 is an important stage-specific transcription factor that directly activates the adult genes and indirectly, via BLC11A, represses the fetal genes [45,52,53]. In addition, a number of proteins have been shown to bind to the γ-*globin* promoter region and to repress transcription of the fetal genes, including the DRED complex, IKAROS, and GATA-1 [20,78,80]. In addition, BCL11A and SOX6 (SRY (sex determining region Y)-box 6) are involved in the silencing of the embryonic ε- and the fetal γ-*globin* genes [70]. These two proteins likely function in the context of a larger protein complex that also involves the corepressor complexes NuRD and GATA-1 [70].

7. CONCLUSIONS

The process of erythropoiesis has been extensively studied. Many fundamental processes involved in the regulation of tissue- and developmental stage-specific alterations in chromatin structure and gene expression have first been described for gene loci expressed in erythroid cells. Recent advances have shown that a complex interplay between cell–type restricted transcription factors, ubiquitously expressed transcription factors, coregulators, and noncoding RNA shape the epigenetic landscape and govern erythroid-specific expression programs. How lncRNAs contribute to expression or repression of transcription, or perhaps post-transcriptional regulation, will be a main focus in the future. Furthermore, it appears that distal regulatory elements come in close proximity to the promoters they regulate during activation of transcription. How this conformational change in gene loci affects transcription is not completely understood and will be another focus in the future.

REFERENCES

[1] Palis J. Primitive and definitive hematopoiesis. Front Physiol 2014;5:3.
[2] Hattangadi SM, Wong P, Zhang L, Flygare J, Lodish HF. From stem cells to red blood cell: regulation of erythropoiesis at multiple levels by multiple proteins, RNAs, and chromatin modifications. Blood 2011;118:6258–68.
[3] Wolff L, Humeniuk R. Concise review: erythroid versus myeloid lineage commitment: regulating the master regulators. Stem Cells 2013;31:1237–44.
[4] Ji P, Murata-Hori M, Lodish HF. Formation of mammalian erythrocytes: chromatin condensation and enucleation. Trends Cell Biol 2011;21:409–15.
[5] Jelkmann W. Physiology and pharmacology of erythropoietin. Transfus Med Hemother 2013;40:302–9.
[6] Sankaran VG, Xu J, Orkin SH. Advances in the understanding of haemoglobin switching. Br J Haematol 2010;149:181–94.
[7] Stamatoyannopoulos G. Control of globin gene expression during development and erythroid differentiation. Exp Hematol 2005;33:259–71.
[8] Higgs DR, Engel JD, Stamatoyannopoulos G. Thalassaemia. Lancet 2012;379:373–83.
[9] Layer G, Reichelt J, Jahn D, Heinz DW. Structure and function of enzymes in heme biosynthesis. Protein Sci 2010;19:1137–61.
[10] Dailey HA, Meissner PN. Erythroid heme biosynthesis and its disorders. Cold Spring Harb Perspect Med 2013;3:a011676.
[11] May BK, Dogra SC, Sadlon TJ, Bhasker CR, Cox TC, Bottomley SS. Molecular regulation of heme biosynthesis in higher vertebrates. Prog Nucleic Acid Res Mol Biol 1995;51:1–51.
[12] Mohandas N, Gallagher PG. Red cell membrane: past, present, and future. Blood 2008;112:3939–48.
[13] Krieger CC, An X, Tang HY, Mohandas N, Speicher DW, Discher DE. Cysteine shotgun-mass spectrometry (CS-MS) reveals dynamic sequence of protein structure changes within mutant and stressed cells. Proc Natl Acad Sci USA 2011;108:8269–74.
[14] Bulger M, Groudine M. Looping versus linking: toward a model for long-distance gene activation. Genes Dev 1999;13:2465–77.
[15] Grosveld F. Activation by locus control regions? Curr Opin Genet Dev 1999;9:152–7.
[16] Higgs D, Wood WG. Long-range regulation of alpha globin gene expression during erythropoiesis. Curr Opin Hematol 2008;15:176–83.
[17] Felsenfeld G, Groudine M. Controlling the double helix. Nature 2003;421:448–53.
[18] Kim A, Dean A. Chromatin loop formation in the β-globin locus and its role in globin gene transcription. Mol Cells 2012;34:1–5.
[19] Kim SI, Bresnick EH. Transcriptional control of erythropoiesis: emerging mechanisms and principles. Oncogene 2007;26:6777–94.
[20] Moriguchi T, Yamamoto M. A regulatory network governing GATA-1 and GATA-2 gene transcription orchestrates erythroid lineage differentiation. Int J Hematol 2014;100:417–24.
[21] Takemoto N, Arai K, Miyatake S. Cutting edge: the differential involvement of the N-terminal finger of GATA-3 in chromatin remodeling and transactivation during Th2 development. J Immunol 2002;169:4103–7.
[22] Tsiftsoglu AS, Vizirianakis IS, Strouboulis J. Erythropoiesis: model systems, molecular regulators, and developmental programs. IUBMB Life 2009;61:800–30.
[23] Devilbiss AW, Sanalkumar R, Johnson KD, Keles S, Bresnick EH. Hematopoietic transcriptional mechanisms: from locus-specific to genome-wide vantage points. Exp Hematol 2014;42:618–29.

[24] Pal S, Cantor AB, Johnson KD, Moran T, Boyer ME, Orkin SH, et al. Coregulator-dependent facilitation of chromatin occupancy by GATA-1. Proc Natl Acad Sci USA 2004;101:980–5.

[25] Burdach J, Funnell AP, Mak KS, Artuz CM, Wienert B, Lim WF, et al. Regions outside the DNA-binding domain are critical for proper in vivo specificity of an archetypal zinc finger transcription factor. Nucleic Acids Res 2014;42:276–89.

[26] Rothbart SB, Strahl BD. Interpreting the language of histone and DNA modifications. Biochim Biophys Acta 2014. Epub ahead of print.

[27] Bresnick EH, Lee HY, Fujiwara T, Johnson KD, Keles S. GATA switches as developmental drivers. J Biol Chem 2010;285:31087–93.

[28] Grass JA, Jing H, Kim SI, Martowicz ML, Pal S, Blobel GA, et al. Distinct functions of dispersed GATA factor complexes at an endogenous gene locus. Mol Cell Biol 2006;26:7056–67.

[29] Rodriguez P, Bonte E, Krijgsveld J, Kolodziej KE, Guyot B, Heck AJ, et al. GATA-1 forms distinct activating and repressive complexes in erythroid cells. EMBO J 2005;24:2354–66.

[30] Miccio A, Wang Y, Hong W, Gregory GD, Wang H, Yu X, et al. NuRD mediates activating and repressive functions of GATA-1 and Fog-1 during blood development. EMBO J 2010;29:442–56.

[31] Blobel GA, Nakajima T, Eckner R, Montminy M, Orkin SH. CREB-binding protein cooperates with transcription factor GATA-1 and is required for erythroid differentiation. Proc Natl Acad Sci USA 1998;95:2061–6.

[32] Papadopoulos GL, Karkoulia E, Tsamardinos I, Porcher C, Ragoussis J, Bungert J, et al. GATA-1 genome-wide occupancy associates with distinct epigenetic profiles in mouse fetal liver erythropoiesis. Nucleic Acids Res 2013;41:4938–48.

[33] McLean CM, Karemaker ID, van Leeuwen F. The emerging roles of Dot1L in leukemia and normal development. Leukemia 2014;28:2131–8.

[34] Feng Y, Yang Y, Ortega MM, Copeland JN, Zhang M, Jacob JB, et al. Early mammalian erythropoiesis requires the Dot1L methyltransferase. Blood 2010;116:4483–91.

[35] Bartholomew B. Regulating the chromatin landscape: structural and mechanistic perspectives. Annu Rev Biochem 2014;83:671–96.

[36] Bultman SJ, Gebuhr TC, Magnuson T. A Brg1 mutation that uncouples ATPase activity from chromatin remodeling reveals an essential role for SWI/SNF-related complexes in beta-globin expression and erythroid development. Genes Dev 2005;19:2849–61.

[37] Kim SI, Bultman SJ, Jing H, Blobel GA, Bresnick EH. Dissecting molecular steps in chromatin domain activation during hematopoietic differentiation. Mol Cell Biol 2007;27:4551–65.

[38] Biterge B, Schneider R. Histone variants: key players of chromatin. Cell Tissue Res 2014;356:457–66.

[39] Wadman IA, Osada H, Grutz GG, Agulnick AD, Westphal H, Forster A, et al. The LIM-only domain protein Lmo2 is a bridging molecule assembling an erythroid, DNA-binding complex which includes TAL1, E47, GATA-1 and Ldb1/NL1 proteins. EMBO J 1997;16(11):3145–57.

[40] Love PE, Warzecha C, Li L. Ldb1 complexes: the new master regulators of erythroid gene transcription. Trends Genet 2014;30:1–9.

[41] Deng W, Lee J, Wang H, Miller J, Reik A, Gregory PD, et al. Controlling long-range genomic interactions at a native locus by targeted tethering of a looping factor. Cell 2012;149:1233–44.

[42] Stumpf M, Waskow C, Krötschel M, van Essen D, Rodriguez P, Zhang X, et al. The mediator complex functions as a coactivator for GATA-1 in erythropoiesis via subunit Med1/TRAPP220. Proc Natl Acad Sci USA 2006;103:18504–9.

[43] Yin JW, Wang G. The mediator complex: a master coordinator of transcription and cell lineage development. Development 2014;141:977–87.

[44] Stumpf M, Yue X, Schmitz S, Luche H, Reddy JK, Borggrefe T. Specific erythroid-lineage defect in mice conditionally deficient for mediator subunit Med1. Proc Natl Acad Sci USA 2010;107:21541–6.

[45] Yien YY, Bieker JJ. EKLF/KLF1, a tissue-restricted integrator of transcription control, chromatin remodeling, and lineage determination. Mol Cell Biol 2013;33:4–13.

[46] Wijgerde M, Gribnau J, Trimborn T, Nuez B, Philipsen S, Grosveld F, et al. The role of EKLF in human beta-globin gene competition. Genes Dev 1996;10:2894–902.

[47] Sengupta T, Cohet N, Morlé F, Bieker JJ. Distinct modes of gene regulation by a cell-specific transcriptional activator. Proc Natl Acad Sci USA 2009;106:4213–8.

[48] Kandiah E, Trowitzsch S, Gupta K, Haffke M, Berger I. More pieces to the puzzle: recent structural insights into class II transcription initiation. Curr Opin Struct Biol 2014;24:91–7.

[49] Compe E, Egly JM. TFIIH: when transcription met DNA repair. Nat Rev Mol Cell Biol 2012;13:343–54.

[50] Schoenfelder S, Sexton T, Chakalova L, Cope NF, Horton A, Andrews S, et al. Preferential associations between co-regulated genes reveal a transcriptional interactome in erythroid cells. Nat Genet 2010;42:53–61.

[51] Bartlett J, Blagojevic J, Carter D, Eskiw C, Fromaget M, Job C, et al. Specialized transcription factories. Biochem Soc Symp 2006;73:67–75.

[52] Zhou D, Liu K, Sun CW, Pawlik KM, Townes TM. KLF1 regulates BCL11A expression and gamma- to beta- globin gene switching. Nat Genet 2010;42:742–4.

[53] Borg J, Papadopoulos P, Georgitsi M, Gutiérrez L, Grech G, Fanis P, et al. Haploinsufficiency for the erythroid transcription factor KLF1 causes hereditary persistence of fetal hemoglobin. Nat Genet 2010;42:801–5.

[54] Pearson RC, Funnell AP, Crossley M. The mammalian zinc finger transcription factor Kruppel-like factor 3 (KLF2/BKLF). IUBMB Life 2011;63:86–93.

[55] Lecuyer E, Hoang T. SCL: from the origin of hematopoiesis to stem cells and leukemia. Exp Hematol 2004;32:11–24.

[56] Anantharaman A, Lin IJ, Barrow J, Liang SY, Masannat J, Strouboulis J, et al. Role of helix-loop-helix proteins during differentiation of erythroid cells. Mol Cell Biol 2011;31:1332–43.

[57] Porcher C, Swat W, Rockwell K, Fujiwara Y, Alt FW, Orkin SH. The T cell leukemia oncoprotein SCL/tal-1 is essential for development of all hematopoietic lineages. Cell 1996;86:47–57.

[58] Lindberg SR, Olsson A, Persson AM, Olsson I. The leukemia-associated ETO homologues are differently expressed during hematopoietic differentiation. Exp Hematol 2005;284:189–98.

[59] Kassouf MT, Chagraoui H, Vyas P, Porcher C. Differential use of SCL/TAL-1 DNA-binding domain in developmental hematopoiesis. Blood 2008;112:1056–67.

[60] Kassouf MT, Hughes JR, Taylor S, McGowan SJ, Soneji S, Green AL, et al. Genome-wide identification of TAL1's functional targets: insight into its mechanisms of action in primary erythroid cells. Genome Res 2010;20:1064–83.

[61] Motohashi H, Shavit JA, Igarashi K, Yamamoto M, Engel JD. The world according to Maf. Nucleic Acids Res 1997;25:2953–9.

[62] Amrolia PJ, Ramamurthy L, Saluja D, Tanese N, Jane SM, Cunningham JM. The activation domain of the enhancer binding protein p45 NF-E2 interacts with TAFII130 and mediates long-range activation of the alpha- and beta-globin gene loci in erythroid cells. Proc Natl Acad Sci USA 1997;94:10051–6.

[63] Brand M, Ranish JA, Kummer NT, Hamilton J, Igarashi K, Francastel C, et al. Dynamic changes in transcription factor complexes during erythroid differentiation revealed by quantitative proteomics. Nat Struct Mol Biol 2004;11:73–80.

[64] Tahara T, Sun J, Nakanishi K, Yamamoto M, Mori H, Saito T, et al. Heme positively regulates the expression of beta-globin at the locus control region via the transcription factor Bach1 in erythroid cells. J Biol Chem 2004;279:5480–7.

[65] Johnson KD, Christensen HM, Zhao B, Bresnick EH. Distinct mechanisms control RNA polymerase II recruitment to a tissue-specific locus control region and a downstream promoter. Mol Cell 2001;8:465–71.

[66] Zhou Z, Li X, Deng C, Ney PA, Huang S, Bungert J. USF and NF-E2 cooperate to regulate the recruitment and activity of RNA polymerase II in the beta-globin gene locus. J Biol Chem 2010;285:15894–905.

[67] Neph S, Vierstra J, Stergachis AB, Reynolds AP, Haugen E, Vernot B, et al. An expansive human regulatory lexicon encoded in transcription factor footprints. Nature 2012;489:83–90.

[68] Shivdasani RA, Orkin SH. Erythropoiesis and globin gene expression in mice lacking the transcription factor NF-E2. Proc Natl Acad Sci USA 1995;92:8690–4.

[69] Kooren J, Palstra RJ, Klous P, Splinter E, von Lindern M, Grosveld F, et al. Beta-globin active chromatin hub formation in differentiating erythroid cells and in p45 NF-E2 knock-out mice. J Biol Chem 2007;282:16544–52.

[70] Medina KL, Singh H. Genetic networks that regulate B lymphopoiesis. Curr Opin Hematol 2005;12:203–9.

[71] Menzel S, Garner C, Gut I, Matsuda F, Yamaguchi M, Heath S, et al. A QTL influencing F cell production maps to a gene encoding a zinc finger protein on chromosome 2p15. Nat Genet 2007;39:1197–9.

[72] Xu J, Peng C, Sankaran VG, Shao Z, Esrick EB, Chong BG, et al. Correction of sickle cell disease in adult mice by interference with fetal hemoglobin silencing. Science 2011;334:993–6.

[73] Eilers M, Eisenmann RN. Myc's broad reach. Genes Dev 2008;22:2755–66.

[74] Eberhardy SR, Farnham PJ. Myc recruits P-TEFB to mediate the final step in the transcription activation of the cad promoter. J Biol Chem 2002;277:40156–62.

[75] Deng C, Li Y, Liang S, Cui K, Salz T, Yang H, et al. USF1 and hSET1A mediated epigenetic modifications regulate lineage differentiation and HoxB4 transcription. PLoS Genet 2013;9:e1003524.

[76] Greig KT, Carotta S, Nutt SL. Critical roles for c-Myb in hematopoietic progenitor cells. Semin Immunol 2008;20:247–56.

[77] Bianchi E, Zini R, Salati S, Tenedini E, Norfo R, Tagliafico E, et al. c-myb supports erythropoiesis through the transactivation of KLF1 and LMO2 expression. Blood 2010;116:e99–110.

[78] Tanabe O, Katsuoka F, Campbell AD, Song W, Yamamoto M, Tanimoto K, et al. An embryonic/fetal beta-type globin gene repressor contains a nuclear receptor TR2/TR4 heterodimer. EMBO J 2002;21:3434–42.

[79] Bottardi S, Zmiri FA, Bourgoin V, Ross J, Mavoungou L, Milot E. Ikaros interacts with P-TEFB and cooperates with GATA-1 to enhance transcription elongation. Nucleic Acids Res 2011;32: 3624–38.

[80] Bank A, O'Neill D, Lopez R, Pulte D, Ward M, Mantha S, et al. Role of intergenic human gamma-delta globin sequences in human hemoglobin switching and reactivation of fetal hemoglobin in adult erythroid cells. Ann N Y Acad Sci 2005;1054:48–54.

[81] Ameres SL, Zamore PD. Diversifying microRNA sequence and function. Nat Rev Mol Cell Biol 2013;14:475–88.

[82] Zhang L, Sankaran VG, Lodish HF. MicroRNAs in erythroid and megakaryocytic differentiation and megakaryocyte-erythroid progenitor lineage commitment. Leukemia 2012;26:2310–6.

[83] Listowski MA, Heger E, Boguslawska DM, Machnicka B, Kuliczkowski K, Leluk J, et al. microRNAs: fine tuning of erythropoiesis. Cell Mol Biol Lett 2013;18:34–46.

[84] Pase L, Layton JE, Kloosterman WP, Carradice D, Waterhouse PM, Lieschke GJ. miR-451 regulates zebrafish erythroid maturation in vivo via its target GATA-2. Blood 2009;113:1794–804.

[85] Patrick DM, Zhang CC, Tao Y, Yao H, Qi X, Schwartz RJ, et al. Defective erythroid differentiation in miR-451 mutant mice mediated by 14-3-3zeta. Genes Dev 2010;24:1614–9.

[86] Vance KW, Ponting CP. Transcriptional regulatory functions of nuclear long noncoding RNAs. Trends Genet 2014;30:348–55.

[87] Tuan D, Kong S, Hu K. Transcription of the hypersensitive site HS2 enhancer in erythroid cells. Proc Natl Acad Sci USA 1992;89:11219–23.

[88] Routledge SJ, Proudfoot NJ. Definition of transcriptional promoters in the human beta globin locus control region. J Mol Biol 2002;323:601–11.

[89] Kim TK, Hemberg M, Gray JM, Costa AM, Bear DM, Wu J, et al. Widespread transcription at neuronal activity-regulated enhancers. Nature 2010;465:182–7.

[90] Bulger M, Groudine M. Functional and mechanistic diversity of distal transcription enhancers. Cell 2011;144:327–39.

[91] Yang L, Lin C, Jin C, Yang JC, Tanasa B, Li W, et al. Lnc-RNA dependent mechanisms of androgen-receptor-regulated gene activation programs. Nature 2013;500:598–602.

[92] Chu C, Qu K, Zhong FL, Artandi SE, Chang HY. Genomic maps of long noncoding RNA occupancy reveal principles of RNA-chromatin interactions. Mol Cell 2011;44:667–78.

[93] Kaneko S, Li G, Son J, Xu CF, Margueron R, Neubert TA, et al. Phosphorylation of the PRC2 component Ezh2 is cell cycle-regulated and up-regulates its binding to ncRNA. Genes Dev 2010;24:2615–20.

[94] Yoo EJ, Cooke NE, Liebhaber SA. An RNA-independent linkage of noncoding transcription to long-range enhancer function. Mol Cell Biol 2012;32:2020–9.

[95] Alvarez-Dominguez JR, Hu W, Yuan B, Shi J, Park SS, Gromatzky AA, et al. Global discovery of erythroid long noncoding RNAs reveals novel regulators of red cell maturation. Blood 2014;123:570–81.

[96] Ghirlando R, Giles K, Gowher H, Xiao T, Xu Z, Yao H, et al. Chromatin domains, insulators, and the regulation of gene expression. Biochim Biophys Acta 2012;1819:644–51.

[97] Chetverina D, Aoki T, Erokhin M, Georgiev P, Schedl P. Making connections: insulators organize eukaryotic chromosomes into independent cis-regulatory networks. Bioessays 2014;36:163–72.

[98] Ong CT, Corces VG. CTCF: an architectural protein bridging genome topology and function. Nat Rev Genet 2014;15:234–46.

[99] Jing H, Vakoc CR, Ying L, Mandat S, Wang H, Zheng X, et al. Exchange of GATA factors mediates transitions in looped chromatin organization at a developmentally regulated gene locus. Mol Cell 2008;29:232–42.

[100] Palstra RJ. Close encounters of the 3C kind: long-range chromatin interactions and transcriptional regulation. Brief Funct Genomic Proteomic 2009;8:297–309.

[101] De Laat W, Grosveld F. Spatial organization of gene expression: the active chromatin hub. Chromosome Res 2003;11:447–59.

[102] Shearstone JR, Pop R, Bock C, Boyle P, Meissner A, Socolovsky M. Global DNA demethylation during mouse erythropoiesis in vivo. Science 2011;290:350–8.

[103] Pronier E, Almire C, Mokrani H, Vasanthakumar A, Simon A, da Costa Reis Monte Mor B, et al. Inhibition of TET2-mediated conversion of 5-methylcytosine to 5-hydroxymethylcytosine disturbs erythroid and granulomonocytic differentiation of human hematopoietic progenitors. Blood 2011;118:2551–5.

[104] Madzo J, Liu H, Rodriguez A, Vasanthakumar A, Sundaravel S, Caces DB, et al. Hydroxymethylation at gene regulatory regions directs stem/early progenitor cell commitment during erythropoiesis. Cell Rep 2014;6:231–44.

[105] Zaret KS, Carroll JS. Pioneer transcription factors: establishing competence for gene expression. Genes Dev 2011;25:2227–41.

[106] Graf T, Enver T. Forcing cells to change lineages. Nature 2009;462:587–94.

[107] Liang S, Moghimi B, Yang TP, Strouboulis J, Bungert J. Locus control region mediated regulation of adult beta globin gene expression. J Cell Biochem 2008;105:9–16.

[108] Weintraub H, Groudine M. Chromosomal subunits in active genes have an altered conformation. Science 1976;193:848–56.

[109] Jiménez G, Griffiths SD, Ford AM, Greaves MF, Enver T. Activation of the beta-globin locus control region precedes commitment to the erythroid lineage. Proc Natl Acad Sci USA 1992;89:10618–22.

[110] Bottardi S, Aumont A, Grosveld F, Milot E. Developmental stage-specific epigenetic control of human beta-globin gene expression is potentiated in hematopoietic progenitor cells prior to their transcriptional activation. Blood 2003;102:3989–97.

[111] Levings PP, Bungert J. The human beta globin locus control region. Eur J Biochem 2002;269:1589–99.

[112] Demers C, Chaturvedi CP, Ranish JA, Juban G, Lai P, Morle F, et al. Activator-mediated recruitment of the MLL2 methyltransferase complex to the beta-globin locus. Mol Cell 2007;27:573–84.

[113] Fan AX, Papadopoulos G, Hossain MA, Lin IJ, Hu J, Tang TM, et al. Genomic and proteomic analysis of transcription factor TFII-I reveals insight into the response to cellular stress. Nucleic Acids Res 2014;42:7625–41.

CHAPTER 12

Genetically altered cancer epigenome

Ming Tang, Huacheng Luo, Jianrong Lu
Department of Biochemistry and Molecular Biology, College of Medicine, University of Florida, Gainesville, FL, USA

Contents

Epigenetic Gene Expression and Regulation
http://dx.doi.org/10.1016/B978-0-12-799958-6.00012-3

1. INTRODUCTION

Cancer arises from abnormal expression and/or function of tumor suppressor genes and oncogenes, which may result from either genetic or epigenetic alterations. Eukaryotic genomic DNA is wrapped around the outside surface of the histone octamer to form nucleosomes, the repeating building blocks of chromatin. Chromatin is a dynamic structure and can be packed lightly or tightly. Chromatin structure impacts gene expression primarily by governing the genome accessibility for transcription factors and the transcriptional machinery. Covalent modifications of histones and DNA, nucleosome positioning, and long-distance chromatin interactions are critical determinants of active and repressive chromatin states. Cancer cells frequently exhibit an aberrant epigenome, notably epigenetic silencing of various tumor suppressor genes with vital functions in cancer-relevant signaling pathways, such as cell proliferation, apoptosis, and DNA repair [1,2].

In 2013 and 2014 large-scale cancer genomic studies (whole-exome and whole-genome sequencing) of nearly 5000 human tumor samples with matched normal tissues across 21 cancer types identified significantly mutated genes that drive malignancy [3,4]. The analyzed somatic mutations include substitutions and small insertions or deletions. Most cancer gene mutations in most patients occur at intermediate frequencies (2–20%) or lower. Statistically significant somatic mutations are found in a wealth of chromatin regulators that are directly involved in histone modifications, DNA methylation, nucleosome remodeling, and long-range chromatin interactions. Genetic alterations of these regulators are expected to disturb global epigenetic patterns and have the potential to deregulate numerous genes genome-wide (including tumor suppressor genes and oncogenes), which may catastrophically contribute to tumor initiation, progression, and metastasis. The findings illustrate a mechanistic link that leads from genetic mutations to altered epigenome [5,6].

2. HISTONE MODIFICATIONS

The amino-terminal tails of core histones protrude outward beyond the gyres of DNA. Many specific residues within the histone tails can undergo a diverse array of post-translational modifications, such as acetylation, methylation, and monoubiquitination, which are established or reverted by corresponding histone-modifying enzymes. These modifications may directly impact chromatin structure and/or serve as docking signals to recruit new chromatin-modifying or remodeling complexes that may further stabilize or reprogram the epigenetic landscape. Therefore, histone modifications are instrumental for the regulation of chromatin accessibility and dynamics, and affect essentially all DNA-templated processes, including gene transcription, replication, and repair. Deregulation of histone modifications may result in abnormal cell proliferation and survival and eventual development of cancer (Table 1).

Table 1 Histone-modifying complexes significantly mutated in cancer

Enzyme type	Mutated subunit	Substrate	Mutation prevalence in cancer
KMT	MLL1/KMT2A	H3K4	1% breast, 13% bladder, 3% combined
	MLL2/KMT2D	H3K4	26% bladder, 21% diffuse large B-cell lymphoma, 20% lung SCC, 15% head and neck, 6% combined, 4% medulloblastoma
	MLL3/KMT2C	H3K4	7% breast, 6% combined, 24% bladder, 15% lung adenocarcinoma, 7% head and neck
	MLL4/KDM2B	H3K4	2% combined, 14% endometrial
	SETD2/KMT3A	H3K36	12% kidney CCC, 4% combined, 2% glioblastoma, 6% bladder
	NSD1/KMT3B	H3K36	10% head and neck, 3% combined, 6% lung SCC
	EZH2/KMT6A[a]	H3K27	10% diffuse large B-cell lymphoma, 2% AML
KDM	UTX/KDM6A	H3K27me2/3	27% bladder, 2% combined, 2% head and neck, 2% AML
	SMCX/JARID1C/KDM5C	H3K4me2/3	7% kidney CCC, 2% combined
	ARID5B (complex with PHF2/JHDM1E/KDM7C)	H3K9me1/2	12% endometrial
DUB	BAP1	H2AK119ub1	11% kidney CCC, 2% combined
	ASXL1 (complex with BAP1)		2% combined, 3% AML, 5% lung SCC, 3% head and neck
	ASXL2		2% combined, 7% bladder
KAT	CBP/CREBBP/KAT3A	Core histones	24% diffuse large B-cell lymphoma, 3% combined, 12% bladder
	p300/EP300/KAT3B	Core histones	3% combined, 9% endometrial, 7% head and neck, 4% lung SCC

Acronyms: AML: acute myeloid leukemia; CCC: clear cell carcinoma; SCC: squamous cell carcinoma.
[a]Gain-of-function mutations.

2.1 Lysine methyltransferases (KMTs)

Histone methylation involves lysine and arginine residues. The major families of SET (**s**u(var)3–9, **e**nhancer-of-zeste (EHZ), **T**rithorax (Trx)) domain-containing proteins act as lysine methyltransferases (KMTs) to catalyze the methylation of particular lysine residues on histones H3 and H4 [7]. Lysine methylation occurs in multiple states, including mono-, di-, or trimethylation. Different lysine methylation marks may be associated with distinct transcriptional readouts. Typically, H3 lysine 4 di- and trimethylation (H3K4me2/3) are marks of active transcription, whereas H3K9me2/3 and H3K27me2/3 indicate repressive chromatin.

2.1.1 MLL1-4

Mixed lineage leukemias (MLLs) are H3K4-specific methyltransferases and play important roles in the regulation of gene transcription, epigenetic modification, and tumorigenesis [8]. In mammal, there exist several MLL proteins, such as MLL1-4, which are capable of mono-, di-, and trimethylating H3K4. As H3K4me2/3 are associated with active transcription, MLLs are transcriptional coactivators [9].

MLL1 and MLL2 share a high degree of structural similarity and are the mammalian homologs of the *Drosophila* Trx protein. In *Drosophila*, the Trithorax group (TrxG) and the Polycomb group (PcG) proteins play a central role in the regulation of homeotic gene expression throughout development [10]. The two families of proteins positively and negatively regulate transcription, respectively. MLL1 is required for the H3K4 trimethylation and transcription of a small subset of genes, including developmental regulators, such as homeobox (Hox) genes [11]. MLL1 is perhaps best known for its involvement in leukemia [12]. Approximately 10% of all human acute leukemias harbor MLL1 chromosomal translocations. MLL1 translocations involve more than 60 different partner genes, many of which seem unrelated. A number of the most common MLL1 translocation partners are involved in transcriptional elongation, suggesting abnormal RNA polymerase II (Pol II) transcription elongation may be fundamental to MLL-induced leukemogenesis [13]. All MLL-fusion proteins retain the amino terminus of MLL1 that is responsible for chromatin association. Furthermore, in-frame fusion of MLL1 to a partner protein is required for cellular transformation. These observations suggest that MLL1 translocation-mediated leukemic pathogenesis is not due to a loss-of-function mechanism. MLL is also affected by other chromosomal abnormalities, including partial tandem duplications and genomic amplifications. All MLL1 rearrangements cause aberrantly increased and sustained expression of Hox genes, which disturbs hematopoietic differentiation and is central to leukemogenesis.

However, inactivating mutations of MLL1 and MLL2 have been uncovered in multiple types of solid tumors, suggesting that MLL1 and MLL2 possess tumor-suppressive function. In fact, the tumor suppressor Menin, which is mutated in multiple endocrine neoplasia type 1, directly interacts with the amino terminus of MLL1/MLL2 and is an essential component of both the MLL1 and MLL2 complexes. Deletion of Menin abolishes the majority of H3K4 trimethylation and gene expression on all Hox gene clusters [11]. Importantly, MLL1 activates expression of cyclin-dependent kinase inhibitors (CKIs) p27 and p18 in a Menin-dependent manner [14]. Inactivating Menin mutations lead to failed induction of these CKIs, which is likely to be an important oncogenic pathway in such endocrine tumors. Induction of cellular senescence by oncogenic signals (e.g., Ras [rat sarcoma], Myc [myelocytomatosis]) functions as a barrier to cellular transformation. MLL1 is required for the induction of p16 (also a CKI and a tumor suppressor) by oncogenic Ras during oncogenic checkpoint response [15]. MLL2 is involved in several important cellular signaling pathways, including the p53 tumor suppressor [16]. These functions of MLL1 and MLL2 are of critical importance in the epigenetic regulation of the cell cycle and senescence, and may explain why they exert tumor suppressor activities.

MLL3 and MLL4 are the mammalian homologs of the *Drosophila* trithorax-related (Trr) gene [8]. In *Drosophila*, functions of Trr and Trx are not redundant, and each has specific genomic targets. Genome-wide binding studies demonstrated that Trr and MLL3/MLL4 are bound to distal regulatory elements and transcription start sites [17,18]. Monomethylated H3K4 (H3K4me1) is a general mark for enhancers [19]. MLL3 and MLL4 are involved in the implementation of H3K4 monomethylation on enhancers. Loss of MLL3 and MLL4 results in decreased H3K4me1 on the majority of enhancers [18]. H3K4me1 and acetylation of H3K27 are histone modifications that are highly enriched on active enhancers, whereas H3K4me1 and H3K27me3 mark inactive/poised enhancers. The MLL3 and MLL4 complexes specifically contain the H3K27 demethylase UTX (lysine-specific demethylase 6A). Impairment of MLL3/MLL4 also alters H3K27 acetylation at enhancer regions [17]. These studies reveal a direct role for MLL3/MLL4 and UTX in the transition of enhancers from inactive/poised to active status.

In *Drosophila*, both Trr mutant cells and UTX mutant cells similarly display an overgrowth phenotype, suggesting Trr and UTX restrict tissue growth [20]. In human, MLL3 has been identified as a tumor suppressor gene frequently altered in various tumors. Mutations of MLL3 and MLL4 are expected to cause malfunction of numerous enhancers. Deregulation of enhancer activity may cause aberrant expression of tumor suppressors and oncogenes, contributing to tumorigenesis. Moreover, the MLL3/MLL4 family members interact with tumor suppressor p53 and are involved in the p53-mediated DNA damage response [21].

2.1.2 SETD2 and NSD1

SETD2 (SET domain containing 2) and NSD1 (nuclear receptor binding SET domain protein 1) are histone H3K36 methyltransferases. SETD2 is mainly responsible for H3K36me3, and NSD1 preferentially for H3K36me1/2 [22]. H3K36 methylation is implicated in diverse processes, including gene transcription and DNA repair and recombination [23]. Altered placement of H3K36 methylation within the chromatin landscape can lead to a range of human diseases, including cancer. Mutations of SETD2 result in a global reduction of H3K36me3 in tumor cells [24]. In collaboration with other genetic lesions (e.g., MLL1 translocation), loss of SETD2 contributes to both initiation and progression of leukemia by enhancing the self-renewal potential of leukemia stem cells [25]. Therefore, the SETD2-H3K36me3 pathway represents a tumor-suppressive mechanism. SETD2 interacts with p53 and regulates its transcription activity [26]. SETD2 is required for activation of ATM (ataxia telangiectasia mutated) by DNA double-strand breaks (DSBs) and for homologous recombination repair of DSBs [27,28]. SETD2-mutant cells exhibit impaired DNA damage signaling and fail to activate p53. Depleting SETD2 causes defective homologous recombination repair. In addition, the H3K36me3 mark is required in vivo to recruit the mismatch recognition protein complex [29]. Cells lacking SETD2 display microsatellite instability (MSI) and an elevated spontaneous mutation frequency, characteristic of DNA mismatch repair (MMR)-deficient cells. Loss-of-function mutations in NSD1 cause the Sotos overgrowth syndrome [30], but their connection to cancer remains poorly understood.

2.1.3 EZH2

Histone methyltransferase EZH2 (enhancer of zeste homolog 2) is the catalytic subunit of the Polycomb repressive complex 2 (PRC2) and is involved in repressing gene expression through di- and trimethylation of H3K27 [31]. EZH2 is a major proto-oncogene. EZH2 is overexpressed in several common solid tumor types, and is associated with advanced stages of disease and poor prognosis [32]. Moreover, the enzyme may be activated by mutations. The Y641 residue within the catalytic domain of EZH2 is a hotspot of mutations, especially in diffuse large B-cell lymphoma and follicular lymphoma [33]. Recurrent missense mutations at this site change substrate preferences of the mutant enzymes and increase H3K27me3, thereby representing gain-of-function mutations [34,35].

PcG proteins (including EZH2) can induce tumorigenesis in part through direct repression of critical tumor suppressor genes, such as p16-ARF (where ARF stands for alternate reading frame) and E-cadherin [36,37]. The p16-ARF locus encodes several well-established tumor suppressors and is frequently mutated or silenced in human tumors [38]. EZH2 mediates epigenetic silencing of p16 and is essential in cancer stem cells [39]. Expression of EZH2 and E-cadherin exhibits an inverse correlation in cancer, and E-cadherin expression is restored when EZH2 is depleted [37,40]. Pharmacological inhibition of EZH2 methyltransferase activity decreases global H3K27me3 levels, reactivates silenced PRC2 target genes, and markedly inhibits the growth of EZH2 mutant tumors [41,42].

Paradoxically, however, inactivating mutations in EZH2 have also been reported to promote myeloid disorders [43,44]. In accordance, recurrent K27 missense mutations in histone H3 and its variant H3.3 have been observed particularly in gliomas [45–48]. K27-mutated H3 or H3.3 tails function as a pseudosubstrate to aberrantly recruit the PRC2 complex and inhibit the enzymatic activity of EZH2. The dominant-negative effect of K27 mutations leads to a global reduction of H3K27me3 and aberrant gene activation. The dual role of EZH2 and H3K27 methylation in cancer suggests EZH2's pro- or anti-tumorigenesis function is probably context dependent.

2.2 Lysine demethylases (KDMs)

The Jumonji C (JmjC)-domain-containing enzymes can demethylate lysine residues and are the major histone demethylases [49]. The demethylases play critical roles in developmental processes and human diseases such as cancer.

2.2.1 UTX

UTX is a H3K27-specific demethylase (opposite to EZH2). UTX is inactivated by recurrent somatic mutations in a number of tumors, leading to increased H3K27 methylation [50]. Inactivation of UTX is considered the same as enhancing EZH2 activity. UTX is a specific subunit of the MLL3 and MLL4 complexes [17,18]. UTX, MLL3, and MLL4 are all frequently mutated in human cancers, suggesting deregulated enhancers may be an important mechanism underlying UTX tumorigenesis. Furthermore, mutations in UTX and MLL2 are mutually exclusive in urothelial carcinoma of the bladder [51], suggesting that mutations in the two genes may have similar downstream effects on carcinogenesis.

2.2.2 SMCX

SMCX (lysine (K)-specific demethylase 5C) is a member of the JARID1 (jumonji, AT-rich interactive domain 1) family of H3K4 demethylases [52] and is frequently mutated in renal cell carcinoma (RCC) [53]. SMCX has been identified as a target gene of hypoxia-inducible factor (HIF) in RCC cells to decrease H3K4Me3 levels [54]. Inactivation of SMCX promotes tumor formation in RCC [54]. Furthermore, another member of the JARID1 family, RBP2 (retinol binding protein 2), was firstly identified as a binding partner of the Rb tumor suppressor [55]. RBP2 also inhibits the oncogenic Notch signaling pathway and Notch-induced tumorigenesis by erasing the H3K4me3 mark at Notch target genes [56]. Given their homology, SMCX may function similarly as RBP2.

2.2.3 ARID5B

ARID5B (AT-rich interactive domain-containing protein 5B) is a DNA-binding protein and forms a complex with the histone H3K9 demethylase PHF2 [57]. The ARID5B-PHF2 complex acts as a transcriptional coactivator. PHF2 (PHD[polyhomeotic distal protein] finger protein 2) recognizes the active H3K4me3 mark. Once the ARID5B-PHF2 complex is recruited to target promoters, PHF2 mediates the demethylation of the repressive H3K9Me2 mark, facilitating transcriptional activation of target genes [57,58]. Defects in ARID5B may be a cause of susceptibility to transformation. However, it remains obscure how the ARID5B-PHF2 complex exhibits tumor-suppressive function.

2.3 Deubiquitinase (DUB)
2.3.1 BAP1-ASXL1/2

H3K27me3-marked repressive chromatin is recognized by Polycomb repressive complex 1 (PRC1), which possesses ubiquitin ligase activity and maintains the transcriptionally silenced state by further monoubiquitinating K119 of histone H2A (H2AK119Ub1) [59]. Therefore, deubiquitination of H2AK119Ub plays a critical role in chromatin modulation and transcriptional regulation. The Polycomb repressive deubiquitinase (PR-DUB) complex specifically mediates deubiquitination of H2AK119ub1 [60]. This deubiquitinating complex is at least composed of BAP1 (BRCA1 [breast cancer 1, early onset] associated protein-1) and ASXL1 (additional sex combs like transcriptional regulator 1), in which the BAP1 enzyme is the catalytic subunit. ASXL1 is frequently mutated in most types of myeloid malignancies [61–63]. Mutations in ASXL1 result in loss of H3K27me3 [64]. ASXL2 may share similar biochemical functions with ASXL1. BAP1 is deleted in some human cancers and is an established tumor suppressor gene [65,66]. Besides ASXL1, BAP1 is associated with other protein complexes. For instance, BAP1 interacts with the BRCA1 tumor suppressor, and exerts growth inhibitory effects in a BRCA1-dependent manner [67].

2.4 Histone acetyltransferases (HATs)

2.4.1 CBP and p300

Histone acetyltransferases (HATs) and deacetylases (HDACs) play a pivotal role in modifying chromatin structure and gene expression. HATs acetylate lysine residues in histone tails and loosen chromatin structure, thereby increasing accessibility of regulatory proteins to DNA. Acetylated lysine may also serve as recruiting marks to attract chromatin regulatory complexes [68]. By contrast, HDACs condense chromatin and repress gene transcription. Aberrant histone acetylation is associated with the initiation and progression of cancer. Histone acetyltransferases CBP (CCAAT enhancer-binding protein) and p300 are close homologs and can acetylate multiple lysine sites of core histones. They function as transcriptional coactivators of various sequence-specific transcription factors that are involved in cell proliferation, differentiation, and apoptosis. Somatic mutations of p300 and CBP have been detected in a number of malignancies. Chromosomal translocations frequently target CBP and p300 in acute leukemia, whereas loss–of-function mutations in the two genes have been identified in solid tumors, including colorectal and breast carcinomas [69]. The inactivating mutations result in truncated protein products or amino-acid substitutions in critical protein domains, and are often associated with inactivation of the second allele. Recurrent mutations at D1399 of the p300 catalytic domain are observed, which are likely to affect the catalytic activity of this HAT [70]. Mutations in CBP and p300 can impair histone acetylation and transcriptional regulation of their targets. These factors are involved in critical tumorigenic pathways (including p53 and Rb) [69,71]. Their tumor-suppressive activity has been confirmed in a mouse model [69].

3. DNA METHYLATION

DNA methylation is a common epigenetic modification in chromatin. In cancer, hypermethylation of CpG islands is a well-recognized epigenetic event. Hypermethylation of CpG island promoters leads to repressive histone modifications and gene silencing, which occurs at numerous tumor suppressor genes [1].The cancer genome is also characterized by global hypomethylation at repetitive and gene-body sequences. Hypomethylation of DNA loosens chromatin and causes chromosomal instability.Therefore, DNA methylation has a profound impact on transcription and genome stability, and is significantly associated with cell growth, differentiation, and transformation.

3.1 DNA methyltransferase (DNMT)

3.1.1 DNMT3A

DNA methylation is established by DNA methyltransferases (DNMTs), which generate 5-methylcytosine (5mC) in CpG dinucleotides. DNMT3A is a de novo DNA methyltransferase. A variety of inactivating mutations in DNMT3A, including missense, nonsense, frame shift, and splice-site mutations, have been documented in human cancer [1]. While these mutations occur throughout various domains of DNMT3A, the majority of mutations in acute myeloid leukemia (AML) are heterozygous and specifically alter a single amino acid, R882, in the catalytic domain [72]. DNMT3A R882 hotspot mutations severely impair the DNA methyltransferase enzymatic activity. Moreover, as DNMT3A forms a tetrameric complex with other DNMT3 [73], the R882 mutants are capable of interacting with wild-type DNMT3, thereby inhibiting their ability to methylate DNA [74,75]. Therefore, DNMT3A R882 mutants, in addition to being hypomorphic, exert dominant-negative effects.These mutations induce focal hypomethylation at specific CGs throughout AML cell genomes and may cause abnormal activation of oncogenic genes. Indeed, DNMT3A R882H mutation was reported to increase the CDK1 (cyclin-dependent kinase 1) expression and enhance cell-cycle activity, thereby contributing to leukemogenesis [76].

3.2 5mC hydroxylase

3.2.1 TET2 and IDH1/2

TET2 is one of the three enzymes of the TET (ten-eleven translocation) family, which are evolutionarily conserved dioxygenases that catalyze the oxidization of 5mC to 5-hydroxymethyl-cytosine (5hmC) and promote DNA demethylation [77]. Loss-of-function mutations in TET2 have been discovered in myeloid malignancies and are implicated in the development of cancers. TET2 mutations reduce the global 5hmC levels [78]. Inactivation of TET2 likely alters genomic 5hmC and 5mC patterns and disrupts gene regulation.Analysis ofTET2-deficient mice demonstrates thatTET2 functions as a dose-dependent tumor suppressor, as TET2 haploinsufficiency initiates myeloid and lymphoid transformations [79].

Many chromatin-modifying enzymes depend on specific metabolites. For example,TET and Jmjc (Jumonji C) family enzymes use α-ketoglutarate (α-KG) as an essential cosubstrate. In this regard, abnormal metabolism may impact epigenome. Isocitrate dehydrogenases (IDHs) are metabolic enzymes that convert isocitrate to α-KG through oxidative decarboxylation. Recurrent hotspot missense mutations in IDH1 and IDH2 are frequently found in glioma and AML. These mutations occur at a single amino acid residue of IDH1 (R132) and IDH2 (R140). Only a single copy of the genes is mutated in tumors. Groundbreaking studies have demonstrated that the tumor-derived IDH mutations are neomorphic: the IDH mutants acquire new enzymatic activity and are able to convert α-KG into (R)-2-hydroxyglutarate (2HG) [80,81]. 2HG inhibits α-KG-dependent epigenetic regulators (e.g., TET, Jmjc) and causes histone and DNA hypermethylation. IDH mutants induce global DNA hypermethylation and a hypermethylator phenotype [82,83]. In AML, IDH mutations are mutually exclusive with mutations in the α-KG-dependent TET2, and TET2 loss-of-function mutations cause similar epigenetic defects as IDH mutants.

Mutant IDHs can significantly increase histone methylation as well, presumably through 2HG-mediated inhibition of Jmjc histone demethylases including KDM4C (lysine (K)-specific demethylase 4C) [84]. Collectively, 2HG functions as an oncometabolite, and its excess accumulation contributes to tumor formation.

4. NUCLEOSOME REMODELING

Nucleosomes block access to DNA and impede the initiation and elongation of transcription. To activate transcription, nucleosomes need to be repositioned or removed to expose DNA sequence for binding of transcription factors and RNA Pol II. Nucleosome remodeling is a major means by which the cell modulates nucleosome mobility. This is accomplished by chromatin-remodeling complexes that use ATP hydrolysis to shuffle nucleosomes around and to replace or remove them from chromatin [85]. There are four classes of chromatin remodelers: SWI/SNF (SWItch/Sucrose NonFermentable), CHD (chromodomain helicase DNA-binding protein), ISWI (imitation SWI), and INO80. SWI/SNF and CHD-type remodelers are frequently mutated in human cancer (Table 2).

4.1 The SWI/SNF complexes

SWI/SNF complexes can remodel the chromatin structures by sliding nucleosomes [86] and can either eject or insert histone octamers [87]. SWI/SNF chromatin remodeling complexes are diverse assemblies of at least 14 subunits (a single subunit may be encoded by multiple genes) [88]. SWI/SNF complexes function as tumor suppressors in human malignancies [89]. Recurrent loss-of-function mutations in genes encoding various subunits of the SWI/SNF complexes have been found in a wide spectrum of cancer types [3,4,88]. The SWI/SNF complex has been identified as the most frequently mutated chromatin regulatory complex in cancer.

4.1.1 BRG1

BRG1 (Brahma-related gene 1) is one of the catalytic ATPase subunits in the SWI/SNF complexes. The other homologous BRM (SMARCA2 SWI/SNF-related,

Table 2 Nucleosome remodeling complexes mutated in cancer

Nucleosome remodeler	Mutated subunit	Mutation prevalence in cancer types
SWI/SNF	BRG1/SMARCA4	9% lung adenocarcinoma, 3% combined, 6% esophageal
	ARID1A/Baf250	33% endometrial, 25% bladder, 9% colorectal, 7% lung, 4% kidney CCC, 2% breast, 5% combined
	ARID2/Baf200	3% combined, 8% melanoma, 6% colorectal, 5% lung adenocarcinoma
	SNF5/SMARCB1/Baf47	7% rhabdoid, 2% esophageal, 1% combined
	PBRM1/Baf180	35% kidney CCC, 4% combined, 5% endometrial
NuRD	CHD4/Mi-2β	3% combined, 15% endometrial
CHD8	CHD8	9% GBM, 2% combined

matrix-associated, Actin-dependent regulator of chromatin, subfamily a, member 2) is present in the SWI/SNF complexes mutually exclusively with BRG1. Both BRG1 and BRM are able to reposition the nucleosome in vitro [90]. Although BRG1- and BRM-containing complexes show some redundancy, they may function distinctively [90]. Notably, BRG1-deficient mice are embryonic lethal, while BRM-deletion mice are viable [91]. In human cancer, BRG1 seems to be one of the most frequently mutated subunit genes, whereas the *BRM* gene is rarely mutated. The BRG1 subunit is either lost or mutated in a significant proportion of human primary non-small-cell lung cancer (NSCLC) samples [92]. It has also been found to be mutated in medullobalstoma [93], pancreatic cancer, breast cancer prostate cancer [94], and rhabdoid tumors [95]. Reduced BRG1 expression level can also drive tumorigenesis as evidenced by the fact that 10% BRG1 heterozygous mice develop mammary tumors [91]. Recently, it was shown that BRG1 binds to Myc (avian myelocytomatosis viral oncogene homolog) and Myc-target promoters to antagonize Myc activity and promotes cell differentiation in cancer cell lines and primary tumors [96]. Inactivation of BRG1 enables cancer cells to maintain the undifferentiated gene expression and prevents its response to environmental stimuli [96]. More recently, loss of BRG1 is shown to cooperate with oncogenic Kras (Kirsten rat sarcoma) to form cystic neoplastic lesions that resemble human intraductal papillary mucinous neoplasia (IPMN) [97]. Mechanistically, depletion of BRG1 leads to nucleosome landscape change genome-wide in vivo, highlighting its role in regulating chromatin structure and transcription [98]. Given its important role in controlling gene expression, inactivation of BRG1 can lead to aberrant expression of tumor suppressor genes and oncogenes.

4.1.2 ARID1A

ARID1A (AT-rich interactive domain 1A) and its paralog ARID1B (AT-rich interactive domain 1B) associate with other proteins to form the BRG1-associated (BAF) complexes in a mutually exclusive manner. ARID1A gene is frequently deleted in cancer [99]. Moreover, mutations in ARID1A were identified in various types of cancers. Fifty percent of ovarian clear cell carcinomas (OCCCs) and 30 percent of endometrioid carcinomas have mutations in ARID1A [100,101]. Mutations were also found in gastric carcinoma [99], esophageal adenocarcinoma [102], pediatric Burkitt lymphoma [103], medulloblastomas [93], breast cancers, and lung cancers, indicating its tumor suppressor activity in various cancer types [104]. ARID1A contains an AT-rich interaction domain (ARID) that binds to DNA without sequence specificity [105] and increases the BAF affinity to chromatin [106]. It is possible that mutations in the ARID domain disrupt the ARID-DNA interaction, thus abolishing its tumor suppression activity. ARID1A interacts with sequence-specific transcription factors, such as nuclear hormone receptors and p53, through its carboxyl-terminal domain and is required for the transcription activity of these transcription factors [107–109]. ARID1A can repress cellular proliferation in ovarian cancer [109], breast cancers [110], and gastric cancer [111]. Functional knockdown of ARID1A in these cancer cells enhances the proliferation rate. ARID1A regulates cell cycle by targeting BAF complexes directly to Myc promoter, whose expression is critical for p21 induction [112]. ARID1A physically interacts with p53 and they co-occupy the promoter of p21. Depletion of ARID1A resulted in reduced BRG1 binding at the p21 promoter and diminished p21 expression [109]. PIK3CA (phosphatidyl-inositol-4,5-bisphosphate 3-kinase, catalytic subunit alpha) mutations and ARID1A mutations are often observed hand-by-hand in ovarian clear cell carcinoma [100,113], suggesting that ARID1A mutations may cooperate with PI3K/Akt (phosphatidyl-inositol-4,5-bisphosphate 3-kinase/v-Akt murine thymoma viral oncogene homolog 1) pathway to promote cancer development. ARID1A can also help to maintain genome stability by ensuring proper chromosome segregation [114] and facilitating DNA damage repair [115].

4.1.3 ARID2

AT-rich interaction domain 2 (ARID2) was initially identified in the PBAF (Polybromo-associated BAF) complex [116]. ARID2 and ARID1A/B are mutually exclusive subunits occupying one position in the SWI/SNF complex. ARID2 is mutated in hepatocellular carcinoma [117]. Recently, ARID2 is identified as a new cancer driving gene in melanoma [118]. ARID2 is also frequently mutated in colorectal cancer as identified by exome sequencing [119]. ARID2 is mutated in 5% of non-small-cell lung cancers, making it one of the most frequently mutated genes in this cancer type [120]. Most of these mutations are inactivating mutations, suggesting that ARID2 is a potential tumor suppressor. ARID2 may help maintain the PBAF complex integrity, as depletion of ARID2 reduces the protein levels of other subunits of the PBAF complex, presumably due to stoichiometric disturbance [121]. Depletion of ARID2 also diminishes the transcription of interferon-α-induced IFITM1 (interferon-induced transmembrane protein 1), whose expression is critical for interferon-induced antiproliferative activity in hepatocellular carcinoma cells [122]. ARID2-deficient cells thus gain uncontrolled cell proliferation.

4.1.4 SNF5

SNF5 (sucrose nonfermenting factor gene number 5) is a highly conserved core subunit in the SWI/SNF complexes [90]. It has been shown to be required for the recruitment of SWI/SNF complexes to specific genes [123]. It is inactivated by biallelic mutations and/or loss of heterozygosity (LOH) in virtually all malignant rhabdoid tumors (RTs), an aggressive set of pediatric malignancies, and in a few other rare cancers [89]. In mouse models, 30% of Snf5-heterozygous mice develop sarcomas that are similar with human RTs [124,125]. Conditional inactivation of the other allele renders a cancer phenotype of full penetrance at a median onset of only 11 weeks [126].

Loss of SNF5 activates genes associated with cell proliferation, reminiscing its tumor suppressor activity [127]. SNF5 is shown to physically interact with p53, and SNF5 is recruited to p53-dependent promoters in vivo and is necessary for p53-mediated transcriptional activation [128]. Loss of p53, in addition to loss of SNF5, accelerates tumor formation [127]. Interestingly, SNF5 interacts with the MLL3 histone methyltransferase and facilitates its recruitment to chromatin [129]. Given that MLL3 is a primary H3K4me enhancer mark writer [18], it is not surprising that SNF5 is associated with gene activation. Consistent with its role in gene activation, SNF5 interacts with RNA Pol II complex containing acetyltransferases CBP and p/CAF(p300/CBP Associated Factor) [130]. As the SWI/SNF complexes and Polycomb play antagonistic roles in *Drosophila* development, mammalian SNF5 and EZH2 also exhibit antagonism toward each other [131]. Loss of the SNF5 leads to elevated expression of EZH2, and inactivation of EZH2 blocks tumor formation driven by Snf5 deficiency in mouse models.

4.1.5 PBRM1

BAF180 is composed of six tandem bromodomains that can recognize acetylated histones, two bromo-adjacent homology (BAH) domains needed for protein–protein interaction, and a high-mobility group (HMG) for binding DNA. BAF180 is a unique subunit in the polybromo/BRG1-associated factor (PBAF) complex, suggesting that it provides selective activity to this class of SWI/SNF remodelers. Indeed, it is implicated in recruiting PBAF complexes to specific loci [132]. Tumor suppressor VHL (von Hippel-Lindau) is the most frequently mutated gene in RCCs. Intriguingly, PBRM1 (polybromo 1), which encodes BAF180, is found to be mutated in 41% of the RCCs, making it a highly mutated gene, second only to VHL [133]. BAF180 is mutated in breast cancer, and it is a critical regulator of CKI p21 induction [134]. BAF180 is shown to be a critical regulator of p53 and p53-induced replicative senescence as well [135]. Recently, a study demonstrated that BAF180 is required for centrometric cohesion. Cells without cohesion undergo genome instability followed by DNA damage [136]. Thus, protecting genome from instability is another mechanism for BAF180's tumor suppressor activity.

4.2 The CHD family of chromatin remodelers

4.2.1 CHD4

The nucleosome remodeling and histone deacetylase (NuRD), also known as Mi-2 complex, is an ATP-dependent chromatin remodeling complex [137]. NuRD complex remodels chromatin structure and thus regulates gene transcription [85]. This complex is composed of six core subunits: CHD3 (also known as Mi-2α) and CHD4 (also known as Mi-2β); HDAC1 and HDAC2; methyl-CpG-binding domain 2 (MBD2) and MBD3; metastasis-associated gene1 (MTA1), MTA2, and MTA3; retinoblastoma-binding protein 4 (RBBP4) and RBBP7; and GATAD2A (GATA zinc finger domain-containing protein 2A) and GATAD2B subunits. Among them, CHD3 and CHD4 have the ATP-dependent chromatin remodeling activity; HDAC1 and HDAC2 possess the protein deacetylation activity. Nonenzymatic MBD and MTA subunits are shown to target the complex to methylated DNA. The rest of the components of the NuRD complex are thought to provide structural support and serve as protein–protein interaction intermediates [138].

Whole-exome sequencing of uterine serous carcinoma revealed frequent mutations in CHD4 [139]. Another study identified somatic mutations in CHD4 in 17% of endometrial tumors [140]. Although most mutations identified are loss-of-function mutations, NuRD complex can either suppress or promote tumorigenesis depending on the cellular context. For example, CHD4 and HDAC1 are shown to interact with DNA methyltransferases 1 (DNMT1) physically and co-occupy hypermethylated tumor suppressor gene promoters. NuRD complex cooperates with DNMT1 to maintain the silencing of several negative regulators of the oncogenic Wnt (wingless-type MMTV integration site family) signaling pathways, thus contributing to tumorigenesis in colon cancer [141]. Recently, CHD4 is shown to interact with ZFHX4 (zinc finger homeobox 4), a transcription factor that is required to maintain tumor-initiating cell phenotype in glioblastoma. ZFHX4 and CHD4 colocalize at many genomic sites, suggesting that CHD4-mediated chromatin remodeling activity is critical to maintain glioblastoma phenotype [142]. Conversely, in breast cancer, ZIP (zinc finger, CCCH-type with G patch), a zinc finger, and G-patch domain-containing protein preferentially interacts with CHD3 and CHD4 subunits of the NuRD complex to suppress genes associated with cell proliferation, survival, and migration [143].

NuRD complex's role in maintaining genome stability is well established. Loss of CHD4 expression has been observed in gastric and colorectal cancer cases that are marked by genome instability [144]. Knockdown of CHD4 leads to cell-cycle arrest at G1/S transition phase, degradation of CDC25A (cell division cycle 25A), and accumulation of p21 [145,146]. NuRD complex is also well known as an integral component of DNA repair machinery. CHD4-deficient cells exhibit hypersensitivity to DNA damage, following ionizing radiation exposure, and display an increased number of unrepaired breaks at DNA damage sites [147].

4.2.2 CHD8

CHD8 is also a member of the chromodomain helicase (CHD) family. CHD8 is composed of two amino-terminal chromodomains, an SNF2-like helicase/ATPase domain, and two uncharacterized BRK domains. Mutations in CHD8 were identified in 35% of the gastric cancers and 28% of the colorectal cancers. These mutations lead to a loss of CHD8 expression [144]. CHD8 is also significantly mutated in glioblastoma [3]. CHD8 was initially shown to interact with β-catenin, a key oncogenic component of the canonical Wnt signaling pathway, and repress β-catenin target gene expression. ATP-dependent chromatin remodeling activity of CHD8 is critical to the repression of β-catenin-targeted gene expression [148]. More recently, CHD8 is shown to promote the association of β-catenin and histone H1, forming a trimeric complex on chromatin that is required for inhibition of β-catenin-dependent transactivation [149]. In a similar mechanism, CHD8 suppresses p53-mediated transactivation and apoptosis through histone H1 recruitment during embryogenesis [150]. CHD8 was copurified with MLL histone-modifying complexes. Depletion of CHD8 resulted in a loss of MLL activity and H3K4me3 at the HOXA2 (homeobox A2) promoter [151]. However, the significance of this interaction in cancer development is still unclear. Another notable partner of CHD8 is CTCF, an insulator protein that plays a critical role in gene regulation and high-order genome organization [152]. CHD8 colocalizes with CTCF at the differentially methylated region (DMR) of H19, the locus control region of β-globin, and the promoter region of BRAC1 and the Myc genes. Ablation of CHD8-affected CTCF-dependent insulator activity at those loci indicates a role of CTCF-CHD8 complex in insulation and epigenetic remodeling [152].

5. GENOME ORGANIZATION

5.1 CTCF genome organizer and the cohesin complex

CTCF (CCCTC-binding factor) is a multiple zinc finger protein that exerts diversified functions under different genomic contexts. CTCF was first isolated and cloned on the basis of its ability to bind to highly divergent 50–60 bp sequences within the promoter region of the chicken Myc gene [153]. Subsequent studies showed that CTCF can act both as a transcriptional repressor and activator [154–156]. In addition, CTCF is well known as an insulator protein that possesses enhancer blocking function and barrier function [157]. Interestingly, recent studies show that CTCF can help to tether distal enhancers to their cognate promoters [158]. Moreover, 79% of long-range interactions between distal elements and promoters are not blocked by the presence of CTCF binding sites sitting in between [159]. This is in contrast to the traditional view of CTCF as an insulator protein.

Apart from its role in transcriptional regulation, perhaps the most famous role of CTCF is its genome organizer function [157,160]. Hi-C experiments revealed that the human genome is organized into megabase-sized topological domains [161]. Interestingly, CTCF is found to be enriched in the domain boundaries. However, CTCF binding alone is not sufficient to demarcate the domain boundaries, as only 15% of CTCF binding sites are located in the domain boundaries [161]. Additional factors such as cohesin might be also required at the topological domain boundaries [162,163]. However, cohesin and CTCF may differentially affect chromatin architecture [164]. Depletion of cohesin generally disrupts local chromatin interactions within the topological domains while keeping outer topological domains intact. In contrast, depletion of CTCF both decreases the intradomain interactions and simultaneously increases the interdomain interactions, suggesting that CTCF plays a dominant role in shaping the topological domain structure [164].

Various findings indicate that CTCF is a major tumor suppressor gene. CTCF is highly mutated in endometrial cancer and breast cancer [3]. Heterozygous deletion or mutation of CTCF was observed in leukemia [165]. Genome-wide ChIP-seq analysis revealed tens of thousands of binding sites for CTCF, indicating its wide-range regulatory function in the genome [166]. Mutations that reside in the 11 zinc finger region affect the binding specificity to DNA sequences [167]. Disruption of CTCF binding at specific gene loci can lead to aberrant expression of cancer-related genes. For example, tumor-derived CTCF mutants lost the binding affinity to growth-regulatory genes such as Myc, ARF and Igf2 (insulin-like growth factor 2) [168]. Deregulation of these genes may contribute to malignant tumor phenotypes. It is proposed that CTCF integrity may be important in controlling tumor angiogenesis [169,170]. Zinc-finger-mutated CTCF lost the binding affinity at the VEGFA (vascular endothelial growth factor A) locus, which encodes a prominent proangiogenic growth factor. Cancer cells harboring mutated CTCF thus are sensitized to hypoxia induction and gain increased angiogenic activity [169]. Conceivably, mutations in CTCF binding sites can also alter CTCF binding affinity and lead to unfavorable outcomes. Mutations in CTCF binding sites at the Igf2/H19 locus have been identified in patients with Beckwith-Wiedemann syndrome, an overgrowth disorder predisposing patients to pediatric cancer, suggesting that functional loss of CTCF binding may contribute to this disease [171]. Reduced expression level of CTCF can cause cancer as well. Hemizygous CTCF knockout mice are susceptible to spontaneous, radiation-, and chemical-induced cancer and exhibit a genome-wide perturbation in DNA methylation [172]. This evidence demonstrates that CTCF haploinsufficiency predisposes to cancer by losing CTCF-mediated epigenetic stability [172].

The cohesin complex consists of SMC1A (structural maintenance of chromosome protein 1A), SMC3 (structural maintenance of chromosome protein 3), RAD21/SCC1 (sister-chromatid cohesin protein 1), and SCC3 (sister-chromatid cohesin protein 3) (encoded by STAG-1, -2, and -3). It mediates sister chromatid cohesion to ensure accurate chromosome segregation during mitosis. Cohesin also collaborates with CTCF to control gene expression through long-distance DNA looping [173,174]. Recurrent mutations in cohesin subunits have been identified in myeloid neoplasms [175] and bladder cancer [176]. While inactivation of *STAG2* (cohesin subunit SA-2) expectedly causes chromatid cohesion defects and aneuploidy in Ewing's sarcoma, glioblastoma, and melanoma [177], it is not associated with aneuploidy in bladder cancer [178]. In addition, acute myeloid leukemias with mutations in *STAG2* and other cohesin genes have completely normal karyotypes [179]. These observations suggest that cohesin may have tumor-suppressive function independent of mitotic regulation, which may involve CTCF-related genome organization and gene regulation.

6. CONCLUDING REMARKS

Cancer genome sequencing studies have identified driver mutations in chromatin regulatory proteins, which cause deregulation of histone modifications, DNA methylation, nucleosome remodeling, and high-order chromatin organization. As most chromatin regulators are normally associated with thousands of target genes and loci throughout the genome, their disruption is likely to have profound effects on global gene expression and contribute to various forms of malignancies. Although in most cases, the precise mechanisms linking mutations in a particular chromatin regulator to corresponding neoplasms have yet to be fully deciphered, these studies demonstrate that genetic alterations give rise to epigenetic aberrations, thereby illustrating a genetic basis for altered cancer epigenome.

Unlike genetic changes, epigenetic alterations are in principle reversible by pharmacologic manipulation. Although the whole-exome analysis uncovers predominantly loss-of-function mutations, oncogenic activating mutations are also detected in a few chromatin regulators (e.g., EZH2, IDH1/2). Other gain-of-function alterations may be achieved by overexpression, amplification, and chromosomal translocations [180]. The epigenetic abnormalities resulting from deregulated chromatin-modifying enzymes offer promising opportunities for cancer therapy [2,181]. Indeed, small molecule inhibitors targeting DNA methylation and certain forms of histone methylation have led to tangible clinical benefits.

However, epigenetic gene regulation is complex. Each chromatin regulator may affect numerous genes genome-wide and is likely to be pleiotropic. It remains a challenging task to understand the many downstream effects when a chromatin factor is mutated or therapeutically targeted in cancer. Presumably depending on cellular context, a particular chromatin regulator or epigenetic modification may function as an oncogene in one setting but a tumor suppressor in other circumstances. This is exemplified by the dual roles of EZH2 and H3K27me3, and inactivating mutations in both DNMT3A and TET2 that may have opposite effects on DNA methylation. It is difficult to predict potential outcomes when a defined drug is systematically administered because different tissues may respond differently. Cautions should be taken for long-term treatment with epigenetic inhibitors.

REFERENCES

[1] Baylin SB, Jones PA. A decade of exploring the cancer epigenome-biological and translational implications. Nat Rev Cancer 2011;11:726–34.
[2] Dawson MA, Kouzarides T. Cancer epigenetics: from mechanism to therapy. Cell 2012;150:12–27.
[3] Lawrence MS, Stojanov P, Mermel CH, et al. Discovery and saturation analysis of cancer genes across 21 tumour types. Nature 2014;505:495–501.
[4] Kandoth C, McLellan MD, Vandin F, et al. Mutational landscape and significance across 12 major cancer types. Nature 2013;502:333–9.
[5] Shen H, Laird PW. Interplay between the cancer genome and epigenome. Cell 2013;153:38–55.
[6] Plass C, Pfister SM, Lindroth AM, Bogatyrova O, Claus R, Lichter P. Mutations in regulators of the epigenome and their connections to global chromatin patterns in cancer. Nat Rev Genet 2013;14:765–80.
[7] Greer EL, Shi Y. Histone methylation: a dynamic mark in health, disease and inheritance. Nat Rev Genet 2012;13:343–57.
[8] Shilatifard A. The COMPASS family of histone H3K4 methylases: mechanisms of regulation in development and disease pathogenesis. Annu Rev Biochem 2012;81:65–95.
[9] Schuettengruber B, Martinez A-M, Iovino N, Cavalli G. Trithorax group proteins: switching genes on and keeping them active. Nat Rev Mol Cell Biol 2011;12:799–814.
[10] Ringrose L, Paro R. Epigenetic regulation of cellular memory by the Polycomb and Trithorax group proteins. Annu Rev Genet 2004;38:413–43.
[11] Wang P, Lin C, Smith ER, et al. Global analysis of H3K4 methylation defines MLL family member targets and points to a role for MLL1-mediated H3K4 methylation in the regulation of transcriptional initiation by RNA polymerase II. Mol Cell Biol 2009;29:6074–85.
[12] Muntean AG, Hess JL. The pathogenesis of mixed-lineage leukemia. Annu Rev Pathol 2012;7:283–301.
[13] Mohan M, Lin C, Guest E, Shilatifard A. Licensed to elongate: a molecular mechanism for MLL-based leukaemogenesis. Nat Rev Cancer 2010;10:721–8.
[14] Milne TA, Hughes CM, Lloyd R, et al. Menin and MLL cooperatively regulate expression of cyclin-dependent kinase inhibitors. Proc Natl Acad Sci USA 2005;102:749–54.
[15] Kotake Y, Zeng Y, Xiong Y. DDB1-CUL4 and MLL1 mediate oncogene-induced p16INK4a activation. Cancer Res 2009;69:1809–14.
[16] Guo C, Chang C-C, Wortham M, et al. Global identification of MLL2-targeted loci reveals MLL2's role in diverse signaling pathways. Proc Natl Acad Sci USA 2012;109:17603–8.
[17] Herz H-M, Mohan M, Garruss AS, et al. Enhancer-associated H3K4 monomethylation by Trithorax-related, the Drosophila homolog of mammalian Mll3/Mll4. Genes Dev 2012;26:2604–20.
[18] Hu D, Gao X, Morgan MA, Herz H-M, Smith ER, Shilatifard A. The MLL3/MLL4 branches of the COMPASS family function as major histone H3K4 monomethylases at enhancers. Mol Cell Biol 2013;33:4745–54.

[19] Heintzman ND, Ren B. Finding distal regulatory elements in the human genome. Curr Opin Genet Dev 2009;19:541–9.

[20] Kanda H, Nguyen A, Chen L, Okano H, Hariharan IK. The Drosophila ortholog of MLL3 and MLL4, trithorax related, functions as a negative regulator of tissue growth. Mol Cell Biol 2013;33:1702–10.

[21] Lee J, Kim D-H, Lee S, et al. A tumor suppressive coactivator complex of p53 containing ASC-2 and histone H3-lysine-4 methyltransferase MLL3 or its paralogue MLL4. Proc Natl Acad Sci USA 2009;106:8513–8.

[22] Li Y, Trojer P, Xu C-F, et al. The target of the NSD family of histone lysine methyltransferases depends on the nature of the substrate. J Biol Chem 2009;284:34283–95.

[23] Wagner EJ, Carpenter PB. Understanding the language of Lys36 methylation at histone H3. Nat Rev Mol Cell Biol 2012;13:115–26.

[24] Edmunds JW, Mahadevan LC, Clayton AL. Dynamic histone H3 methylation during gene induction: HYPB/Setd2 mediates all H3K36 trimethylation. EMBO J 2008;27:406–20.

[25] Zhu X, He F, Zeng H, et al. Identification of functional cooperative mutations of SETD2 in human acute leukemia. Nat Genet 2014;46:287–93.

[26] Xie P, Tian C, An L, et al. Histone methyltransferase protein SETD2 interacts with p53 and selectively regulates its downstream genes. Cell Signal 2008;20:1671–8.

[27] Carvalho S, Vítor AC, Sridhara SC, et al. SETD2 is required for DNA double-strand break repair and activation of the p53-mediated checkpoint. Elife 2014;3:e02482.

[28] Pfister SX, Ahrabi S, Zalmas L-P, et al. SETD2-dependent histone H3K36 trimethylation is required for homologous recombination repair and genome stability. Cell Rep 2014;7:2006–18.

[29] Li F, Mao G, Tong D, et al. The histone mark H3K36me3 regulates human DNA mismatch repair through its interaction with MutSα. Cell 2013;153:590–600.

[30] Tatton-Brown K, Rahman N. The NSD1 and EZH2 overgrowth genes, similarities and differences. Am J Med Genet C Semin Med Genet 2013;163C:86–91.

[31] Margueron R, Li G, Sarma K, et al. Ezh1 and Ezh2 maintain repressive chromatin through different mechanisms. Mol Cell 2008;32:503–18.

[32] Kleer CG, Cao Q, Varambally S, et al. EZH2 is a marker of aggressive breast cancer and promotes neoplastic transformation of breast epithelial cells. Proc Natl Acad Sci USA 2003;100:11606–11.

[33] Morin RD, Johnson NA, Severson TM, et al. Somatic mutations altering EZH2 (Tyr641) in follicular and diffuse large B-cell lymphomas of germinal-center origin. Nat Genet 2010;42:181–5.

[34] Yap DB, Chu J, Berg T, et al. Somatic mutations at EZH2 Y641 act dominantly through a mechanism of selectively altered PRC2 catalytic activity, to increase H3K27 trimethylation. Blood 2011;117:2451–9.

[35] Sneeringer CJ, Scott MP, Kuntz KW, et al. Coordinated activities of wild-type plus mutant EZH2 drive tumor-associated hypertrimethylation of lysine 27 on histone H3 (H3K27) in human B-cell lymphomas. Proc Natl Acad Sci USA 2010;107:20980–5.

[36] Gil J, Peters G. Regulation of the INK4b-ARF-INK4a tumour suppressor locus: all for one or one for all. Nat Rev Mol Cell Biol 2006;7:667–77.

[37] Cao Q, Yu J, Dhanasekaran SM, et al. Repression of E-cadherin by the polycomb group protein EZH2 in cancer. Oncogene 2008;27:7274–84.

[38] Bracken AP, Kleine-Kohlbrecher D, Dietrich N, et al. The Polycomb group proteins bind throughout the INK4A-ARF locus and are disassociated in senescent cells. Genes Dev 2007;21:525–30.

[39] Sauvageau M, Sauvageau G. Polycomb group proteins: multi-faceted regulators of somatic stem cells and cancer. Cell Stem Cell 2010;7:299–313.

[40] Fujii S, Ochiai A. Enhancer of zeste homolog 2 downregulates E-cadherin by mediating histone H3 methylation in gastric cancer cells. Cancer Sci 2008;99:738–46.

[41] McCabe MT, Ott HM, Ganji G, et al. EZH2 inhibition as a therapeutic strategy for lymphoma with EZH2-activating mutations. Nature 2012;492:108–12.

[42] Tan J, Yang X, Zhuang L, et al. Pharmacologic disruption of Polycomb-repressive complex 2-mediated gene repression selectively induces apoptosis in cancer cells. Genes Dev 2007;21:1050–63.

[43] Nikoloski G, Langemeijer SMC, Kuiper RP, et al. Somatic mutations of the histone methyltransferase gene EZH2 in myelodysplastic syndromes. Nat Genet 2010;42:665–7.

[44] Ernst T, Chase AJ, Score J, et al. Inactivating mutations of the histone methyltransferase gene EZH2 in myeloid disorders. Nat Genet 2010;42:722–6.

[45] Schwartzentruber J, Korshunov A, Liu X-Y, et al. Driver mutations in histone H3.3 and chromatin remodelling genes in paediatric glioblastoma. Nature 2012;482:226–31.

[46] Lewis PW, Müller MM, Koletsky MS, et al. Inhibition of PRC2 activity by a gain-of-function H3 mutation found in pediatric glioblastoma. Science 2013;340:857–61.

[47] Chan K-M, Fang D, Gan H, et al. The histone H3.3K27M mutation in pediatric glioma reprograms H3K27 methylation and gene expression. Genes Dev 2013;27:985–90.

[48] Bender S, Tang Y, Lindroth AM, et al. Reduced H3K27me3 and DNA hypomethylation are major drivers of gene expression in K27M mutant pediatric high-grade gliomas. Cancer Cell 2013;24:660–72.

[49] Pedersen MT, Helin K. Histone demethylases in development and disease. Trends Cell Biol 2010;20:662–71.

[50] Van Haaften G, Dalgliesh GL, Davies H, et al. Somatic mutations of the histone H3K27 demethylase gene UTX in human cancer. Nat Genet 2009;41:521–3.

[51] Comprehensive molecular characterization of urothelial bladder carcinoma. Nature 2014;507:315–22.

[52] Iwase S, Lan F, Bayliss P, et al. The X-linked mental retardation gene SMCX/JARID1C defines a family of histone H3 lysine 4 demethylases. Cell 2007;128:1077–88.

[53] Dalgliesh GL, Furge K, Greenman C, et al. Systematic sequencing of renal carcinoma reveals inactivation of histone modifying genes. Nature 2010;463:360–3.

[54] Niu X, Zhang T, Liao L, et al. The von Hippel-Lindau tumor suppressor protein regulates gene expression and tumor growth through histone demethylase JARID1C. Oncogene 2012;31:776–86.

[55] Benevolenskaya EV, Murray HL, Branton P, Young RA, Kaelin WG. Binding of pRB to the PHD protein RBP2 promotes cellular differentiation. Mol Cell 2005;18:623–35.

[56] Liefke R, Oswald F, Alvarado C, et al. Histone demethylase KDM5A is an integral part of the core Notch-RBP-J repressor complex. Genes Dev 2010;24:590–601.

[57] Baba A, Ohtake F, Okuno Y, et al. PKA-dependent regulation of the histone lysine demethylase complex PHF2-ARID5B. Nat Cell Biol 2011;13:668–75.

[58] Wen H, Li J, Song T, et al. Recognition of histone H3K4 trimethylation by the plant homeodomain of PHF2 modulates histone demethylation. J Biol Chem 2010;285:9322–6.

[59] Wang H, Wang L, Erdjument-Bromage H, et al. Role of histone H2A ubiquitination in Polycomb silencing. Nature 2004;431:873–8.

[60] Scheuermann JC, de Ayala Alonso AG, Oktaba K, et al. Histone H2A deubiquitinase activity of the Polycomb repressive complex PR-DUB. Nature 2010;465:243–7.

[61] Katoh M. Functional and cancer genomics of ASXL family members. Br J Cancer 2013;109:299–306.

[62] Carbuccia N, Murati A, Trouplin V, et al. Mutations of ASXL1 gene in myeloproliferative neoplasms. Leukemia 2009;23:2183–6.

[63] Gelsi-Boyer V, Trouplin V, Adélaïde J, et al. Mutations of polycomb-associated gene ASXL1 in myelodysplastic syndromes and chronic myelomonocytic leukaemia. Br J Haematol 2009;145:788–800.

[64] Abdel-Wahab O, Adli M, LaFave LM, et al. ASXL1 mutations promote myeloid transformation through loss of PRC2-mediated gene repression. Cancer Cell 2012;22:180–93.

[65] Testa JR, Cheung M, Pei J, et al. Germline BAP1 mutations predispose to malignant mesothelioma. Nat Genet 2011;43:1022–5.

[66] Wiesner T, Obenauf AC, Murali R, et al. Germline mutations in BAP1 predispose to melanocytic tumors. Nat Genet 2011;43:1018–21.

[67] Jensen DE, Proctor M, Marquis ST, et al. BAP1: a novel ubiquitin hydrolase which binds to the BRCA1 RING finger and enhances BRCA1-mediated cell growth suppression. Oncogene 1998;16:1097–112.

[68] Yang X-J, Seto E. HATs and HDACs: from structure, function and regulation to novel strategies for therapy and prevention. Oncogene 2007;26:5310–8.

[69] Iyer NG, Ozdag H, Caldas C. p300/CBP and cancer. Oncogene 2004;23:4225–31.

[70] Liu X, Wang L, Zhao K, et al. The structural basis of protein acetylation by the p300/CBP transcriptional coactivator. Nature 2008;451:846–50.

[71] Wang F, Marshall CB, Ikura M. Transcriptional/epigenetic regulator CBP/p300 in tumorigenesis: structural and functional versatility in target recognition. Cell Mol Life Sci 2013;70:3989–4008.

[72] Ley TJ, Ding L, Walter MJ, et al. DNMT3A mutations in acute myeloid leukemia. N Engl J Med 2010;363:2424–33.

[73] Jia D, Jurkowska RZ, Zhang X, Jeltsch A, Cheng X. Structure of Dnmt3a bound to Dnmt3L suggests a model for de novo DNA methylation. Nature 2007;449:248–51.

[74] Kim SJ, Zhao H, Hardikar S, Singh AK, Goodell MA, Chen T. A DNMT3A mutation common in AML exhibits dominant-negative effects in murine ES cells. Blood 2013;122:4086–9.

[75] Russler-Germain DA, Spencer DH, Young MA, et al. The R882H DNMT3A mutation associated with AML dominantly inhibits wild-type DNMT3A by blocking its ability to form active tetramers. Cancer Cell 2014;25:442–54.

[76] Xu J, Wang Y-Y, Dai Y-J, et al. DNMT3A Arg882 mutation drives chronic myelomonocytic leukemia through disturbing gene expression/DNA methylation in hematopoietic cells. Proc Natl Acad Sci USA 2014;111:2620–5.

[77] Kohli RM, Zhang Y. TET enzymes, TDG and the dynamics of DNA demethylation. Nature 2013;502:472–9.

[78] Ko M, Huang Y, Jankowska AM, et al. Impaired hydroxylation of 5-methylcytosine in myeloid cancers with mutant TET2. Nature 2010;468:839–43.

[79] Solary E, Bernard OA, Tefferi A, Fuks F, Vainchenker W. The Ten-Eleven Translocation-2 (TET2) gene in hematopoiesis and hematopoietic diseases. Leukemia 2014;28:485–96.

[80] Dang L, White DW, Gross S, et al. Cancer-associated IDH1 mutations produce 2-hydroxyglutarate. Nature 2009;462:739–44.

[81] Ward PS, Patel J, Wise DR, et al. The common feature of leukemia-associated IDH1 and IDH2 mutations is a neomorphic enzyme activity converting alpha-ketoglutarate to 2-hydroxyglutarate. Cancer Cell 2010;17:225–34.

[82] Figueroa ME, Abdel-Wahab O, Lu C, et al. Leukemic IDH1 and IDH2 mutations result in a hypermethylation phenotype, disrupt TET2 function, and impair hematopoietic differentiation. Cancer Cell 2010;18:553–67.

[83] Turcan S, Rohle D, Goenka A, et al. IDH1 mutation is sufficient to establish the glioma hypermethylator phenotype. Nature 2012;483:479–83.

[84] Lu C, Ward PS, Kapoor GS, et al. IDH mutation impairs histone demethylation and results in a block to cell differentiation. Nature 2012;483:474–8.

[85] Clapier CR, Cairns BR. The biology of chromatin remodeling complexes. Annu Rev Biochem 2009;78:273–304.

[86] Bowman GD. Mechanisms of ATP-dependent nucleosome sliding. Curr Opin Struct Biol 2010;20:73–81.

[87] Saha A, Wittmeyer J, Cairns BR. Chromatin remodelling: the industrial revolution of DNA around histones. Nat Rev Mol Cell Biol 2006;7:437–47.

[88] Kadoch C, Hargreaves DC, Hodges C, et al. Proteomic and bioinformatic analysis of mammalian SWI/SNF complexes identifies extensive roles in human malignancy. Nat Genet 2013;45:592–601.

[89] Wilson BG, Roberts CWM. SWI/SNF nucleosome remodellers and cancer. Nat Rev Cancer 2011;11:481–92.

[90] Phelan ML, Sif S, Narlikar GJ, Kingston RE. Reconstitution of a core chromatin remodeling complex from SWI/SNF subunits. Mol Cell 1999;3:247–53.

[91] Bultman S, Gebuhr T, Yee D, et al. A Brg1 null mutation in the mouse reveals functional differences among mammalian SWI/SNF complexes. Mol Cell 2000;6:1287–95.

[92] Medina PP, Romero OA, Kohno T, et al. Frequent BRG1/SMARCA4-inactivating mutations in human lung cancer cell lines. Hum Mutat 2008;29:617–22.

[93] Parsons DW, Li M, Zhang X, et al. The genetic landscape of the childhood cancer medulloblastoma. Science 2011;331:435–9.

[94] Wong AK, Shanahan F, Chen Y, et al. BRG1, a component of the SWI-SNF complex, is mutated in multiple human tumor cell lines. Cancer Res 2000;60:6171–7.

[95] Schneppenheim R, Frühwald MC, Gesk S, et al. Germline nonsense mutation and somatic inactivation of SMARCA4/BRG1 in a family with rhabdoid tumor predisposition syndrome. Am J Hum Genet 2010;86:279–84.

[96] Romero OA, Setien F, John S, et al. The tumour suppressor and chromatin-remodelling factor BRG1 antagonizes Myc activity and promotes cell differentiation in human cancer. EMBO Mol Med 2012;4:603–16.

[97] Von Figura G, Fukuda A, Roy N, et al. The chromatin regulator Brg1 suppresses formation of intraductal papillary mucinous neoplasm and pancreatic ductal adenocarcinoma. Nat Cell Biol 2014;16:255–67.

[98] Tolstorukov MY, Sansam CG, Lu P, et al. Swi/Snf chromatin remodeling/tumor suppressor complex establishes nucleosome occupancy at target promoters. Proc Natl Acad Sci USA 2013;110: 10165–70.

[99] Bagchi A, Mills AA. The quest for the 1p36 tumor suppressor. Cancer Res 2008;68:2551–6.

[100] Jones S, Wang T-L, Shih I-M, et al. Frequent mutations of chromatin remodeling gene ARID1A in ovarian clear cell carcinoma. Science 2010;330:228–31.

[101] Wiegand KC, Shah SP, Al-Agha OM, Zhao Y, Al E. ARID1A mutations in endometriosis-associated ovarian carcinomas. N Engl J Med 2011;363:1532–43.

[102] Wang X, Nagl NG, Flowers S, Zweitzig D, Dallas PB, Moran E. Expression of p270 (ARID1A), a component of human SWI/SNF complexes, in human tumors. Int J Cancer 2004;112:255.

[103] Giulino-Roth L, Wang K, MacDonald TY, et al. Targeted genomic sequencing of pediatric Burkitt lymphoma identifies recurrent alterations in antiapoptotic and chromatin-remodeling genes. Blood 2012;120:5181–4.

[104] Huang J, Zhao Y-L, Li Y, Fletcher JA, Xiao S. Genomic and functional evidence for an ARID1A tumor suppressor role. Genes Chromosom Cancer 2007;46:745–50.

[105] Dallas PB, Pacchione S, Wilsker D, Bowrin V, Kobayashi R, Moran E. The human SWI-SNF complex protein p270 is an ARID family member with non-sequence-specific DNA binding activity. Mol Cell Biol 2000;20:3137–46.

[106] Chandler RL, Brennan J, Schisler JC, Serber D, Patterson C, Magnuson T. ARID1a-DNA interactions are required for promoter occupancy by SWI/SNF. Mol Cell Biol 2013;33:265–80.

[107] Nie Z, Xue Y, Yang D, et al. A specificity and targeting subunit of a human SWI/SNF family-related chromatin-remodeling complex. Mol Cell Biol 2000;20:8879–88.

[108] Inoue H, Furukawa T, Giannakopoulos S, Zhou S, King DS, Tanese N. Largest subunits of the human SWI/SNF chromatin-remodeling complex promote transcriptional activation by steroid hormone receptors. J Biol Chem 2002;277:41674–85.

[109] Guan B, Wang T-L, Shih I-M. ARID1A, a factor that promotes formation of SWI/SNF-mediated chromatin remodeling, is a tumor suppressor in gynecologic cancers. Cancer Res 2011;71: 6718–27.

[110] Mamo A, Cavallone L, Tuzmen S, et al. An integrated genomic approach identifies ARID1A as a candidate tumor-suppressor gene in breast cancer. Oncogene 2012;31:2090–100.

[111] Zang ZJ, Cutcutache I, Poon SL, et al. Exome sequencing of gastric adenocarcinoma identifies recurrent somatic mutations in cell adhesion and chromatin remodeling genes. Nat Genet 2012;44: 570–4.

[112] Nagl Jr NG, Zweitzig DR, Thimmapaya B, Beck Jr GR, Moran E. The c-myc gene is a direct target of mammalian SWI/SNF-related complexes during differentiation-associated cell cycle arrest. Cancer Res 2006;66:1289–93.

[113] Huang H-N, Lin M-C, Huang W-C, Chiang Y-C, Kuo K-T. Loss of ARID1A expression and its relationship with PI3K-Akt pathway alterations and ZNF217 amplification in ovarian clear cell carcinoma. Mod Pathol 2014;27:983–90.

[114] Dykhuizen EC, Hargreaves DC, Miller EL, et al. BAF complexes facilitate decatenation of DNA by topoisomerase IIα. Nature 2013;497:624–7.

[115] Park J-H, Park E-J, Lee H-S, et al. Mammalian SWI/SNF complexes facilitate DNA double-strand break repair by promoting gamma-H2AX induction. EMBO J 2006;25:3986–97.

[116] Wang W, Côté J, Xue Y, et al. Purification and biochemical heterogeneity of the mammalian SWI-SNF complex. EMBO J 1996;15:5370–82.

[117] Li M, Zhao H, Zhang X, et al. Inactivating mutations of the chromatin remodeling gene ARID2 in hepatocellular carcinoma. Nat Genet 2011;43:828–9.

[118] Hodis E, Watson IR, Kryukov GV, et al. A landscape of driver mutations in melanoma. Cell 2012;150:251–63.

[119] Cajuso T, Hänninen UA, Kondelin J, et al. Exome sequencing reveals frequent inactivating mutations in ARID1A, ARID1B, ARID2, and ARID4A in microsatellite unstable colorectal cancer. Int J Cancer 2013;2:1–13.

[120] Manceau G, Letouzé E, Guichard C, et al. Recurrent inactivating mutations of ARID2 in non-small cell lung carcinoma. Int J Cancer 2013;132:2217–21.

[121] Yan Z, Cui K, Murray DM, et al. PBAF chromatin-remodeling complex requires a novel specificity subunit, BAF200, to regulate expression of selective interferon-responsive genes. Genes Dev 2005;19:1662–7.

[122] Yang G, Xu Y, Chen X, Hu G. IFITM1 plays an essential role in the antiproliferative action of interferon-gamma. Oncogene 2007;26:594–603.

[123] Oruetxebarria I, Venturini F, Kekarainen T, et al. P16INK4a is required for hSNF5 chromatin remodeler-induced cellular senescence in malignant rhabdoid tumor cells. J Biol Chem 2004;279:3807–16.

[124] Roberts CW, Galusha SA, McMenamin ME, Fletcher CD, Orkin SH. Haploinsufficiency of Snf5 (integrase interactor 1) predisposes to malignant rhabdoid tumors in mice. Proc Natl Acad Sci USA 2000;97:13796–800.

[125] Klochendler-Yeivin A, Fiette L, Barra J, Muchardt C, Babinet C, Yaniv M. The murine SNF5/INI1 chromatin remodeling factor is essential for embryonic development and tumor suppression. EMBO Rep 2000;1:500–6.

[126] Roberts CWM, Leroux MM, Fleming MD, Orkin SH. Highly penetrant, rapid tumorigenesis through conditional inversion of the tumor suppressor gene Snf5. Cancer Cell 2002;2:415–25.

[127] Isakoff MS, Sansam CG, Tamayo P, et al. Inactivation of the Snf5 tumor suppressor stimulates cell cycle progression and cooperates with p53 loss in oncogenic transformation. Proc Natl Acad Sci USA 2005;102:17745–50.

[128] Lee D, Kim JW, Seo T, Hwang SG, Choi E-J, Choe J. SWI/SNF complex interacts with tumor suppressor p53 and is necessary for the activation of p53-mediated transcription. J Biol Chem 2002;277:22330–7.

[129] Lee S, Kim D-H, Goo YH, Lee YC, Lee S-K, Lee JW. Crucial roles for interactions between MLL3/4 and INI1 in nuclear receptor transactivation. Mol Endocrinol 2009;23:610–9.

[130] Cho H, Orphanides G, Sun X, et al. A human RNA polymerase II complex containing factors that modify chromatin structure. Mol Cell Biol 1998;18:5355–63.

[131] Wilson BG, Wang X, Shen X, et al. Epigenetic antagonism between polycomb and SWI/SNF complexes during oncogenic transformation. Cancer Cell 2010;18:316–28.

[132] Thompson M. Polybromo-1: the chromatin targeting subunit of the PBAF complex. Biochimie 2009;91:309–19.

[133] Varela I, Tarpey P, Raine K, et al. Exome sequencing identifies frequent mutation of the SWI/SNF complex gene PBRM1 in renal carcinoma. J Urol 2011;186:1150.

[134] Xia W, Nagase S, Montia AG, et al. BAF180 is a critical regulator of p21 induction and a tumor suppressor mutated in breast cancer. Cancer Res 2008;68:1667–74.

[135] Burrows AE, Smogorzewska A, Elledge SJ. Polybromo-associated BRG1-associated factor components BRD7 and BAF180 are critical regulators of p53 required for induction of replicative senescence. Proc Natl Acad Sci USA 2010;107:14280–5.

[136] Brownlee PM, Chambers AL, Cloney R, Bianchi A, Downs JA. BAF180 promotes cohesion and prevents genome instability and aneuploidy. Cell Rep 2014;6:973–81.

[137] Tong JK, Hassig CA, Schnitzler GR, Kingston RE, Schreiber SL. Chromatin deacetylation by an ATP-dependent nucleosome remodelling complex. Nature 1998;395:917–21.

[138] Lai AY, Wade PA. Cancer biology and NuRD: a multifaceted chromatin remodelling complex. Nat Rev Cancer 2011;11:588–96.

[139] Zhao S, Choi M, Overton JD, et al. Landscape of somatic single-nucleotide and copy-number mutations in uterine serous carcinoma. Proc Natl Acad Sci USA 2013;110:2916–21.

[140] Le Gallo M, O'Hara AJ, Rudd ML, et al. Exome sequencing of serous endometrial tumors identifies recurrent somatic mutations in chromatin-remodeling and ubiquitin ligase complex genes. Nat Genet 2012;44:1310–5.

[141] Cai Y, Geutjes E-J, de Lint K, et al. The NuRD complex cooperates with DNMTs to maintain silencing of key colorectal tumor suppressor genes. Oncogene 2014;33:2157–68.

[142] Chudnovsky Y, Kim D, Zheng S, et al. ZFHX4 interacts with the NuRD core member CHD4 and regulates the glioblastoma tumor-initiating cell state. Cell Rep 2014;6:313–24.

[143] Li R, Zhang H, Yu W, et al. ZIP: a novel transcription repressor, represses EGFR oncogene and suppresses breast carcinogenesis. EMBO J 2009;28:2763–76.

[144] Kim MS, Chung NG, Kang MR, Yoo NJ, Lee SH. Genetic and expressional alterations of CHD genes in gastric and colorectal cancers. Histopathology 2011;58:660–8.

[145] Polo SE, Kaidi A, Baskcomb L, Galanty Y, Jackson SP. Regulation of DNA-damage responses and cell-cycle progression by the chromatin remodelling factor CHD4. EMBO J 2010;29:3130–9.

[146] Larsen DH, Poinsignon C, Gudjonsson T, et al. The chromatin-remodeling factor CHD4 coordinates signaling and repair after DNA damage. J Cell Biol 2010;190:731–40.

[147] Chou DM, Adamson B, Dephoure NE, et al. A chromatin localization screen reveals poly (ADP ribose)-regulated recruitment of the repressive polycomb and NuRD complexes to sites of DNA damage. Proc Natl Acad Sci USA 2010;107:18475–80.

[148] Thompson BA, Tremblay V, Lin G, Bochar DA. CHD8 is an ATP-dependent chromatin remodeling factor that regulates beta-catenin target genes. Mol Cell Biol 2008;28:3894–904.

[149] Nishiyama M, Skoultchi AI, Nakayama KI. Histone H1 recruitment by CHD8 is essential for suppression of the Wnt-β-catenin signaling pathway. Mol Cell Biol 2012;32:501–12.

[150] Nishiyama M, Oshikawa K, Tsukada Y, et al. CHD8 suppresses p53-mediated apoptosis through histone H1 recruitment during early embryogenesis. Nat Cell Biol 2009;11:172–82.

[151] Yates JA, Menon T, Thompson BA, Bochar DA. Regulation of HOXA2 gene expression by the ATP-dependent chromatin remodeling enzyme CHD8. FEBS Lett 2010;584:689–93.

[152] Ishihara K, Oshimura M, Nakao M. CTCF-dependent chromatin insulator is linked to epigenetic remodeling. Mol Cell 2006;23:733–42.

[153] Klenova EM, Nicolas RH, Paterson HF, et al. CTCF, a conserved nuclear factor required for optimal transcriptional activity of the chicken c-myc gene, is an 11-Zn-finger protein differentially expressed in multiple forms. Mol Cell Biol 1993;13:7612–24.

[154] Baniahmad A, Steiner C, Kohne AC, Renrkawitz R. Modular structure of a chicken lysozyme silencer: involvement of an unusual thyroid hormone receptor binding site. Cell 1990;61:505–14.

[155] Burcin M, Arnold R, Lutz M, et al. Negative protein 1, which is required for function of the chicken lysozyme gene silencer in conjunction with hormone receptors, is identical to the multivalent zinc finger repressor CTCF. Mol Cell Biol 1997;17:1281–8.

[156] Vostrov AA, Quitschke WW. The zinc finger protein CTCF binds to the APBbeta domain of the amyloid beta-protein precursor promoter. Evidence for a role in transcriptional activation. J Biol Chem 1997;272:33353–9.

[157] Phillips JE, Corces VG. CTCF: master weaver of the genome. Cell 2009;137:1194–211.

[158] Shen Y, Yue F, McCleary DF, et al. A map of the *cis*-regulatory sequences in the mouse genome. Nature 2012;488:116–20.

[159] Sanyal A, Lajoie BR, Jain G, Dekker J. The long-range interaction landscape of gene promoters. Nature 2012;489:109–13.

[160] Ong C-T, Corces VG. CTCF: an architectural protein bridging genome topology and function. Nat Rev Genet 2014;15:234–46.

[161] Dixon JR, Selvaraj S, Yue F, et al. Topological domains in mammalian genomes identified by analysis of chromatin interactions. Nature 2012;485:376–80.

[162] Phillips-Cremins J, Sauria MG, Sanyal A, et al. Architectural protein subclasses shape 3D organization of genomes during lineage commitment. Cell 2013;153:1281–95.

[163] DeMare LE, Leng J, Cotney J, et al. The genomic landscape of cohesin-associated chromatin interactions. Genome Res 2013;23:1224–34.

[164] Zuin J, Dixon JR, van der Reijden MIJA, et al. Cohesin and CTCF differentially affect chromatin architecture and gene expression in human cells. Proc Natl Acad Sci USA 2014;111:996–1001.

[165] Yoshida K, Toki T, Okuno Y, et al. The landscape of somatic mutations in Down syndrome-related myeloid disorders. Nat Genet 2013;45:1293–9.

[166] Barski A, Cuddapah S, Cui K, et al. High-resolution profiling of histone methylations in the human genome. Cell 2007;129:823–37.

[167] Ciriello G, Miller ML, Aksoy BA, Senbabaoglu Y, Schultz N, Sander C. Emerging landscape of oncogenic signatures across human cancers. Nat Genet 2013;45:1127–33.

[168] Filippova GN, Qi C-F, Ulmer JE, et al. Tumor-associated zinc finger mutations in the CTCF transcription factor selectively alter tts DNA-binding specificity. Cancer Res 2002;62:48–52.

[169] Tang M, Chen B, Lin T, et al. Restraint of angiogenesis by zinc finger transcription factor CTCF-dependent chromatin insulation. Proc Natl Acad Sci USA 2011;108:15231–6.

[170] Lu J, Tang M. CTCF-dependent chromatin insulator as a built-in attenuator of angiogenesis. Transcription 2012;3:73–7.

[171] Sparago A, Russo S, Cerrato F, et al. Mechanisms causing imprinting defects in familial Beckwith-Wiedemann syndrome with Wilms' tumour. Hum Mol Genet 2007;16:254–64.

[172] Kemp CJ, Moore JM, Moser R, et al. CTCF haploinsufficiency destabilizes DNA methylation and predisposes to cancer. Cell Rep 2014;7:1020–9.

[173] Nasmyth K, Haering CH. Cohesin: its roles and mechanisms. Annu Rev Genet 2009;43:525–58.

[174] Remeseiro S, Losada A. Cohesin, a chromatin engagement ring. Curr Opin Cell Biol 2013;25:63–71.

[175] Kon A, Shih L-Y, Minamino M, et al. Recurrent mutations in multiple components of the cohesin complex in myeloid neoplasms. Nat Genet 2013;45:1232–7.

[176] Solomon DA, Kim J-S, Bondaruk J, et al. Frequent truncating mutations of STAG2 in bladder cancer. Nat Genet 2013;45:1428–30.

[177] Solomon DA, Kim T, Diaz-Martinez LA, et al. Mutational inactivation of STAG2 causes aneuploidy in human cancer. Science 2011;333:1039–43.

[178] Balbás-Martínez C, Sagrera A, Carrillo-de-Santa-Pau E, et al. Recurrent inactivation of STAG2 in bladder cancer is not associated with aneuploidy. Nat Genet 2013;45:1464–9.

[179] Walter MJ, Payton JE, Ries RE, et al. Acquired copy number alterations in adult acute myeloid leukemia genomes. Proc Natl Acad Sci USA 2009;106:12950–5.

[180] Zack TI, Schumacher SE, Carter SL, et al. Pan-cancer patterns of somatic copy number alteration. Nat Genet 2013;45:1134–40.

[181] Helin K, Dhanak D. Chromatin proteins and modifications as drug targets. Nature 2013;502:480–8.

CHAPTER 13

Long noncoding RNAs and carcinogenesis

Félix Recillas-Targa

Instituto de Fisiología Celular, Departamento de Genética Molecular, Universidad Nacional Autónoma de México, Ciudad de México, México

Contents

1. INTRODUCTION

More than five decades ago, Scherrer and Darnell proposed the existence of the "giant RNAs" would correspond to large transcripts covering noncoding sequences and represent a major part of primary transcripts with a high molecular weight [1–4]. At that time, pure biochemical methods were used to identify these types of molecules, mainly sedimentation, but also pulse labeling and spectrophotometry. With these strategies, Scherrer, Darnell, and other biochemists started to define not only processed RNAs but also the identification of other subfamilies of RNAs, among them the so-called giant RNAs with no-specific defined function at that time [1–4]. Some years later, Darnell discovered the nuclear presence of heterogeneous RNAs that are not assembled into polyribosomes; this may reflect what we rediscovered in more recent times as noncoding RNAs [5]. Moreover, Penman and collaborators started to characterize and visualize another subfamily of RNAs, the heterogeneous nuclear RN-proteins (hnRNPs), as a component of fibers contributing to cell nucleus substructures. They demonstrated that hnRNA is a structural component of the nuclear matrix and chromatin architecture [6,7]. These observations have been, to a certain extent, controversial, but show that between the 1960s and 1980s, we can find the seminal works that started to suggest that the RNA world was not simply restricted to premessenger RNAs and messenger RNAs with coding potential.

The following years were critical for progress into the functional description of relatively small and specialized subsets of RNAs that included ribosomal RNAs, small nuclear and nucleolar RNAs, and transfer RNAs linked to the translation of RNA information into proteins. In recent years, another subfamily of RNAs has emerged with regulatory functions at distinct levels. For example, the microRNAs (miRNAs or miRs) have been identified that induce the degradation of target transcripts or the blockage of translation of those mRNAs; the piwi (P-element induced wimpy testis)-associated RNA (piRNA) was identified and found to be involved in the formation of heterochromatin mainly in the male germline; and the small interfering RNAs (siRNAs) were shown to induce the degradation of a perfectly complementary target RNA [8]. The majority of all of the above-mentioned RNAs were less than 200 nucleotides in length and considered (as part of the frequently cited and arbitrarily assigned definition) as small noncoding RNAs [9]. Genome-wide studies, in particular the massive RNA sequencing (RNA-seq) strategies, the Functional Annotation of the Mammalian Genome (FANTOM) Consortium [10] and the ENCODE consortium [11] revealed that between 70% and 80% of the human genome is transcribed, leaving 2% of the genome with the possibility of being translated into a functional protein. These observations at the scale of entire genomes led to the conclusion that a significant number of noncoding transcripts are larger than 200 nucleotides, and, therefore, are considered as long noncoding RNAs (lncRNAs).

In this chapter, the relationship between lncRNAs and carcinogens will be addressed. First, a brief overview of lncRNAs' features and mechanisms of action will be given to provide a better understanding of the different levels in which lncRNAs act. With this background in mind, the most relevant and recent lncRNAs involved in cancer development will be described. Therefore, future directions and possible therapeutic applications should consider targeting lncRNAs, especially in diverse cellular processes such as cancers.

2. GENERAL FEATURES AND MECHANISMS OF ACTION OF THE LONG NONCODING RNA

2.1 Major subtypes of lncRNAs

Based on their location relative to coding sequences, the lncRNAs are classified as intragenic and intergenic, with intergenic being designated as long intergenic noncoding RNAs or lincRNAs. One of the most abundant classes of lncRNAs is the so-called natural antisense transcripts (NATs) that correspond to noncoding RNAs that are complementary, with different degrees of sequence complementation, to another transcript that frequently correspond to a coding sequence [12]. This type of noncoding RNA raises an interesting question that has to do with its transcriptional regulation, or coregulation, depending on the location and degree of sequence complementarity. Another type of lncRNA is the sense overlapping RNA that corresponds to an RNA that is transcribed on the same DNA strand but with a different sequence, and includes the sense intronic RNA, which is transcribed exclusively within an intron of a coding gene (Figure 1(A)). And finally, there are examples in which a combination of these subtypes exist and can be processed generating different isoforms [13,14]. There are even cases of lncRNAs, divergent between them or with respect to a coding gene, that apparently share their promoter regions (Figure 1(A)). All these locations may imply a large variety of functions, as for example their regulatory action in *cis* over adjacent regulatory elements or genes (see below) or the need of a tight coregulation in a tissue- and time-specific manner.

Interestingly, as part of the Encyclopedia of DNA Element (ENCODE) Project, the Encyclopædia of Genes and Gene Variants (GENCODE) Consortium has performed one of the most complete human lncRNA annotations [13]. In this study, they found 14,880 lncRNAs, among which 9519 correspond to genic and 9518 to intergenic lncRNAs. It would not be surprising if, in the future, the total number of lncRNAs continued to rise, based on novel strategies used to define them.

2.2 Detailed features of the lncRNAs

The genome-wide studies, along with the coupling of the generated databases, have shown the distribution of RNA polymerases and chromatin-associated modifications and proteins. This demonstrates that the majority of the lncRNAs are transcribed by the RNA polymerase II, and that the distribution of histone modifications is similar to the ones found in coding genes. Some histone marks include H3K4me3 in the promoter region of the lncRNAs and H3K36me3 associated with the transcriptional elongation of lncRNAs (Figure 1(B)). In addition, the lncRNAs often present with a 5′-terminal methylguanosine cap and are 3′-polyadenylated. Importantly, the genome-wide survey of lncRNAs has shown that approximately 42% of the lncRNA transcripts have two exons.

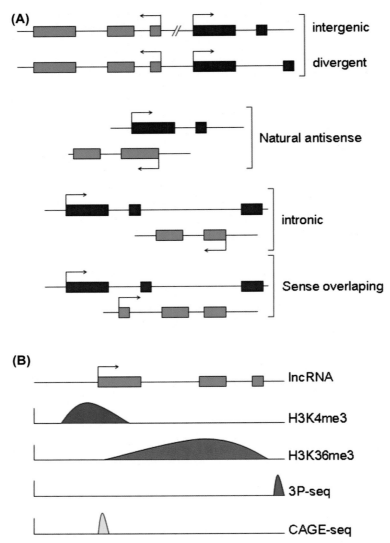

Figure 1 Subtypes of long noncoding RNAs. (A) The distribution of the main subtypes of lncRNAs (shown in green), which are frequently defined in relation to their location relative to the protein-coding genes (shown in blue). Note that, the lncRNAs possess exons that can be processed and even spliced differentially to generate different isoforms [13] (B) Genome-scale distribution of chromatin-associated features of noncoding RNAs, as well as polyadenylated ends sequencing (3P-seq) and capped RNA sequencing (CAGE-seq) fragments.

The great majority are spliced with canonical splice site signals, and, furthermore, some lncRNAs are subject to alternative splicing [13]. Finally, one relevant aspect of lncRNAs' function is linked to their relative abundance. The lncRNAs are, in general orders of magnitude, less abundant than protein encoding genes. This is a very enigmatic feature of the lncRNAs that supports a regulatory role that requires significantly fewer transcripts in order to translate their function into regulatory signals. Of note, particular caution should be paid to the fact that, recently, it has been demonstrated that some lncRNAs have the potential to encode small polypeptides with emerging new roles [15].

2.3 The lncRNA multifunctional capacity based on their physiological properties

One of the most evident aspects of the lncRNA is the intrinsic capacity of the RNA to acquire multiple secondary and tertiary structures. This certainly has direct consequences on their multifunctional capacities. The most evident functional interaction is the one between RNA and proteins [16]. Such interactions may be highly variable, in particular due to the flexibility of the RNA molecules to acquire different stem-loop structures. These interactions can also bring together or into proximity different proteins, including chromatin components. These types of multiple interactions have led to the proposal of a model of modular interaction for the function of lncRNAs [17]. Studies have shown that such interactions are difficult to define, and, in addition, there are frequent unspecific associations between RNA and a large myriad of proteins [18]. Therefore, the prediction of complex RNA structures interacting with proteins is not an easy task. The capacity of the lncRNAs is not restricted to the RNA–protein interactions. There is a growing list of examples showing that RNA molecules can interact with DNA, and recruit in *trans*-regulatory proteins [19,20]. We also cannot disregard RNA–RNA interactions and more recently, tripartite DNA–protein–RNA interactions, which have been shown to have functional consequences such as in the case of the p53–CTCF–Wrap53 association [21].

In conclusion, a large amount of research is needed to better understand how RNA secondary and tertiary structures are adopted in response to signals or molecular contacts. This work will be critical in order to validate or discard some of the models proposed for the mechanisms of action of lncRNAs.

2.4 Mechanisms of action of the lncRNAs

One of the most enigmatic aspects of lncRNAs is their way of action. The most direct mechanism is associated with the ability of lncRNAs to base pair with diverse types of RNAs, such as mRNA, miRNA (see below), and other lncRNAs (including NATs) [9,12,22]. Another aspect that is less explored is the role of lncRNAs as precursors to generate different isoforms, as well as small RNAs that may serve as effector molecules [23,24].

Another mechanism of action of lncRNAs is the one that takes into consideration their interaction with proteins. As mentioned before, the diversity of RNA structures allows for a regulated contact with proteins to direct different interactions [18,25]. One of the most studied mechanisms, in particular through the interaction with chromatin remodeling complexes, is the guide model [26]. This model accounts for the recruitment of regulatory proteins to their target sequences, in *cis* or *trans*. In contrast, in the decoy model, lncRNAs compete or titrate proteins (similar to, for example, transcription factors), instead of guiding the regulatory proteins to their site of action, thereby attenuating their regulatory action. The scaffold model is, in some ways, related to the previous one, and is based on the ability of an lncRNA to attract one or more proteins into a complex and/or guide them to specific genomic locations. Based on such models, it turns out that the need to develop new experimental strategies to decipher RNA–protein interactions at the entire genome scale is critical. One of the most reliable strategies is the use of photoactivatable ribonucleoside-enhanced cross-linking and immunoprecipitation (PAR-CLIP). This method provides a way that reduces unspecific interactions in order to define RNA–protein interactions [21,27]. These kinds of experimental strategies will provide novel insights concerning the functional properties of lncRNAs.

Interestingly, there are many models for lncRNAs, but one that recently came to light is based on the one associated with transcriptional enhancers, known as enhancer RNAs (eRNAs) [28]. These kind of lncRNAs act at different levels by either blocking or attracting cofactors to the enhancer elements, which favor the formation of chromatin loops to contact long-distance regulatory elements, like other enhancers, to their target promoter [29–31]. It is important to comment that, presently, the generality of this mechanism is difficult to estimate. However, genome-wide studies have demonstrated transcription of a large portion of these enhancers.

Together, these are the models associated with the mechanisms of action of lncRNAs. In this field, it is clear that the vision of lncRNA is incomplete; therefore, based on the number and structural properties of lncRNAs, many other mechanisms may exist.

3. DIFFERENT FUNCTIONAL STRATEGIES OF lncRNAs

The general features of lncRNAs, as well as their structural properties and some of their mechanisms of action, have been described previously. Now, it is time to describe the biological functions associated to lncRNAs. To a certain extent, it is assumed that lncRNAs are involved in chromatin remodeling. Historically, the noncoding maternally expressed transcript H19 was the first to be isolated and characterized [32], and over the years it has been shown to be involved in genomic imprinting, as are several other lncRNAs. LncRNAs also participate in diverse regulatory mechanisms such as dosage compensation through X-chromosome inactivation, pluripotency control, and differentiation of embryonic stem cells (ESCs), the regulation of *Hox* genes, organism development, importantly in central nervous system (CNS) development and neural plasticity, and, of course, in human diseases. Next, an overview of each one of these nuclear and cellular processes in which the lncRNAs act will be briefly presented.

3.1 The lncRNAs and genomic imprinting

Genomic imprinting has to do with the monoallelic expression of a particular subset of genes involved in different aspects of early metazoan development. The imprinted locus *IGF2/H19* (insulin–like growth factor 2/H19), where IGF2 and H19 are both located on human chromosome 11, but IGF2 is only expressed from the paternally inherited chromosome, and H19 is only expressed from the maternally expressed chromosome without having a direct role in the imprinting mechanism. Importantly, H19 is associated with different pathologies including Wilm's disease and rhabdomyosarcoma tumors [33]. For years, no clear function was assigned to the lncRNA H19 until recently, when a detailed analysis of the H19 primary sequence confirmed the absence of open reading frames (ORFs) and revealed the presence of miRNAs, in particular miR-675-3p and miR-675-5p, which promote skeletal muscle differentiation [33].

Alternatively, among other imprinted loci, there is a growing list of lncRNAs that are mainly associated with silencing of imprinted genes [34]. The lncRNA KCNQ1OT1 (potassium voltage-gated channel, subfamily Q, member 1 overlapping transcript 1, human) guides DNA methylation of overlapping and surrounding genes during embryonic development. In contrast, in the placenta, the lncRNA KCNQ1OT1 induces silencing of its target region in an allelic-specific manner by recruiting the Polycomb repressive complex (PRC2) and the G9a histone methyltransferase (*Drosophila*; EHMT2 euchromatic histone-lysine N-methyltransferase 2, in humans) to incorporate the histone mark H3K9me3 [35]. In another example, the lncRNA AIRN (antisense IGF2R [insulin-like growth factor 2 receptor] RNA, in humans) is paternally expressed and also silences the genes located in the IGF2R locus. AIRN is a NAT to the *IGF2R* gene, and in early embryonic stages, AIRN lncRNA interferes with the recruitment of the RNA polymerase II [36]. But in the placenta, AIRN follows a similar mechanism as the lncRNA KCNQ1OT1 by recruiting HMTs and silencing autosomal imprinted genes [37]. Therefore, these are two examples of lncRNAs, which are expressed in an allelic-specific way that contributes to genomic imprinting through chromatin mechanisms.

3.2 Dosage compensation and X-chromosome inactivation

X-chromosome inactivation involves the epigenetic silencing of the majority of the female X-chromosomes, inducing the formation of heterochromatin by the action and propagation of Xist RNA, one of the most well-studied lncRNAs [34,38]. Xist is a 15–17-kb noncoding and nuclear RNA that is polyadenylated, capped, spliced, and devoid of any ORF. Xist is required for the initiation of the X-inactivation and for triggering basically all the epigenetic processes associated with the generation of facultative heterochromatin at the scale of an entire chromosome. These processes include, in a sequential order: *Xist* transcription, histone acetylation, additional histone modifications, recruitment and actions of the PRC2 complex, incorporation of the histone variant macroH2A, DNA methylation, and modifications to the intranuclear matrix of the nucleus through the formation of the Barr body in the periphery of the nucleus [39]. Finally, many aspects of the modes of action remain unsolved, in particular, the process(es) related to the mechanism(s) of coating the entire X-chromosome [40].

3.3 The lncRNAs in pluripotency and differentiation of embryonic stem cells

ESCs are characterized by their potential to proliferate in culture, maintain their undifferentiated state (self-renewal), and capacity to differentiate into most of the body cell types. These latter two processes are coordinated through the differential expression of specific genes, epigenetic modulation of the ESC genome, and post-translational regulation [41]. In addition, a well-defined subset of transcription factors, like SOX2 (sex determining region Y-Box 2), OCT4 (octamer-binding transcription Factor 4), NANOG (NANOG homeobox), KLF4 (Krüpple-like factor 4), and c-Myc (avian myelocytomatosis viral oncogene homolog), participate in the regulation of genes required for the self-renewal capacity of ESCs and suppress cell differentiation [42]. Several recent studies revealed the existence and expression of hundreds of lncRNAs in human and mouse ESCs [43,44]. Interestingly, some of these lncRNAs seem to have expression profiles that correlate with the expression patterns of pluripotency markers like SOX2, OCT4, and NANOG. As previously mentioned, there are many lncRNAs expressed in ESCs; the following are only three examples.

The first example is the lncRNA named Braveheart (*Bvht*) that has been described in mouse. Taking advantage of different strategies to differentiate ESCs, it was demonstrated that the lncRNA Bvht was needed for progression of mesoderm toward a cardiac commitment [45]. Interestingly, Bvht interacts with Suz12 (suppressor of zeste 12 homolog, a member of the PRC2 complex) during the process of cardiomyocyte differentiation.

More recently, Chang and collaborators described the binding of the WDR5 (WD repeat-containing protein 5) protein to hundreds of lncRNAs in mouse ESCs [46]. WDR5 is considered a multifunctional adaptor protein that binds to the MLL (mixed lineage leukemia) complex, which possesses an H3K4 histone methyltransferase activity and regulates gene transcription positively [47]. Furthermore, WDR5 is necessary in mammalian ESCs for the maintenance of an active chromatin configuration of genes associated with the ESC pluripotency state [48]. In fact, it has been demonstrated that WDR5 binds with the lncRNA HOTTIP (HOXA transcript at the distal tip) that regulates the human HOXA locus (see below), and together they recruit the histone H3K4me3-associated MLL1 complex [49]. Since then, it has been demonstrated that lncRNA-binding to the WDR5 protein is essential for the histone mark H3K4me3 in ESCs [46]. In conclusion, around 100 lncRNAs present in ESCs bind WDR5, and such binding is critical to maintaining the histone H3K4me3 mark associated with chromatin regions where transcription of genes for ESC self-renewal is needed.

The third example is the long intergenic nonprotein-coding RNA regulator of reprogramming (lincRNA-RoR), which is associated to ESC pluripotency [50]. Based on its mechanism of action, the lincRNA-RoR is considered as a competing endogenous RNA (ceRNA), or sponge noncoding, RNA since it competes for the binding of key miRNAs that directly regulate the key pluripotent transcription factors SOX2, OCT4, and NANOG [51]. During self-renewing of ESCs, lincRNA-RoR shows high levels of expression that titrate different families of miRNAs, in particular the miR-145. This activity turns out to be a critical step in maintaining the pluripotency of ESCs and the integrity of the pluripotent transcription factors. Finally, in the absence of lincRNA-RoR, the ESCs are prone to differentiate [51].

Together, these examples demonstrate how different lncRNA strategies contribute to the molecular mechanisms associated with ESC self-renewal and differentiation.

3.4 Regulation of the homeotic genes by lncRNAs

The *HOX* genes represent a multigene family in vertebrates that is highly conserved and organized in gene clusters located on different chromosomes [52]. *HOX* genes encode a family of transcriptional regulators that participate in several different developmental programs associated with the regulation of the head-to-tail body axis and cell specification in some adult differentiation processes [52]. In mammals, there are 39 *HOX* genes that are grouped into four different genomic cluster domains (*HOXA*, *HOXB*, *HOXC*, and *HOXD*). Studies have shown the existence of hundreds of lncRNAs revealing, in the majority of the cases, a similar pattern of expression as their neighboring genes [19].

In a 2007 seminal work, Rinn and collaborators characterized one of the most relevant lncRNAs studies to date on the lncRNA HOTAIR (HOX transcript antisense RNA) [19]. Among the 231 identified ncRNAs, HOTAIR is a 2158-nucleotide antisense transcript residing in the *HOXC* locus on human chromosome 12 [19]. HOTAIR mediates epigenetic silencing by physically attracting the PRC2 complex and LSD1-CoREST (lysine (K)-specific demethylase 1 -[co]repressor for element 1 silencing transcription factor) complex via its 5′ and 3′ region, respectively [10]. By comparison, the EZH2 (enhancer of zeste protein 2), subunit of the PRC2 complex, has histone methyltransferase activity that incorporates the histone H3K27me3 repressive mark, whereas LSD1 demethylates H3K4me2 and H3K4me1. HOTAIR loss of function decreases H3K27me3 and SUZ12 occupancy across 40 kb of the *HOXD* locus in the human chromosome 2, demonstrating that HOTAIR represses transcription via Polycomb in *trans* by acting as a scaffold for histone-repressive modification complexes [10,18]. Importantly, abnormal HOTAIR overexpression has been associated with cancer (see below).

Among the lncRNAs found in the *HOX* gene clusters, the *cis*-acting lncRNA *HOTTIP* has been identified at the 5′ end of the human *HOXA* locus upstream of the *HOXA13* gene in primary fibroblasts [49]. In contrast to HOTAIR, HOTTIP regulates

HOXA genes in *cis* by interacting with MLL1, a member of the Trithorax complex. Regulation occurs through the gradual formation of chromatin loops with HOTTIP transcripts and HOXA promoters with increasing frequency and intensity toward the 5′-end concurrently with gradual decreasing toward the 3′-side of the *HOXA* locus [49]. A similar mode of action has been seen in the mouse with the lncRNA *MISTRAL*, also known as *Mira*, which is transcribed from the *HOXA* locus [53]. Differentiation of mouse ESCs with retinoic acid identified *Mira* as being involved in regulating in *cis* the expression of *HOXAC* and *HOXA7* genes. *Mira* loss-of-function experiments affected the activation of germ layer specification genes, suggesting a role for *Mira* in early mouse ESC differentiation [53].

In conclusion, there are only three known examples of lncRNAs acting in the regulation of the human and mouse *HOX* genes. It is impressive to realize that just in the human *HOX* loci, there are more than 200 different lncRNAs [19,49], where the majority of these are yet to be studied. Certainly, the potential exists for several new mechanisms of action, which is yet to be discovered.

3.5 The lncRNAs and their influence during development

The role of lncRNAs in the development of an organism is probably one of the primary topics where there is a greatest lack of knowledge. However, there are some examples that have revealed links between the development and formation of different tissues and organs, aspects that could have relevance in organogenesis and human diseases. For example, lncRNAs have been implicated in haematopoiesis through the action of lincRNA-EPS (lncRNA erythroid prosurvival), an linc RNA that was characterized during erythroid differentiation in mouse fetal liver progenitor cells [54]. In the case of lincRNA-EPS, loss of function causes the blockage of differentiation of the mouse erythroid progenitor cells and, consequently, an increase in apoptosis (or programmed cell death) through the inhibition of *PYCARD* (apoptosis-associated speck-like protein containing a CARD) gene expression [54]. Thus, the normal expression of lincRNA-EPS can regulate erythroid apoptosis, to some extent through the repression of PYCARD. Therefore, the role of lncRNAs in hematopoiesis is important since defects can be associated with major defects and even blood cell cancers such as leukemias.

Two of the lncRNAs that clearly participate in development are *Braveheart* (*Bvht*) and *Fendrr* (FOXF1 adjacent noncoding developmental regulatory RNA); both of these lncRNAs were identified from the mesoderm, from which the myogenic lineage specializes and therefore the heart originates [45,55]. RNA interference (RNAi) experiments against *Bvht* in mouse ESCs, and neonatal cardiomyocyte cell cultures demonstrated an aberrant development of cardiomyocytes [45]. These results suggested that *Bvht* could participate in cardiac tissue regeneration in case of an injury. As discussed previously, *Bvht* interacts with the PRC2 complex whose actions impart to *Bvht* a repressive role over sets of genes involved in cardiac commitment. In the case of *Fendrr* loss of function in a knockdown context resulted in embryonic lethality with direct defects in heart function and body wall development in the mouse, and as seen with *Bvht*, *Fendrr* also acts through the recruitment of the PRC2 complex [55].

Another example linked to myogenesis is the lncRNA linc-MD1 (muscle differentiation 1). Linc-MD1 possesses a well-defined temporal pattern of expression that has been associated *in vitro* with muscle differentiation of mouse myoblasts and the transit between early to late stages of muscle differentiation based on the ceRNA mechanism of action of the linc-MD1 [56,57]. Of note, linc-MD1 expression is reduced in myoblasts of patients with Duchenne muscular dystrophy (DMD) [57]. Mechanistically, the linc-MD1 competes for the binding of miR-133 and miR-135, which are post-transcriptional regulators of the mastermind-like protein 1 (MAML1) and myocyte-specific enhancer factor 2C (MEF2C), respectively. Finally, and as mentioned earlier, lncRNA H19 also participates in skeletal muscle differentiation through the action of the miR-675-3p and miR-675-5p [33].

These are only a few examples of lncRNAs participating in the regulation of cell differentiation and organogenesis. Certainly, the role of lncRNAs in these processes probably represents one of the less explored aspects of the biology and function of lncRNAs.

3.6 The lncRNAs and their role in the central nervous system

Starting with the fact that the majority of the cell types forming the CNS are postmitotic arrested cells, several questions arise when trying to understand the ability of the CNS to respond to environmental stimuli and the epigenetic mechanisms associated with such requirements. Furthermore, it has been shown that within the CNS, there is a higher frequency of noncoding transcripts in comparison to other tissues and cell types [58]. In particular, there is a growing list of evidence suggesting that lncRNAs play a key role in the evolutionary complexity of the CNS, its plasticity, and in neurodegenerative diseases.

Therefore, lncRNAs have been identified in the brain and mapped to specific anatomical regions [59,60]. Two functional groups of lncRNAs have been proposed, one involved in neural stem cell self-renewal capacity and neural differentiation, and the other is associated with functioning of the CNS, such as regulation of synaptic activity. For the first group, the lncRNA-ES1, lncRNA-ES2, and lncRNA-ES3 have been associated with pluripotent stem cells, since within their regulatory elements, they have binding sites for the plutipotency transcription factors OCT4 and NANOG [44]. In addition, other lncRNAs were identified with a role in neuronal differentiation: RMST (rhabdomyosarcoma 2 associated transcript), lncRNA-N1, lncRNA-N2, and lncRNA-N3 [44]. Loss-of-function experiments involving these lncRNAs resulted in a reduction of the number of neurons in culture and decreased expression of neuronal markers, and instead an increase of glial markers was observed. As an example of the second group of lncRNAs in the CNS, BC1 (brain cytoplasmic RNA 1) is a cytoplasmic lncRNA, found in dendrites and regulates postsynaptic signaling [61]. Interestingly, the loss of lncRNA BC1 induces hyperexcitation of neurons. Furthermore, BC1 has been implicated in a mouse model of epilepsy [62]. Another example in this group is the antisense lncRNA BDNF-AS, which represses the brain-derived neurotrophic factor (BDNF) with the concomitant restriction of neurite growth through the recruitment of the PRC2 complex [63]. In conclusion, these are only a few examples among the many lncRNAs involved in the development and function of the CNS. There are many more lncRNAs that remain uncharacterized, which might have unexpected functions, in particular with regard to brain plasticity and cognitive abilities of the CNS.

In summary, consider these few examples of lncRNAs as a mere demonstration of the vast scope of roles and functions with which they are associated across a broad diversity of cellular processes. The variety of functions and targets of lncRNAs clearly reveals their relevance but, more importantly, their diversification and tissue-specificity reveal a novel level of regulation exhibited by lncRNAs as regulatory signal amplifiers that have yet to be fully understood (see the following section, and Figure 2 as an example).

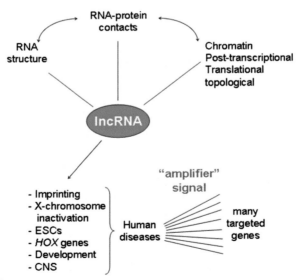

Figure 2 Diagram showing lncRNAs (in the center) function based on RNA secondary structure, potential to interact with a large diversity of proteins, and the processes associated with such configurations and interactions (top of the diagram). Derived from these ways of action, different cellular and nuclear processes are dependent on the lncRNA way of action. In normal and abnormal conditions, this figure proposes to envision lncRNAs as amplifiers that can amplify regulatory signals within sets of targets in a cell-specific manner.

4. THE LONG NONCODING RNAs IN CANCER

Due to the large number of lncRNAs identified and their diverse roles in many different cellular processes, it is not surprising that aberrant lncRNAs expression has increasingly been linked to numerous and diverse human diseases and disorders. Numerous studies showed that lncRNAs contribute to breast, lung, and liver cancers, as well as glioma carcinogenesis [64–67]. It is not possible to describe all of the lncRNAs associated with, nor their complex roles in cancer within the limits of this chapter; therefore, a limited number of examples will be presented. These examples include three of the first discovered and better characterized lncRNAs: HOTAIR, MALAT-1 (metastasis–associated lung adenocarcinoma transcript 1), and lincRNA-p21 (tumor protein p53 pathway corepressor 1 [nonprotein coding]), as well as some other general examples.

4.1 Some general examples of lncRNAs in cancer

With an ever-growing list of this type of lncRNAs with direct relationships with cancer, it is surprising to realize that we still have not deciphered the rules and mechanisms of action of these lncRNAs in cancer. The first example is the antisense noncoding RNA in the cyclin–dependent kinase inhibitor 2a (INK4) locus known as ANRIL, which is an antisense lncRNA that is transcribed within the INK4b–ARF–INK4a locus. This locus is an important genomic location with a role in the control of the cell cycle, senescence, and apoptosis [68]. ANRIL is transcribed by RNA polymerase II and processed into different splice isoforms [69]. Interestingly, the unspliced p15AS isoform is able to silence the tumor suppressor gene *p15*, but until now no mechanism has been elucidated [70]. In addition, different single nucleotide polymorphisms have been linked to an increased susceptibility to several diseases, including cancer [71]. Mechanistically, ANRIL interacts with the Polycomb-associated PRC2 and PRC1 complexes to induce heterochromatin formation in the vicinity of the INK4b–ARF–INK4a gene cluster by inducing its repression [68]. This is in part due to the physical contacts between ANRIL and CBX7 (chromobox homolog 7), and SUZ12, respectively [72,73]. Under normal conditions, ANRIL may bypass senescence to allow cell proliferation of stem or progenitor cells, but in aberrant situations, ANRIL can be involved in tumorigenesis by misregulation of the INK4b–ARF–INK4a locus [68,70].

A second example of an antisense lncRNA is the C-terminal binding protein 1 (CTBP1-AS), which has been related to the epigenetic upregulation of a set of androgen-responsive genes in prostate cancer [74]. CTBP1-AS acts both in *cis* and in *trans*; in *cis* CTBP1-AS acts to negatively regulate the *CTBP1* gene, whose peptide product is involved in the stimulation of cell proliferation. This repressive action of the lncRNA CTBP1-AS is mediated by the complex formed by the PSF repressor (phosphotyrosine-binding-associated splicing factor) and the recruitment of the HDAC-Sin3A (histone deacetylase-SIN3 transcription regulator family member A) chromatin repressive complex acting specifically over the *CTBP1* gene promoter [74]. In contrast, the CTBP1-AS activity in *trans* acts to recruit the PSF repressor to genes involved in the control of the cell cycle [74]. It is worth mentioning that other hormone-responsive lncRNAs have been identified and can act similarly to CTBP1-AS, which promotes prostate cancer [74,75].

A third lncRNA associated with cancer that was identified by high-throughput sequencing of poly(A$^+$) RNA (RNA-seq) [76] is PCAT1 (prostate cancer-associated ncRNA transcript 1). PCAT1 is transcribed from human chromosome 8q24 and is involved in prostate cancer cell proliferation by acting in *trans* over *BRCA2* (breast cancer 2, early onset), *CENPF* (centromere protein F), and *CENPE* (centromere protein E)-coding genes, among others [76]. As with other lncRNAs, the lncRNA PCAT1 regulates its target genes through the PRC2 complex. From a clinical point of view, it has been shown that after analyzing 108 tumor samples with their corresponding non-tumor controls, PCAT1 had a tendency to be overexpressed. Therefore, PCAT1 has been suggested to act as a potential biomarker for the poor prognosis of patients with colorectal cancer [77].

The final example of this group of lncRNAs is the antisense lncRNA named ANRASSF1 (antisense noncoding RASSF1). ANRASSF1 is an intronic noncoding RNA located in and transcribed from the RASSF1A (RAS-association domain family member 1A) locus [78], on chromosome 3p21.3 in the antisense orientation in relation to the tumor suppressor gene *RASSF1*, which has been implicated in many tumor types [79]. Mechanistically, ANRASSF1 forms an RNA–DNA hybrid that facilitates the recruitment of the PRC2 complex. Analysis of expression levels revealed that the lncRNA ANRASSF1 is overexpressed in breast and prostate tumor cancer cell lines, and in addition, one of its specific targets is the *RASSF1A* gene promoter [78]. Importantly, ectopic overexpression of ANRASSF1 induces an increase in HeLa cells proliferation, whereas loss-of-function experiments cause a decrease in cell proliferation [78].

In conclusion, it is clear that lncRNAs are involved in cancer. It is also obvious that their mechanisms of action remain poorly understood, but what seems to further complicate understanding their mechanisms of action is that, in many cases, they seem to be transcriptionally deregulated. Consequently, in the future, it will be necessary to analyze in detail how lncRNAs are transcriptionally regulated. Thus, in the next section will describe in more detail, three of the most studied lncRNAs involved in tumorigenesis.

4.2 The lncRNA: HOTAIR and its role in cancer

As mentioned before, HOTAIR is a 2.2-kb antisense transcript originally identified in an intergenic region of the *HOXC* locus [19]. In a 2010 seminal work, Chang and collaborators determined that an abnormal increase in HOTAIR expression in primary breast tumors and metastases was linked to HOTAIR as a predictor of eventual metastasis and death [80]. Overexpression of HOTAIR in different breast cancer cells, covering different degrees of tumorigenesis, demonstrated that the cell line more prone to metastasis (MDA-MB-231) promoted high levels of cancer cell invasion and metastasis in matrix invasion assays. Importantly, overexpression of HOTAIR fused to the luciferase reporter gene in the nonmetastatic cell line SK-BR3, which showed lung colonization after tail vein xenograft that disappeared after 1 week. In contrast, when the same experiment was performed using the breast metastatic MDA-MB-231 cells, overexpressing HOTAIR resulted in a drastic increase of lung tumor formation, thus confirming the capacity of HOTAIR to regulate metastatic progression [80]. As part of the HOTAIR overexpression effect, there was a generalized retargeting of the PRC2 complex to more than 800 genes, which then induced their repression by the incorporation of the histone mark H3K27me3. Detailed analysis of these genes showed that a large proportion of them were related to cell–cell signaling pathways and development [80]. Further studies at the whole-genome scale concluded the existence of a functional interdependency between HOTAIR and PRC2 in promoting cancer invasiveness. Furthermore, aberrant HOTAIR expression established an abnormal chromatin state for hundreds of genes that led to drastic changes in gene expression that promoted tumor metastasis. It is also important to note that HOTAIR is not only upregulated in breast cancer, it has been shown to also be involved in colorectal, hepatocellular, gastrointestinal, pancreatic carcinomas, and several other tumor types. For some of these cancers, upregulation of HOTAIR expression has been correlated with poor prognosis [80–84].

4.3 The lncRNA: MALAT-1 and lung cancer

The metastasis-associated lung carcinoma transcript 1 (MALAT1, also known as NEAT2) was one of the first lncRNAs discovered and has been associated with lung cancer metastasis. It also serves as a prognostic marker for patient survival in early stages of lung adenocarcinoma and squamous cell carcinoma [85,86]. Furthermore, MALAT1 has effects in the regulation of cell proliferation, cell migration, and invasion associated to colorectal cancer metastasis [87]. MALAT1 is also overexpressed in cell lines and tissue samples of patients with hepatocellular carcinoma and bladder cancer tissues [88,89]. In addition, MALAT1 promotes invasion and cell proliferation of cervical cancer cells by regulating BAX (BCL2-associated X protein), BCL2 (B-cell CLL/lymphoma), Caspase-8, Caspase-3, and BCL-XL (BCL2-like isoform 1) [86]. Therefore, from the cancer development perspective, MALAT1 is frequently abnormally overexpressed in many tissues, where its function is related to different cellular functions such as cell migration, invasion, proliferation, and apoptosis.

But, what is the mechanism of action of the lncRNA MALAT1? Studies have shown the lncRNA MALAT1 interacts with the unmethylated form of the Polycomb 2 protein (Pc2), promoting E2F1 (E2F transcription factor 1) sumoylation as a signal for activation of the cell growth program [90]. It is important to recall that Pc2 regulates relocation of growth-control genes between Polycomb bodies and interchromatin granules in response to different growth signals. These observations lead to the proposal that the lncRNA

MALAT1 participates in the organization of subnuclear components that participate in the coordination of different gene-expression programs through the relocation of transcription units in the nuclear environment [90]. Consequently, MALAT1 is retained in the nucleus. Using PAR–CLIP experiments, MALAT1 has been shown to have the capacity to bind different RNA-binding proteins. It also controls splicing by modulating the phosphorylation levels of serine/arginine splicing factors [91,92]. Finally, MALAT1 has been shown to be highly conserved throughout evolution across many different mammalian species, further supporting its functional relevance. Despite its conservation, null MALAT1 mice are viable and display no apparent signs of severe developmental defects [86].

In conclusion, there is an evident gap between all the effects of MALAT1 deregulation in cancer and what we know of its mechanisms of action. There are no clear correlations and connections between these processes, thus rendering MALAT1 an attractive challenge for future and more detailed investigations.

4.4 The intergenic lncRNA: lincRNA-p21 and the p53 pathway

Only recently has it become clear that lncRNAs show diverse roles, not only in gene expression, but also in many other cellular processes that, when perturbed, can directly or indirectly cause diverse human diseases and disorders [93]. This is certainly the case for the lincRNA-p21 that was revealed through differential high-throughput screens. In particular, one screen focused in detecting lincRNAs involved in the p53 pathway using MEF (mouse embryonic fibroblasts) cells and human lung tumor cells. This screen determined that the lincRNA-p21 was regulated by the human *p53* tumor suppressor gene, where its role was to induce epigenetic silencing of the p53 pathway target genes and trigger apoptosis [14,94]. The lincRNA-p21 was found to be an intergenic noncoding RNA expressed from the human genomic region located between the *CDKN1A* (cyclin-dependent kinase inhibitor 1A) and *SFSR3* (serine/arginine-rich splicing factor 3) genes. In HeLa cells, it has been characterized as a 3 kb in length intergenic noncoding RNA. The lincRNA-p21 can modulate the localization of repressive complexes to target genes linked to the p53 pathway and induce apoptosis via an interaction with the heterogeneous nuclear ribonucleoprotein K (hnRNP-K), a protein that plays various roles in the p53 pathway [94]. The hnRNP-K protein interacts with 780 nucleotides from the 5′ region of the lincRNA-p21. This contact is required to induce the interphase between the lincRNA and transcriptional repression of target genes shared between p53 and lincRNA-p21 [9]. Interestingly, a similar occurrence within the human *p53* locus takes place, where an RNA–protein interaction contact is involved between the natural antisense WRAP53 (WD40 repeat-containing protein-encoding RNA antisense to P53) and the chromatin associated CTCF (CCCTC-binding factor protein). Disruption of this interaction affects the transcriptional expression of the *p53* gene and its response to DNA damage [21]. Therefore, the combination of p53 and lincRNA-p21 can influence many transcriptional programs downstream of the p53 pathway in a sort of amplification of signal that allows the targeting of many and varied genes. Consequently, *p53* deregulation can have, through this amplification role with lincRNA-p21, an effect over many genes that can induce different disease pathways.

Consistent with these diverse roles of lncRNAs, a more recent study by Yoon and colleagues in 2012 determined that there was a post-transcriptional function of lincRNA-p21, as an inhibitor of translation [95]. These authors found that the RNA-binding protein HuR (human antigen R) interacts with lincRNA-p21, and that the HuR–lincRNA-p21 complex subsequently recruits the let-7/Ago2 (lethal-7/Argonaute 2) complex, thereby inducing instability of lincRNA-p21 and facilitating the derepression of both JunB (jun B proto-oncogene) and CTNNB1 (catenin beta 1) mRNA translation.

As expected, the lincRNA-p21 was found to be involved in several diseases, including cancer. In clinical trials, it was shown that abnormal variations in lincRNA-p21 expression levels were in accordance with the patient's tumor stage in colon cancer cells [96].

As with other lncRNAs, the lincRNA-p21 is yet another interesting lncRNA with many different functions and, as in the previous cases, several questions concerning its mode of action remain unsolved.

5. CONCLUDING REMARKS AND PERSPECTIVES

After describing the general features of the lncRNAs and the diversity of their function, including those related to diseases, we reach a clear conclusion: we are confronted with a novel field of research even though the existence of lncRNA was discovered several decades ago [1]. This new research field will have unsuspected consequences, that is as yet difficult to anticipate today. We are confronted with a novel paradigm in biology that requires an important effort from the research community and the development of new tools to systematically identify lncRNAs, but more importantly uncover and better elucidate their mechanisms of action. At this point, it is important to note that lncRNA function does not rely exclusively in chromatin-related mechanisms. There are many other modes of action that have been elucidated based on the intrinsic properties of the RNA molecule itself, including chemical modifications of RNA [97].

Another interesting aspect of lncRNA research is the possibility to design selective targets via strand-specific oligonucleotides or antagoNAT (oligonucleotides designed against NATs) in order to therapeutically derepress protein-coding gene expression [12]. Therefore, this strategy envisions a highly selective targeting for each particular lncRNA. This seems both very attractive and highly feasible. Unfortunately, the suggested amplifier role of lncRNA implies the possibility of more than one target gene, thus rendering this strategy indirect (Figure 2). Due to the relevance and importance of key RNA protein contacts, the design and synthesis of selective drugs that recognize particular structural features of lncRNAs and/or their RNA protein contacts might be of use to interfere with the recruitment of chromatin-remodeling enzymes and complexes to specific gene loci.

Finally, due to their genetic characteristics, a challenge in the field is to begin to understand how the lncRNAs, if considered as genes, are transcriptionally regulated at a particular moment, stage of development, and/or in a particular cell type.

ACKNOWLEDGMENTS

I thank Ricardo Saldaña-Meyer for critical reading of the manuscript. This work was supported by the DGAPA-PAPIIT, UNAM (IN209403, IN203811 and IN201114), and CONACyT (42653-Q, 128464 and 220503).

REFERENCES

[1] Scherrer K. Historical review: the discovery of "giant" RNA and RNA processing: 40 years of enigma. Trends Biochem Sci 2003;28:566–71.
[2] Scherrer K. Regulation of gene expression and the transcription factor cycle hypothesis. Biochimie 2012;94:1057–68.
[3] Scherrer K, Darnell JE. Sedimentation characteristics of rapidly labeled RNA from HeLa cells. Biochem Biophys Res Commun 1962;7:486–90.
[4] Warner JR, Soeiro R, Birnboim HC, Girard M, Darnell JE. Rapidly labeled HeLa cell nuclear RNA. Identification by zone sedimentation of a heterogeneous fraction separate from ribosomal precursor RNA. J Mol Biol 1966;19:349–61.
[5] Salditt-Georgieff M, Harpold MM, Wilson MC, Darnell JE. Large heterogeneous nuclear ribonucleic acid has three time as many 5′ caps as polyadenylic acid segments, and most caps do not enter polyribosomes. Mol Cell Biol 1981;1:179–87.
[6] Herman R, Weymouth L, Penman S. Hetergeneous nuclear RNA-protein fibers in chromatin-depleted nuclei. J Cell Biol 1978;78:663–74.
[7] Nickerson JA, Krochmalnic G, Wan KM, Penman S. Chromatin architecture and nuclear RNA. Proc Natl Acad Sci USA 1989;86:177–81.
[8] Cech TR, Steitz JA. The noncoding RNA revolution-trashing old rules to forge new ones. Cell 2014;157:77–94.
[9] Mercer TR, Mattick JS. Structure and function of long noncoding RNAs in epigenetic regulation. Nat Struct Mol Biol 2013;(3):300–7.
[10] Tsai MC, Manor O, Wan Y, et al. Long noncoding RNA as modular scaffold of histone modification complexes. Science 2010;329:689–93.
[11] Djebali S, Davis CA, Merkel A, et al. Lanscape of transcription in human cells. Nature 2012;489:101–8.
[12] Wahlestedt C. Targeting long non-coding RNA to therapeutically upregulate gene expression. Nat Rev Drug Discovery 2013;12:433–46.
[13] Derrien T, Johnson R, Bussotti G, et al. The GENCODE v7 catalog of human long noncoding RNAs: analysis of their gene structure, evolution and expression. Genome Res 2012;22:1775–89.
[14] Geisler S, Coller J. RNA in unexpected places: long non-coding RNA functions in diverse cellular contexts. Nat Rev Mol Cell Biol 2013;14:699–712.
[15] Cohen SM. Everything old is new again: (linc)RNAs make proteins. EMBO J 2014;33:937–8.
[16] Chu C, Qu K, Zhong FL, Artandi SE, Chang H. Genomic maps of long noncoding RNA occupancy reveal principles of RNA-chromatin interactions. Mol Cell 2011;44:667–78.
[17] Guttman M, Rinn JL. Modular regulatory principles of large non-coding RNAs. Nature 2012;482:339–46.
[18] Lunde BM, Moore C, Varani G. RNA-binding proteins: modular design for efficient function. Nat Rev Mol Cell Biol 2007;8:479–90.

[19] Rinn JL, Kertesz M, Wang JK, et al. Functional demarcation of active and silent chromatin domains in human HOX loci by noncoding RNAs. Cell 2007;129:1311–23.

[20] Schmitz KM, Mayer C, Postepska A, Grummt I. Interaction of noncoding RNA with the rDNA promoter mediates recruitment of DNMT3b and silencing of rRNA genes. Genes Dev 2010;24:2264–9.

[21] Saldaña-Meyer R, González-Buendía E, Guerrero G, et al. CTCF regulated the human p53 gene through direct interaction with its natural antisense transcript, Wrap53. Genes Dev 2014;28:723–34.

[22] Carrieri C, Cimatti L, Bialioli M, et al. Long non-coding antisense RNA controls *Uchl1* translation through an embedded SINEB2 repeat. Nature 2012;491:454–7.

[23] Wilsuz JE, Freier SM, Spector DL. 3′ end processing of a long nuclear-retained noncoding RNA yields a tRNA-like cytoplasmic RNA. Cell 2008;135:919–32.

[24] Arriaga-Canon C, Fonseca-Guzmán Y, Valdes-Quezada C, Arzate-Mejía R, Guerrero G, Recillas-Targa F. A long non-coding RNA promotes full activation of adult gene expression in the chicken α-globin domain. Epigenetics 2014;9:173–81.

[25] Hogan DJ, Riordan DP, Gerber AP, Herschlag D, Brown PO. Diverse RNA-binding proteins interact with functionally related sets of RNAs, suggesting an extensive regulatory system. PLoS Biol 2008;6:e255.

[26] Rinn JL, Chang HY. Genome regulation by long noncoding RNAs. Annu Rev Biochem 2012;81:145–66.

[27] Scheibe M, Butter F, Hafner M, Tuschl T, Mann M. Quantitative mass spectrometry and PAR-CLIP to identify RNA-protein interactions. Nucleic Acids Res 2012;40:9897–902.

[28] Ørom UA, Shiekhattar R. Long noncoding RNA usher in a new era in the biology of enhancers. Cell 2013;154:1190–3.

[29] Melo CA, Drost J, Wijchers PJ, et al. eRNAs are required for p53-dependent enhancer activity and gene transcription. Mol Cell 2013;49:524–35.

[30] Li W, Notani D, Ma Q, et al. Functional roles of enhancer RNAs for oestrogen-dependent transcriptional activation. Nature 2013;498:516–20.

[31] Lam MTY, Cho H, Lesch HP, et al. Rev-Erbs repress macrophage gene expression by inhibiting enhancer-directed transcription. Nature 2013;498:511–5.

[32] Pachnis V, Brannan CI, Tilghman SM. The structure and expression of a novel gene activated in early mouse embryogenesis. EMBO J 1988;7:673–81.

[33] Dey BK, Pfeifer K, Dutta A. The H19 long noncoding RNA gives rise to microRNAs miR-675-3p and miR.675-5p to promote skeletal muscle differentiation and regeneration. Genes Dev 2014;28:491–501.

[34] Lee JT, Bartolomei MS. X-inactivation, imprinting and long noncoding RNAs in health and disease. Cell 2013;152:1308–23.

[35] Mancini-Dinardo D, Steele SJ, Levorse JM, Ingram RS, Tilghman SM. Elongation of the *Kcnq1ot1* transcript is required for genomic imprinting of neighboring genes. Genes Dev 2006;20:1268–82.

[36] Latos PA, Pauler FM, Koerner MV, et al. Airn transcriptional overlap, but not its lncRNA products, induces imprinted *Igf2r* silencing. Science 2012;338:1469–72.

[37] Sleutels F, Zwart R, Barlow DP. The non-coding *Air* RNA is required for silencing autosomal imprinted genes. Nature 2002;415:810–3.

[38] Wutz A. Gene silencing in X-chromosome inactivation: advances in understanding facultative heterochromatin formation. Nat Rev Genet 2011;12:542–53.

[39] Escamilla-Del-Aranal M, Teixeira de Rocha S, Heard E. Evolutionary diversity and developmental regulation of X-chromosome inactivation. Hum Genet 2011;130:307–27.

[40] Engreitz JM, Pandya-Jones A, McDonel P, et al. The Zist lncRNA exploits three-dimensional genome architecture to spread across the X chromosome. Science 2013;341:1237973.

[41] Chen T, Dent SYR. Chromatin modifiers and remodelers: regulators of cellular differentiation. Nat Rev Genet 2014;15:93–106.

[42] Boyer LA, Lee TI, Cole MF, et al. Core transcriptional regulatory circuitry in human embryonic stem cells. Cell 2005;122:947–56.

[43] Dinger ME, Amaral PP, Mercer TR, et al. Long noncoding RNAs in mouse embryonic stem cell pluripotency and differentiation. Genome Res 2008;18:1433–45.

[44] Ng S-Y, Johnson R, Stanton LW. Human long non-coding RNAs promote pluripotency and neuronal differentiation by association with chromatin modifiers and transcription factors. EMBO J 2012;31:522–33.

[45] Klattenhoff CA, Scheuermann JC, Surface LE, et al. *Braveheart*, a long noncoding RNA required for cardiovascular lineage commitment. Cell 2013;152:1–14.

[46] Yang YW, Flynn RAN, Chen Y, et al. Essential role of lncRNA binding for WDR5 maintenance of active chromatin and embryonic stem cell pluripotency. Elife 2014;3:e02046.

[47] Wysocka J, Swigut T, Milne TA, et al. WDR5 associates with histone H3 methylates at K4 and is essential for H3K4 methylation and vertebrate development. Cell 2005;121:859–72.

[48] Ang YS, Tsai SY, Leed DF, et al. Wdr5 mediates self-renewal and reprogramming via the embryonic stem cell core transcriptional network. Cell 2011;145:183–97.

[49] Wang KC, Yang YW, Lius B, et al. A long noncoding RNA maintains active chromatin to coordinate homeotic gene expression. Nature 2011;472:120–4.

[50] Loewer S, Cabili MN, Guttman M. Large intergenic non-coding RNA-RoR modulated reprogramming of human induces pluripotent stem cells. Nat Genet 2012;42:1113–7.

[51] Wang Y, Xu Z, Jian J, et al. Endogenous miRNAs sponge lincRNA-RoR regulates Oct4, Nanog and Soc2 in human embryonic stem cell self-renewal. Dev Cell 2013;25:69–80.

[52] Mallo M, Alonso CR. The regulation of the Hox gene expression during animal development. Development 2013;140:3951–63.

[53] Bertani S, Sauer S, Bolotin E, Sauer F. The noncoding RNA *Mistral* activates Hoxa6 and Hoxa7 expression and stem cell differentiation by recruiting MLL1 to chromatin. Mol Cell 2011;43:1040–6.

[54] Hu W, Yuan B, Flygare J, Lodisch HF. Long noncoding RNA-mediated anti-apoptotic activity in murine erythroid terminal differentiation. Genes Dev 2011;25:2573–8.

[55] Grote P, Wittler L, Hendrix D, et al. The tissue-specific lncRNA *Fendrr* is an essential regulator of heart and body wall development in the mouse. Dev Cell 2013;24:206–14.

[56] Cesana M, Cacchiarelli D, Legnini I, et al. A long noncoding RNA controls muscle differentiation by functioning as a competing endogenous RNA. Cell 2011;147:358–69.

[57] Twayana S, Legnini I, Cesana M, Cacchiarelli D, Morlando M, Bozzoni I. Biogenesis and function of non-coding RNAs in muscle differentiation and in Duchenne muscular dystrophy. Biochem Soc Trans 2013;41:844–9.

[58] Qureshi IA, Mehler MF. Emerging roles of non-coding RNAs in brain evolution, development, plasticity and disease. Nat Rev Neurosci 2012;13:528–54.

[59] Mercer T, Qureshi IA, Gokhan S, et al. Long noncoding RNAs in neuronal-glial fate specification and oligodendrocyte lineage maturation. BMC Neurosci 2010;11:14.

[60] Iyengar BR, Choudhry A, Sarangdhar MA, Veukatesh KV, Gadgil CJ, Pillai B. Non-coding RNA interact to regulate neuronal developmental and function. Front Cell Neurosci 2014;8:47.

[61] Wang H, Iacoangli A, Lin D. Dendritic BC1 RNA in translational control mechanism. J Cell Biol 2005;171:811–21.

[62] Gitaí DL, Fachin AL, Mello SS, et al. The non-coding RNA BC1 is down-regulated in the hippocampus of Wistar-Audiogenic Rat (WAR) strain after audiogenic kindling. Brain Res 2011;1367:114–21.

[63] Modarresi F, Faghihi MA, Lopez-Toledano MA, et al. Inhibition of natural antisense transcripts *in vivo* results in gene-specific transcriptional upregulation. Nat Biotechnol 2012;30:453–9.

[64] Yang F, Zhang L, Huo X, et al. Long noncoding RNA high expression in hepato cellular carcinoma facilitates tumor growth through enhancer of zeste homolog 2 in humans. Hepatology 2011;54:1679–89.

[65] Han L, Zhang K, Shi Z, et al. LncRNA profile of glioblastoma reveals the potential role of lncRNAs in contributing to glioblastoma patogénesis. Int J Oncol 2012;40:2004–12.

[66] Shore AN, Herschkowitz JI, Rosen JM. Noncoding RNAs involved in mammary gland development and tumorigenesis: there's a long way to go. J Mammary Gland Biol Neoplasia 2012;17:43–58.

[67] Qiu M, Xu Y, Yang X, Wang J, Hu J, Xu L, et al. CCAT2 is a lung adenocarcinoma-specific long non-coding RNA and promotes invasion of non-small cell lung cancer. Tumour Biol 2014;36:5375–80.

[68] Aguilo F, Zhou MN, Walsh MJ. Long noncoding RNA, polycomb, and the ghosts haunting INK4b-ARF-INK4a expression. Cancer Res 2011;71:5365–9.

[69] Folkersen L, Kyriakou T, Goel A, et al. Relationship between CAD risk phenotype in the chromosome 9p21 locus and gene expression identification of eight new ANRIL splice variants. PLoS One 2009;4:e7677.

[70] Yu W, Gius D, Onyango P, Muldoon-Jacobs K, Karp J, Feinberg AP, et al. Epigenetic silencing of tumour suppressor gene p15 by its antisense RNA. Nature 2008;451:202–6.

[71] Popov N, Gil J. Epigenetic regulation of the INK4b-ARF-INK4a locus; in sickness and in health. Epigenetics 2010;5:685–90.

[72] Yap KL, Li S, Muñoz-Cabello AM, et al. Molecular interplay of the noncoding RNA ANRIL and methylated histone H3 lysine 27 by Polycomb cbx7 in transcriptional silencing of INK4a. Mol Cell 2010;38:662–74.

[73] Kotake Y, Nakagawa T, Kitagawa K, Suzuki S, Liu N, Kitagawa M, et al. Long non-coding RNA ANRIL is required for the PRC2 recruitment to and silencing of p15(INK4b) tumor suppressor gene. Oncogene 2011;30:1956–62.

[74] Takayama KI, Horie-Inoue K, Katayama S, et al. Androgen-response long noncoding RNA CTBP1-as promotes prostate cancer. EMBO J 2013;32:1665–80.

[75] Louro R, Nakaya HI, Amaral PP, et al. Androgen responsive intronic non-coding RNAs. BMC Biol 2007;5:4.

[76] Prensener JR, Iyer MK, Balbin OA, et al. Transcriptome sequencing across a prostate cancer cohort identifies PCAT-1, an unannotated lincRNA implicated in disease progression. Nat Biotechnol 2011;29:742–9.

[77] Ge X, Chen Y, Liao X, Liu D, Li F, Ruan H, et al. Overexpression of long noncoding RNA PCAT-1 is a novel biomarker of poor prognosis in patients with colorectal cancer. Med Oncol 2013;30:588.

[78] Beckedorff FC, Ayupe AC, Crocci-Souza R, et al. The intronic long noncoding RNA ANRASSF1 recruits PRC2 to the RASSF1A promoter, reducing the expression of RASSF1A and increasing cell proliferation. PLoS Genet 2013;9:e1003705.

[79] Agathanggelou A, Cooper WN, Latif F. Role of the Ras-association domain family 1 tumour suppressor gene in human cancers. Cancer Res 2005;65:3497–508.

[80] Gupta RA, Shah N, Wang KC, et al. Long non-coding RNA HOTAIR reprograms chromatin state to promote cancer metastasis. Nature 2010;464:1072–6.

[81] Yang Z, Zhou L, Wu LM, Lai MC, Xie HY, Zhang F, et al. Overexpression of long non-coding RNA HOTAIR predicts tumor recurrence in hepatocellular carcinoma patients following liver transplantation. Ann Surg Oncol 2011;18:1243–50.

[82] Kim K, Jutooru I, Chadalapaka G, et al. HOTAIR is a negative prognostic factor and exhibits pro-oncogenic activity in pancreatic cancer. Oncogene 2013;32:1616–25.

[83] Ishibashi M, Kogo R, Shibata K, et al. Clinical significance of the expression of long non-coding RNA HOTAIR in primary hepatocellular carcinoma. Oncol Rep 2013;29:946–50.

[84] Svoboda M, Slyskova J, Schneiderova M, et al. HOTAIR long non-coding RNA is a negative prognostic factor not only in primary tumors, but also in the blood of colorectal cancer patients. Carcinogenesis 2014;35:1510–5.

[85] Ji P, Diederichs S, Wang W, Böing S, et al. MALAT-1, a novel noncoding RNA, and thymosin beta4 predict metastasis and survival in early-stage non-small cell lung cancer. Oncogene 2003;22:8031–41.

[86] Eißmann M, Gutschner T, Hämmerle M, et al. Loss of the abundant nuclear non-coding RNA *MALAT-1* is compatible with life and development. RNA Biol 2012;9:1076–87.

[87] Xu C, Yang M, Tian J, Wang X, Li Z. MALAT-1 enhances cell motility of lung carcinoma cells by influencing the expression of motility-related genes. FEBS Lett 2011;584:4575–80.

[88] Ying L, Chen Q, Wang Y, Zhou Z, Huang, Qiu F. Upregulated MALAT-1 contributes to bladder cancer cell migration by inducing epithelial-to-mesenchynal transition. Mol Biosyst 2012;8: 2289–94.

[89] Lai M, Yang Z, Zhou L, et al. Long non-coding RNA MALAT-1 overexpression predicts tumor recurrence of hepatocellular carcinoma after liver implantation. Med Oncol 2012;29:1810–6.

[90] Yang L, Lin C, Liu W, et al. ncRNA and Pc2 methylation-dependent gene relocation between nuclear structures mediates gene activation programs. Cell 2011;147:773–88.

[91] Hafner M, Landthaler M, Burger L, et al. Transcriptome-wide identification of RNA-binding protein and microRNA target sites by PAR-CLIP. Cell 2010;141:129–41.

[92] Tripathi V, Ellis JD, Shen Z, Song DY, Pan Q, Watt AT, et al. The nuclear-retained noncoding RNA MALAT-1 regulated alternative splicing by modulating SR binding factor phosphorylation. Mol Cell 2010;39:925–38.

[93] Shi X, Sun M, Liu H, Yao Y, Song Y. Long non-coding RNAs: a new frontier in the study of human diseases. Cancer Lett 2013;339:159–66.

[94] Huarte M, Guttman M, Feldser D, et al. A large intergenic noncoding RNA induced by p53 mediates global gene expression in the p53 response. Cell 2010;142:409–19.

[95] Yoon J-H, Abdelmohsen K, Srikantan S, et al. LincRNA-p21 suppresses target mRNA translation. Mol Cell 2012;47:648–55.

[96] Zhai H, Fesler A, Schee K, Fodstad O, Flatmark K, Ju J. Clinical significance of long intergenic non-coding RNA-p21 in colorectal cancer. Clin Colorectal Cancer 2013;12:261–6.

[97] Fu Y, Dominissini D, Rechavi G, He C. Gene expression regulation mediated through reversible m^{6}A RNA methylation. Nat Rev Genet 2014;15:293–306.

CHAPTER 14

Epigenetics of physiological and premature aging

Shrestha Ghosh[1], Zhongjun Zhou[1,2]
[1]Department of Biochemistry, Li Ka Shing Faculty of Medicine, The University of Hong Kong, Hong Kong, China; [2]Shenzhen Institute of Research and Innovation, The University of Hong Kong, Shenzhen, China

Contents

1. INTRODUCTION

Aging refers to the gradual decline of proper body functioning, which eventually leads to collapse of the entire living organism. It can be broadly classified into physiological aging and premature aging. The former is defined as the outcome of progressive occurrence of deterioration in physical and mental abilities in an individual over time, while the latter refers to an early onset of the aging phenotypes, such as loss of subcutaneous fat, wrinkling of skin, muscular atrophy, alopecia, and many more [1]. So far, numerous studies have been carried out to understand, not only the underlying mechanisms of

Epigenetic Gene Expression and Regulation
http://dx.doi.org/10.1016/B978-0-12-799958-6.00014-7

aging, but also the age-related pathologies. Some of the well-established mechanisms resulting in aging are accumulating DNA damage, telomere shortening and dysfunction, apoptosis, atrophy of stem cells, somatic mutations, aberration in hormonal pathways, alteration in epigenetic marks, metabolic dysregulation, and cellular senescence as well [2]. Among these pathways, epigenetic regulation of gene expression is one of the crucial mechanisms, which explains the imbalance observed in gene expression profiles during aging [3].

Epigenetics is defined as the modifications occurring in the genome without any alteration in the underlying DNA sequence. Such modifications are termed heritable, as they can last throughout a cell's lineage and can also propagate to several generations, once occurring in the germline [4]. These alterations, occurring at the molecular level, are potent enough to cause a range of physiological and behavioral changes in organisms. Epigenetic modifications impart another layer of dynamism to the chromatin structure in response to transcription or DNA damage repair signaling. Multiple lines of study have advocated the existence of two basic principles in the causation of epigenetic modifications, namely, the environmental factors and nondeterministic pathways [5]. Till date, epigenomic studies have focused extensively on cancer, neurodegenerative diseases, aging, and metabolic syndromes [3,6]. In this study, we will discuss the epigenetic modifications occurring in both physiological and premature aging, with a special focus on the recent advancements in the field and also convergence and divergence of various epigenetic modifications in the process of aging.

2. EPIGENETICS IN AGING: A BROAD OUTLOOK

In the recent years, epigenetics has gained much limelight in explaining the molecular mechanisms resulting in the predisposition of aging. Although most pathways that cause phenotypic changes in an individual without affecting the genetic makeup are classified under epigenetics, special attention needs to be taken to determine whether those alterations are heritable or not. There remains a fine line between epigenetics and alterations at the DNA or histone levels, which do not run down the generations. So far, several epigenetic mechanisms have been identified that cause changes in the expression patterns of genes. Some of the widely studied epigenetic mechanisms are DNA methylation, histone modifications, miRNAs, noncoding RNAs (ncRNAs), and prions [7], as shown in Figure 1. The majority of these mechanisms congregate at the level of chromatin. The chromatin can be broadly classified into euchromatin (the more transcriptionally active form) and heterochromatin (the more transcriptionally repressed form). Heterochromatin can be further subdivided into constitutive and facultative heterochromatin. Facultative heterochromatin, upon specific signaling, can choose to undergo transcriptional activation, while constitutive heterochromatin remains silent almost permanently [8]. X-chromosomal inactivation in mammalian females and pericentromeric or telomeric DNA present excellent examples of

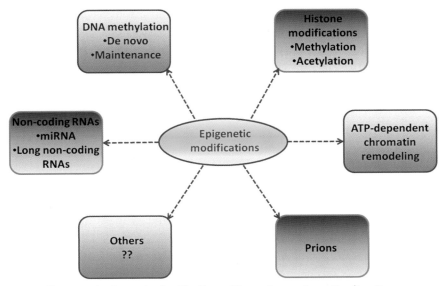

Figure 1 A schematic classification of the various epigenetic alterations.

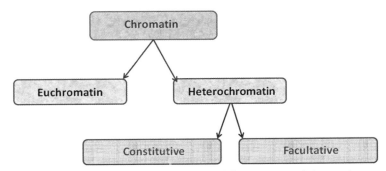

Figure 2 A schematic classification of different types of chromatin.

the facultative and constitutive heterochromatin, respectively [9]. A schematic classification of the chromatin is presented in Figure 2.

DNA and histones are wrapped around each other to form nucleosomes. Each nucleosome constitutes 146 base pairs (bps) of DNA wound around an octamer of histone proteins (two copies each of H2A, H2B, H3, and H4). The histone H1 lies at the base of nucleosome, at the entry and exit point of DNA. H1 is known as the linker histone, since it binds to the linker region of DNA and thus forms a compact chromatin conformation [10]. Most of the epigenetic mechanisms (as classified in Figure 1) have been known to impart a dynamic state to the chromatin by altering its structure via both covalent and noncovalent alterations in the DNA and histones. Hence, DNA methylation and histone modifications are the two most broadly studied epigenetic patterns.

However, in the past few years, miRNAs, ncRNAs, and prions have also started attracting attention in the field of aging epigenetics [7,11]. Apart from these, transposable element positioning and also telomere attrition have been classified under epigenetics in several organisms [12].

3. EPIGENETIC MECHANISMS

The different mechanisms of epigenetic regulation in aging have been discussed in details below.

3.1 DNA methylation

DNA methylation is the most extensively studied epigenetic alteration. It occurs mostly on the cytosine residues of CpG dinucleotide sequences, an aggregation of which forms CpG islands [13]. In addition, CpA methylation has been rendered important in sensory neurons for the function of allelic exclusion [14]. Other non-CpG methylations, like CNN and CNG, have also been reported in early mouse embryos and embryonic stem cells (ESCs), although their expression levels are largely reduced in somatic tissues [15]. These findings insinuate at an important part played by the non-CpG methylation patterns in early developmental phase, which possibly becomes redundant in the later stages. However, further experimentation needs to be done to establish the same. DNA methylation has been implicated in a wide variety of cellular functions, such as X-chromosome inactivation in mammalian females, centromeric chromatin silencing, formation of heterochromatin, and also in mammalian imprinting [15]. Noteworthy is the point that cytosine methylation has long been established to play pivotal roles in both augmentation and repression of gene expression [16]. Hence, this mechanism presents a very sensitive and finely balanced chromatin mark, which is specifically recognized by protein complexes to mediate the corresponding response signaling.

The loss of DNA methylation patterns across the genome has been strongly implicated in the aging process. During aging, there is a prominent drift occurring in the distribution of methylated cytosines globally. These alterations have been linked to a decrease in the 5-methyl-cytosine concentration on a global basis, while CpG islands in the promoters of some specific genes become abnormally hypermethylated [7,17]. Intriguingly, several of those promoters regulate expression of some important tumor-suppressor genes, such as *RUNX* (runt-related transcription factor 3), *CDKN2A* (cyclin-dependent kinase (CDK) inhibitor 2A), *LOX* (lysyl oxidase), and *TIG1* (tazarotene-induced gene 1). This explains the higher risk of cancer predisposition observed in aged individuals. Some other genes repressed by promoter hypermethylation during aging are insulin-like growth factor 2 (*IGF2*), c-Fos, estrogen receptor (*ER*), and myogen differentiation antigen 1 [see Ref. [7] for a review]. In addition, hypermethylation has also been reported in the ribosomal DNA (rDNA) clusters in aged rat livers, which further explains the downregulation of ribosomal RNA (rRNA) observed during aging [18]. On the other hand, a number of proteins involved in DNA damage repair have been observed to be downregulated with replicative senescence [19]. Thus, it remains to be investigated whether downregulation of rDNA clusters in turn results in the reduced expression of several proteins deemed important for maintaining genomic integrity or not. In a report by Raddatz et al., the authors employed transcriptome and whole genome bisulphate sequencing on human skin to demonstrate that aging is narrowly associated with epigenome destabilization [20]. However, since aging has been known to affect a multitude of tissues and organ systems and has varied consequences, it would be inappropriate to extrapolate the observations on only skin to the entire human body. On the other hand, age-related hypermethylation of CpG islands in breast tissues has been linked to the ultimate occurrence of breast cancer [21]. This study, if applied, can have potential benefits in the early diagnosis of a possible risk of breast tumorigenesis in aged women. Another group has claimed to have identified three specific loci for aberrant methylation as a mark of aging in blood [22]. Several independent studies have identified numerous other CpG sites which are aberrantly methylated in aging [23,24]. Additionally, differential DNA methylation patterns have also been observed in the two most severe forms of premature aging disorders, namely Hutchinson–Gilford progeria syndrome (HGPS) and Werner syndrome (WS). They, like several other laminopathies, result from mutations in the lamin A/C (*LMNA*) and WS, RecQ helicase-like (*WRN*) genes, respectively (see Ref. [25] for a comprehensive review on laminopathies). Also, recent work has reported a strong association between hypomethylation of *GSK*-gene body (responsible for encoding glucokinase) and the occurrence of coronary heart disease (CHD) [26]. This study has claimed hypermethylated GSK loci to be a positive biomarker for healthy aging. Since mortality of HGPS patients mostly occurs due to heart attack, it is tempting to speculate that a possible hypomethylation at the CpG sites of GSK gene could be one of the underlying reasons.

DNA methylation is catalyzed by the enzymes known as DNA methyltransferases (DNMTs). These enzymes use S-adenosylmethionine (SAM) as the donor of methyl group. In mammals, three active DNMTs have been identified as of now, which are DNMT1, DNMT3a, and DNMT3b [27]. Although two other forms have also been identified in mammals, namely DNMT2 and DNMT3L, they are reported to be of less significance in aging till date [13]. DNA methylation can be broadly categorized into de novo methylation and maintenance methylation. The former refers to formation of new methylation marks on DNA, whereas the latter corresponds to the copying of methylation patterns from parental strands of DNA onto the daughter strands after undergoing DNA replication. DNMT1 has been known to function in maintenance methylation while DNMT3a and DNMT3b are responsible for de novo methylation [28]. The huge importance of methyltransferases in the system can be understood from the observation that mice lacking DNMT activity become lethal at the embryonic stage [29].

The aberrant functioning of DNA methylases has been strongly implicated in the process of aging. Casillas et al. identified significant differences in the expression patterns of DNA methylases in aged and neoplastically transformed WI-38 human fetal lung fibroblasts [30]. The authors demonstrated a prominent decrease in the DNMT1 mRNA and protein expression with aging, while an opposite trend was observed in case of DNMT3b. These results are suggestive of the possibility that lack of sufficient DNMT1 expression could be the underlying cause of global loss of methylation patterns observed in aged cells and tissues. Further, increased DNMT3b expression is explanatory to the hypermethylation observed on some selected promoters in aged tissues [11]. Although these findings answer several questions, they also open up a plethora of other intriguing questions. It still remains to be understood how the DNMT1 gene undergoes repression in aged cells, what triggers this downregulation, which upstream regulatory pathways are involved in this, are there specific miRNAs or ubiquitination machinery coming into play, and whether the enzyme is erroneously targeted to nonspecific sites by binding to some inhibitory protein complexes. Apart from these, a recent study has revealed that maternal aging results in impaired cytoplasm-to-nucleus translocation of DNMTs in mouse oocytes [31], thus providing a mechanistic explanation to the increased infertility observed with maternal aging. On the whole, these findings suggest a very important role played by DNA methylation patterns in the disposition of aging phenotypes and also that these methylation patterns hold great potential in serving as biomarkers for age-related pathologies.

3.2 Histone modifications

While DNA methylation mostly relates to repression of gene expression, histone modifications can be responsible for either gene repression or activation. These modifications provide a local chromatin conformation, which facilitates the required level of gene expression. They can be either covalent or noncovalent. The covalent modifications largely comprise of arginine (R) methylation, lysine (K) methylation, acetylation, ubiquitination, ADP-ribosylation. sumoylation, and serine (S) and threonine (T) phosphorylation [32]. On the other hand, noncovalent modifications are brought about by the ATP-dependent protein complexes, which function in chromatin remodeling [33]. Although histones undergo several modifications, their methylation and acetylation are largely considered as epigenetic marks. It still remains to be understood whether other histone modifications like phosphorylation, sumoylation, or ADP-ribosylation could also be heritable changes or not. A large number of studies have demonstrated the inter-relationship of histone modifications and the process of aging.

3.2.1 Histone methylation

Histone methylation has been involved in several cellular functions like transcriptional repression, DNA damage repair, heterochromatin formation, X-chromosome inactivation, and many more [34]. The property of lysine residues to be either mono-, di-, or trimethylated makes histone methylation more complex than DNA methylation. In addition, the ability of arginine residues to be either monomethylated or symmetrically or asymmetrically dimethylated adds an extra layer of complicacy and sensitivity to the lysine methylation marks [35]. The enzymes which catalyze the transfer of methyl groups from SAM to the lysine and arginine residues are termed as the histone methyltransferases (HMTs). Most of them constitute a catalytic SET domain (suppressor of variegation 3-9 and enhancer of zeste and trithorax) [36]. Another study has identified that Polycomb group (PcG) and Trithorax group complexes catalyze the methylation of histones H3K27 and H3K4, respectively [37]. This study revealed the importance of PcG group of proteins in the maintenance of heritable lineage-specific gene expression during mitosis. Although the authors have claimed that the association between PcG complexes and methylated histones guides the former back into its position after a detachment during mitosis, it still remains to be illustrated whether other unknown protein complexes and signaling pathways also play a part in the process. Since both PcG complexes and DNMT1 are involved in the maintenance methylation process, it would not be completely inappropriate to hypothesize a possible interdependence and collaboration between these two major players in maintaining the heritability of the methylation patterns in both DNA and histones.

So far, several epigenetic markers have been identified, which can direct either activation or repression of specific genes. For example, H3K9me3, H3K27me3, and H4K20me3 have been found to be associated with transcriptional repression, whereas H3K4me3 and H3K36me3 are seen to constitute the genomic loci that are transcriptionally active [38,39]. In this regard, worth mentioning are the roles played by lysine methyltransferases (KMTs) and lysine demethylases (KDMs). KMTs are broadly classified into catalytic SET domain containing and noncontaining groups. On the other hand, several KDMs have also been identified to modulate gene expression by demethylating histones on their N-terminal tails [see Ref. [38] for a review]. The first ever KMT identified was SUV39H1 (suppressor of variegation 3-9 homolog 1, also known as KMT1a in humans and mouse), which is found to be conserved from yeast to humans [40]. SUV39H1 is classically known to result in the trimethylation of histone H3 at lysine 9, H3K9me3 in the heterochromatin.

As observed for DNA methylation, histone methylation patterns have also been reported to be grossly affected in aged individuals. Of special mention, is the work by Sarg et al. which identified that a trimethylated form of histone H4, H4K20me3 does exist in mammals and that this mark shows a remarkable increase in aging tissues [41]. This study also identified differential trends of histone methylation patterns in various mice tissues and organs. For example, dimethylated H4K20 (H4K20me2) was reported to be more dominant in rat kidney and liver than H4K20me (the monomethylated form). Further, H4K20me3 mark significantly increases with age in animals, while the mono- and dimethylated form of H4K20 remains almost constant throughout the process of aging. Similarly, in HGPS patients, H4K20me3 mark is significantly increased. On the contrary, histone alterations like H3K27me3 and H3K9me3 are known to be downregulated in those patient cell lines [42]. The decrease in H3K27 methylation could be explained by the downregulation observed in the corresponding methyltransferase, EZH2 (enhancer of zeste protein 2) in HGPS patients [43]. On the other hand, our study identified SUV39H1 levels to be stabilized in *ZMPSTE24−/−* (zinc metallopeptidase STE24) mice (a mouse model recapitulating many progeroid phenotypes) by increased binding between prelamin A/progerin with SUV39H1, thus resulting in enhanced H3K9Me3 levels [44]. Interestingly, upon depletion of SUV39H1, DNA damage repair machinery could be restored, and cellular senescence could be delayed in the mice. The difference in methylation patterns observed in HGPS patients and *ZMPSTE24−/−* mice models could be explained by the relative differences in expression levels of proteins upon passage of cells. This calls for further investigation to shed light on the differential methylation patterns (if any) of histones as the cells and tissues undergo passage. This study can give mechanistic explanations to the drift observed in histone methylation trends in physiologically aging individuals. A comparison of the epigenetic drift observed in physiologically aging fibroblasts and HGPS patient-derived fibroblasts has been diagrammatically represented in Figure 3.

3.2.2 Histone acetylation

Histone acetylation is one of the most extensively studied epigenetic modifications. It refers to the transfer of an acetyl moiety from acetyl-coenzyme A to a lysine residue, which is catalyzed by the enzymes known as histone acetyltransferases (HATs). This acetylation can also be reversed by the enzymes collectively known as the histone deacetylases (HDACs). Acetylated histones are a universal marker for transcriptionally active genes, since addition of an acetyl group weakens the binding between histones and DNA, thus resulting in chromosome loosening for giving access to transcriptional machinery, the DNA damage repair system, and several other signaling complexes [45].

There are several HATs which have been identified till date. Some of the most studied are Gcn5-related N-acetyltransferases (GNATs), CREB-binding protein (CBP), p300, and MYST family of HATs (consisting of MOZ (monocytic leukemia zinc finger protein), MOF (males-absent on the first protein), Tip60 (tat-interactive protein 60 kD), and others). Out of them, mostly CBP, p300, and MOF have been linked to the process of aging. Several studies have pointed out a downregulation in CBP and p300 expression levels during mammalian aging [46]. A significant drop in CBP/p300 levels has also been observed in the motor neurons of rat brains with aging [47]. This can ultimately lead to reduction in the expression of some proteins, which might be important to maintain brain functioning. This could also be one of the contributing reasons to the development of several neurological pathologies in aged individuals like dementia, Alzheimer's disease (AD), and many more. In addition to

Histone modifications in physiological and premature aging of humans

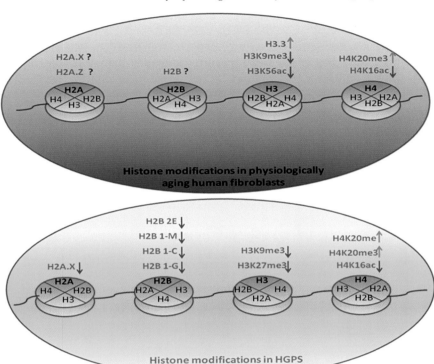

Figure 3 Comparison of the histone modifications observed in physiologically aging vs. accelerated aging (Hutchinson–Gilford Progeria Syndrome, HGPS) models.

downregulation in expression, the HAT activity of CBP/p300 has also been reported to decline in the liver, muscles, and testes of old mice [48]. In a recent study, MOF has been identified as a vital constituent of the ESC core transcriptional network and helps maintain the self-renewing potential of ESCs [49]. Since in progeroid conditions, hematopoietic stem cell exhaustion has been implicated in the occurrence of diseased state [50], it would be tempting to speculate that MOF functioning might be affected in the progeria patient cells. In our recent study, we reported a significant hypomethylation of H4K16 in the *ZMPSTE24−/−* mice which resulted from mislocalization of MOF to the nuclear matrix [51]. A gradual H4K16 hypomethylation was also observed in aging wild type mice. In addition to that, administration of sodium butyrate (NAB, a histone deacetylase inhibitor) further led to rescue of the aging phenotypes in the mutant mice. As observed in HGPS cells, the HAT, Tip60 has been reported to have an affinity for histones H2A and H4, which could also have further implications [52].

HDACs are the enzymes responsible for catalyzing the removal of acetyl groups from the ε-N-acetyl lysine amino acid moieties. So far, there have been four major classes of HDACs identified based on the sequence homology with yeast enzymes. Of special interest in aging are the class III HDACs, widely known as the sirtuin proteins [53]. Although Hdac3 and Hdac5 have been recently claimed to have a role in the aging of brain and neurodegeneration [54], the sirtuins have been more directly implicated in the epigenetic regulation of aging.

3.2.3 Sirtuins: key players in the epigenomics of aging

The sirtuin family of proteins has been widely associated with metabolism, longevity, and aging in multiple organisms. The yeast sir2 was the first ever sirtuin protein to be discovered. Since then, several other sirtuin proteins have been identified across different species based on sequence homology. Sirtuins are typically NAD$^+$ (nicotinamide adenine dinucleotide)–dependent deacetylases, which catalyze the removal of an acetyl group from the lysine residues of histones and other protein targets, thus resulting in the formation of *O-acetyl-ADP-ribose* units. Loss of sir2 in yeast has been demonstrated to reduce longevity, and its activation or overexpression has increased lifespan considerably [55]. Sir2 has been reported to regulate yeast lifespan and mediate anti-aging effects via different mechanisms and pathways. Firstly, sir2 is known to deacetylate histones H3 and H4 at the heterochromatin loci of rDNA, silenced-mating type loci, and also telomeres, and thus establish chromatin silencing. Secondly, rDNA repeats can form extrachromosomal rDNA circles (ERCs) having poignant impact on reducing longevity of yeast cells. The sir2 protein is known to get recruited to the nucleolar rDNA repeats from telomeres and form a complex termed as regulator of nucleolar silencing and telophase exit (RENT). This complex prevents ECR formation and thus results in expansion of yeast lifespan. Thirdly, sir2 protein has been reported to deacetylate H4K16 at the subtelomeric regions and thus modulate replicative lifespan of yeast cells. Since sir2 protein expression level declines with replicative aging in yeast, this results in hyperacetylation of H4K16 at the subtelomeric regions. However, this increase in H4K16 acetylation could also be attributed to the functioning of the HAT Sas2 (something about silencing 2). This suggests that sir2 and Sas2 possibly work in balance to maintain chromatin silencing and thus regulate the process of aging in yeast [7,55].

The functions of sirtuin proteins are found to be evolutionarily conserved across species in terms of exerting their anti-aging effects. Till date, seven sirtuin proteins have been identified in mammals namely, SIRT1-7. Out of them, SIRT1 and SIRT2 have been extensively studied in the process of aging, followed by SIRT6, whose role in premature aging is being increasingly revealed [56]. SIRT1 expression is reported to decrease with age in mouse as well as human fibroblasts. A similar phenomenon was also observed in mice with premature aging. SIRT1 is a well-known modulator of heterochromatin formation and maintenance. It mediates its gene silencing effects by regulating the epigenetic modifications of histones. SIRT1 is widely known to deacetylate histones H4K16, H3K9, and also H3K56. It also has a multitude of nonhistone protein targets, such as p53, FOXO3A (forkhead box O3), PGC-1α (proliferator-activated receptor gamma coactivator-1 alpha), and many others. In addition, SIRT1 also associates with linker histone H1b and deacetylates it at lysine 26. Apart from directly deacetylating histones, SIRT1 has also been reported to indirectly regulate histone methylation patterns. SIRT1 induces reduction of transcriptionally active chromatin marks like dimethylated H3K79 (H3K79me2) and promotes the formation of repressed chromatin marks like H3K9me3 and also H4K20me1 [7,57]. The underlying molecular mechanism can be understood from the targeting of mammalian HMT SUV39H1 by SIRT1. SIRT1 is observed to directly interact with SUV39H1 and deacetylate it at lysine 266, leading to downregulation of its activity, which eventually results in increased H3K9me3 levels in the cell [58]. Although the role of SIRT1 in aging and longevity has been established mainly via histone deacetylation and indirect regulation of histone methylation, several other pathways also reinforce SIRT1's anti-aging effects. For example, in SIRT1's functioning in DNA damage repair, p53 or FOXO3A-mediated pathways also have significant implications in organismal aging [56]. Recently, sirtuins' lifespan extending function has been reported to be dependent on methylation of NAM (nicotinamide) [59]. The authors identified a novel methyltransferase for NAM, now known as anmt-1, which generates 1-methylnicotinamide (MNA) from NAM. In yeast cells, it is observed that the sirtuin-mediated lifespan extension gets remarkably downregulated in complete absence of anmt-1.

Apart from SIRT1, SIRT2 has also been linked to interconnecting aging and tumor suppression. SIRT2 is predominantly found in the cytoplasm, but is also seen to translocate to the nucleus during G2/M transition phase of mitosis and deacetylate histone H4K16. Apart from histones, it is also known to deacetylate α-tubulin at lysine 40. Because of its varied roles in cellular metabolism, neurodegenerative diseases and also cancer progression, it has been deemed to be a potential target for therapeutic intervention in age-related pathologies [60]. In this regard, SIRT3 has also been identified in providing protection against several neurological disorders like Huntington's disease, AD, and Parkinson's disease, and thus holds promise in mediating anti-aging effects in the brain [61]. Lastly, SIRT6 can be termed as an emerging player in the epigenetic regulation of aging. Mostoslavsky et al. [109] first observed that mice lacking SIRT6 develop premature aging phenotypes with severely compromised DNA damage repair machinery. With increasing evidence, SIRT6's roles in mediating anti-aging effects are becoming clearer. SIRT6 is known to globally deacetylate histone H3 at lysine 9 and 56. As discussed later, SIRT6 also deacetylates histone H3K9 at the telomeric regions to maintain a repressed heterochromatin state. Additionally, SIRT6 has been established as a key regulator of DNA damage repair machinery. Moreover, SIRT6 has been reported to suppress NF-kB (nuclear factor kappa-light-chain-enhancer of activated B cells) signaling to mediate anti-aging effects. Recently, SIRT6 has also been identified as a tumor suppressor. Thus, these data are clearly suggestive of SIRT6's multifaceted functions in preventing aging [62]. Recently, SIRT7 has also emerged as a histone modifier, which has been reported to specifically deacetylate histone H3 at lysine 18 and thus promote maintenance of oncogenic transformation [63]. The authors observed that loss of SIRT7 remarkably inhibits the tumorigenicity of human cancer cell xenografts in mice, since SIRT7 mediated H3K18 deacetylation is required for the tumor cells to maintain cancerous properties.

Apart from the epigenetic regulation of sirtuins via maintenance of telomere integrity, DNA damage repair, and histone modifications, sirtuin proteins have been extensively associated with caloric restriction (CR) as well. CR is by far the only dietary intervention that has been widely implicated in extension of lifespan in various organisms. CR has been demonstrated to induce the expression levels of SIRT1, SIRT3, and also SIRT5. In this regard, resveratrol has been identified as an activator of SIRT1, which mediates effects similar to CR. However, several studies have also refuted the activating effect of resveratrol on SIRT1 [64,65]. While addressing this question, we identified lamin A (the nuclear lamina protein) to be the connecting link between resveratrol, SIRT1, and lifespan extension [66]. We reported that lamin A directly binds to and activates SIRT1. Resveratrol enhances this association and therefore its administration (at low doses) partially rescues progeroid phenotypes in *ZMPSTE24−/−* mice (a progeria mouse model). Apart from SIRT1, SIRT2 has also been associated with CR and oxidative stress signaling. On the other hand, loss of mitochondrial sirtuin, SIRT3 results in decrease of CR-mediated protection against hearing loss [64]. However, it still remains to be clarified whether the other known sirtuins also play a part in mediating effects of CR. Other intriguing questions in this regard are that whether sirtuins mediate the effects of CR, or CR-mediated effects regulate sirtuin functioning, or these two events are mutually dependent to bring about healthy aging.

3.3 ATP-dependent chromatin remodelers

The ATP (adenosine triphosphate)-dependent chromatin remodeling complexes mediate epigenetic modifications noncovalently. These protein complexes hydrolyze ATP to generate the energy that is required for modulating histone-DNA binding. Since maintenance of chromatin structure and assembly are extremely essential in providing the required accessibility of protein complexes to the DNA, these chromatin modifiers hold immense importance in determining the expression level of genes, heterochromatin formation, and several downstream effects like cell cycle control, DNA damage response, apoptosis, and senescence [67]. A majority of these multisubunit complexes contain an ATPase catalytic domain. These subunits can be broadly classified into three families, all of which contain a conserved catalytic core domain. These families are the SWI/SNF (SWItch/Sucrose Nonfermentable), Mi-2/CHD (the chromodomain-helicase-DNA binding complexes), and the ISWI (imitation SWI) ATPases [68]. An emerging number of studies are pointing out the significance of these complexes in the process of aging.

In a recent study on *Caenorhabditis elegans* (*C. elegans*), SWI/SNF has been identified as an interacting partner of the transcription factor DAF-16/FOXO (forkhead box, subgroup O) [69]. DAF-16/FOXO, via physical interaction, recruits the SWI/SNF complex to the target gene promoters of the former. The SWI/SNF complex then, via chromatin remodeling, activates transcription of the target genes, which leads to extension of longevity in the organism. The authors have thus established a novel mechanistic link between the role of ATP-dependent chromatin remodelers and lifespan extension, which could also have further implications in the aging process. BRG1 (Brahma-related gene 1, or smarca4), an important constituent of the SWI/SNF family, has been implicated several times to modulate senescence. Upon overexpression in a carcinoma-derived human cell line, BRG1 has been reported to upregulate the process of cellular senescence [70]. This phenotype was rescued by the administration of cyclin E, which is responsible for G1/S transition. This finding suggests a critical role of BRG1 in the regulation of cellular senescence via controlling the cell cycle. In addition to this, Napolitano et al. demonstrated the involvement of BRG1 in apoptosis, cellular growth arrest, and also senescence in the mesenchymal stem cells (MSCs) of rats [71]. The authors overexpressed BRG1 by adenoviral transduction in the MSCs. This led to a remarkable upregulation in apoptosis in the cells mediated by the p53 and Rb (retinoblastoma) pathways. Additionally, upregulation of cellular senescence was also observed upon BRG1 overexpression. This study has shed light on the underlying mechanism of BRG1-induced senescence and has corelated the roles of p53 and Rb in the process. In this scenario, the NURD complex (nucleosome remodeling and histone deacetylation) deserves a special mention. This ATP-dependent chromatin remodeling complex, containing seven subunits, also possesses HDAC and methyl-CpG binding activity and is hence mostly responsible for transcriptional repression [72]. In HGPS and also normally aging cells, the expression level of NURD components, such as HDAC1 and histone chaperones RBBP4 (retinoblastoma binding protein 4), has been reported to be downregulated [73]. This could be one of the contributing factors in the loss of heterochromatinization in aging cells and tissues. The study also identified aging related DNA damage and chromatin defects upon knockdown of NURD components, thus further reinforcing the strong involvement of ATP-dependent chromatin remodeling complexes in both physiological and premature aging.

3.4 Noncoding RNAs

Noncoding RNAs (ncRNAs) are referred to the transcripts which do not get translated into proteins, but they retain their functionality of regulating gene expression by various mechanisms. Since ncRNAs do not translate into proteins, the gene sequence coding for ncRNAs is termed as RNA genes. They are typically less than 300 nucleotides, which is why they are also sometimes termed as small RNAs. On the basis of size, biogenesis, and functionality, ncRNAs can be classified into several groups namely, microRNAs (miRNAs), long noncoding RNAs (lncRNAs), Piwi-interacting RNAs (piRNAs, where Piwi is named for the *Drosophila* P-element induced wimpy testis), short noncoding RNAs (sncRNAs), and also circular RNAs (circRNAs) (see Ref. [74] for a review). Since the ncRNAs regulate gene expression and phenotype without affecting the genome, they are widely categorized as epigenetic modulators. So far, ncRNAs have been associated with a spectrum of functions, such as epigenetic regulation, cellular senescence and aging, synaptic responses, germ cell development and also apoptosis [75]. Among the various types of ncRNAs, miRNAs are the best characterized ones, followed by lncRNAs, which have been implicated in the regulation of cellular senescence and aging by multiple independent studies.

3.4.1 MicroRNAs (miRNAs)

MicroRNAs (miRNAs) are typically ~18–25 nucleotides long, single-stranded RNAs. They form an integral part of RNA interference strategy employed by the cellular machinery to silence particular gene expressions. They are complementary to the target mRNA sequences. The genes encoding for miRNAs transcribe into long pri-miRNAs, which are then processed by Drosha (a type of double-stranded RNA-specific endoribonuclease type III) to result in stem-loop structured pre-miRNAs. This moiety then enters the nuclear compartment via exportin-5 protein and is cleaved by Dicer (a type of double-stranded RNA-specific endoribonuclease type III) to result in a double-stranded RNA complex with two nucleotide overhangs on the 3′ ends. This entity is then recruited into the RISC (RNA-induced silencing complex) where the sense strand is cleaved off, while the antisense strand remains attached to the RISC. This activates the RISC complex, which then binds to target mRNA and causes its silencing via degradation or prevention of ribosomal attachment because of steric hindrance [76]. A number of miRNAs have been identified which play pivotal roles in aging.

The importance of miRNAs in the regulation of aging can be prominently understood from the finding that overexpression of the miRNA, *lin-4* increased lifespan in *C. elegans* and its knockdown significantly reduced longevity of the organism [77]. However, for a majority of the miRNAs, the underlying pathway being regulated has been traced down to either insulin/IGF signaling, mTOR (mammalian target of rapamycin) signaling or DAF-12 (the nuclear receptors found in *C. elegans*) signaling. Insulin/IGF signaling has been repeatedly correlated with longevity and senescence. Hence, identification of a number of miRNAs targeting various components of insulin/IGF signaling clearly establishes the vastly important role played by the miRNAs in regulating senescence. Some of the well-known examples are miR-1, miR-7, miR-320, and many more. In addition to insulin/IGF signaling, mTOR and DAF-12 signaling are also known to modulate senescence and organismal longevity. These signaling pathways are also found to be regulated by the miRNAs, such as miR-30A, miR-21, miR-14, and others (see Ref. [78] for a review). Recently, miR-449a has been reported to cause cell-cycle arrest and also senescence in the prostate cancer cells [79]. The authors employed in silico analysis to reveal that miR-449a targets cyclin D1, and thus controls cellular growth and senescence via Rb-dependent pathway. This study opens up a possibility of therapeutic intervention in prostate cancer cells by inducing senescence via targeting miR-449a. In another recent study, miR-141-3p has been established to directly target ZMPSTE24 (the enzyme responsible for cleaving prelamin A into mature lamin A) and thus regulate aging of human MSCs [80]. In this study, 3′UTR luciferase reporter assay was performed to identify the 3′UTR of ZMPSTE24 as the direct target of miR-144-3p. Since miR-144-3p levels are also found to be upregulated in senescent cells, this study provided a mechanistic explanation to the accumulation of prelamin A upon aging. Additionally, miRNAs have also been implicated in the regulation of synaptic responses, learning, and memory and are thus associated with bringing about senescence in the brain cells [74]. However, it still remains largely unclear as to how upstream epigenetic regulation can affect the expression of the RNA genes. This is indicative of the presence of a multilayer of epigenetic pathways, which, if not efficiently coordinated, results in aging.

3.4.2 Long noncoding RNAs
The long ncRNAs or lncRNAs are an emerging group of ncRNAs whose functional regulation in aging is being increasingly identified. In a recent study, the lncRNA *SAL-RNA1* (senescence-associated lncRNA 1) has been reported to be downregulated in senescent cells [81]. In the human diploid WI-38 fibroblasts, knockdown of *SAL-RNA1*-induced senescence associated phenotypic traits like increased β-galactosidase activity, induced p53 signaling, and also enlarged cellular morphology. This suggests a direct involvement of lncRNAs in the regulation of aging. Further, lncRNAs are also reported to be involved in age-related osteoarthritis [82]. In another study, the androgen-responsive lncRNA, CTBP1-AS (C-terminal-binding protein 1-antisense) has been found to induce prostate cancer progression by repressing CTBP1 activity [83]. This could also have further implications in the process of aging. Moreover, lncRNAs have been reported to regulate synaptic plasticity and memory [74]. This raises the possibility of an involvement of lncRNAs in the aging-associated synaptic malfunctioning.

3.5 Prions
Prions are the proteins which can convert themselves catalytically into an infectious conformation. This altered conformation also results in a different function of the protein. Since they are capable of affecting an organism's phenotype without altering the underlying DNA sequence, they are often referred to as epigenetic alterations. The

diseases caused by these proteins are termed as prion diseases. The prion proteins (PrP) are mostly associated with neural degenerative disorders [84].

PrP^C, the cellular form of the prion protein, is the most extensively studied prion protein in mammals, which can transform into a more β-sheet rich disease causing form known as PrP^{SC}. PrP^C has been reported to be downregulated in the aging neurons of the hippocampus [85]. Since PrP^C inhibits amyloid β production, which results in the pathogenesis of AD, this study provides a mechanistic link between reduced PrP^C expression and increased risk of developing AD in aged individuals. Moreover, the glycosylation on PrP^C has also been reported to increase during aging in mouse brain [86]. Since increased glycosylation is also a characteristic of the disease form PrP^{SC}, this study presents a possible link between increased glycosylation of prion proteins with aging to render them disease causing. Another study has reported that prion proteins in aged individuals can lead to differential neural processing [87]. Although the field of prion proteins is still in its developing phase, it holds potential in explaining several pathologies observed during aging.

3.6 Histone variants: potential candidates for epigenetic biomarkers

Histone variants are the different forms of histones which may or may not differ in their molecular structure from the classical histone proteins, although they have very different biological functioning. The association of histone variants with senescence and aging has been identified for over a decade back. It has been reported that the ratio of histone variant proteins, that is, H2A.1/H2A.2, declines linearly in the human dermal fibroblasts during aging [88]. The observed linear decrease in the ratio is a function of cumulative population doublings (CPDs). The same group of researchers also reported that the histone variants of H2A and H3 could also be regulated in in vitro aging, and this regulatory pattern is similar in differentiation as well [89]. They measured the biogenesis of H2A and H3 variants as a function of CPDs using the same cellular system. In another study, it was reported that the expression level of H2A variants was downregulated upon drug-induced senescence [90]. This study also identified phosphorylation of the histone variant H2A.X (ɣH2A.X) as a biomarker for DNA damage foci. This finding has a great significance because antibodies against ɣH2A.X are widely used nowadays to identify DNA damage foci in the cells. In addition, age-related alterations in the histone variants H1 and H1(0) have also been reported in the tissues of mice [91]. Although there are reports describing downregulation of some histone variants upon aging, there are studies showing an upregulation of other histone variants as cells undergo senescence. In a recent study, the expression level of histone variants H3.3 and macroH2A was reported to increase as the cells underwent senescence [92]. The authors explained this phenomenon as an important event for the cells to maintain chromosomal integrity as they senesce. However, this study raises another possibility that these histone variants might be produced to provide an extra layer of protection to DNA damage as aging cells become more susceptible to genomic insults. On the contrary, another possibility cannot be neglected that the accumulation of these histone variants might result from some underlying genetic mutations as the cells gradually undergo senescence. Apart from histone variants, the chaperones associated with them also show alterations in an age-dependent manner. For example, the histone chaperone, histone cell cycle regulation defective homolog A (HIRA), which deposits histone variant H3.3 into nucleosomes, has been reported to have altered expression in an age-dependent manner in baboon dermal fibroblasts [93]. Taken together, these histone variants could be employed as important biomarkers to identify cells undergoing senescence and also age-related pathologies.

4. EPIGENETIC REGULATION OF CELLULAR SENESCENCE

Cellular senescence can be defined as the irreversible arrest of the cell division. Multiple independent studies have opined in favor of accumulation of cellular senescence as one of the causative factors of organismal aging. A range of cellular events have been implicated in causing cellular senescence, such as oncogenic activation, oxidative stress, telomere attrition, genomic instability, dysfunctioning of the key cell cycle regulators like p53 and Rb pathways, and many more. Cellular senescence mostly results in the formation of senescence-associated heterochromatin foci (SAHF). The above-mentioned cellular events have been reported to act both independently and also in conjuncture to bring upon the pathologies associated with aging [7]. A diagrammatic comparison of young and senescent cells has been shown in Figure 4.

4.1 Telomere epigenetics

Telomeres can be defined as the heterochromatic regions found at the ends of chromosomes which essentially constitute repetitive DNA sequences (TTAGGG repeats). This nucleoprotein structure functions in the protection of chromosomal ends from untimely degradation. Telomeres remain associated with the shelterin protein complex, comprising of six proteins. This configuration shields the chromosomal ends from being identified as potential DNA double-strand breaks by the cell's DNA repair machinery. During replication, DNA polymerase remains unable to replicate the linear genome completely. This is why the cells undergo telomere shortening after every round of replication. This telomere attrition gets accrued to induce a DNA damage response, which eventually results in cellular senescence and is termed as replicative senescence [94]. To restore this terminal DNA loss, the telomerase enzyme comes into play and adds newly synthesized telomeric repeats to the shortened ends. Interestingly, this rescuing activity of telomerase is only seen in germlines and stem cells and not in the somatic cells. This is clearly suggestive of the fact that telomere maintenance in stem cells is one of the underlying mechanisms which restores their regenerative capacity. Several independent studies have

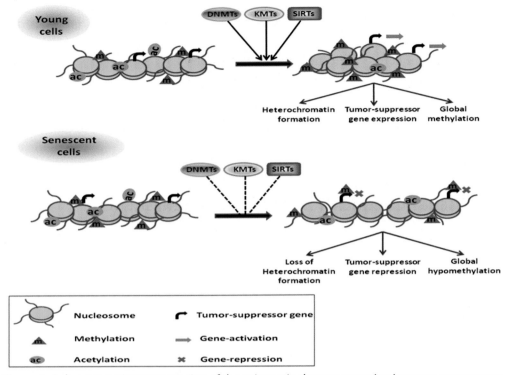

Figure 4 A diagrammatic representation of the epigenetic changes occurring in young versus senescent cells.

rendered replicative senescence as a contributing factor to the development of physiological aging and other age-related pathologies (see Ref. [95] for a review).

A large number of epigenetic modifications have been associated with telomere length maintenance. In human as well as mouse cells, it has been reported that loss of epigenetic alterations in the heterochromatin, such as DNA methylation, can potentially lead to telomere elongation and also heighten telomere recombination. On the other hand, reduction of the H3K9me3 mark at telomeric and subtelomeric regions has been observed in mammalian cells lacking telomerase function. As a mechanistic link, the lack of HMTs (SUV39H1 or SUV39H2) has also been demonstrated to result in deficient telomeric histone methylation, which ultimately leads to aberrantly long telomeres [95,96]. These studies indicate that a decrease in both DNA and histone methylation can lead to abruptly long telomeres, which can cause age-related disease phenotypes. It is also possible that the HMTs and DNMTs act in coordination to maintain the heterochromatin foci in telomeres. Another possibility can also not be ignored that there might exist a fine balance between the HMTs and DNMTs in the maintenance of telomere integrity, the disruption of which results in age-related pathologies. However, this model requires further experimentation for establishment.

Apart from viewing telomeres as the subjects of epigenetic regulation, one can also observe them as the epigenetic regulatory elements. For example, in telomere position effect (TPE), genes placed adjacently to telomeres have been seen to have a transcriptionally repressed state in yeast. This telomere-induced silencing effect has also been reported for subtelomeric genes [96]. This phenomenon raises an interesting question as to whether subtelomeric genes become expressed upon telomeric attrition in the cells undergoing senescence. Although several studies speak in favor of an interrelationship between epigenetics and telomere biology in the process of aging, some studies have also observed opposing results. Roberts et al. provided evidence that reduced dosage of the epigenetic modifiers like Dnmt1, Dnmt3L, SmcHD1 (structural maintenance of chromosomes flexible hinge domain containing sRNA), and also FOXO3A had no significant effect on the telomere length in mouse [97].

Nevertheless, the role of epigenetics in the maintenance of telomere has been established time and again. Telomeres constitute some of the classical epigenetic marks of pericentromeric chromatin like H3K9me3, H4K20me3, and also HP1 protein binding. Apart from these, sirtuins have also been reported to modulate telomere biology. The mammalian sirtuin, SIRT6, has been demonstrated to deacetylate histone H3K9 at the telomeric regions and thus facilitate binding of the WS protein, WRN. Upon SIRT6 depletion from cells, telomere dysfunction and premature cellular senescence become increasingly evident. The SIRT6 knockout mice show severe signs of premature aging phenotypes [7,96]. In yeast telomeric heterochromatin, methylation of histone H3 at lysine 4 and 79 and ubiquitination of histone H2B increase upon cellular aging. In addition, it has been demonstrated that loss of Set2 HMT, responsible for methylation of histone H3 at lysine 36, increases longevity in yeast cells [98]. These data clearly indicate that the epigenetic regulation of the telomeric DNA has a profound impact on the occurrence of cellular senescence and aging.

4.2 Epigenetics of SAHF formation

SAHF can be defined as the foci containing facultative heterochromatin, which leads to silencing of several proliferation promoting genes in the senescent cells. As studied in human diploid fibroblasts (HDFs), the formation of SAHF is readily observable upon induction of senescence. Till date, several epigenetic biomarkers have been identified, which serve as the biomarkers of SAHF in cells. For example, SAHF are enriched in the heterochromatin markers like H3K9me3, H3K27me3, core histone macroH2A, heterochromatin protein 1 (HP1), and also HMGA proteins (nonhistone chromatin architectural proteins). Similarly, SAHF show absence of the euchromatin marks such as H3K9 acetylation and H3K4me3, and they also lack linker Histone H1 [99]. Recently, it has been demonstrated that these heterochromatin marks segregate from each other in the

SAHF to form layered high-order chromatin structures (HOCS) [100]. This layered conformation of HOCS results from the spatial rearrangement of the existing heterochromatin and not from heterochromatin scattering. Formation of SAHF has also been demonstrated to be affected by the Rb protein [101]. In this study, overexpression of HMG2A (high mobility group A2 protein) is reported to sufficiently induce SAHF formation in primary WI38 cells (human dermal fibroblast cell lines). Although Rb protein has been established to play a crucial role in HMG2A-induced SAHF formation, it has been illustrated that the Rb-mediated pathway is not an indispensible event in the process. This opens up the possibilities for unidentified protein cascades to function in SAHF formation upon senescence. On the other hand, Kosar et al. reported that SAHF are not indispensible for the occurrence of cellular senescence [102]. The authors have demonstrated that SAHF formation is not a universal characteristic of cells undergoing senescence. Rather, SAHF formation is cell-type specific and also occurs in a genotoxic insult-dependent manner. On the whole, these studies suggest that epigenetic marks are important determinants of SAHF formation. However, it still remains to be clearly understood as to how the upstream signaling protein complexes direct the cells to form SAHF and whether SAHF have other important functions apart from repression of cell cycle-controlling proteins like E2F transcription factor.

5. CONVERGENCE AND DIVERGENCE OF EPIGENETIC ALTERATIONS TO CAUSE OR CONTROL AGING

The epigenetic modifications act both independently and also in an interdependent manner. In a massive number of instances, these epigenetic modifications, and also the epigenetic modifiers, have been observed to form a complicated and delicately balanced meshwork, which results in proper functioning of the cellular system. While considering aging, a similar phenomenon has been observed. However interestingly, these modifications sometimes function as a group to bring about the same effect, and in other cases they act antagonistically to nullify or repress each other's function. For example, the methyltransferase SUV39H1 has been shown to be regulated by the sirtuin SIRT1 during the formation of heterochromatin [58]. Since SUV39H1 promotes the formation of heterochromatin mark H3K9me3, and SIRT1 expression gets downregulated with senescence, this could be explanatory to the loss of heterochromatin mark with the progression of aging in individuals. Moreover, SUV39H1-mediated H3K9 trimethylation is reported to employ DNMT3b-dependent DNA methylation to the pericentromeric DNA [103]. This presents an example of intricate intersection among the epigenetic modifiers. In addition, miRNAs have been found to target both HDACs and DNMTs. On the other hand, alterations in DNA methylation and histone acetylation levels have also been reported to affect the transcription of miRNA genes [7]. The KMTs and the KDMs have also been observed to act in concert for maintaining genomic integrity [39].

Additionally, histone modifiers and DNMTs have been reported to crosstalk in order to regulate the expression of rRNA genes [104]. Hence, an age-related disorder may not be the consequence of functioning of one particular epigenetic modifier, but an accumulation of consequences brought about by a multitude of epigenetic modifiers.

6. EPIGENETICS AND DRUG DEVELOPMENT

The reversibility of epigenetic modifications like DNA methylation and histone modifications make them excellent targets for introducing therapeutic interventions. Since histone methylation patterns have emerged as important biomarkers in determining senescent cells, studies are being carried out to develop small molecular compounds in order to modulate HMTs and also histone demethylases and regulate aging. These small compounds can act as potential drugs to increase lifespan in individuals and may also be able to alleviate age-related disorders. For example, several small molecule compounds have been identified as specific inhibitors for G9a/GLP proteins (the homologous methyltransferases responsible for H3K9 mono- or dimethylation), such as BIX-01294, UNC0224, UNC0638, E72, and others. Similarly, chaetocin (a fungal metabolite) was screened out as the inhibitor of SUV39H1. In addition, EPZ004777 was identified to specifically inhibit DOT1L and DOT1-like methyltransferase for H3K79 methylation [105]. Interestingly, cytosine methylation patterns are also used in forensic departments to identify the age group of samples [106], further reinstating the importance of epigenetic patterns in identifying cells undergoing senescence. Recently, methylation of mitochondrial DNA (mtDNA) has also been established as a potential biomarker in determining aging [107]. Moreover, 4-phenylbutyrate (PBA, an HDAC inhibitor) has been shown to extend lifespan in *Drosophila* [108]. In addition, we have shown that administration of sodium butyrate (NAB, also an HDAC inhibitor) extends longevity in a progeria mouse model [51]. Thus, the area of epigenetics presents very important tools in not only diagnosing age and age-related pathologies, but also formulating drugs to extend longevity.

7. CONCLUSIONS

As discussed above, numerous pathways and protein cascades have been identified till date to understand the underlying mechanisms of the process of aging. Epigenetics stands out as one of the most prominent players in regulating aging. Epigenetics is a broad term per se, constituting a plethora of proteins and signaling pathways, which act individually and also as a group to bring about the desired modifications and eventually result in the required phenotype. Aging, on the other hand, is also very complex, since not just one, but a multitude of signaling pathways come into play, such as DNA damage repair mechanisms, oxidative stress signaling, somatic mutation, replicative senescence, and many more. If analyzed carefully, the role of epigenetic regulation in controlling a majority of the above-mentioned pathways become evident. However, what remains as a large void in our understanding is the question of whether epigenetic regulation acts upstream of these pathways, or these pathways cause epigenetic alterations as a result. Is it possible that the metabolic state within the cellular environment of aging cells signals the epigenetic modifiers to bring about the desired chromatin conformation? Or is it that the epigenetic modifications cause activation or repression of required genes to bring about the desired cellular environment? Till date, it remains quite unclear whether epigenetics is a top-down or bottom-up approach in the aging process. It is also possible that both these approaches come into play and maintain a balance for optimum cellular functioning.

Another aspect that has enthralled researchers across the globe is the difference (if any) in epigenetic modifications between physiological aging and premature aging. While considering centenarians (people aged 100 or above), who present excellent examples of healthy aging, the questions that often arise are how and why their cellular signaling is different from the rest, which leads to lifespan extension. Are their epigenetic modifications any different from the others? Much has been known so far regarding the epigenetic alterations occurring in premature aging, like upregulation and downregulation of DNA methylation patterns, histone modifications, and others. But what remains to be seen is that whether such modifications are slowed down or completely inhibited in the centenarians. If so, which molecular pathways and signaling cascades come into play to affect the same? The answer to these questions can open up new vistas in our current understanding of epigenetic regulations and their impact on aging. With emerging studies and reports, the role of epigenetics in aging are getting revealed more and more to enhance our knowledge. The day is not far when epigenetic biomarkers would become the most sought after signs of aging, and therapeutic interventions can be designed to increase not only longevity but also healthspan of individuals.

REFERENCES

[1] Kirkwood TB. A systematic look at an old problem. Nature 2008;451(7179):644–7.
[2] Moskalev AA, Aliper AM, Smit-McBride Z, Buzdin A, Zhavoronkov A. Genetics and epigenetics of aging and longevity. Cell Cycle 2014;13(7):0–14.
[3] Callaway E. Epigenomics starts to make its mark. Nature 2014;508(7494):22.
[4] Bird A. Perceptions of epigenetics. Nature 2007;447(7143):396–8.
[5] Flanagan JM, Popendikyte V, Pozdniakovaite N, Sobolev M, Assadzadeh A, Schumacher A, et al. Intra-and interindividual epigenetic variation in human germ cells. Am J Hum Genet 2006;79(1):67–84.
[6] Neumeister P, Albanese C, Balent B, Greally J, Pestell RG. Senescence and epigenetic dysregulation in cancer. Int J Biochem Cell Biol 2002;34(11):1475–90.
[7] Munoz-Najar U, Sedivy JM. Epigenetic control of aging. Antioxid Redox Signal 2011;14(2):241–59.
[8] Grewal SI, Jia S. Heterochromatin revisited. Nat Rev Genet 2007;8(1):35–46.
[9] Heard E. Delving into the diversity of facultative heterochromatin: the epigenetics of the inactive X chromosome. Curr Opin Genet Dev 2005;15(5):482–9.
[10] Luger K, Mäder AW, Richmond RK, Sargent DF, Richmond TJ. Crystal structure of the nucleosome core particle at 2.8 Å resolution. Nature 1997;389(6648):251–60.

[11] Fraga MF, Esteller M. Epigenetics and aging: the targets and the marks. Trends Genet 2007;23(8):413–8.

[12] Wood JG, Helfand SL. Chromatin structure and transposable elements in organismal aging. Front Genet 2013;4.

[13] Goll MG, Bestor TH. Eukaryotic cytosine methyltransferases. Annu Rev Biochem 2005;74:481–514.

[14] Lomvardas S, Barnea G, Pisapia DJ, Mendelsohn M, Kirkland J, Axel R. Interchromosomal interactions and olfactory receptor choice. Cell 2006;126(2):403–13.

[15] Bernstein BE, Meissner A, Lander ES. The mammalian epigenome. Cell 2007;128(4):669–81.

[16] Bird A. DNA methylation patterns and epigenetic memory. Genes Dev 2002;16(1):6–21.

[17] Issa JP. Aging and epigenetic drift: a vicious cycle. J Clin Invest 2014;124(1):24–9.

[18] Oakes CC, Smiraglia DJ, Plass C, Trasler JM, Robaire B. Aging results in hypermethylation of ribosomal DNA in sperm and liver of male rats. Proc Natl Acad Sci 2003;100(4):1775–80.

[19] Mao Z, Tian X, Van Meter M, Ke Z, Gorbunova V, Seluanov A. Sirtuin 6 (SIRT6) rescues the decline of homologous recombination repair during replicative senescence. Proc Natl Acad Sci 2012;109(29):11800–5.

[20] Raddatz G, Hagemann S, Aran D, Söhle J, Kulkarni PP, Kaderali L, et al. Aging is associated with highly defined epigenetic changes in the human epidermis. Epigenetics Chromatin 2013;6(1):36.

[21] Johnson KC, Koestler DC, Cheng C, Christensen BC. Age-related DNA methylation in normal breast tissue and its relationship with invasive breast tumor methylation. Epigenetics 2013;9(2):0–7.

[22] Weidner CI, Lin Q, Koch CM, Eisele L, Beier F, Ziegler P, et al. Aging of blood can be tracked by DNA methylation changes at just three CpG sites. Genome Biol 2014;15(2):R24.

[23] Zykovich A, Hubbard A, Flynn JM, Tarnopolsky M, Fraga MF, Kerksick C, et al. Genome–wide DNA methylation changes with age in disease–free human skeletal muscle. Aging Cell 2013;13(2):360–66.

[24] Florath I, Butterbach K, Müller H, Bewerunge-Hudler M, Brenner H. Cross-sectional and longitudinal changes in DNA methylation with age: an epigenome-wide analysis revealing over 60 novel age-associated CpG sites. Hum Mol Genet 2013;23(5):1186–201.

[25] Schreiber KH, Kennedy BK. When lamins go bad: nuclear structure and disease. Cell 2013;152(6): 1365–75.

[26] Xu L, Zheng D, Wang L, Jiang D, Liu H, Xu L, et al. GCK gene-body hypomethylation is associated with the risk of coronary heart disease. Bio Med Res Int 2014;2014.

[27] Okano M, Xie S, Li E. Cloning and characterization of a family of novel mammalian DNA (cytosine-5) methyltransferases. Nat Genet 1998;19(3):219–20.

[28] Okano M, Bell DW, Haber DA, Li E. DNA methyltransferases Dnmt3a and Dnmt3b are essential for de novo methylation and mammalian development. Cell 1999;99(3):247–57.

[29] Li E, Bestor TH, Jaenisch R. Targeted mutation of the DNA methyltransferase gene results in embryonic lethality. Cell 1992;69(6):915–26.

[30] Casillas MA, Lopatina N, Andrews LG, Tollefsbol TO. Transcriptional control of the DNA methyltransferases is altered in aging and neoplastically-transformed human fibroblasts. Mol Cell Biochem 2003;252(1–2):33–43.

[31] Zhang L, Lu DY, Ma WY, Li Y. Age-related changes in the localization of DNA methyltransferases during meiotic maturation in mouse oocytes. Fertil Steril 2011;95(4):1531–4.

[32] Kouzarides T. Chromatin modifications and their function. Cell 2007;128(4):693–705.

[33] Smith CL, Peterson CL. ATP-dependent chromatin remodeling. Curr Top Dev Biol 2004;65:115–48.

[34] Lachner M, Jenuwein T. The many faces of histone lysine methylation. Curr Opin Cell Biol 2002;14(3):286–98.

[35] Bannister AJ, Schneider R, Kouzarides T. Histone methylation: dynamic or static? Cell 2002;109(7): 801–6.

[36] Ng SS, Yue WW, Oppermann U, Klose RJ. Dynamic protein methylation in chromatin biology. Cell Mol Life Sci 2009;66(3):407–22.

[37] Ringrose L, Paro R. Epigenetic regulation of cellular memory by the Polycomb and Trithorax group proteins. Annu Rev Genet 2004;38:413–43.

[38] Rivera C, Gurard-Levin ZA, Almouzni G, Loyola A. Histone lysine methylation and chromatin replication. Biochimica Biophysica Acta 2014;1839(12):1433–39.

[39] Black JC, Van Rechem C, Whetstine JR. Histone lysine methylation dynamics: establishment, regulation, and biological impact. Mol Cell 2012;48(4):491–507.

[40] Aagaard L, Laible G, Selenko P, Schmid M, Dorn R, Schotta G, et al. Functional mammalian homologues of the Drosophila PEV–modifier Su (var) 3–9 encode centromere–associated proteins which
complex with the heterochromatin component M31. EMBO J 1999;18(7):1923–38.

[41] Sarg B, Koutzamani E, Helliger W, Rundquist I, Lindner HH. Postsynthetic trimethylation of histone
H4 at lysine 20 in mammalian tissues is associated with aging. J Biol Chem 2002;277(42):39195–201.

[42] Arancio W, Pizzolanti G, Genovese SI, Pitrone M, Giordano C. Epigenetic involvement in Hutchinson-
Gilford Progeria Syndrome: a Mini-review. Gerontology 2014;60(3):197–203.

[43] Shumaker DK, Dechat T, Kohlmaier A, Adam SA, Bozovsky MR, Erdos MR, et al. Mutant nuclear
lamin A leads to progressive alterations of epigenetic control in premature aging. Proc Natl Acad Sci
2006;103(23):8703–8.

[44] Liu B, Wang Z, Zhang L, Ghosh S, Zheng H, Zhou Z. Depleting the methyltransferase Suv39h1
improves DNA repair and extends lifespan in a progeria mouse model. Nat Commun 2013;4:1868.

[45] Clayton AL, Hazzalin CA, Mahadevan LC. Enhanced histone acetylation and transcription: a dynamic
perspective. Mol Cell 2006;23(3):289–96.

[46] Chan HM, La Thangue NB. p300/CBP proteins: HATs for transcriptional bridges and scaffolds. J
Cell Sci 2001;114(13):2363–73.

[47] Matsumoto A. Age-related changes in nuclear receptor coactivator immunoreactivity in motoneurons
of the spinal nucleus of the bulbocavernosus of male rats. Brain Res 2002;943(2):202–5.

[48] Li Q, Xiao H, Isobe KI. Histone acetyltransferase activities of cAMP-regulated enhancer-binding protein and p300 in tissues of fetal, young, and old mice. J Gerontol A Biol Sci Med Sci 2002;57(3):B93–8.

[49] Li X, Li L, Pandey R, Byun JS, Gardner K, Qin Z, et al. The histone acetyltransferase MOF is a key
regulator of the embryonic stem cell core transcriptional network. Cell Stem Cell 2012;11(2):163–78.

[50] Wang Y, Schulte BA, Zhou D. Extra view hematopoietic stem cell senescence and long-term bone
marrow injury. Cell Cycle 2006;5(1):35–8.

[51] Krishnan V, Chow MZY, Wang Z, Zhang L, Liu B, Liu X, et al. Histone H4 lysine 16 hypoacetylation
is associated with defective DNA repair and premature senescence in Zmpste24-deficient mice. Proc
Natl Acad Sci 2011;108(30):12325–30.

[52] Arancio W. A bioinformatics analysis of lamin-a regulatory network: a perspective on epigenetic
involvement in Hutchinson–Gilford progeria syndrome. Rejuvenation Res 2012;15(2):123–7.

[53] Longo VD, Kennedy BK. Sirtuins in aging and age-related disease. Cell 2006;126(2):257–68.

[54] Cosín-Tomás M, Alvarez-López MJ, Sanchez-Roige S, Lalanza JF, Bayod S, Sanfeliu C, et al. Epigenetic alterations in hippocampus of SAMP8 senescent mice and modulation by voluntary physical
exercise. Front Aging Neurosci 2014;6:51.

[55] Wierman MB, Smith JS. Yeast sirtuins and the regulation of aging. FEMS Yeast Res 2013;14(1):73–88.

[56] Verdin E. The many faces of sirtuins: coupling of NAD metabolism, sirtuins and lifespan. Nat Med
2014;20(1):25–7.

[57] Rehan L, Laszki-Szcząchor K, Sobieszczańska M, Polak-Jonkisz D. SIRT1 and NAD as regulators of
ageing. Life Sci 2014;105(1):1–6.

[58] Vaquero A, Scher M, Erdjument-Bromage H, Tempst P, Serrano L, Reinberg D. SIRT1 regulates the
histone methyl-transferase SUV39H1 during heterochromatin formation. Nature 2007;450(7168):
440–4.

[59] Schmeisser K, Mansfeld J, Kuhlow D, Weimer S, Priebe S, Heiland I, et al. Role of sirtuins in lifespan
regulation is linked to methylation of nicotinamide. Nat Chem Biol 2013;9(11):693–700.

[60] de Oliveira RM, Sarkander J, Kazantsev AG, Outeiro TF. SIRT2 as a therapeutic target for age-
related disorders. Sirtuins Biol Dis 2012;92.

[61] Kincaid B, Bossy-Wetzel E. Forever young: SIRT3 a shield against mitochondrial meltdown, aging,
and neurodegeneration. Front Aging Neurosci 2013;5:48.

[62] Tennen RI, Chua KF. Chromatin regulation and genome maintenance by mammalian SIRT6. Trends
Biochem Sci 2011;36(1):39–46.

[63] Barber MF, Michishita-Kioi E, Xi Y, Tasselli L, Kioi M, Moqtaderi Z, et al. SIRT7 links H3K18
deacetylation to maintenance of oncogenic transformation. Nature 2012;487(7405):114–8.

[64] Someya S, Yu W, Hallows WC, Xu J, Vann JM, Leeuwenburgh C, et al. SIRT3 mediates reduction of
oxidative damage and prevention of age-related hearing loss under caloric restriction. Cell
2010;143(5):802–12.

[65] Park S, Mori R, Shimokawa I. Do sirtuins promote mammalian longevity? A critical review on its relevance to the longevity effect induced by calorie restriction. Mol Cells 2013;35(6):474–80.

[66] Liu B, Ghosh S, Yang X, Zheng H, Liu X, Wang Z, et al. Resveratrol rescues SIRT1-dependent adult stem cell decline and alleviates progeroid features in laminopathy-based progeria. Cell Metab 2012;16(6):738–50.

[67] Varga-Weisz P. ATP-dependent chromatin remodeling factors: nucleosome shufflers with many missions. Oncogene 2001;20(24):3076–85.

[68] Eisen JA, Sweder KS, Hanawalt PC. Evolution of the SNF2 family of proteins: subfamilies with distinct sequences and functions. Nucleic Acids Res 1995;23(14):2715–23.

[69] Riedel CG, Dowen RH, Lourenco GF, Kirienko NV, Heimbucher T, West JA, et al. DAF-16/FOXO employs the chromatin remodeller SWI/SNF to promote stress resistance and longevity. Nat Cell Biol 2013;15(5):491–501.

[70] Shanahan F, Seghezzi W, Parry D, Mahony D, Lees E. Cyclin E associates with BAF155 and BRG1, components of the mammalian SWI-SNF complex, and alters the ability of BRG1 to induce growth arrest. Mol Cell Biol 1999;19(2):1460–9.

[71] Napolitano MA, Cipollaro M, Cascino A, Melone MA, Giordano A, Galderisi U. Brg1 chromatin remodeling factor is involved in cell growth arrest, apoptosis and senescence of rat mesenchymal stem cells. J Cell Sci 2007;120(16):2904–11.

[72] Xue Y, Wong J, Moreno GT, Young MK, Côté J, Wang W. NURD, a novel complex with both ATP-dependent chromatin-remodeling and histone deacetylase activities. Mol Cell 1998;2(6):851–61.

[73] Ibrahim MX, Sayin VI, Akula MK, Liu M, Fong LG, Young SG, et al. Targeting isoprenylcysteine methylation ameliorates disease in a mouse model of progeria. Science 2013;340(6138):1330–3.

[74] Earls LR, Westmoreland JJ, Zakharenko SS. Non-coding RNA regulation of synaptic plasticity and memory: implications for aging. Ageing Res Rev 2014;17:34–42.

[75] Peschansky VJ, Wahlestedt C. Non-coding RNAs as direct and indirect modulators of epigenetic regulation. Epigenetics 2013;9(1):3–12.

[76] Wang Y, Liang Y, Lu Q. MicroRNA epigenetic alterations: predicting biomarkers and therapeutic targets in human diseases. Clin Genet 2008;74(4):307–15.

[77] Boehm M, Slack F. A developmental timing microRNA and its target regulate life span in C. elegans. Science 2005;310(5756):1954–7.

[78] Grillari J, Grillari-Voglauer R. Novel modulators of senescence, aging, and longevity: small non-coding RNAs enter the stage. Exp Gerontol 2010;45(4):302–11.

[79] Noonan EJ, Place RF, Basak S, Pookot D, Li LC. miR-449a causes Rb-dependent cell cycle arrest and senescence in prostate cancer cells. Oncotarget 2010;1(5):349.

[80] Yu KR, Lee S, Jung JW, Hong IS, Kim HS, Seo Y, et al. MicroRNA-141-3p plays a role in human mesenchymal stem cell aging by directly targeting ZMPSTE24. J Cell Sci 2013;126(23):5422–31.

[81] Abdelmohsen K, Panda A, Kang MJ, Xu J, Selimyan R, Yoon JH, et al. Senescence-associated lncRNAs: senescence-associated long noncoding RNAs. Aging Cell 2013;12(5):890–900.

[82] Ukai T, Sato M, Akutsu H, Umezawa A, Mochida J. MicroRNA-1-9a-3p, microRNA-193b, and microRNA-320c are correlated to aging and regulate human cartilage metabolism. J Orthop Res 2012;30(12):1915–22.

[83] Takayama KI, Horie-Inoue K, Katayama S, Suzuki T, Tsutsumi S, Ikeda K, et al. Androgen–responsive long noncoding RNA CTBP1-AS promotes prostate cancer. EMBO J 2013;32(12):1665–80.

[84] Prusiner SB. Prions. Proc Natl Acad Sci 1998;95(23):13363–83.

[85] Whitehouse IJ, Jackson C, Turner AJ, Hooper NM. Prion protein is reduced in aging and in sporadic but not in familial Alzheimer's disease. J Alzheimer's Dis 2010;22(3):1023–31.

[86] Goh AXH, Li C, Sy MS, Wong BS. Altered prion protein glycosylation in the aging mouse brain. J Neurochem 2007;100(3):841–54.

[87] Gouras GK. Convergence of synapses, endosomes, and prions in the biology of neurodegenerative diseases. Int J Cell Biol 2013;2013.

[88] Rogakou EP, Sekeripataryas KE. A biochemical marker for differentiation is present in an in vitro aging cell system. Biochem Biophys Res Commun 1993;196(3):1274–9.

[89] Rogakou EP, Sekeri–Pataryas KE. Histone variants of H2A and H3 families are regulated during in vitro aging in the same manner as during differentiation. Exp Gerontol 1999;34(6):741–54.

[90] Lopez MF, Tollervey J, Krastins B, Garces A, Sarracino D, Prakash A, et al. Depletion of nuclear histone H2A variants is associated with chronic DNA damage signaling upon drug-evoked senescence of human somatic cells. Aging (Albany NY) 2012;4(11):823.

[91] Medvedev ZA, Medvedeva MN. Age-related changes of the H1 and H1° histone variants in murine tissues. Exp Gerontol 1990;25(2):189–200.

[92] Rai TS, Adams PD. Lessons from senescence: chromatin maintenance in non-proliferating cells. Biochim Biophys Acta 2012;1819(3):322–31.

[93] Jeyapalan JC, Ferreira M, Sedivy JM, Herbig U. Accumulation of senescent cells in mitotic tissue of aging primates. Mech Ageing Dev 2007;128(1):36–44.

[94] Galati A, Micheli E, Cacchione S. Chromatin structure in telomere dynamics. Front Oncol 2013;3.

[95] Fojtová M, Fajkus J. Epigenetic regulation of telomere maintenance. Cytogenet Genome Res 2014.

[96] García-Cao M, O'Sullivan R, Peters AH, Jenuwein T, Blasco MA. Epigenetic regulation of telomere length in mammalian cells by the Suv39h1 and Suv39h2 histone methyltransferases. Nat Genet 2004;36(1):94–9.

[97] Roberts AR, Blewitt ME, Youngson NA, Whitelaw E, Chong S. Reduced dosage of the modifiers of epigenetic reprogramming Dnmt1, Dnmt3L, SmcHD1 and Foxo3a has no detectable effect on mouse telomere length in vivo. Chromosoma 2011;120(4):377–85.

[98] Ryu HY, Rhie BH, Ahn SH. Loss of the Set2 histone methyltransferase increases cellular lifespan in yeast cells. Biochem Biophys Res Commun 2014;446(1):113–8.

[99] Funayama R, Ishikawa F. Cellular senescence and chromatin structure. Chromosoma 2007;116(5):431–40.

[100] Chandra T, Narita M. High-order chromatin structure and the epigenome in SAHFs. Nucleus 2013;4(1):23–8.

[101] Shi X, Tian B, Liu L, Gao Y, Ma C, Mwichie N, et al. Rb protein is essential to the senescence-associated heterochromatic foci formation induced by HMGA2 in primary WI38 cells. J Genet Genomics 2013;40(8):391–8.

[102] Kosar M, Bartkova J, Hubackova S, Hodny Z, Lukas J, Bartek J. Senescence-associated heterochromatin foci are dispensable for cellular senescence, occur in a cell type-and insult-dependent manner and follow expression of p16 (ink4a). Cell Cycle 2011;10(3):457–68.

[103] Lehnertz B, Ueda Y, Derijck AA, Braunschweig U, Perez-Burgos L, Kubicek S, et al. *Suv39h*-Mediated histone H3 lysine 9 methylation directs DNA methylation to major satellite repeats at pericentric heterochromatin. Curr Biol 2003;13(14):1192–200.

[104] Grummt I. Different epigenetic layers engage in complex crosstalk to define the epigenetic state of mammalian rRNA genes. Hum Mol Genet 2007;16(R1):R21–7.

[105] Liu Y, Liu K, Qin S, Xu C, Min J. Epigenetic targets and drug discovery: Part 1: histone methylation. Pharmacol Ther 2014;143(3):275–94.

[106] Yi SH, Xu LC, Mei K, Yang RZ, Huang DX. Isolation and identification of age-related DNA methylation markers for forensic age-prediction. Forensic Sci Int Genet 2014;11:117–25.

[107] Iacobazzi V, Castegna A, Infantino V, Andria G. Mitochondrial DNA methylation as a next-generation biomarker and diagnostic tool. Mol Genet Metab 2013;110(1):25–34.

[108] Kang HL, Benzer S, Min KT. Life extension in Drosophila by feeding a drug. Proc Natl Acad Sci 2002;99(2):838–43.

[109] Mostoslavsky R, Chua KF, Lombard DB, Pang WW, Fischer MR, Gellon L, et al. Genomic instability and aging-like phenotype in the absence of mammalian SIRT6. Cell 2006;124(2):315–29.

CHAPTER 15

Epigenetic effects of environment and diet

Jason O. Brant, Thomas P. Yang

Department of Biochemistry and Molecular Biology, Center for Epigenetics, Genetics Institute, University of Florida College of Medicine, Gainesville, FL, USA

Contents

1. INTRODUCTION

Eukaryotic cells, tissues, and organisms constantly respond to external influences of the environment to maximize survival and fitness. These responses to the environment typically do not occur directly in the genome but rather at the level of the epigenome. Epigenetic regulation is defined as stable and mitotically heritable changes in gene expression that are not due to changes in DNA sequence. More generally, epigenetics can be regarded as mechanisms that stably regulate DNA function, as epigenetic mechanisms can also govern DNA replication, repair, recombination, and even RNA processing, in addition to transcription. Accumulating evidence indicates that mechanisms mediating the response and adaptation to environmental conditions include ones that are epigenetically based. Epigenetic changes and adaptation to conditions at early stages of development, particularly in utero, may have effects on disease risk and susceptibility later in life. Thus, understanding epigenetic alterations and mechanisms associated with responses to environmental conditions may provide insight into underlying mechanisms of adult onset diseases. These epigenetic regulatory mechanisms include DNA methylation, histone modification, chromatin remodeling, and noncoding RNAs [1]. Most studies to date have examined the effects of environment and diet on DNA methylation

Epigenetic Gene Expression and Regulation
http://dx.doi.org/10.1016/B978-0-12-799958-6.00015-9

patterns. For this reason, and because mechanisms of mitotic transmission of epigenetic information are best understood for DNA methylation (via the maintenance methyltransferase DNMT1) and less well-characterized for other mechanisms, this chapter will focus on the effects of environment and diet on DNA methylation. In mammals, the stable and mitotically heritable aspect of epigenetic regulation and the epigenetic responses to environmental influences, is likely to have the greatest effect and long-term consequences in utero. Therefore, this chapter also will emphasize the epigenetic effects of environment and diet on gene expression in mammals and the epigenetic effects of the environment in utero.

During early mammalian development, the preimplantation embryo undergoes a genome-wide erasure of DNA methylation [2–5]. Subsequently, upon implantation and differentiation, DNA methylation patterns are re-established that will eventually reflect terminal cell-type-specific methylation patterns (reviewed in Ref. [6]). This period of dynamic changes of erasure and resetting of DNA methylation patterns in the embryonic and fetal epigenomes is likely to be a developmental window that is particularly sensitive to perturbation by environmental influences with potential long-term consequences on disease susceptibility and health outcomes. The idea that conditions in utero and early life may impact health in adulthood has led to the developmental origins hypothesis (or fetal programming) which posits that early life experiences or conditions can affect adult disease risk and health outcomes. Accumulating evidence indicates that this process can be mediated, at least in part, by epigenetic mechanisms such as DNA methylation.

2. DEVELOPMENTAL ORIGINS OF HEALTH AND HUMAN DISEASE

As early as the time of Hippocrates, the concept that adult diseases may originate during the process of development has been proposed. Since then, epidemiological studies and experimental data have contributed to an ever-changing hypothesis. As recently as 50 years ago, Rose published observations of familial patterns of coronary heart disease (CHD) [7]. Shortly thereafter, Fordsdahl showed that poor living conditions in early life were important risk factors for arteriosclerotic heart disease [8]. These hypotheses were greatly expanded by the work of Barker and Osmond [9,10], which would later lead to the concept of the fetal origins of adult disease. Barker termed this the thrifty phenotype, which states that environmental cues during development may influence development in such a way as to prepare the fetus for a predicted future environment in adult life. This concept of developmental plasticity also suggested that these adaptations may alter metabolism in a way as to be detrimental in later life. At the same period as Baker and colleagues were performing their work, Trichopoulos proposed a similar hypothesis, the fetal origins of cancer, for the origination of breast cancer in utero [11]. Realization that developmental plasticity extends into the postnatal period led to the term developmental origins hypothesis [12]. Waterland and Garza demonstrated that epigenetic changes can occur in a limited period of opportunity during development and that these changes persist into adulthood; this hypothesis was termed metabolic imprinting [13]. All of these refinements in the nomenclature of these hypotheses were intended to highlight specific biological mechanisms that could be unified into a single hypothesis, now termed the developmental origins of health and disease (DOHaD) [14]. Thus, when conducting epidemiological studies of adult disease, it is important to recognize that both the genome and the epigenome interactively influence sensitivity to disease in adult life [15].

However, the mechanism by which early environmental and nutritional insults leads to an increase in disease susceptibility and altered metabolism in adult life remains poorly understood. One possibility is that epigenetic changes in the placenta and/or fetus due to exposure to maternal malnutrition during development may lead to stable, heritable, aberrant changes in the regulation of genes important for proper placental and fetal development. Methylation of CpG dinucleotides is a well-studied epigenetic modification, typically associated with transcriptionally repressed chromatin. As genome-wide DNA methylation patterns are reprogrammed during prenatal and early postnatal development and then maintained throughout adult life, it is possible that aberrant nutritional conditions, environmental contaminants/toxins, stress exposure, etc. during this period of epigenetic reprogramming could alter the establishment of normal methylation patterns [13].

In particular, this perturbation of the normal establishment of DNA methylation patterns could affect the normal regulation of imprinted genes. Differential expression of the two alleles of imprinted genes is dependent upon the parent of origin of each allele, with some imprinted genes expressing only the paternally-inherited allele while other imprinted genes express only the maternally-inherited allele [16–18]. Many imprinted genes are responsible for proper placental and fetal development in placental mammals [19]. Imprinted genes are commonly associated with differentially methylated regions (DMRs) that govern parent-of-origin expression patterns [20–22]. Because parent-of-origin DNA methylation patterns are established during gametogenesis [23], it has been proposed that exposure to nutritional or environmental insults during gametogenesis and placental and fetal development can lead to alterations in the epigenetic states of imprinted genes, specifically aberrant DNA methylation at DMRs [13,24].

3. EPIGENETIC EFFECTS OF ENVIRONMENTAL CONTAMINANTS AND TOXINS

A rapidly growing literature on the epigenetic effects of various environmental contaminants and toxins provides strong evidence that many such compounds have the potential to change normal patterns of DNA methylation. These substances and compounds include, but are not limited to, endocrine disruptors such as bisphenol A (BPA), vinclozilin and methyoxychlor, herbicides and insecticides [25–27], phytochemicals [28], air pollution [29–31], arsenic [32,33] and heavy metals (reviewed in [34]), polycyclic aromatic hydrocarbons (PAHs) [35,36], fetal alcohol exposure [37–39], nanoparticles (reviewed in [40]), maternal cigarette smoking [41] and secondhand cigarette smoke [42], etc.

A groundbreaking study in the field of environmental epigenetics was published by the laboratory of Michael Skinner in 2005 [26]. They demonstrated that exposure of rats to the endocrine disruptor vinclozolin in utero resulted in decreased spermatogenic capacity in adulthood that was correlated with altered DNA methylation patterns in germ cells. Significantly, the germ cell phenotype, as well as the alterations in DNA methylation patterns, persisted through the F3 generation even though the exposure to vinclozolin occurred only in the F0 mother. Persistence of the phenotype and altered methylation patterns through to the F3 generation is particularly significant because the F3 generation is the first generation in which vinclozolin exposure of the original F0 mother also did not include potential exposure of subsequent generations. For example, the F1 offspring were in utero during exposure of the F_0 mother and therefore susceptible to vinclozolin exposure, while the developing germ cells that would give rise to the F2 offspring were developing within the vinclozolin-exposed F1 offspring in utero. Therefore, the F3 generation would be the first generation in which no contact, direct or indirect, with vinclozolin was possible. Because the F3 offspring in this study exhibited altered DNA methylation after vinclozolin exposure only to the original F0 dam, this was the first molecular evidence for transgenerational transmission of altered epigenetic states (Figure 1). However, this initial study only examined two loci in any detail for DNA methylation changes and transgenerational transmission (regions within the lysophospholipase gene, and within 1 kb of the transcription initiation site of the gene encoding the cytokine-inducible SH2 protein). Subsequently, the Skinner lab has gone on to use technologies for genome-wide DNA methylation analyses to interrogate a broader representation of the genome. Using methylated DNA immunoprecipitation in conjunction with microarrays (meDIP-ChIP) of rat sperm genome promoters (interrogating ~15,000 promoters), 52 promoter regions of altered methylation (RAMs) were detected in the F3 generation, of which 16 were confirmed by an alternative method [43]. DNA sequence analysis of these 16 validated promoter regions identified a consensus DNA sequence motif associated with the transgenerational epigenetic changes elicited by vinclozolin. This sequence contained potential binding sites for several known eukaryotic transcription factors. Curiously, this study also identified a copy number variation (CNV) in the *Fam111a* promoter, with an increase in copy number in the vinclozolin samples. The significance of this CNV is unclear, particularly with regard to the germ cell-associated phenotype and transgenerational inheritance; the possibility that epigenetic changes in the F1 generation somehow fostered the *Fam111a* CNV identified in the F3 generation is proposed.

The Skinner group also has examined the potential transgenerational epigenetic effects in rats of other environmental contaminants and toxins such as dioxin, a

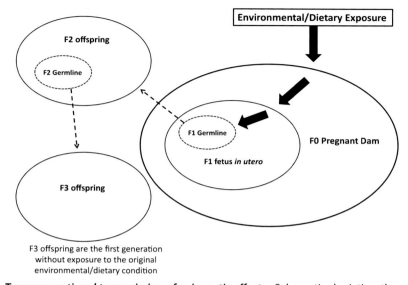

F3 offspring are the first generation
without exposure to the original
environmental/dietary condition

Figure 1 *Transgenerational transmission of epigenetic effects.* Schematic depicting the mechanism of transgenerational transmission of epigenetic effects due to exposure of the F0 pregnant dam to environmental or dietary conditions. The F0 pregnant dam is directly exposed to the environmental/ dietary condition, with the F1 offspring in utero and the F1 germline also potentially exposed to the condition. Exposure of the F1 germline to the environmental/dietary condition may directly transmit epigenetic alterations in the F1 germline to the resulting F2 offspring. Therefore, the F3 generation is the first generation without direct exposure or direct consequences of the environmental/dietary condition, and any epigenetic alterations in this generation must be due to transgenerational transmission of the initial epigenetic effect(s). Thick black arrows depict direct exposure while dotted arrows depict germline inheritance.

pesticide/repellant mixture of permethrin; the insect repellant DEET, a plastics–associated mixture of BPA and phthalates; and jet fuel, all of which have been shown to have toxicological effects [25,44–47]. These compounds were shown to be associated with early pubertal onset, increased sperm apoptosis, and/or reduction in follicle numbers in the ovary transgenerationally. Again, genome-wide DNA methylation analysis of ~15,000 promoters in rat sperm of the F3 generation by meDIP-ChIP showed altered methylation patterns for each of the contaminants analyzed. Interestingly, most regions that showed altered DNA methylation in the F3 generation were unique to the contaminant; no RAM was common to all four compounds or mixtures. The pesticide/repellant mixture and plastics mixture exhibited the most overlap in RAMs, and relatively few other RAMs were shared by two or more contaminants. Analysis of the RAMs associated with each of these compounds/mixtures showed a common feature to be low CpG density, regions termed CpG deserts [48]. Similar studies were also performed by the Skinner lab on additional compounds commonly found in the environment [45,49].

An interesting consequence of the transgenerational epigenetic changes induced by exposure to environmental contaminants is the potential effect on the response to other environmental conditions. For example, Crews et al. showed that after a single exposure to vinclozolin in F0 dams, male F3 offspring responded differently to chronic restraint stress with associated changes in brain physiology, metabolic activity, and gene expression patterns [50]. Thus, transgenerational epigenetic effects of environmental contaminants, and perhaps other environmental conditions, may have the potential to affect the normal response to other environmental conditions several generations after the initial (and perhaps only) exposure. This would suggest that our constant and ongoing response to our environment may be, in part, molded by the cumulative response of our ancestors to their environment, particularly in utero.

Similar studies of transgenerational epigenetic inheritance in mice has produced the interesting observation that this phenomenon is specific to outbred mouse strains and not observed in inbred strains [51]. The reason for this is currently unclear.

Several studies have also investigated the epigenetic effects of another endocrine disruptor, BPA. BPA is a widely used chemical for the production of polycarbonate plastic and expoxy resins. It is used in a variety of everyday consumer items such as baby bottles, water bottles, dental composites, CDs and DVDs, thermal printing paper used for sales receipts, the lining of many food and beverage containers, and the lining of water pipes. BPA exposure in rodents has been shown to be associated with breast and prostate cancer, increased body weight, and altered reproduction [52–55]. Disturbingly, 95% of human urine samples contain detectable levels of BPA [56], making environmental contamination by this compound a concern for human health. An early study of the potential epigenetic effects of BPA in rodents used the agouti mouse model system [57,58]. The agouti gene encodes a signaling protein that induces hair follicles to produce a yellow hair pigment in mice; this gene is normally expressed only in hair follicles in the skin and contributes to the normal brown coat color. The *Avy* agouti allele carries an insertion of an intracisternal A particle (IAP) transposable element upstream of the normal agouti gene, which aberrantly drives constitutive ectopic expression to the agouti gene from a functional IAP promoter. This results in yellow fur, obesity, diabetes, and cancer. However, expression of the *Avy* allele is regulated by DNA methylation of the IAP promoter, leading to a wide variation in coat color from yellow to mottled to brown, depending upon the level of methylation of the IAP promoter in the skin of *Avy* mice. This demonstrates the *Avy* allele is a so-called metastable epiallele. Thus, the agouti mouse system allows for a readout of relative methylation levels of the IAP promoter based on the degree of yellow color in the coat, with a fully yellow coat indicating an unmethylated IAP, a brown coat indicating full methylation and silencing of the IAP, and intermediate degrees of yellow-brown mottling of coat color indicating various intermediate levels of IAP methylation in the population of follicles in the skin (Figure 2). Dolinoy et al. [57] used the agouti mouse system to examine the epigenetic effects of

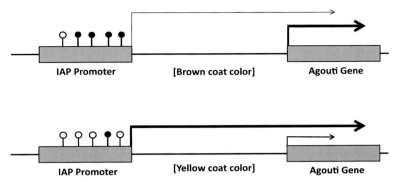

Figure 2 *Epigenetic regulation of coat color in the agouti mouse.* Schematic representation of the agouti gene under epigenetic regulation of the IAP element. Circles (lollipops) represent CpG dinucleotides; open circles depict unmethylated CpG sites and black-filled circles depict hypermethylated sites. Arrows represent gene expression, with thicker arrows representing higher levels of transcription. Methylation of the IAP element results in low levels of transcription from the IAP promoter and normal levels of transcription from the agouti gene resulting in a brown coat color. Hypomethylation of the IAP element yields constitutive transcription from the IAP promoter and lower levels of transcription from the agouti promoter. This results in elevated constitutive agouti expression and a yellow coat color.

prenatal BPA exposure via maternal diet. Females were given normal diets and diets supplemented with BPA for two weeks before mating and throughout pregnancy and lactation. Their offspring at day 22 were assessed for coat color and subjected to DNA methylation analysis of the IAP by sodium bisulfite genomic sequencing. Offspring of BPA-fed mothers showed a shift in coat color toward yellow coat color, compared to non-BPA exposed control offspring. This indicated that BPA exposure in utero was decreasing methylation levels at the IAP promoter. This was confirmed by high resolution sodium bisulfite genomic sequencing of nine CpG sites in the IAP promoter of tail DNA; BPA-exposed offspring showed ~27% overall methylation at these sites, whereas non-BPA-exposed controls showed ~39% methylation levels. A similar reduction in DNA methylation was obtained upon analysis of a second metastable epiallele, the $Cabp^{IAP}$ (CDK5 activator-binding protein gene, also regulated by an IAP insertion) epiallele, after gestational BPA exposure. DNA methylation levels at this locus were also observed to decrease after BPA exposure in utero. Thus, gestational BPA exposure has the potential to stably alter DNA methylation levels in somatic tissues. Interestingly, when the BPA-supplemented diet was co-supplemented with a cocktail of potential methyl donor compounds consisting of folic acid, vitamin B_{12}, betaine, and choline, the effect of BPA-exposure on DNA methylation in offspring was suppressed (i.e., normal agouti coat color distribution and DNA methylation levels at the IAP were restored). And though it is not a methyl donor compound, the phytoestrogen genistein showed the same counteractive effect on BPA-induced hypomethylation in this system. Thus, the methylome in utero may be undergoing dynamic combinatorial responses to various environmental conditions and stimuli during the course of development. The stable changes in DNA methylation patterns resulting from these gestational epigenetic responses (and their corresponding changes in gene expression levels/patterns) have the potential to affect long-term disease susceptibility and risk.

Another study investigating the effect of BPA on DNA methylation used imprinted genes rather than metastable epialleles associated with retroviral insertions, which may not be typical or representative of the response of normal endogenous genes [59]. The differential expression of the maternal and paternal alleles of imprinted genes is typically regulated at some level by differential DNA methylation. Many imprinted genes play a role in normal embryonic, postnatal, and placental development. Thus, Susiarjo et al. examined the effect of prenatal BPA exposure on DNA methylation in a subset of imprinted genes. BPA was administered by diet to female mice two weeks prior to mating through pregnancy until either E9.5 or E12.5, at which time F1 offspring were sacrificed for analysis. Analysis of allele-specific gene expression at E9.5 and E12.5 in development showed the normally repressed allele of the small nuclear ribonucleoprotein polypeptide N (*Snrpn*) gene in placentas, insulin-like growth factor (*Igf2*) gene in embryos, and potassium channel, voltage-gated KQT-like subfamily Q, member 1, overlapping transcript 1 (*Kcnq1ot1*) gene in placentas were detectably expressed after exposure to BPA during gestation. The failure to establish and/or maintain strict imprinted gene expression was indicative of altered epigenetic regulation in response to BPA exposure in utero. This loss of imprinting (LOI) was shown to be tissue-specific as *Snrpn* in embryos, *Igf2* in placentas and *Kcnq1ot1* in embryos did not show LOI. Furthermore, LOI occurred only when BPA exposure was administered during the latter stage of oocyte development and during the developmental period of genome-wide epigenetic reprogramming. These imprinted genes were then examined to correlate LOI with altered patterns of DNA methylation. Analysis by pyrosequencing and sodium bisulfite genomic sequencing showed that DNA methylation levels in the imprinting control region (ICR) of the *Snrpn* gene in E9.5 placentas was reduced upon BPA exposure in utero. Altered methylation was also detected at an ICR located between the *Igf2* and *H19* genes in BPA-exposed embryos and in the *Igf2* promoter region in embryos. Analysis of global DNA methylation showed that gestational BPA exposure reduced global DNA methylation levels in E9.5 placentas compared to controls, but not in E9.5 embryos.

Recent studies also investigated genome-wide effects of maternal BPA consumption on mouse DNA methylation patterns in F1 offspring [60]. Liver DNA from *Avy* offspring of mothers fed BPA was subjected to genome-wide DNA methylation profiling using the Sequenom Epi-TYPER MassARRAY platform. Nonmonotonic dose-dependent changes in DNA methylation in response to prenatal BPA exposure were detected in so-called RAMs. Approximately 2000–5000 RAMs were identified upon comparison of control and BPA-exposed animals, depending upon BPA dosage. RAMs were preferentially located in CpG island shores (regions flanking CpG islands), compared to CpG islands, and ~100 to ~200 RAMs were identified in promoter regions. Thus, taken together, these studies of BPA and DNA methylation clearly demonstrate an epigenetic effect of maternal BPA exposure on F1 offspring. Given the known toxicological effects of BPA exposure in mammals, it now is important to consider aberrant epigenetic regulation as a significant component in BPA-associated diseases.

A number of recent studies have examined the effects of another environmental contaminant, air pollutants, on DNA methylation patterns of exposed individuals. For example, airborne pollutants from the Map Ta Phut industrial estate (MIE) in Thailand were investigated for potential epigenetic effects on blood leukocytes in MIE workers (high exposure), Map Ta Phut residents (intermediate exposure), and rural resident controls (low exposure). Using bisulfite polymerase chain reaction (PCR) and pyrosequencing, altered methylation was detected at multiple candidate loci/sequences. LINE-1 repetitive DNA sequences and *IL-6* (*interleukin-6*) showed reduced methylation and elevated *HIC1* (*hypermethylated-in-cancer-1*) methylation in MIE workers compared to rural residents (though the magnitude of the differences were <5%). Differences in methylation levels at these loci/sequences in Map Ta Phut residents compared to rural controls was intermediate to those between MIE workers and rural residents. Thus, environmental exposures from steel, oil refinery, and petrochemical complexes have the potential to alter DNA methylation levels in the blood of humans.

To examine the effect of particulate matter (PM; a ubiquitous urban airborne pollutant associated with lung cancer) on DNA methylation of human tandemly repeated DNA sequences, truck drivers and office workers in Beijing, China were subjected to PM monitoring and PCR-pyrosequencing of blood DNA [29]. The results show PM exposure is correlated with reduced DNA methylation of certain tandem repeat DNA families (e.g., alpha-satellite DNA). Similar studies were performed on Chinese truck drivers and office workers to assess epigenetic effects of individual components of PM such as silicone, calcium, sulfur, aluminum, titanium, etc. on tandem DNA repeats [61]. Statistically significant effects were detected with silicone, calcium, and sulfur. In addition to specific loci/sequences and candidate genes, the effect of air pollutants also was examined on global DNA methylation levels in healthy non-smoking adults [62]. Exposure to NO_2, different sizes of particulate matter, and ozone showed decreased global DNA methylation in blood cells. Another component of air pollution is PAHs, which may be associated with increased incidence of asthma [63]. To assess potential epigenetic effects of PAH exposure in humans during pregnancy, levels of maternal PAH exposure were monitored during the third trimester of non-smoking subjects from high traffic areas of New York City. Then cord blood samples were collected at delivery and analyzed for effects of PAH exposure on DNA methylation of the *IFN*γ (interferon γ) promoter region. Maternal PAH exposure correlated with hypermethylation of the *IFNγ* promoter in cord blood, supporting a possible role for altered epigenetic regulation in childhood asthma due to maternal PAH exposure. Epigenetic effects of indoor air pollution from the burning of solid fuels have also been investigated recently [30]. These solid fuels include coal, wood, dung, and crop residues used for cooking and heating the home. Tao et al. reported that global DNA methylation levels in non-smoking women exposed to both coal and wood were decreased 6–7% compared to control women in peripheral blood, but no association with exposure to solid fuels and methylation at the imprinted *Igf2/H19 ICR*.

Not surprisingly, these studies collectively demonstrate the potential for epigenetic damage (or at least change) by numerous environmental contaminants and toxins, some of which, like vinclozolin, may transmit their effects on long-term health through multiple generations.

4. EPIGENETIC EFFECTS OF ETHANOL AND TOBACCO EXPOSURE

Ethanol and tobacco are two substances of common human consumption. Exposure to each has now been shown to be associated with epigenetic changes. Ethanol is a common chemical encountered and consumed by humans that clearly has teratogenic effects on embryos when exposure occurs in utero. The teratogenic consequences of prenatal ethanol exposure leads to a broad spectrum of disorders, collectively termed fetal alcohol spectrum disorders (FASD), the most severe form of which is fetal alcohol syndrome (FAS) [64,65]. FAS is characterized by pre- and postnatal growth deficiencies, craniofacial abnormalities, and CNS dysfunction. The high incidence of FASD/FAS is a major public health problem worldwide and the leading cause of preventable birth defects and mental retardation [66,67]. However, since its original description as a syndrome in 1973, the underlying molecular mechanism(s) by which prenatal ethanol exposure causes FASD/FAS remains unclear. Several studies have investigated the potential contribution of epigenetic dysregulation to the mechanism of FASD/FAS using rodent models. Using the *Avy* agouti mouse model, Kaminen-Ahola et al. demonstrated that maternal ethanol consumption between E0.5 and E8.5 altered DNA methylation at the IAP and the distribution of coat color in offspring [37]. Pregnant females fed ethanol yielded offspring with an increase in the proportion of brown coat color compared to non-ethanol exposed controls at weaning; this is consistent with an increase in DNA methylation (and concomitant increase in gene silencing) of the IAP in response to prenatal ethanol exposure. These results were verified by methylation analysis of the IAP by sodium bisulfite genomic sequencing, which also showed hypermethylation of the IAP in response to maternal ethanol consumption. This was one of the earliest studies to demonstrate that early embryonic exposure to ethanol could result in stable changes in DNA methylation later in life. Using genome-wide profiling technologies, Feng Zhou and colleagues have shown the ability of ethanol exposure to alter DNA methylation at loci across the genome in mouse embryos and cultured neural stem cells (NSCs) [68,69]. In cultured whole mouse embryos, Liu et al. examined the effect of ethanol on DNA methylation at the stage of early embryonic neurulation using methylated DNA immunoprecipitation (meDIP) in conjunction with microarrays (meDIP-Chip) that interrogated promoters and CpG islands. They observed both increases and decreases in DNA methylation levels in response to ethanol exposure, with promoters containing high CpG density tending to show increased methylation, while promoters with low CpG density tending to show decreased methylation. Changes in DNA methylation was correlated with changes in expression of 84 genes, and imprinted genes showed notable changes in methylation in response to ethanol exposure. Treatment of cultured NSCs with ethanol retarded NSC differentiation and altered the normal genome-wide pattern of DNA methylation changes that occur during NSC differentiation (as determined by meDIP-Chip). More recently, Laufer et al. have used meDIP-Chip (of CpG islands and promoter regions) to characterize genome-wide changes in DNA methylation in adult mouse brains in response to prenatal ethanol exposure [39]. Voluntary maternal consumption during the course of pregnancy (from mating through postnatal day 10) yielded adult male offspring with altered DNA methylation at ≥6660 promoter regions assayed at postnatal day 70 (compared to control offspring of mothers provided with water not supplemented with ethanol). Among these genes, a significant proportion of imprinted genes showed altered DNA methylation in response to prenatal ethanol exposure. Thus, these studies (and additional studies not described here) provide increasingly strong evidence for a component of epigenetic dysregulation in the etiology of FASD/FAS. Interestingly, recent studies have shown that prenatal supplementation with choline can suppress the phenotypic effects of prenatal ethanol exposure [70]. Simultaneous administration of choline with ethanol to pregnant female rats during gestation diminished the effects of ethanol on offspring such that birth and brain weight, incisor emergence, and behavioral effects of prenatal ethanol were unchanged from controls. This finding is intriguing because choline is a component of methyl donor diets that have been shown to affect DNA methylation in offspring when administered to pregnant females (e.g., [57,71,72]). Because the components of a methyl donor diet, including choline, are involved the biochemical synthesis of S-adenosylmethionine (SAM, the universal biological methyl donor for DNA, RNA, and protein methylation), it is tempting to conclude that the effects of these dietary supplements on phenotypes and DNA methylation are occurring via SAM and DNA methylation pathways. However, choline also is involved in membrane synthesis and is a precursor to the neurotransmitter acetylcholine, and it is conceivable that effects of methyl donor diets are indirect via these other functions of choline. Currently, it is unclear if the effects of choline on prenatal ethanol exposure are through SAM or other pathways.

Another common consumer product that appears to induce epigenetic changes is tobacco via cigarette smoking. The effects of maternal smoking during pregnancy on placenta, cord blood, and buccal epithelium tissue of offspring in humans has recently been reviewed and detailed by Nielsen et al. and will not be recapitulated here [41]. Human genome-wide DNA methylation profiling has shown a correlation between a decrease in methylation at a CpG site in the human protease-activated receptor 4 (*F2RL3*) gene in smokers versus non-smokers in blood and lung tissue [73,74]. Shenker et al. [74] also identified decreases in methylation levels at CpG sites in the aryl hydrocarbon receptor repressor (*AHRR*) gene, in 2q37, and in 6p21.33 in blood DNA of smokers compared to non-smokers. The decrease in methylation in the *AHRR* gene was correlated with an increase in *AHRR* mRNA expression levels and was also detected in human lung tissue DNA. This finding of decreased methylation at a CpG site in the *AHRR* gene was confirmed in a study of genome-wide DNA methylation profiling of blood DNA from African American males using the Illumina HumanMethylation450 beadchip; it was proposed that the *AHRR* locus could serve as a sensitive biomarker of smoking history [75]. In contrast to these studies, Tommasi et al. [42] examined the effect of secondhand smoke on genome-wide patterns of DNA methylation in mouse lung tissue using the methylated CpG island recovery (MIRA) method in conjunction with Roche NimbleGen CpG Island Plus RefSeq Promoter Microarrays (interrogating ~20,000 RefSeq promoters and 16,000 CpG islands). They found no statistically significant methylation differences between samples exposed to secondhand smoke and controls. They also found no difference in methylation of LINE1 (long interspersed element 1), B1 (a member of the SINE (short interspersed element) family of repetitive elements), and IAP-long terminal repeat repetitive DNA families between lung samples exposed to secondhand smoke and control samples. This is somewhat surprising, given the accumulating evidence from other studies of DNA methylation changes in response to smoking, as well as the strong connection between smoking, lung cancer, and the significant alterations in DNA methylation found in many tumors. No mention was made of the *AHRR* gene found to be differentially methylated in smoke-exposed and control samples in other studies (see above), though it is unclear if the CpG site identified in these other studies was included in the NimbleGen microarray used. The methodology used for methylation profiling in this study was confined to promoters and CpG islands, and therefore may have missed methylation changes in other important regulatory regions of the genome (e.g., enhancers, insulators, etc.); the authors offer an alternative explanation for their findings. More recent DNA methylation profiling studies have extended the earlier findings of altered DNA methylation patterns in response to smoking. Analysis of human blood DNA by Tsaprouni et al. confirmed loci of altered methylation in previous reports and extended these results by identifying additional loci at which DNA methylation changes occurred in response to smoking [76]. Interestingly, most of these loci were associated with regions of open chromatin (e.g., DNase I hypersensitive sites) that are commonly correlated with regulatory regions. Also, these changes in DNA methylation were partially reversible upon cessation of smoking for at least three months. More recently, Guida et al. used the Illumina Infinium HumanMethylation450 platform to investigate the epigenetic effects of smoking on blood cells of two human study groups [77]. They identified 461 CpG sites that exhibited altered DNA methylation in response to smoking, most of which were hypomethylated, including three sites in the *AHRR* gene and one site in the *F2RL3* gene (both genes identified in earlier studies of DNA methylation and smoking). Interestingly, in an analysis of the effects of cessation of smoking on DNA methylation, two distinct groups of CpG sites were identified, one in which DNA methylation changes in response to smoking remained altered even many years (>35 years) after smoking ceased, and a group in which aberrant methylation levels correlated with smoking eventually returned to normal non-smoking levels after cessation of smoking. The CpG sites at which smoking alters DNA methylation the most were predominantly those sites which do not return to normal after quitting smoking. This subset of epigenetic changes triggered by smoking that persist after cessation of smoking may be potentially useful as lifetime biomarkers of smoking history.

5. NUTRITION AND DNA METHYLATION

Throughout most western countries, cardiovascular and respiratory diseases and cancer account for fully three-quarters of all mortalities in adults [78]. For this reason, chronic diseases are becoming the major focus of healthcare related problems. There is a growing body of evidence suggesting that chronic diseases may originate in response to nutritional and/or environmental insults during in utero development [24,79–85]. For example, in human and animal studies, both epidemiological and experimental, evidence suggests nutrient deprivation in utero can have adverse long-term effects on the metabolic and physiological states of offspring. Epidemiological studies looking at standard obstetric birth records show that low birth weight has been linked with increased incidence of hypertension [86], noninsulin-dependent diabetes [87], and chronic bronchitis and CHD [88]. Cancer has been linked with a high birth weight in similar epidemiological studies [89–91]. Unfortunately, these types of studies cannot determine the intrauterine conditions that led to these results, and many factors could contribute to low birth weight. The association between small birth weight and CHD has been studied in several countries [9,88,92–94]. These studies depended on size at gestational age rather than prematurity [95–97]. Death rates among men who were thin at birth, as recorded by a low ponderal index, but had accelerated weight gain during childhood, had the highest death rates from CHD. The ponderal index determines an individual's leanness and is similar to the body mass index (BMI). These men had a fivefold increase in mortality compared to men with high birth weights that were lean in childhood. This effect rate is the highest observed in cardiovascular epidemiology. The authors suggested that the increased death rate may have been due to poor prenatal nutrition followed by improved postnatal nutrition [97]. The results of these epidemiological studies have led to the Barker hypothesis, or the fetal origins hypothesis, which states that the fetus adapts to a limited supply of nutrients in utero, which in turn permanently alters its physiology and metabolism, leading to an increased risk of adult onset disease [10].

6. THE DUTCH FAMINE

A cohort used to study the effects of prenatal nutrition on long-term health and disease risk in a human population is the survivors of the Dutch Hunger Winter (Dutch famine) of 1944–45. The Dutch famine of 1944–45 has provided a wealth of epidemiological data relating malnutrition during early development to adult onset chronic degenerative diseases. During World War II, due to a ban on all food transport in the Netherlands, and an unusually harsh winter which blocked passage from the rural east to the urban west, the daily food rations fell below 1000 calories per day, and at the height of the famine, between 400 and 800 calories per day in the western cities of the Netherlands. During this disastrous famine, women were still able to conceive and give birth, and offspring of these pregnancies have been studied throughout their lives. This has provided an enormous opportunity to study the effects of malnutrition during gestation and correlate this to health-related issues in adult life [79,80,98]. For the study, the obstetric records of 2424 offspring were included, and 741 adults agreed to attend the clinic for extensive measurements. Three periods of 16 weeks were used to distinguish between babies exposed during early, mid- and late gestation. Babies born before the famine, or those conceived after the famine, were termed unexposed and used as comparison. People exposed to famine in early gestation appeared to have a higher risk of CHD [99], had a higher BMI [100] and a more atherogenic lipid profile [80], and those exposed to famine in later stages of development showed increases in the occurrence of obstructive airways disease [101]. People who were exposed to famine in early gestation were also more likely to rate their overall health as poor [79]. Critics have pointed out that many variables, besides extreme shortage of food, must also be taken into account [102,103]. The famine coincided with an unusually harsh winter, which, combined with the stress of war, absence of their spouses, and a general decline in basic services, coinciding with widespread infection, are confounding factors that need to be considered. However, the mechanism by which early nutritional insult leads to an increase in disease susceptibility and altered metabolism in adult life, remains, at best, poorly understood. One possibility is that epigenetic changes in the placenta and/or fetus due to exposure to maternal malnutrition during early development may lead to stable, heritable, aberrant changes in the regulation of genes important for proper placental and fetal development. As discussed earlier, methylation of CpG dinucleotides is a well-characterized epigenetic modification generally associated with silent chromatin. As DNA methylation patterns are reprogrammed in the early embryo and maintained throughout adult life, it is possible that early deprivation of nutrients could lead to alterations within the establishment of proper methylation imprints [13]. Many genes responsible for proper placental and fetal development are imprinted in placental mammals [19], and since methylation imprints are established during gametogenesis [23], it has been proposed that exposure to nutritional insults during gametogenesis and placental and fetal development will lead to alterations in the epigenetic states of imprinted genes, specifically aberrant DNA methylation at DMRs [13,24].

Recently, Tobi et al. performed a genome-wide DNA methylation analysis of blood from subjects of the Dutch Hunger Winter using reduced representation bisulfite sequencing [104]. The study interrogated methylation at 1.2 million CpG sites in 24 subjects exposed to the famine and 24 unexposed controls, the most extensive analysis of methylation to date of this cohort. They identified 181 RAMs associated with famine exposure, with most regions showing hypermethylation and a median change in methylation of a modest 4.6%. A majority of these DMRs mapped to gene bodies, and preferentially localized to regulatory regions, transcribed gene bodies, and expressed exons. These regions also were preferentially associated with genes differentially expressed during preimplantation development. Though this study did not examine the entire methylome, the results provide important insight into the epigenetic effects of the Dutch Hunger Winter and potential molecular mechanisms and pathways for the long-term health consequences of prenatal exposure to famine.

7. MATERNAL NUTRITION AND FETAL HEALTH

In order to test the hypothesis of fetal origins of adult disease, a study by Rees et al., in which female rats were fed either a normal protein diet containing 18% casein or a low protein diet (LPD) containing 8% casein for two weeks prior to mating, was performed. Female rats continued on their respective diet throughout pregnancy until a day before natural birth (21 days post coitum (dpc)) when they were euthanized along with the fetuses, at which time fetal liver, heart, and kidneys were removed [105]. To ensure that the researchers were only investigating changes in protein consumption and not a calorie difference between groups, these diets were isocaloric and the amount of food consumed during pregnancy did not differ significantly between experimental groups. At 21 dpc, the fetuses from the dams on the LPD were smaller (13.7%) than the fetuses from dams on the normal protein diet, and fetal livers from the LPD group were approximately 24% smaller. In addition to weights of fetal tissue, as well as overall fetal weights, the authors examined the global methylation levels in the fetal tissues using a methyl acceptance assay [106]. The results showed a greater than 25% increase in global DNA methylation levels in the livers of the fetuses from the LPD. No significant changes were observed in the kidney or heart. The authors summarize by stating that exposure to protein restriction in utero caused genome-wide changes in DNA methylation in the fetal liver, and suggests that these changes have the potential to alter the regulation of important genes in the offspring [105]. While the study by Rees et al. looked at global levels of DNA methylation, another set of experiments performed by Lillycrop et al. using a similar animal model of protein restriction in utero, looked at locus-specific changes in DNA methylation in postnatal liver tissue [24]. For this study, timed pregnant rats were put into one of three diet groups at time of conception, a LPD and a LPD that has been fortified with folic acid, as well as a normal protein diet control. The dams from each group were fed this diet throughout pregnancy until spontaneous birth at day 21 dpc, at which point the dams were switched onto a lactating diet until the pups were weaned onto normal chow at postnatal day 28. At approximately five weeks after birth, the pups were sacrificed and the livers were harvested. It is important to be able to distinguish acute effects, that is, changes in gene expression that are caused by direct, active exposure to protein restriction, from long-term effects that persist long after the nutritional insult has ended. In this particular study, the pups were on a normal diet for about five weeks after birth and were only exposed to protein restriction while in utero. Since epigenetic marks are, by definition, stable and heritable, any changes observed at this later point after exposure are most likely due to epigenetic changes. DNA methylation and gene expression levels were examined for two hepatic genes, glucocorticoid receptor (GR) and peroxisomal proliferator-activated receptor (PPARα) through the use of methylation-specific PCR and RT-PCR. The results indicated that offspring exposed to a restricted diet in utero had DNA methylation levels 26% lower at the PPARα promoter, and 23% less methylation at the GR promoter, as compared to offspring in the normal protein diet group, whereas methylation levels of PPARα and GR promoters were essentially the same for offspring of both the normal protein and protein-restricted fortified diet groups. The authors also examined whether the observed changes in DNA methylation led to a corresponding change in expression levels and found that mRNA levels of the PPARα gene were increased by 945% and the expression of GR was increased by 300% in the protein-restricted diet group as compared to the control diet group. No statistical difference in the expression levels of PPARα or GR between the control and restricted fortified diet groups was observed [24]. This study was important as it demonstrated that early life events, such as protein restriction in utero, could cause long-term stable changes to the epigenome, resulting in aberrant gene expression in the adult offspring.

In a subsequent study, it was shown through high-resolution sodium bisulfite genomic sequencing that the hypomethylation that was observed at the promoter of the PPARα gene occurred at specific CpG dinucleotides [84], and that the CpG sites that were affected were within the putative binding site of various transcription factors. The authors suggested that changes in DNA methylation at these sites could alter transcription factor binding, ultimately leading to changes in gene expression [84]. These results provide a potential mechanism of how changes to the epigenome in early life, which alter the ability of transcription factors to bind to chromatin, can alter gene expression patterns later in life. It is possible that these changes in gene expression could lead to increased disease susceptibility in adulthood. In another follow-up study, using the same rodent model of maternal nutrition, Lillycrop et al. were able to examine changes in gene expression in the livers of adult male offspring exposed to protein restriction in utero in a whole genome manner, as opposed to looking at individual loci, through the use of microarrays [85]. Pathway analysis of genes with altered expression levels reveals altered pathways with diverse metabolic and physiological functions consistent with previously reported phenotypic changes due to maternal diet [107–110]. Taken together, these studies suggest that nutritional insults during the period of gestation can produce stable epigenetic changes, such as altering DNA methylation levels, as well as changes in gene expression, that persist throughout the lifetime of adult offspring. Additionally, these changes in gene expression result in an altered phenotype that affects the health of adult offspring.

Most recently, an extensive and important study by Radford et al. examined connections between maternal undernourishment, DNA methylation, and transgenerational transmission in mice [111]. F1 offspring of dams exposed to severe caloric restriction showed low birth weight, early life adiposity, reduced muscle stem cell number and function, reduced pancreatic function, and glucose intolerance. The offspring (F2 generation) from sperm of the F1 generation exhibited low birth weight and glucose intolerance without further exposure to caloric restriction, thus demonstrating paternal transmission of a disease phenotype to the next generation. Genome-wide DNA methylation profiling of F1 sperm by MeDIP-Seq showed altered methylation at 166 genomic regions (most of which were hypomethylated) relative to normally fed controls, suggesting a mechanism for transgenerational transmission of the phenotype to the F2 generation.

However, the alterations in DNA methylation identified in F1 sperm DNA were not maintained in somatic tissues of the F2 offspring, indicating that transgenerational transmission of the phenotypic effects of maternal caloric restriction was not due simply to direct transmission of altered DNA methylation patterns in sperm to somatic cells of the next generation. Nonetheless, the altered methylation in F1 sperm appeared to have an effect in the F2 generation because gene expression was altered in F2 somatic cells in genomic regions, which exhibited altered DNA methylation in F1 sperm. The authors suggest that the epigenetic alterations in F1 sperm affected F2 somatic cell gene expression patterns via alterations in chromatin architecture, transcriptional regulatory networks, differentiation, or tissue structure.

8. FOLIC ACID AND DNA METHYLATION

Folate is a water soluble B vitamin that is essential for the synthesis of nucleotides and plays a major role in one-carbon metabolism [112]. Folate, in its various coenzyme forms, acts as both a methyl donor and acceptor in one-carbon metabolism, with its most prominent role being involved in the remethylation of homocysteine to methionine. Methionine is a precursor of SAM, which serves as the universal one-carbon donor involved in methylation of DNA, RNA, lipids, and proteins [113,114]. Because of this essential role, perturbations in the levels of folate can have profound effects on both nucleotide synthesis and methylation of DNA, both of which are critical in maintaining the integrity of DNA and the proper regulation of gene expression [115,116]. Folate plays an important role in the pathogenesis of several disorders, including anemia, cardiovascular disease, and developmental abnormalities such as neural tube defects (NTDs) [117], and has been linked to an increased risk of developing several types of cancer, including colon, pancreatic, and possibly breast cancers [118–120].

The benefits of periconceptional supplementation with folic acid, a synthetic, oxidized form of folate, in the reduction in NTDs have been well studied, and it is recommended that women of child-bearing age consume 400 µg of folate daily [121]. As this level of intake was generally not being achieved in the United States population, a program of mandatory, nationwide fortification of flour and uncooked cereal grains with folic acid was implemented in 1998 in the US, and shortly thereafter in Canada. After the fortification program began, plasma folate concentration increased by 100%, along with a reduction in homocysteine levels, resulting in a 20–50% decrease in the incidence of NTDs [122–124]. Interestingly, it was observed that following the implementation of this program, there was a temporal association between folic acid fortification and an increase in colorectal cancer rates in both the US and Canada [125]. Although it is impossible to draw a causal link between these two events, it has been suggested that an excess of folic acid could have promoted the growth of previously undetected preneoplasia [126,127]. Although the success in prevention of NTDs has been dramatic, and folic acid is generally considered to be safe, there are concerns that the levels of folic acid intake may be far higher than recommended [128]. As the long-term consequences of increased folate levels have yet to be extensively studied in long-term clinical trials, this is becoming an area of increased concern and research.

In human studies conducted in metabolic units involving folate depletion of volunteers, global DNA hypomethylation was observed in circulating peripheral blood lymphocytes. After repletion of folate levels, DNA methylation levels returned to normal, indicating that the effect on DNA methylation was transitory and persisted only during the folate depletion period [129]. Studies involving folate depletion in rodents indicate similar effects. Folate-deficient rats had significantly lower global DNA methylation levels in liver tissue compared to folate replete control rodents [130]. It has also been demonstrated that rats fed a diet supplemented with folic acid showed increased levels of global DNA methylation in liver tissue [131]. Perhaps the best studied example of how prenatal folate can affect DNA methylation in the offspring is in the *agouti* mouse system described in Sec. 3. Feeding pregnant dams a diet fortified with methyl donors resulted in hypermethylation of the agouti repetitive element, leading to decreased expression [81].

More recently, the effects of maternal and postweaning folic acid supplementation on colorectal cancer risk were analyzed in rats [132]. The authors found that maternal folic acid supplementation significantly reduced the occurrence of colorectal cancer in adult offspring. No reduction was noted for postweaning folic acid supplementation. Additionally, the levels of colorectal global DNA methylation were increased in the maternal folic acid supplemented group and were reduced in the postweaning supplemented group [132]. These studies highlight the fact that changes to the intrauterine environment can alter the epigenome of the offspring, resulting in increased risk of cancer and other diseases throughout adulthood.

To examine epigenetic effects of postnatal folate exposure, Crider and Hao examined the effects of folic acid on DNA methylation in women of childbearing age in a large population-based, double blind study of folic acid supplementation [133,134]. This large population-based study set out to determine the effects of folic acid supplementation and withdrawal on homocysteine levels in a folic acid deplete population, based on dosages of folic acid ranging from 100 to 4000 µg daily. Blood samples were collected at baseline (0 months), 1, 3, and 6 months, at which time folic acid supplementation ceased. Blood samples were again collected 3 months after cessation of supplementation. In this study, no statistically significant changes in DNA methylation were observed in circulating blood at any dose or time point. However, when the authors examined methylation levels in coagulated blood samples, there were significant time-, dose- and genotype-dependent changes observed. The authors suggest that folate may play a here-to-fore unknown role in the clotting and coagulation processes, and that the whole blood samples may more accurately represent circulating blood in the subjects.

9. MATERNAL NURTURING AND STRESS

It has been shown that early life events can alter responses to stress in such a way as to predispose adults to various diseases, including ischemic heart disease and an increased risk of obesity, as well as several psychological diseases [135–140]. In addition to changes in the intrauterine environment such as altered nutrition and environmental toxins, early life experiences can also alter the epigenome. Maternal nurturing has been shown to affect the neural and endocrine responses to stress [139,141–149]. In the rat, there are naturally occurring variations in maternal behavior in the nurturing of pups. During lactation, the mother arches her back to allow easy suckling by pups, termed arched-back nursing (ABN), while also occasionally licking and grooming pups (LG). Researchers use rat dams that exhibit either high LG-ABN or low LG-ABN to study how maternal nurturing can affect stress responses in the adult offspring. These variations in grooming and nursing result in offspring with altered behavior and hypothalamic-pituitary-adrenal responses to stress in the adult offspring. Weaver et al. demonstrated that the observed altered behavior may be due in part to changes in DNA methylation and histone acetylation in the hippocampus. Offspring of low LG-ABN dams had lower levels of DNA methylation and histone H3K9 acetylation at the GR promoter compared to high LG-ABN-exposed offspring [141]. Cross-fostering of the offspring between high and low LG-ABN suggests that these changes to the epigenome were not due to genetic differences between high and low LG-ABN dams. The changes to the epigenome were established as early as postnatal day 7, and cross-fostering after this point reversed the effect, indicating that these epigenetic marks are dynamic [141]. The results of this study demonstrated for the first time that maternal behavior could program gene expression patterns in the adult by altering epigenetic marks in the offspring.

In a follow-up study, the effects of early life experiences were examined in a genome-wide manner in the hippocampus of offspring of high and low LG-ABN dams. This study showed that more than 900 genes appear to be stably altered by maternal care. Treatment of the offspring with a histone deacetylase inhibitor (TSA) or with the methyl-donor L-methionine reversed the altered expression of a portion of these genes and reversed the observed behavioral changes. This study demonstrated that early life experiences can program gene expression patterns and adult stress response behavior by altering the epigenome, and these changes are in part reversible by treatment with TSA or L-methionine [142]. Additional studies in rats, using the same model of maternal care, have shown that genes involved in various psychopathologies, such as schizophrenia and depression, as well as genes involved in learning and cognition, are also altered by maternal care [144,147]. In an alternate model of early life stress, Blaze et al. demonstrated that maternal rough handling and dragging and dropping of pups, or maternal nurturing administered outside of the home cage, resulted in locus-specific changes in DNA methylation in the medial prefrontal cortex. The effect on DNA methylation was complex, with observed alterations varying depending on the treatment group, as well as the sex, time after treatment and locus being analyzed [148]. It has even been demonstrated that indirect stress, that is stress that is administered to a cage mate, can also alter DNA methylation and behavior in adult offspring [146].

The effects of early life experiences on the epigenome can also be seen in humans. Childhood adversity, such as abuse and neglect, is strongly correlated with suicide, and it has been shown that hippocampal GR is downregulated in psychological disorders associated with suicide [150]. To investigate this link, the authors analyzed DNA methylation and expression of the neuronal specific promoter of the glucocorticoid receptor

(NR3C1) in post-mortem hippocampal tissue from suicide victims. It was found that methylation was increased, and the expression was decreased in suicide victims that were abused as children as compared to non-abused suicide victims or control hippocampal tissue samples (non-abused, sudden, or accidental death) [143].

It has also been shown that exposure to prenatal stress can induce stable alterations in DNA methylation. The Quebec Ice Storm of 1998 was one of Canada's worst natural disasters in history. After a particularly large ice storm, power was out for three million people for as long as 6 weeks, with temperatures between −10 and −20°C, causing tremendous hardship for citizens. Offspring of women who were pregnant during the ice storm were analyzed for changes in DNA methylation in T cells at 13 years of age. It was observed that over 900 genes showed changes in methylation as compared to controls [151]. Genes related to immune function seemed to be the most affected by prenatal stress. This study was the first to show that prenatal maternal stress exposure can produce stable changes in DNA methylation. This natural disaster allowed researchers to separate objective and subjective maternal stress, making analysis of the results easier to interpret. As with the Dutch famine of 1944–45, the aberrant changes to the epigenome of offspring appears to be quite long-lasting, persisting well into adulthood.

REFERENCES

[1] Berger SL, et al. An operational definition of epigenetics. Genes Dev 2009;23(7):781–3.
[2] Li E. Chromatin modification and epigenetic reprogramming in mammalian development. Nat Rev Genet 2002;3(9):662–73.
[3] Santos F, Dean W. Epigenetic reprogramming during early development in mammals. Reproduction 2004;127(6):643–51.
[4] Morgan HD, et al. Epigenetic reprogramming in mammals. Hum Mol Genet 2005;14(1):R47–58.
[5] Leitch HG, Tang WW, Surani MA. Primordial germ-cell development and epigenetic reprogramming in mammals. Curr Top Dev Biol 2013;104:149–87.
[6] Bergman Y, Cedar H. DNA methylation dynamics in health and disease. Nat Struct Mol Biol 2013;20(3):274–81.
[7] Rose G. Familial patterns in ischaemic heart disease. Br J Prev Soc Med 1964;18:75–80.
[8] Forsdahl A. Are poor living conditions in childhood and adolescence an important risk factor for arteriosclerotic heart disease? Br J Prev Soc Med 1977;31(2):91–5.
[9] Barker DJ, Osmond C. Infant mortality, childhood nutrition, and ischaemic heart disease in England and Wales. Lancet 1986;1(8489):1077–81.
[10] Barker DJ. Fetal origins of coronary heart disease. BMJ 1995;311(6998):171–4.
[11] Trichopoulos D. Hypothesis: does breast cancer originate in utero? Lancet 1990;335(8695):939–40.
[12] Gluckman PD, Hanson MA. Developmental origins of disease paradigm: a mechanistic and evolutionary perspective. Pediatr Res 2004;56(3):311–7.
[13] Waterland RA, Garza C. Potential mechanisms of metabolic imprinting that lead to chronic disease. Am J Clin Nutr 1999;69(2):179–97.
[14] Waterland RA, Michels KB. Epigenetic epidemiology of the developmental origins hypothesis. Annu Rev Nutr 2007;27:363–88.
[15] Dolinoy DC, Weidman JR, Jirtle RL. Epigenetic gene regulation: linking early developmental environment to adult disease. Reprod Toxicol 2007;23(3):297–307.
[16] Surani MA. Imprinting and the initiation of gene silencing in the germ line. Cell 1998;93(3):309–12.

[17] Bartolomei MS, Tilghman SM. Genomic imprinting in mammals. Annu Rev Genet 1997;31: 493–525.

[18] Wood AJ, Oakey RJ. Genomic imprinting in mammals: emerging themes and established theories. PLoS Genet 2006;2(11):e147.

[19] Constancia M, et al. Placental-specific IGF-II is a major modulator of placental and fetal growth. Nature 2002;417(6892):945–8.

[20] Tremblay KD, Duran KL, Bartolomei MS. A 5′ 2-kilobase-pair region of the imprinted mouse H19 gene exhibits exclusive paternal methylation throughout development. Mol Cell Biol 1997;17(8): 4322–9.

[21] Thorvaldsen JL, Duran KL, Bartolomei MS. Deletion of the H19 differentially methylated domain results in loss of imprinted expression of H19 and Igf2. Genes Dev 1998;12(23):3693–702.

[22] Feil R, Khosla S. Genomic imprinting in mammals: an interplay between chromatin and DNA methylation? Trends Genet 1999;15(11):431–5.

[23] Reik W, Dean W, Walter J. Epigenetic reprogramming in mammalian development. Science 2001;293(5532):1089–93.

[24] Lillycrop KA, et al. Dietary protein restriction of pregnant rats induces and folic acid supplementation prevents epigenetic modification of hepatic gene expression in the offspring. J Nutr 2005;135(6): 1382–6.

[25] Manikkam M, et al. Plastics derived endocrine disruptors (BPA, DEHP and DBP) induce epigenetic transgenerational inheritance of obesity, reproductive disease and sperm epimutations. PLoS One 2013;8(1):e55387.

[26] Anway MD, et al. Epigenetic transgenerational actions of endocrine disruptors and male fertility. Science 2005;308(5727):1466–9.

[27] Collotta M, Bertazzi PA, Bollati V. Epigenetics and pesticides. Toxicology 2013;307:35–41.

[28] Guerrero-Bosagna CM, Skinner MK. Environmental epigenetics and phytoestrogen/phytochemical exposures. J Steroid Biochem Mol Biol 2014;139:270–6.

[29] Guo L, et al. Effects of short-term exposure to inhalable particulate matter on DNA methylation of tandem repeats. Environ Mol Mutagen 2014;55(4):322–35.

[30] Tao MH, et al. Indoor air pollution from solid fuels and peripheral Blood DNA methylation: findings from a population study in Warsaw, Poland. Environ Res 2014;134:325–30.

[31] Baccarelli A, et al. Rapid DNA methylation changes after exposure to traffic particles. Am J Respir Crit Care Med 2009;179(7):572–8.

[32] Mass MJ, Wang L. Arsenic alters cytosine methylation patterns of the promoter of the tumor suppressor gene p53 in human lung cells: a model for a mechanism of carcinogenesis. Mutat Res 1997;386(3): 263–77.

[33] Reichard JF, Schnekenburger M, Puga A. Long term low-dose arsenic exposure induces loss of DNA methylation. Biochem Biophys Res Commun 2007;352(1):188–92.

[34] Cheng TF, Choudhuri S, Muldoon-Jacobs K. Epigenetic targets of some toxicologically relevant metals: a review of the literature. J Appl Toxicol 2012;32(9):643–53.

[35] Herbstman JB, et al. Prenatal exposure to polycyclic aromatic hydrocarbons, benzo[a]pyrene-DNA adducts, and genomic DNA methylation in cord blood. Environ Health Perspect 2012;120(5):733–8.

[36] Perera F, et al. Relation of DNA methylation of 5'-CpG island of ACSL3 to transplacental exposure to airborne polycyclic aromatic hydrocarbons and childhood asthma. PLoS One 2009;4(2):e4488.

[37] Kaminen-Ahola N, et al. Maternal ethanol consumption alters the epigenotype and the phenotype of offspring in a mouse model. PLoS Genet 2010;6(1):e1000811.

[38] Kaminen-Ahola N, et al. Postnatal growth restriction and gene expression changes in a mouse model of fetal alcohol syndrome. Birth Defects Res A Clin Mol Teratol 2010;88(10):818–26.

[39] Laufer BI, et al. Long-lasting alterations to DNA methylation and ncRNAs could underlie the effects of fetal alcohol exposure in mice. Dis Model Mech 2013;6(4):977–92.

[40] Stoccoro A, et al. Epigenetic effects of nano-sized materials. Toxicology 2013;313(1):3–14.

[41] Nielsen CH, Larsen A, Nielsen AL. DNA methylation alterations in response to prenatal exposure of maternal cigarette smoking: a persistent epigenetic impact on health from maternal lifestyle? Arch Toxicol 2014;12.

[42] Tommasi S, et al. Whole DNA methylome profiling in mice exposed to secondhand smoke. Epigenetics 2012;7(11):1302–14.

[43] Guerrero-Bosagna C, et al. Epigenetic transgenerational actions of vinclozolin on promoter regions of the sperm epigenome. PLoS One 2010;5(9).

[44] Manikkam M, et al. Transgenerational actions of environmental compounds on reproductive disease and identification of epigenetic biomarkers of ancestral exposures. PLoS One 2012;7(2):e31901.

[45] Manikkam M, et al. Pesticide methoxychlor promotes the epigenetic transgenerational inheritance of adult-onset disease through the female germline. PLoS One 2014;9(7):e102091.

[46] Tracey R, et al. Hydrocarbons (jet fuel JP-8) induce epigenetic transgenerational inheritance of obesity, reproductive disease and sperm epimutations. Reprod Toxicol 2013;36:104–16.

[47] Manikkam M, et al. Pesticide and insect repellent mixture (permethrin and DEET) induces epigenetic transgenerational inheritance of disease and sperm epimutations. Reprod Toxicol 2012;34(4): 708–19.

[48] Skinner MK, Guerrero-Bosagna C. Environmental signals and transgenerational epigenetics. Epigenomics 2009;1(1):111–7.

[49] Skinner MK, et al. Ancestral dichlorodiphenyltrichloroethane (DDT) exposure promotes epigenetic transgenerational inheritance of obesity. BMC Med 2013;11:228.

[50] Crews D, et al. Epigenetic transgenerational inheritance of altered stress responses. Proc Natl Acad Sci USA 2012;109(23):9143–8.

[51] Guerrero-Bosagna C, et al. Epigenetic transgenerational inheritance of vinclozolin induced mouse adult onset disease and associated sperm epigenome biomarkers. Reprod Toxicol 2012;34(4): 694–707.

[52] Murray TJ, et al. Induction of mammary gland ductal hyperplasias and carcinoma in situ following fetal bisphenol A exposure. Reprod Toxicol 2007;23(3):383–90.

[53] Ho SM, et al. Developmental exposure to estradiol and bisphenol A increases susceptibility to prostate carcinogenesis and epigenetically regulates phosphodiesterase type 4 variant 4. Cancer Res 2006;66(11):5624–32.

[54] Williams SA, et al. Effects of developmental bisphenol A exposure on reproductive-related behaviors in California mice (*Peromyscus californicus*): a monogamous animal model. PLoS One 2013;8(2):e55698.

[55] Fernandez M, et al. Neonatal exposure to bisphenol a and reproductive and endocrine alterations resembling the polycystic ovarian syndrome in adult rats. Environ Health Perspect 2010;118(9): 1217–22.

[56] Calafat AM, et al. Urinary concentrations of bisphenol A and 4-nonylphenol in a human reference population. Environ Health Perspect 2005;113(4):391–5.

[57] Dolinoy DC, Huang D, Jirtle RL. Maternal nutrient supplementation counteracts bisphenol A–induced DNA hypomethylation in early development. Proc Natl Acad Sci USA 2007;104(32): 13056–61.

[58] Dolinoy DC. The agouti mouse model: an epigenetic biosensor for nutritional and environmental alterations on the fetal epigenome. Nutr Rev 2008;66(Suppl. 1):S7–11.

[59] Susiarjo M, et al. Bisphenol a exposure disrupts genomic imprinting in the mouse. PLoS Genet 2013;9(4):e1003401.

[60] Kim JH, et al. Perinatal bisphenol A exposure promotes dose-dependent alterations of the mouse methylome. BMC Genomics 2014;15:30.

[61] Hou L, et al. Altered methylation in tandem repeat element and elemental component levels in inhalable air particles. Environ Mol Mutagen 2014;55(3):256–65.

[62] De Prins S, et al. Influence of ambient air pollution on global DNA methylation in healthy adults: a seasonal follow-up. Environ Int 2013;59:418–24.

[63] Tang WY, et al. Maternal exposure to polycyclic aromatic hydrocarbons and 5′-CpG methylation of interferon-gamma in cord white blood cells. Environ Health Perspect 2012;120(8):1195–200.

[64] Jones KL, et al. Pattern of malformation in offspring of chronic alcoholic mothers. Lancet 1973;1(7815): 1267–71.

[65] Jones KL, Smith DW. Recognition of the fetal alcohol syndrome in early infancy. Lancet 1973;302(7836):999–1001.

[66] May PA, Gossage JP. Estimating the prevalence of fetal alcohol syndrome. A summary. Alcohol Res Health 2001;25(3):159–67.

[67] Committee on Substance Abuse and Committee on Children With Disabilities. Fetal alcohol syndrome and alcohol-related neurodevelopmental disorders. Pediatrics 2000;106(2 Pt 1):358–61.

[68] Liu Y, et al. Alcohol exposure alters DNA methylation profiles in mouse embryos at early neurulation. Epigenetics 2009;4(7):500–11.

[69] Zhou FC, et al. Alcohol alters DNA methylation patterns and inhibits neural stem cell differentiation. Alcohol Clin Exp Res 2011;35(4):735–46.

[70] Thomas JD, Abou EJ, Dominguez HD. Prenatal choline supplementation mitigates the adverse effects of prenatal alcohol exposure on development in rats. Neurotoxicol Teratol 2009;31(5):303–11.

[71] Wolff GL, et al. Maternal epigenetics and methyl supplements affect agouti gene expression in Avy/a mice. FASEB J 1998;12(11):949–57.

[72] Cooney CA, Dave AA, Wolff GL. Maternal methyl supplements in mice affect epigenetic variation and DNA methylation of offspring. J Nutr 2002;132(8 Suppl):2393S–400S.

[73] Breitling LP, et al. Tobacco-smoking-related differential DNA methylation: 27K discovery and replication. Am J Hum Genet 2011;88(4):450–7.

[74] Shenker NS, et al. Epigenome-wide association study in the European Prospective Investigation into Cancer and Nutrition (EPIC-Turin) identifies novel genetic loci associated with smoking. Hum Mol Genet 2013;22(5):843–51.

[75] Philibert RA, et al. Changes in DNA methylation at the aryl hydrocarbon receptor repressor may be a new biomarker for smoking. Clin Epigenet 2013;5(1):19.

[76] Tsaprouni LG, et al. Cigarette smoking reduces DNA methylation levels at multiple genomic loci but the effect is partially reversible upon cessation. Epigenetics 2014;9(10):1382–96.

[77] Guida F, et al. Dynamics of smoking-induced genome-wide methylation changes with time since smoking cessation. Hum Mol Genet 2015;24(8):2349–59.

[78] Murray CJ, Lopez AD. Global and regional cause-of-death patterns in 1990. Bull World Health Organ 1994;72(3):447–80.

[79] Roseboom TJ, et al. Adult survival after prenatal exposure to the Dutch famine 1944–45. Paediatr Perinat Epidemiol 2001;15(3):220–5.

[80] Roseboom TJ, et al. Effects of prenatal exposure to the Dutch famine on adult disease in later life: an overview. Mol Cell Endocrinol 2001;185(1–2):93–8.

[81] Waterland RA, Jirtle RL. Transposable elements: targets for early nutritional effects on epigenetic gene regulation. Mol Cell Biol 2003;23(15):5293–300.

[82] Burdge GC, et al. Epigenetic regulation of transcription: a mechanism for inducing variations in phenotype (fetal programming) by differences in nutrition during early life? Br J Nutr 2007;97(6):1036–46.

[83] Lillycrop KA, et al. Induction of altered epigenetic regulation of the hepatic glucocorticoid receptor in the offspring of rats fed a protein-restricted diet during pregnancy suggests that reduced DNA methyltransferase-1 expression is involved in impaired DNA methylation and changes in histone modifications. Br J Nutr 2007;97(6):1064–73.

[84] Lillycrop KA, et al. Feeding pregnant rats a protein-restricted diet persistently alters the methylation of specific cytosines in the hepatic PPAR alpha promoter of the offspring. Br J Nutr 2008;100(2):278–82.

[85] Lillycrop KA, et al. Maternal protein restriction with or without folic acid supplementation during pregnancy alters the hepatic transcriptome in adult male rats. Br J Nutr 2010;103(12):1711–9.

[86] Law CM, et al. Maternal and fetal influences on blood pressure. Arch Dis Child 1991;66(11):1291–5.

[87] Phillips DI, et al. Thinness at birth and insulin resistance in adult life. Diabetologia 1994;37(2):150–4.

[88] Barker DJ, et al. Fetal and placental size and risk of hypertension in adult life. BMJ 1990;301(6746):259–62.

[89] Michels KB, et al. Birthweight as a risk factor for breast cancer. Lancet 1996;348(9041):1542–6.

[90] Michels KB, Xue F. Role of birthweight in the etiology of breast cancer. Int J Cancer 2006;119(9):2007–25.

[91] Hjalgrim LL, et al. Birth weight as a risk factor for childhood leukemia: a meta-analysis of 18 epidemiologic studies. Am J Epidemiol 2003;158(8):724–35.

[92] Barker DJ, et al. Weight in infancy and death from ischaemic heart disease. Lancet 1989;2(8663):577–80.

[93] Barker DJ, et al. The relation of small head circumference and thinness at birth to death from cardiovascular disease in adult life. BMJ 1993;306(6875):422–6.

[94] Hales CN, et al. Fetal and infant growth and impaired glucose tolerance at age 64. BMJ 1991;303(6809):1019–22.

[95] Osmond C, et al. Early growth and death from cardiovascular disease in women. BMJ 1993;307(6918):1519–24.

[96] Leon DA, et al. Reduced fetal growth rate and increased risk of death from ischaemic heart disease: cohort study of 15 000 Swedish men and women born 1915-29. BMJ 1998;317(7153):241–5.

[97] Eriksson JG, et al. Catch-up growth in childhood and death from coronary heart disease: longitudinal study. BMJ 1999;318(7181):427–31.

[98] Heijmans BT, et al. Persistent epigenetic differences associated with prenatal exposure to famine in humans. Proc Natl Acad Sci USA 2008;105(44):17046–9.

[99] Ravelli AC, et al. Obesity at the age of 50 y in men and women exposed to famine prenatally. Am J Clin Nutr 1999;70(5):811–6.

[100] Roseboom TJ, et al. Plasma lipid profiles in adults after prenatal exposure to the Dutch famine. Am J Clin Nutr 2000;72(5):1101–6.

[101] Lopuhaa CE, et al. Atopy, lung function, and obstructive airways disease after prenatal exposure to famine. Thorax 2000;55(7):555–61.

[102] Huxley R, Neil A, Collins R. Unravelling the fetal origins hypothesis: is there really an inverse association between birthweight and subsequent blood pressure? Lancet 2002;360(9334):659–65.

[103] Huxley RR, Neil HA. Does maternal nutrition in pregnancy and birth weight influence levels of CHD risk factors in adult life? Br J Nutr 2004;91(3):459–68.

[104] Tobi EW, et al. DNA methylation signatures link prenatal famine exposure to growth and metabolism. Nat Commun 2014;5:5592.

[105] Rees WD, et al. Maternal protein deficiency causes hypermethylation of DNA in the livers of rat fetuses. J Nutr 2000;130(7):1821–6.

[106] Balaghi M, Wagner C. DNA methylation in folate deficiency: use of CpG methylase. Biochem Biophys Res Commun 1993;193(3):1184–90.

[107] Bellinger L, Lilley C, Langley-Evans SC. Prenatal exposure to a maternal low-protein diet programmes a preference for high-fat foods in the young adult rat. Br J Nutr 2004;92(3):513–20.

[108] Bellinger L, Sculley DV, Langley-Evans SC. Exposure to undernutrition in fetal life determines fat distribution, locomotor activity and food intake in ageing rats. Int J Obes (Lond) 2006;30(5):729–38.

[109] Langley-Evans SC, Sculley DV. Programming of hepatic antioxidant capacity and oxidative injury in the ageing rat. Mech Ageing Dev 2005;126(6–7):804–12.

[110] Lucas A, et al. Nutrition in pregnant or lactating rats programs lipid metabolism in the offspring. Br J Nutr 1996;76(4):605–12.

[111] Radford EJ, et al. In utero effects. In utero undernourishment perturbs the adult sperm methylome and intergenerational metabolism. Science 2014;345(6198):1255903.

[112] Mackenzie R. Biogenesis and interconversion of substituted tetrahydofolates. Folate Pterins 1984;1:255–306.

[113] Lucock M. Folic acid: nutritional biochemistry, molecular biology, and role in disease processes. Mol Genet Metab 2000;71(1–2):121–38.

[114] Lamprecht SA, Lipkin M. Chemoprevention of colon cancer by calcium, vitamin D and folate: molecular mechanisms. Nat Rev Cancer 2003;3(8):601–14.

[115] Dolinoy DC, Jirtle RL. Environmental epigenomics in human health and disease. Environ Mol Mutagen 2008;49(1):4–8.

[116] Robertson KD. DNA methylation and human disease. Nat Rev Genet 2005;6(8):597–610.

[117] Kim YI. Role of folate in colon cancer development and progression. J Nutr 2003;133(11 Suppl 1):3731S–9S.

[118] Kim YI. Folate and colorectal cancer: an evidence-based critical review. Mol Nutr Food Res 2007;51(3):267–92.

[119] Ulrich CM. Folate and cancer prevention: a closer look at a complex picture. Am J Clin Nutr 2007;86(2):271–3.

[120] Larsson SC, Giovannucci E, Wolk A. Folate intake, MTHFR polymorphisms, and risk of esophageal, gastric, and pancreatic cancer: a meta-analysis. Gastroenterology 2006;131(4):1271–83.

[121] Berry RJ, et al. Prevention of neural-tube defects with folic acid in China. China-U.S. Collaborative Project for Neural Tube Defect Prevention. N Engl J Med 1999;341(20):1485–90.

[122] Honein MA, et al. Impact of folic acid fortification of the US food supply on the occurrence of neural tube defects. JAMA 2001;285(23):2981–6.

[123] Jacques PF, et al. The effect of folic acid fortification on plasma folate and total homocysteine concentrations. N Engl J Med 1999;340(19):1449–54.

[124] Ray JG. Folic acid food fortification in Canada. Nutr Rev 2004;62(6 Pt 2):S35–9.

[125] Mason JB, et al. A temporal association between folic acid fortification and an increase in colorectal cancer rates may be illuminating important biological principles: a hypothesis. Cancer Epidemiol Biomarkers Prev 2007;16(7):1325–9.

[126] Song J, et al. Effects of dietary folate on intestinal tumorigenesis in the apcMin mouse. Cancer Res 2000;60(19):5434–40.

[127] Song J, et al. Chemopreventive effects of dietary folate on intestinal polyps in Apc+/-Msh2-/- mice. Cancer Res 2000;60(12):3191–9.

[128] Pfeiffer CM, et al. Biochemical indicators of B vitamin status in the US population after folic acid fortification: results from the National Health and Nutrition Examination Survey 1999-2000. Am J Clin Nutr 2005;82(2):442–50.

[129] Jacob RA, et al. Moderate folate depletion increases plasma homocysteine and decreases lymphocyte DNA methylation in postmenopausal women. J Nutr 1998;128(7):1204–12.

[130] Balaghi M, Horne DW, Wagner C. Hepatic one-carbon metabolism in early folate deficiency in rats. Biochem J 1993;291(Pt 1):145–9.

[131] Choi SW, et al. Folate depletion impairs DNA excision repair in the colon of the rat. Gut 1998;43(1):93–9.

[132] Sie KK, et al. Effect of maternal and postweaning folic acid supplementation on colorectal cancer risk in the offspring. Gut 2011;60(12):1687–94.

[133] Hao L, et al. Folate status and homocysteine response to folic acid doses and withdrawal among young Chinese women in a large-scale randomized double-blind trial. Am J Clin Nutr 2008;88(2):448–57.

[134] Crider KS, et al. Genomic DNA methylation changes in response to folic acid supplementation in a population-based intervention study among women of reproductive age. PLoS One 2011;6(12):e28144.

[135] Seckl JR, Meaney MJ. Early life events and later development of ischaemic heart disease. Lancet 1993;342(8881):1236.

[136] Lissau I, Sorensen TI. Parental neglect during childhood and increased risk of obesity in young adulthood. Lancet 1994;343(8893):324–7.

[137] Mirescu C, Peters JD, Gould E. Early life experience alters response of adult neurogenesis to stress. Nat Neurosci 2004;7(8):841–6.

[138] Heim C, et al. Pituitary-adrenal and autonomic responses to stress in women after sexual and physical abuse in childhood. JAMA 2000;284(5):592–7.

[139] Szyf M, et al. Maternal programming of steroid receptor expression and phenotype through DNA methylation in the rat. Front Neuroendocrinol 2005;26(3–4):139–62.

[140] Kazl C, et al. Early-life experience alters response of developing brain to seizures. Brain Res 2009;1285:174–81.

[141] Weaver IC, et al. Epigenetic programming by maternal behavior. Nat Neurosci 2004;7(8):847–54.

[142] Weaver IC, Meaney MJ, Szyf M. Maternal care effects on the hippocampal transcriptome and anxiety-mediated behaviors in the offspring that are reversible in adulthood. Proc Natl Acad Sci USA 2006;103(9):3480–5.

[143] McGowan PO, et al. Epigenetic regulation of the glucocorticoid receptor in human brain associates with childhood abuse. Nat Neurosci 2009;12(3):342–8.

[144] Zhang TY, et al. Maternal care and DNA methylation of a glutamic acid decarboxylase 1 promoter in rat hippocampus. J Neurosci 2010;30(39):13130–7.

[145] Mychasiuk R, Gibb R, Kolb B. Prenatal stress produces sexually dimorphic and regionally specific changes in gene expression in hippocampus and frontal cortex of developing rat offspring. Dev Neurosci 2011;33(6):531–8.

[146] Mychasiuk R, et al. Prenatal bystander stress alters brain, behavior, and the epigenome of developing rat offspring. Dev Neurosci 2011;33(2):159–69.

[147] Bagot RC, et al. Variations in postnatal maternal care and the epigenetic regulation of metabotropic glutamate receptor 1 expression and hippocampal function in the rat. Proc Natl Acad Sci USA 2012;109(Suppl. 2):17200–7.

[148] Blaze J, Scheuing L, Roth TL. Differential methylation of genes in the medial prefrontal cortex of developing and adult rats following exposure to maltreatment or nurturing care during infancy. Dev Neurosci 2013;35(4):306–16.

[149] Roth TL, et al. Bdnf DNA methylation modifications in the hippocampus and amygdala of male and female rats exposed to different caregiving environments outside the homecage. Dev Psychobiol 2014;56(8):1755–63.

[150] Schatzberg AF, et al. A corticosteroid/dopamine hypothesis for psychotic depression and related states. J Psychiatr Res 1985;19(1):57–64.

[151] Cao-Lei L, et al. DNA methylation signatures triggered by prenatal maternal stress exposure to a natural disaster: project ice storm. PLoS One 2014;9(9):e107653.

CHAPTER 16

Epigenetic control of stress-induced apoptosis

Lei Zhou

Department of Molecular Genetics and Microbiology, UF Health Cancer Center, College of Medicine, University of Florida, Gainesville, Florida, USA

Contents

1. TRANSCRIPTIONAL REGULATION AND APOPTOSIS

Apoptosis is a genetically controlled cell suicidal mechanism that serves to eliminate obsolete or damaged cells in multicellular organisms. The executioner of apoptosis, the caspases, are synthesized in all cells as dormant enzymes. The cleavage and activation of caspases is tightly controlled by multiple mechanisms (reviewed in [1]). Key regulators controlling caspase activation, especially those at the upstream, are mostly regulated at the transcription level. For instance, the BH3-only proapoptotic genes *bim* (Bcl2-interacting mediator of cell death), *puma* (p53 upregulated modulator of apoptosis), and *noxa* (phorbol-12-myristate-13-acetate-induced protein 1) are the transcriptional targets of P53 [2] (Figure 1). The expression of these upstream regulators is thus subject to epigenetic control.

The importance of transcriptional regulation in mediating programmed cell death during animal development has been fully demonstrated by genetic analysis in model organisms such as *Caenorahbditis elegans* and *Drosophila* [3,4]. During the development of *C. elegans*, 131 somatic cells are to be eliminated strictly based on cell lineage. Of those cells destined to die during development, most show apoptotic morphology within 30 min of their generation. In these short-lived cells, transcriptional upregulation of the BH3-only gene *egl-1* (egg-laying defective) plays an essential role in initiating the demise of these cells [3]. Egl-1 releases the inhibition of Ced-9 (cell death protein 9) on Ced-4 (cell death protein 4) and leads to the activation of the procaspase Ced-3 (cell death protein 3). Interestingly, a few somatic cells, such as the tail spike cells, do not become committed to cell death right after generation. Rather, they will survive for more than

Stress -Induced Apoptosis Mediated by Transcriptional Activation of Upstream Pro-apoptotic Genes

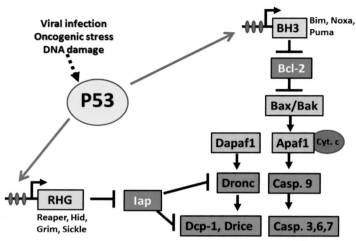

Figure 1 Transcriptional regulation of proapoptotic genes plays an important role in mediating stress-induced apoptosis.

300 min, undergo differentiation, and perform physiological function before committing to apoptosis. In those cells that survive for a prolonged period before committing to apoptosis, the transcriptional upregulation of the caspase gene *ced-3* is also required for inducing them into cellular suicide [5].

In *Drosophila*, the elimination of obsolete neuroblast cells at the end of *Drosophila* embryogenesis requires the expression of proapoptotic genes *reaper*, *grim* (CG4345 gene product from transcript CG4345-RA), and *sickle*. These three genes, together with *hid* (head involution defect), are clustered within a genomic region of about 280 kb. Their transcriptional activation in the neuroblasts is coordinated by the neuroblast regulatory region (NBRR) located in the intergenic region between *grim* and *reaper*. In animals lacking the NBRR but with the transcribed regions of the proapoptotic genes, many neuroblasts fail to undergo apoptosis [6]. In contrast, animals lacking the coding region of *reaper* or *grim* alone, but with an intact NBRR, have minor phenotypes comparing to animals lacking the NBRR. This difference highlights the importance of coordinated transcriptional regulation of not one but multiple proapoptotic genes in controlling developmental cell death.

Unlike *C. elegans*, wherein the programmed cell death of somatic cells is strictly determined by cell lineage, developmental cell death in *Drosophila* and mammals is sensitive to growth factors and environmental perturbation. In many cases, the appropriate level of cell death in a particular tissue is achieved through the competition for limited amount of growth factors. For instance, the live-or-die fate of the midline glia in the developing central nervous system of *Drosophila* is determined by their contact with the axons, from which the growth factor spitz is synthesized [7]. This sensitivity to environmental stress is a key feature of cell death regulation in most animals, ranging from *Drosophila* to mammals. Besides stress response induced by limited growth factors, massive cell death could be induced in response to hormone, nutrient deprivation, or cytotoxic agents, etc. [8,9].

2. EPIGENETIC CONTROL OF PROAPOPTOTIC GENES DURING ANIMAL DEVELOPMENT

An intriguing phenomenon of cell death regulation during development is that cellular sensitivity to death-inducing signals can vary dramatically at different development stages. For instance, *Drosophila* embryos at embryonic stage 9–11 are very sensitive to irradiation-induced apoptosis. However, embryos past development stage 13 become strongly resistant to irradiation. Ionizing irradiation of embryos between stages 9 and 11 induces three proapoptotic genes, *reaper*, *hid*, and *sickle*. The levels of their transcripts are upregulated by five- to tenfold within 15–30 min to mediate a rapid induction of apoptosis [10,11]. The function of the *Drosophila* P53 (dP53) is required for the transcriptional activation of these three genes in response to DNA damage [10]. However, none of the proapoptotic genes can be induced following irradiation in embryos past development stage 13, even though the level of P53 remains about the same. This is despite the fact that the DNA damage responsive signal transduction pathway is fully functional in later stage embryos, since other P53 targeted genes, such as the Ku70 and Ku80 DNA repair genes, are still induced [11].

Genetic analysis indicated that a ~33 kb intergenic region between *reaper* and *sickle* is required for the DNA damage-induced transcription of all of the three proapoptotic genes in stage 9–11 embryos [11]. This region was blocked, even though cell lineage-specific expression of these genes remains intact. In addition to ionizing irradiation, IRER is also responsible for mediating the induction of the proapoptotic genes in cells infected by viruses [12], or following overexpression of the oncogene dMyc (*Drosophila* avian myelocytomatosis viral oncogene homolog). In this sense, IRER seems to function as a centralized regulatory region to control the transcriptional regulation of multiple proapoptotic genes in response to a variety of stresses.

The sensitive-to-resistant transition of the three proapoptotic genes to DNA damage turned out to be due to epigenetic regulation of IRER (Figure 2). In stage 9–11 embryos, IRER adopts a euchromatin-like conformation. However, starting in stage 12, this region became enriched for H3K27me3, and subsequently, also enriched for H3K9me3. ChIP analysis indicated that in embryos past stage 13, this region is bound by HP1 (heterochromatin protein 1), Polycomb group (PcG) proteins Pc (Polycomb), and PSC (posterior sex combs). Following the histone modifications, IRER transforms into a heterochromatin-like conformation that is no longer permeable to DNase I. The epigenetic repression of

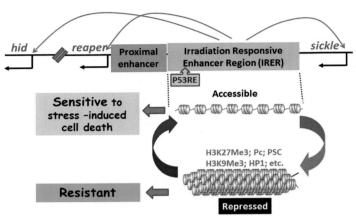

Figure 2 Epigenetic regulation of the IRER controls the responsiveness of proapoptotic genes (*reaper*, *sickle*, and *hid*) to cytotoxic stress.

IRER requires the function of PcG proteins as well as the H3K9 methyltransferase Su(var)3-9 (suppressor of variegation 3-9). In animals mutated for *HDAC1* (histone deacetylase: *rpd3* in *Drosophila*) and *SUV12* (suppressor of zeste 12 protein), the sensitive-to-resistant transition was significantly delayed [11].

Why should the accessibility of the regulatory region IRER be regulated by histone modifications during animal development? Several observations provided clues for us to speculate about the significance of this epigenetic regulation. First, it may serve to alter cellular sensitivity to DNA damage in accordance to cellular differentiation and/or proliferation status. The aforementioned open-to-closed transformation of IRER occurs in the majority of cells in the developing embryo. It results in the sensitive-to-resistant transition of the proapoptotic genes to DNA damage (and other stresses) at a development stage when most cells exit the mitotic cycle and enter into postmitotic differentiation. Prior to stage 12, most embryonic cells are lowly differentiated, and the time between mitotic cycle is very short. Conceivably, there is little time for DNA repair before next mitotic cycle. Presumably, damaged cells eliminated at this stage could be replaced relatively easily by dividing neighboring cells. In post stage 13 embryos, however, most cells enter into postmitotic differentiation. Cells will have more time to repair the damage. And since many development cues for cellular differentiation, such as the cues for axon pathway finding and cell migration, are transient during the development process, it will be to the benefit of the organism to adjust the DNA-damage response from death to repair. An evidence supporting this argument is that following the same X-ray treatment (20–40 Gy), the induction of DNA repair genes

Epigenetic regulation of IRER underlies differential sensitivity to stress

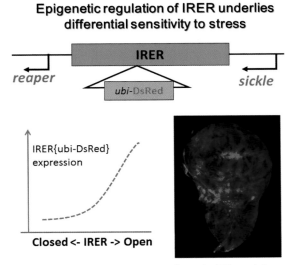

Figure 3 Epigenetic regulation of the IRER generates cells with differential openness of IRER and plasticity of stress response phenotype (for detailed information, refer to Zhang et al., 2014) [14].

Ku70 and Ku80 at 2 h post irradiation is actually much higher in later stage embryos than in early stage embryos [11]. Since the level of DNA repair gene expression at 30 min post X-ray is very similar between young and old embryos, the difference seen at 2 h is likely due to that fact that cells in old embryos did not undergo apoptosis as did cells in young embryos. By suppressing the proapoptotic response in old stage embryos, cells were allowed to accumulate more products of DNA repair genes.

Another important functional significance of epigenetic regulation of IRER might be to establish differential sensitivity among cells in the same tissue compartment. In wild-type flies, ectopic expression of the oncogene dMyc induces over proliferation, but the tissue will remain about the same size because of compensatory cell death. This compensatory induction of cell death following Myc overexpression is also very documented in the mammalian system [13]. How could a tissue compartment, composed of cells with identical genomes and similar lineage and differentiation status, select an appropriate number of cell to die to achieve the right size of the compartment? By inserting an ubiquitin-DsRed (red-fluorescent) reporter into IRER, we were able to monitor the epigenetic status of IRER in individual cells [14]. We found that there is a varying degree of the openness of IRER, reflected by the fluorescent reporter, in otherwise seemingly homogeneous cell populations (Figure 3). Under the condition of dMyc overexpression, cells with relatively more open IRER (higher level of DsRed) were eliminated versus those with relatively closed IRER survived [14]. The requirement of IRER in mediating over proliferation-induced cell death is verified with Df(IRER) (defective IRER) homozygous mutant animals, which have reduced viability but are nonetheless viable. The compensatory cell death following dMyc-induced over-proliferation is significantly inhibited in Df(IRER) homozygous animals, and many superfluous cells would survive [14].

The role of IRER in cell number control is not limited to the context of over-proliferation caused by oncogene expression. Under normal conditions, Df(IRER) mutant flies have about 10% more cells on the wing as compared to their wild-type or heterozygous littermates [14]. When clones of Df(IRER) mutant cells were made in heterozygous animals, the Df(IRER) mutant clones grow significantly larger than their respective twin spot clones that had intact IRER on both chromosomes. The difference between the mutant clones and their wild-type twin spot clones became much more dramatic if the animals were subjected to low dose irradiation during the early larval stage. When the tissues were analyzed for apoptosis following irradiation, cells without IRER were resistant, while their sister cells with two copies of IRER were sensitive. The above evidences strongly indicate that epigenetic regulation of IRER serves to generate a stochastic distribution of cellular sensitivity to stress, and such a mechanism contributes to the maintenance of appropriate cell number and tissue size. It has long been proposed by Feinberg et al. that the main function of epigenetic regulation is to generate phenotypic plasticity within cell population [15,16]. Our study of IRER supports this theory by showing that epigenetic regulation of proapoptotic genes produces varying sensitivity to stress-induced cell death. This phenotypic plasticity plays an important role in controlling cell numbers during tissue development.

Similarly, epigenetic regulation of proapoptotic genes plays an important role in mammalian development. For instance, epigenetic silencing of the BH3-only gene *noxa* is required for the maturation of memory CD4+ T-helper cells [17]. Bmi (B lymphoma Mo-MLV insertion region 1 homolog), a component of the mammalian PRC1 complex, binds to the *noxa* promoter, which becomes enriched for H3K27me3 during the maturation of the memory T-helper cells. This epigenetic silencing is required for maintaining the viability of mature T-helper cells. Mutant *bmi−/−* T cells, and even the heterozygous *bmi+/−* memory T cells, are much more prone to cell death than *bmi+/+* T cells. While Bmi-mediated silencing of *noxa* is required for T-cell development, abnormal silencing of another BH3-only gene, *bim*, by Bmi in plasma cells is responsible for promoting multiple myeloma [18]. This Dr Jekyll versus Mr Hyde role, played by the same epigenetic suppression mechanism, highlights the fact that epigenetic control of proapoptotic genes, while required for animal development, will promote tumorigenesis when misregulated.

3. EPIGENETIC SUPPRESSION OF STRESS-RESPONSIVE PROAPOPTOTIC GENES DURING TUMORIGENESIS

P53-induced apoptosis and cellular senescence play an important role in tumor suppression. However, evidence has shown that P53-induced transcriptional response can be blocked by epigenetic suppression of target genes.

P53 target genes can be specifically silenced by oncogenic viruses. For instance, latency-associated viral gene products from the Epstein–Barr virus can specifically silence the P53 target *bim* in B cells. This silencing process begins with enrichment of H3K27me3 modification on the *bim* promoter, which is then followed by DNA hyper-methylation [19]. This suppression is epigenetically stable, that is, can be propagated to progeny of the infected B cell, and contributes to the development of EBV-positive Burkitt's lymphoma.

A seminal study with adenovirus indicated that tumorigenic virus can hijack cellular epigenetic mechanisms to selectively repress P53 target genes [20]. A small adenovirus protein, E4-ORF3, can block P53-induced expression of tumor-suppressor genes such as *p21* and *puma*. This is achieved via de novo formation of repressive heterochromatin domains at the promoter of P53 target genes. In adenovirus-infected humans, primary small airway epithelial cells, SUV39H1 and SUV39H2, two orthologs of the *Drosophila* H3K9 methyltransferase su(var)3-9, were bound to P53-targeted stress-responsive genes. These genes became enriched for H3K9me3, and the binding of P53 to their promoters was suppressed. There were more than 1000 genes whose expression profile were altered in adenovirus-infected cells. Among those, ~200 genes were suppressed by E4-ORF3. The majority of E4-ORF3-suppressed genes are P53 targets as indicated by the presence of P53 binding site and/or responsiveness to nutlin treatment [20].

How can the small viral protein E4-ORF3 lead to the repression of P53-targeted stress-responsive genes without affecting the global histone modification pattern? It is likely that there exists a cellular system dedicated to regulating the epigenetic status of P53-targeted and stress-responsive proapoptotic genes. Similar to what was observed for the epigenetic regulation of the IRER region during *Drosophila* embryogenesis (see above), such a system could be responsible for regulating cellular sensitivity to stress-induced cell death during mammalian development. Adenovirus, through E4-ORF3, can activate this mechanism to prevent the infected cells from undergoing apoptosis. It is interesting to note that, besides P53 target genes, adenovirus E4-ORF3 also suppresses immune-modulating genes [20]. This theme of P53-mediated transcription and virus infection has a resonance in insects. The epigenetically regulated dP53 target region, IRER, besides being necessary for DNA damage response, is also required for mediating dP53-dependent rapid induction of proapoptotic genes following viral infection [12]. In *Drosophila*, with preliminary evidence indicating in mosquitoes as well, P53-mediated apoptosis is an important innate immune mechanism to block viral proliferation following the primary infection [12,21].

Another example of epigenetic suppression of proapoptotic genes during tumorigenesis came from the study of oncogenic K-Ras (Kirsten rat sarcoma)-mediated silencing of *fas* (Fas (TNF receptor superfamily member 6)). Fas is a membrane death receptor required for the negative selection of T cells during development [22]. Mice with mutation in *fas* display a lupus-like autoimmune phenotype. Silencing of *fas* in cancer cells directly correlates with their resistance to chemotherapy and their ability to escape immune suppression [23]. Expression of *K-ras* in NIH3T3 cells lead to the silencing of *fas* expression. A RNAi–based screen identified 28 genes which are required for K-Ras induced silencing of *fas* in NIH3T3 cells. These include major PRC1/2 components such as Bmi, EZH2 (enhancer of zeste protein 2), EED (embryonic ectoderm development protein), HDAC9 (histone deactylase 9), etc., as well as the DNA methyltransferase DNMT1 [24]. Both DNMT1 and the histone modifying enzymes were found to directly interact with the *fas* promoter during Ras-mediated transformation, which is enriched for H3K27me3 and contains hypermethylated CpG islands after the transformation.

Epistasis analysis indicated that the binding of DNMT1 to the *fas* promoter was dependent on the function of the other 27 factors required for silencing *fas*. This correlates with the observation that DNA methylation was the last event in this silencing process. The first factor to interact with the *fas* promoter is the potential transcription repressor ZFP354B. ZFP354B (Zinc Finger Protein 354B) is required for the binding of down stream players, including CTCF (CCCTC-binding factor), PRC2, and then PRC1, to the *fas* promoter. Overexpression of ZFP354B could induce the silencing of *fas*, but it requires the function of downstream players such as CTCF, EZH2, and Bmi. Thus, ZFP354B acts as the master initiator of this silencing process, starting a cascade of events including recruitment of PRC2, histone methylation at H3K27, recruitment of PRC1 and other DNA binding proteins, and eventually, binding of DNMT1 and hypermethyation of the CpG islands [25].

A genomic and epigenomic analysis indicated that among the more than 1000 genes whose expression levels were significantly changed following K-Ras-mediated transformation, 115 genes became enriched for H3K27me3 in the promoter as well as the gene body [26]. For these genes, the level of average H3K27me3 enrichment over the gene body negatively correlates with the expression level of the gene. However, a thorough time course analysis indicated that the formation of H3K27me3 was a few days after the transcription of these genes had been significantly decreased. Knocking down key PRC2 components, such as Suz12, resulted in depletion of the H3K27me3 marks associated with these genes, but failed to restore their expression. These two observations argue that the enrichment of the suppressive H3K27me3 modification is the consequence but not the cause of decreased gene expression. The same study also found that the suppressive histone modification (H3K27me3) induced by Ras in this context is reversible. Once the cellular level of Ras returned to normal, expression of the 115 genes was restored and the level of H3K27me3 enrichment returned to pre-transformation levels. Thus, strictly speaking, Ras-induced histone modification at the *fas* locus is not a bona fide epigenetic regulation. It is interesting to note that for both IRER suppression during development and the adenovirus-mediated silencing of P53 target genes in primary human lung epithelial cells, enrichment of H3K9me3 was observed at suppressed loci and appears to be able to propagate to progeny cells. The fact that Ras-mediated enrichment of H3K27me3 is not epigenetically stable invites the question about the significance of this histone regulation. It is possible that other endogenous event(s) is(are) necessary to make the silencing inheritable during oncogenesis.

Besides Ras, other oncogenes, such as Myc family members, can also lead to repression of tumor suppressor genes. For instance, MYCN (v-myc myelocytomatosis viral related oncogene, neuroblastoma-derived (avian)) directly interacts with EZH2 to bring the PRC2 complex to tumor suppressor genes [27]. EZH2, as the key component of PRC2, is often overexpressed in malignant cancers. The level of EZH2 expression in cancer cells positively correlates with their resistance to chemotherapy and inversely correlates with the survival of patients [28]. Conversely, knocking down EZH2 increases cancer cell sensitivity to chemotherapy.

4. EPIGENETIC REGULATION OF ANTI-APOPTOTIC GENES

While proapoptotic genes are the target of epigenetic suppression in cancers, anti-apoptotic genes are epigenetically activated to promote cancer cell survival and increase resistance to chemo and radiation therapy. For instance, in T-cell acute lymphoblastic leukemia (T-ALL), the enhancers of anti-apoptotic gene Bcl-2 (B-cell CLL/lymphoma) are enriched for H3K27ac. More importantly, Bcl-2 enhancers in T-ALL cells are occupied by BRD4 (bromodomain containing 4), which helps to preserve the acetylated chromatin through mitosis [29]. Removal of the BRD4 binding by a small molecule inhibitor greatly increased T-ALL sensitivity to chemotherapy [30].

In estrogen-dependent breast cancer cells, the enhancers of Bcl-2 is initially suppressed, and the underlying nucleosomes are enriched for H3K27me3. This suppression could be removed upon estrogen induced activation of JMJD3 (jumonji domain containing 3, lysine-specific demethylase), a H3K27 demethylase [31]. However, the oncogene HER2 (human epidermal growth factor receptor 2) can inactivate EZH2 and thus release the poised Bcl-2 enhancers. Consequently, breast cancer cells with overexpression of HER2 become independent of estrogen [31].

5. EPIGENETICS-BASED THERAPEUTIC STRATEGIES AND DEREPRESSION OF PROAPOPTOTIC GENES

The ultimate goal of radiation therapy, and that of most chemotherapy regimens, are to induce apoptosis of the cancer cell. Under ideal condition, the cellular damage inflicted by radiation or chemotherapeutic agents would initiate the stress-responsive pathways leading to the activation of the apoptosis machinery. However, this process could be hindered by epigenetic regulation of key apoptotic regulatory genes as we outline above.

Since in most cancers the major epigenetic misregulation appears to be silencing/suppression of proapoptotic genes, the therapeutic value of agents that can block DNA or histone methylation has been extensively explored in the last decade [32]. Several DNMT inhibitors, such as Azacitidine (5-azacytidine), and histone deacetylase inhibitors (HDIs), such as SAHA (suberoylanilide hydroxamic acid), have been approved by FDA for certain types of cancer. A major mechanism of the efficacy of DNMT inhibitors and HDIs is reactivation of proapoptotic genes or by removing epigenetic suppression. This reactivation then restores their responsiveness of cancer cells to cellular stresses induced by other chemotherapy agents. For instance, treatment of the HDI vorinostat to lymphoma cells led to the upregulation of both *Bad* (BCL2-associated agonist of cell death) and *Bid* (BH3-interacting domain death agonist), two key proapoptotic genes, and apoptotic cell death [33]. While these enzymatic inhibitors demonstrated considerable efficacy in treating hematological malignancies, their value against solid tumors remains to be fully addressed. More importantly, the toxicity and side effects associated with these inhibitors remain a major hurdle. This problem can be partially alleviated by simultaneous application of low doses of both DNMT inhibitors and HDIs [34].

The lack of specificity of enzymatic inhibitors-based approaches is apparent even at the theoretic level. Global inhibition of DNA and histone methylation could reactivate proapoptotic and other tumor suppressor genes. However, by the same token, they could also derepress many other genes, including potentially oncogenes, since the same enzymes are used for silencing all genes. Indeed, treatment with DNMT inhibitor and HDIs often leads to the activation of a multitude of genes [35]. Thus, it is only logic to conclude that we should strive to understand the mechanisms controlling the specific silencing of P53-targeted genes, which was revealed by the effect of the adenovirus E4-ORF3 and the epigenetic regulation of IRER during *Drosophila* development. Maybe then we could develop means to alleviate the suppression of proapoptotic and other tumor suppressors genes without risking global gene activation.

REFERENCES

[1] Mace PD, Riedl SJ. Molecular cell death platforms and assemblies. Curr Opin Cell Biol 2010;22:828–36.
[2] Villunger A, et al. p53- and drug-induced apoptotic responses mediated by BH3-only proteins puma and noxa. Science 2003;302:1036–8.
[3] Conradt B. Genetic control of programmed cell death during animal development. Annu Rev Genet 2009;43:493–523.
[4] Fuchs Y, Steller H. Programmed cell death in animal development and disease. Cell 2011;147:742–58.
[5] Maurer CW, Chiorazzi M, Shaham S. Timing of the onset of a developmental cell death is controlled by transcriptional induction of the *C. elegans* ced-3 caspase-encoding gene. Development 2007;134: 1357–68.
[6] Tan Y, et al. Coordinated expression of cell death genes regulates neuroblast apoptosis. Development 2011;138:2197–206.
[7] Bergmann A, Tugentman M, Shilo BZ, Steller H. Regulation of cell number by MAPK-dependent control of apoptosis: a mechanism for trophic survival signaling. Dev Cell 2002;2:159–70.
[8] McCall K. Eggs over easy: cell death in the Drosophila ovary. Dev Biol 2004;274:3–14.
[9] Jiang C, Lamblin AF, Steller H, Thummel CS. A steroid-triggered transcriptional hierarchy controls salivary gland cell death during Drosophila metamorphosis. Mol Cell 2000;5:445–55.

[10] Brodsky MH, et al. *Drosophila melanogaster* MNK/Chk2 and p53 regulate multiple DNA repair and apoptotic pathways following DNA damage. Mol Cell Biol 2004;24:1219–31.

[11] Zhang Y, et al. Epigenetic blocking of an enhancer region controls irradiation-induced proapoptotic gene expression in Drosophila embryos. Dev Cell 2008;14:481–93.

[12] Liu B, et al. P53-Mediated rapid induction of apoptosis conveys resistance to viral infection in *Drosophila melanogaster*. PLoS Pathog 2013;9:e1003137.

[13] Meyer N, Kim SS, Penn LZ. The Oscar-worthy role of Myc in apoptosis. Semin Cancer Biol 2006;16:275–87.

[14] Zhang C, et al. An intergenic regulatory region mediates Drosophila Myc-induced apoptosis and blocks tissue hyperplasia. Oncogene 2014:1–13. http://dx.doi.org/10.1038/onc.2014.160.

[15] Feinberg AP. Phenotypic plasticity and the epigenetics of human disease. Nature 2007;447:433–40.

[16] Pujadas E, Feinberg APP. Regulated noise in the epigenetic landscape of development and disease. Cell 2012;148:1123–31.

[17] Yamashita M, et al. Bmi1 regulates memory CD4 T cell survival via repression of the Noxa gene. J Exp Med 2008;205:1109–20.

[18] Jagani Z, et al. The Polycomb group protein Bmi-1 is essential for the growth of multiple myeloma cells. Cancer Res 2010;70:5528–38.

[19] Paschos K, et al. Epstein-barr virus latency in B cells leads to epigenetic repression and CpG methylation of the tumour suppressor gene Bim. PLoS Pathog 2009;5:e1000492.

[20] Soria C, Estermann FE, Espantman KC, O'Shea CC. Heterochromatin silencing of p53 target genes by a small viral protein. Nature 2010;466:1076–81.

[21] Liu B, Becnel JJ, Zhang Y, Zhou L. Induction of reaper ortholog mx in mosquito midgut cells following baculovirus infection. Cell Death Differ 2011;18:1337–45.

[22] Watanabe-Fukunaga R, Brannan CI, Copeland NG, Jenkins NA, Nagata S. Lymphoproliferation disorder in mice explained by defects in Fas antigen that mediates apoptosis. Nature 1992;356:314–7.

[23] Maecker HL, Yun Z, Maecker HT, Giaccia AJ. Epigenetic changes in tumor Fas levels determine immune escape and response to therapy. Cancer Cell 2002;2:139–48.

[24] Gazin C, Wajapeyee N, Gobeil S, Virbasius CM, Green MR. An elaborate pathway required for Ras-mediated epigenetic silencing. Nature 2007;449:1073–7.

[25] Wajapeyee N, Malonia SK, Palakurthy RK, Green MR. Oncogenic RAS directs silencing of tumor suppressor genes through ordered recruitment of transcriptional repressors. Genes Dev 2013;27:2221–6.

[26] Hosogane M, Funayama R, Nishida Y, Nagashima T, Nakayama K. Ras-induced changes in H3K27me3 occur after those in transcriptional activity. PLoS Genet 2013;9:e1003698.

[27] Corvetta D, et al. Physical interaction between MYCN oncogene and polycomb repressive complex 2 (PRC2) in neuroblastoma: functional and therapeutic implications. J Biol Chem 2013;288:8332–41.

[28] Varambally S, et al. The polycomb group protein EZH2 is involved in progression of prostate cancer. Nature 2002;419:624–9.

[29] Nishiyama A, Dey A, Miyazaki J-I, Ozato K. Brd4 is required for recovery from antimicrotubule drug-induced mitotic arrest: preservation of acetylated chromatin. Mol Biol Cell 2006;17:814–23.

[30] Knoechel B, et al. An epigenetic mechanism of resistance to targeted therapy in T cell acute lymphoblastic leukemia. Nat Genet 2014;46:364–70.

[31] Svotelis A, et al. H3K27 demethylation by JMJD3 at a poised enhancer of anti-apoptotic gene BCL2 determines ERα ligand dependency. EMBO J 2011;30:3947–61.

[32] Piekarz RL, Bates SE. Epigenetic modifiers: basic understanding and clinical development. Clin Cancer Res 2009;15:3918–26.

[33] Kretzner L, et al. Combining histone deacetylase inhibitor vorinostat with aurora kinase inhibitors enhances lymphoma cell killing with repression of c-Myc, hTERT, and microRNA levels. Cancer Res 2011;71:3912–20.

[34] Tsai H-C, Baylin SB. Cancer epigenetics: linking basic biology to clinical medicine. Cell Res 2011;21:502–17.

[35] Kortenhorst MSQ, et al. Analysis of the genomic response of human prostate cancer cells to histone deacetylase inhibitors. Epigenetics 2013;8:907–20.

CHAPTER 17

Structure, regulation, and function of TET family proteins

Xin Hu[1], Yun Chen[2], Zhizhuang Joe Zhao[1,2]

[1]Edmond H. Fischer Signal Transduction Laboratory, School of Life Sciences, Jilin University, Changchun, China; [2]Department of Pathology, University of Oklahoma Health Sciences Center, Oklahoma City, OK, USA

Contents

1. INTRODUCTION

The Ten–eleven translocation (TET) family of proteins consists of three members, namely TET1, TET2, and TET3. The founding member of this family, TET1, was initially identified as a fusion partner of the mixed lineage leukemia gene *MLL* in acute myeloid leukemia (AML) in 2002 [1]. These proteins have become hot research topics in recent years since they were found to oxidize 5-methylcytosine (5mC) to 5-hydroymethylcytosine (5hmC), 5-formylcytosine (5fC), and 5-carboxylcytosine (5caC), as part of the DNA demethylation process [2–5]. Among the three members of TET proteins, TET2 appears to attract the most attention in basic and clinical research largely because it is frequently mutated in various hematological malignancies [6]. TET2 plays an important role in regulating hematopoiesis, and its loss–of–function mutations have major pathological

Epigenetic Gene Expression and Regulation
http://dx.doi.org/10.1016/B978-0-12-799958-6.00017-2

consequences. Major progress has been made in recent years on the study of the TET family enzymes. The present review focuses on the structure, regulation, and function of TET2 and its implication for hematological malignancies. Basic information about TET1 and TET3 is also summarized for comparison.

2. STRUCTURE OF TET2

The TET family enzymes are evolutionarily conserved in the animal kingdom. Most animals contain at least one TET orthologue, characterized by an amino-terminal CXXC zinc-finger domain of ~60 amino acids and a carboxyl-terminal catalytic dioxygenase domain (Figure 1). The catalytic domain consists of a Cys-rich segment and a double-stranded β helix (DSBH) domain with a large low-complexity nonconserved insert. Taking the nonconserved insert out of consideration, the catalytic domains of human TET family proteins share 60–70% amino acid sequence identity. Interestingly, TET2 does not contain a CXXC zinc finger. Apparently, during vertebrate evolution, a chromosomal inversion event occurred to the original *TET2* gene, which splits it into two genes encoding CXXC/IDAX and TET2 in opposite directions [7]. The CXXC zinc-finger domain is found in a variety of chromatin-associated proteins, and it binds to nonmethylated CpG dinucleotides [8]. The domain contains eight conserved cysteine residues that chelate two zinc ions. TET1 and TET3 each contain a CXXC zinc finger, which may be responsible for binding to CpG-rich DNA. TET2, on the other hand, containing no N-terminal CXXC finger, may need CXXC4/IDAX for targeting. While the mechanism by which the CXXC zinc finger of CXXC4/IDAX helps to target TET2 to specific DNA is yet to be defined, studies have shown that CXXC4/IDAX is able to activate caspase, thereby downregulating TET2 protein levels [7].

Recently, the crystal structure of the catalytic domain of TET2 in complex with methylated DNA has been determined at a 2.02 Å resolution [9]. The structure reveals 5mC-DNA substrate recognition with bound Fe(II) and N-oxalylglycine, a 2-oxoglutarate (α-ketoglutarate) analog. The overall structure shows a compact globular fold. The DSBH domain forms a central core comprised a DSBH (also known as jelly-roll motif). The

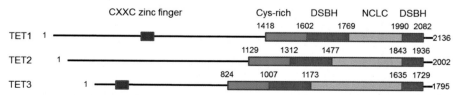

Figure 1 *Schematic structure of the TET family proteins.* Positions of CXXC zinc-finger domain, Cys-rich segment, double-stranded β helix domain (DSBH), and nonconserved low-complexity insert (NCLC) are indicated. The three latter elements form the catalytic domain. Note that TET2 lack a CXXC zinc-finger domain.

Cys-rich domain does not form an independent structural unit but instead, wraps around the DSBH core and stabilizes the DNA above the DSBH core. TET2 specifically recognizes the CpG dinucleotide with substrate preference for 5mC. The catalytic cavity does not seem to discriminate among 5mC, 5hmC, 5fC, and 5caC, thereby allowing further oxidation of 5mC derivatives. The structure thus provides further insight into the mechanism by which TET family enzymes catalyze oxidation of 5mC in DNA.

3. ENZYMATIC ACTIVITY OF TET FAMILY ENZYMES

The TET family proteins were initially discovered to have dioxygenase activity based on their sequence homology with *Trypanosoma brucei* base J-binding protein 1 (JBP1), a dioxygenase that hydroxylates thymine in DNA [3,10]. In vitro biochemical assays demonstrated that recombinant TET catalytic domains and full-length TET proteins can efficiently convert 5mC to 5hmC in the presence of two essential cofactors, namely 2-oxoglutarate and Fe(II), and ectopic expression of TET proteins in cultured cells reduced 5mC levels and produced 5hmC [2,3]. This catalytic activity was abolished by mutation of the signature His-Xaa-Asp motif of these dioxygenases. Subsequent studies revealed that TET proteins can further oxidize 5hmC to 5fC and 5caC in DNA [4,5]. Therefore, the TET family enzymes are able to produce three distinct derivatives of cytosine in DNA, which may represent important fine tuning of epigenetic status in the cells (Figure 2).

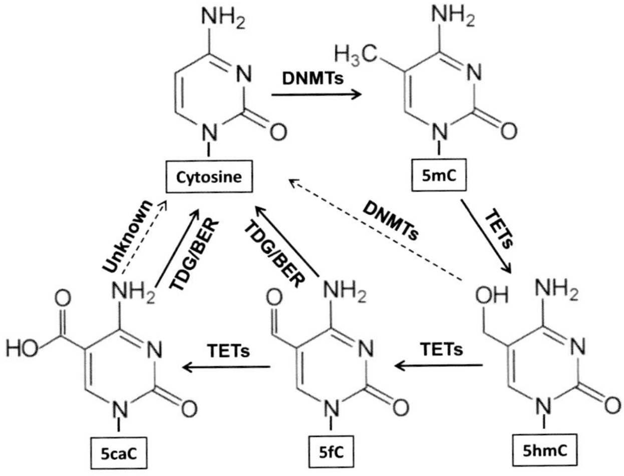

Figure 2 TET-mediated demethylation of methylcytosine. DNA methyltransferases (DNMTs) catalyze methylation of cytosine to 5-methylcytosine (5mC). TET proteins sequentially oxidize 5mC to 5-hydroxymethylcytosine (5hmC), 5-formylcytosine (5fC), and 5-carboxylcytosine (5caC). 5fC and 5caC can then be removed by thymine DNA glycosylase (TDG) and replaced by cytosine via base excision repair (BER). Decarboxylation by an unknown enzymatic activity can convert 5caC to cytosine, and DNMTs can potentially remove the hydroxymethyl group of 5hmC.

The level of 5hmC varies greatly in different mammalian cells, ranging from 5% to 10% of the level of 5mC in embryonic stem cells to as high as 40% of 5mC in Purkinje neurons [3,11]. However, 5fC and 5caC are present in mammalian cells at much lower levels than 5hmC (0.03% and 0.01% of the level of 5mC in mouse embryonic stem cells, respectively) [4,5]. While the physiological functions of these cytosine derivatives are yet to be defined, studies have shown that 5fC and 5caC can be excised by thymine DNA glycosylase and replaced by unmodified cytosine through the base excision repair pathway [4]. In addition, decarboxylation of 5caC by unknown enzymes has been found in embryonic stem cell lysates [12]. Furthermore, in vitro biochemical assays demonstrated that the mammalian de novo DNA methyltransferases DNMT3A and DNMT3B also have redox-dependent DNA dehydroxymethylase activity to remove the hydroxymethyl group of 5hmC, directly converting 5hmC to cytosine [13,14]. Taken together, the TET family enzymes are involved in the initial step for demethylation of DNA (Figure 2). Finally, efforts trying to identify cellular 5hmC readers have led to the identification of UHRF2 as a specific binder of 5hmC in neuronal progenitor cells, and the SRA domain of UHRF2 is responsible for the binding [15,16]. Functional implications of this binding remain to be defined.

4. SPECIFICITY OF TET2

As methylcytosine dioxygenases, the TET family proteins can interact with DNA through their catalytic domains. The cocrystal structure of TET2 catalytic domain with DNA demonstrates how this may occur. The structure shows that TET2 specifically recognizes the CpG dinucleotide with substrate preference for 5mC [9]. In fact, all of the three TET proteins are strongly enriched at gene promoters, especially those that are CpG-rich. In the case of TET1 and TET3, this is at least partly mediated by their CXXC domains, which preferentially bind CpG sequences [17,18]. TET2, on the other hand, may be mediated by CXXC4/IDAX as discussed above [7]. In mouse embryonic stem cells, the level of 5hmC is ~10% the level of 5mC, and TET1 and TET2 are responsible for essentially all of the 5hmC, because TET3 mRNA is expressed at much lower levels than TET1 or TET2 mRNA [19]. However, TET1 and TET2 appear to have distinct roles in regulating 5hmC in these cells. A recent study demonstrated that while knockdown of TET1 diminished 5hmC levels in the region of gene promoters and transcription start sites, knockdown of TET2 caused a predominant decrease of 5hmC in gene bodies [20]. Generally, the enrichment of 5hmC in the promoter and gene body is correlated with low and high gene expression in mouse embryonic stem cells, respectively [21]. However, approximately equal numbers of genes were up- and downregulated upon knockdown of TET1 and TET2 in mouse embryonic stem cells [20]. This apparent contradiction may be explained by the fact that many of the observed changes in gene expression are not directly caused by 5hmC alteration but rather indirectly through other transcriptional repressors or activators. Furthermore, while depletion of TET2 significantly reduced the level of 5hmC in the exon boundaries of highly expressed genes that are otherwise enriched in 5hmC in the region, it increased 5hmC at regions corresponding to the promoters and transcription start site [20]. The later effects may also be mediated indirectly. Overall, TET1 and TET2 have distinct roles in regulating 5hmC, gene expression, and exon usage in embryonic stem cells. The mechanism underlying this specificity and its biological implications remain to be defined.

5. POSTTRANSLATIONAL MODIFICATION OF TET2 AND ITS INTERACTION WITH OTHER PROTEINS

In addition to modifying DNA directly, the TET proteins exert epigenetic regulation by recruiting O-linked β-d-N-acetylglucosamine transferase (OGT) to chromatin [22–24]. OGT transfers the N-acetylglucosamine sugar group to seryl and threonyl residues of numerous proteins, including histones [25]. This glycosylation may directly or indirectly affect other modifications of histones and their stability and interactions with other proteins [25,26]. Importantly, OGT, like TET proteins, is also enriched at gene promoter regions, and TET proteins are apparently responsible for their recruitment to the regions because depletion of TET proteins diminishes the accumulation of OGT at promoters [22–24]. It has been shown that GlcNAcylation of histone H2B at Ser112 facilitates H2B monoubiquitination at Lys120, leading to transcriptional activation [26]. This was diminished upon TET2 depletion, indicating that TET2 may increase transcription through OGT [23]. The interaction of TET2 with OGT also promotes O-GlcNAcylation of host cell factor 1 (HCF1), a component of the H3K4 methyltransferase SET1 complex associated with active RNA polymerase II, and depletion of TET2 results in decreased H3K4 trimethylation and reduced transcription [22]. All three members of TET family proteins are highly phosphorylated, and this phosphorylation is regulated via GlcNAcylation by OGT [27]. Considering that OGT target many proteins, the impact of OCT recruitment by TET proteins to chromatin is likely more profound. Finally, TET proteins bind to VprBP and are monoubiquitylated by the VprBP-DDB1-CUL4-ROC1 E3 ubiquitin ligase (CRL4^{VprBP}) on a highly conserved lysine residue, and this VprBP-mediated monoubiquitylation promotes TET binding to chromatin [28]. It should be noted that many recurrent TET2 mutations found in hematological malignancies affect either the monoubiquitylation site (K1299) or residues essential for VprBP binding.

6. REGULATION OF TET2 ACTIVITY

The structural features of TET2 suggest different modes of regulation for TET2 enzymatic activity. While the catalytic domain is responsible for the enzymatic activity of TET2, full-length TET2 was found to be more active than various truncated versions of

TET2 [4,29], suggesting a regulatory role of the non-catalytic segments. Except for the CXXC zinc finger found in TET1 and TET3, the non-catalytic segments are highly diverse in structure among the TET proteins. They may confer different interactions of TET proteins with DNA substrate and other regulatory or functional proteins. Likewise, the low-complexity nonconserved insert in the catalytic domain may play a similar role. Note, this insert was deleted for crystallization of the TET2 catalytic domain/DNA complex [9]. The catalytic activity of TET enzymes requires two essential cofactors, namely 2-oxoglutarate and Fe(II) [3]. The requirement for 2-oxoglutarate is pathologically significant because 2-hydroxyglutarate, the product resulting from mutations of the *IDH1/2* genes, serves as a competitive inhibitor, causing effects equivalent to the loss-of-function mutations of the *TET2* gene [30,31]. In addition, like many nonheme Fe(II) dioxygenases, the activity of TET1 and TET2 is stimulated by ascorbic acid, which interacts with their catalytic domains [32–34]. Furthermore, treatment of mouse embryonic stem cells with ascorbic acid rapidly elevated the total level of 5hmC. This effect was abolished when both TET1 and TET2 were knocked out from the cells [32,33]. Therefore, ascorbic acid also acts as a cofactor of TET enzymes, and it may stabilize the TET family enzymes by preventing Fe(II) from oxidation and/or reducing Fe(III).

7. REGULATION OF TET2 EXPRESSION

The three TET enzymes exhibit distinct expression patterns. *TET2* gene is widely distributed, but its expression is predominant in a variety of differentiated tissues, especially in hematopoietic and neuronal lineages [35]. In fact, *TET2* is the most expressed *TET* gene in hematopoietic stem cells. In contrast, the *TET1* gene is mainly expressed in embryonic stem cells, while the *TET3* gene is more ubiquitous [36]. It has been well documented that expression of TET proteins is tightly regulated at the transcriptional level. For example, in mouse embryonic stem cells, both *TET1* and *TET2* are positively regulated by Oct4, and their mRNA levels decrease dramatically upon differentiation. In contrast, the *TET3* gene is significantly upregulated during differentiation [17].

In addition to transcriptional control, microRNAs also play a major role in regulation of TET protein expression. In fact, it was reported that miR-22 targets the *TET2* tumor suppressor to promote hematopoietic stem cell self-renewal and transformation and regulates breast cancer stemness and metastasis via TET-family-dependent chromatin remodeling in leukemia and breast cancers [37,38]. Furthermore, a high-throughput 3′ UTR activity screen led to identification of over 30 miRNAs that inhibited *TET2* gene expression and decreased cellular 5hmC [39]. These miRNAs include miR-125b, miR-29b, miR-29c, miR-101, and miR-7. Significantly, forced expression of these *TET2*-targeting miRNAs disrupts normal hematopoiesis, leading to hematopoietic expansion and biased myeloid differentiation, which can be corrected by overexpression of TET2 [39].

TET proteins are also regulated at the posttranslational level. One recent study suggests that CXXC4/IDAX interacts with TET2 and regulates its stability through caspase-dependent degradation [7]. It is not clear whether TET1 and TET3 are subjected to a similar regulation at the protein level. The TET proteins are also direct substrates of calpains, a family of calcium-dependent proteases [40]. Specifically, calpain-1 mediates TET1 and TET2 turnover in mouse embryonic stem cells, and calpain-2 regulates TET3 level during differentiation. Therefore, the TET proteins are subject to calpain-mediated degradation. The TET proteins are differentially regulated by OGT [41]. OGT catalyzes the O-GlcNAcylation of TET3, promotes TET3 nuclear export, and, consequently, inhibits the formation of 5-hydroxymethylcytosine catalyzed by TET3. Although TET1 and TET2 also interact with and can be O-GlcNAcylated by OGT, their subcellular localization does not seem to be affected [41].

Downregulation of *TET2* gene expression at the mRNA and proteins levels may have important pathological consequences. It has been found that a substantial fraction of AML patients with wild-type *TET2* gene shows similarly decreased 5hmC levels as found in *TET2* mutation cases [29]. In addition, several *TET2*-targeting microRNAs are preferentially overexpressed in *TET2*-wild-type AML [39]. Therefore, in addition to genetic mutations and inhibition of enzyme activity, molecular pathways that regulate *TET2* expression can serve as an important alternative mechanism in hematopoietic malignancies. Components of these pathways should be exploited in cancer diagnosis.

8. BIOLOGICAL FUNCTION OF TET2

8.1 Mouse model

It is now well accepted that TET2 has a pleiotropic role in hematopoiesis by regulating stem cell self-renewal, lineage commitment, and terminal differentiation. Much of the information came from analyses of *TET2* knockout mice. Four groups have developed *TET2* knockout mouse models by using conventional or conditional gene-targeting strategies [42–45]. *TET2* knockout mice are born in Mendelian ratios and are fertile, but display clear hematological abnormalities. These mice have increased numbers of hematopoietic stem cells with enhanced self-renewal and proliferative potential in culture and in competitive transplantation assays. Extramedullary hematopoiesis has also been observed in the spleen and liver of these mice. As the mice aged, many of them developed a myelomonocytic leukemia (CMML) phenotype with increased monocytes in the peripheral blood, bone marrow, and spleen. This occurs to both heterozygous $TET2^{+/-}$ and homozygous $TET2^{-/-}$ mice, indicating that *TET2* haplo-insufficiency can initiate myeloid and lymphoid transformations. The phenotype is consistent with the high prevalence of *TET2* mutation in CMML [6]. In addition to the predominant CMML-like phenotype, diseases resembling myelodysplasia (MDS), myeloproliferative neoplasms (MPNs), and aberrant lymphocyte development have also been observed in some $TET2^{-/-}$ mice. It should also be mentioned that a fifth $TET2$-deficient mouse model ($TET2^{trap/trap}$) generated by a gene trap insertion in intron 2 of the $TET2$ gene mostly died by postnatal day three, although they were born in expected Mendelian frequency [46]. Genetic background may play a major role in the postnatal lethality, because this occurred to mice with C57BL/6 background while mice with a C57BL/6 and Balb/c mixed background survived more than 12 weeks. Nonetheless, the liver of $TET2^{trap/trap}$ mice also had an increase in the number of hematopoietic stem cells/progenitors with increased self-renewal ability in vitro and in vivo. Considered together, $TET2$ has a crucial role in regulating hematopoiesis. The age-dependent induction of diverse phenotypes in $TET2$-deficient mice suggests that additional genetic lesions may cooperate with $TET2$ loss to induce disease and influence the generation of distinct types of blood neoplasms in the myeloid or in the lymphoid lineage.

In comparison, various models of $TET1$- and $TET3$-deficient mice have also been generated. $TET1^{-/-}$ mice on a mixed C57BL/6 × 129 background are born with the expected Mendelian ratio and appeared healthy despite a low birth weight [47]. However, $TET1$-deficient mice generated by gene-trapping showed embryonic lethality on a 129P2/OlaHsd background but are viable and fertile on a C57BL/6 background [48]. $TET1$-gene-trap mutant mice with the C57BL/6 background had smaller ovaries and produced smaller litters than control mice. Homozygous knockout of $TET3$ in mice led to neonatal lethality for undetermined reasons [49]. Female conditional $TET3$ knockout mice generated through germline-specific deletion of $TET3$ from primordial germ cells were normal in growth and morphology but displayed much reduced fecundity. About half of the embryos derived from $TET3$-null oocytes arrested around day 11.5, regardless of the sperm genotype, and the data demonstrated that maternal TET3 is essential for conversion of 5mC to 5hmC in developing zygotes [49]. More recently, $TET1/2/3$ triple knockout mouse embryonic fibroblasts were derived and used to study their capacity for reprogramming into induced pluripotent stem cells. The data demonstrated that these cells cannot be reprogrammed because of a block in the mesenchymal-to-epithelial transition step [50]. In summary, the TET family enzymes have a crucial role in stem cell development. The distinct phenotypes observed with mice deficient in the TET proteins are presumably attributed not only to different tissue distributions of these proteins but also their targets, substrate specificity, and regulations. Clearly, genetic background of mice also had a major role in determining the phenotypes of mice deficient in the TET family enzymes.

8.2 Human studies

The role of TET2 in stem cell development has also been demonstrated by using human cells [51]. $TET2$ gene expression is low in human embryonic stem cells and increases during hematopoietic differentiation. Small hairpin RNA–mediated knockdown of TET2 caused no effect on the pluripotency of various embryonic stem cells, but skewed their differentiation into neuroectoderm at the expense of endoderm and mesoderm. Furthermore, knockdown of TET2 in hematopoietic cells increased hematopoietic development. Finally, TET2-driven differentiation is dependent on the NANOG transcriptional factor whose expression is regulated by TET2. Therefore, TET2 is apparently involved in different stages of human embryonic development.

9. *TET2* GENE MUTATIONS IN HEMATOLOGICAL MALIGNANCIES

Somatic mutations of the *TET2* gene were initially discovered in 10–26% of MPN and MDS patients [52,53]. Subsequent studies revealed *TET2* mutations in other myeloid malignancies, including CMML, AML, systemic mastocytosis, and blastic plasmacytoid dendritic cell neoplasm with a frequency of 7.6–42% [54,55]. It should be pointed out that occurrence in CMML is particularly high, reaching nearly 60% [6]. *TET2* mutations have also been detected in lymphoid malignancies including mature B-cell, mature T-cell, angioimmunoblastic T-cell, mantle cell, and diffuse large B-cell lymphoma [56–59]. The prevalence of *TET2* mutations in lymphoid malignancies ranges from 2% to 33%, with angioimmunoblastic T-cell lymphoma carrying the highest mutation rate [6]. The mutations associated with hematological malignancies appear to be restricted to *TET2*, since somatic mutations of *TET1* and *TET3* have not been found [54]. The *TET2* mutations include missense, nonsense, and frameshift-causing insertion/deletion mutations spread out throughout entire molecules (Figure 3). Most missense *TET2* mutations are in the catalytic domain. Structural analyses predict that these mutations likely result in loss-of-function products. Both heterozygous and homozygous *TET2* mutations have been found in the patients, suggesting that *TET2* haplo-insufficiency may cause disease phenotypes.

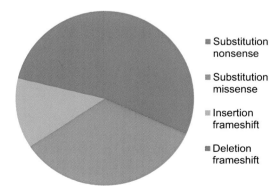

Legend:
- Substitution nonsense
- Substitution missense
- Insertion frameshift
- Deletion frameshift

Figure 3 *Distribution of various TET2 alterations found in cancers. Data were from the COSMIC database (Catalog of Somatic Mutations in Cancer,* http://www.sanger.ac.uk/cosmic/*June 2014). The total number of cases is 1309.*

TET2 mutations were identified in all of the three subtypes of classical MPNs, including polycythemia vera, essential thrombocytosis and primary myelofibrosis [6]. They have been found in JAK2V617F-positive and JAK2V617F-negative cases of MPNs. In JAK2V617F-positive polycythemia vera patients, *TET2* mutations have been identified in both JAK2V617F-positive and negative clones, suggesting that *TET2* mutations represent an early event in the pathogenesis of MPNs [52]. In other cases, however, *TET2* mutations appear only when the disease progresses to an acute phase [60,61]. In MDS, *TET2* alterations appear to be the most prevalent genetic abnormality, occurring with a frequency of 19–35% [6,62–65]. Interestingly, biallelic *TET2* mutations have been identified in some cases of MDS. These patients carry two distinct *TET2* mutations with one mutant allele usually present in the majority of cells, but the allelic burden of the second mutant *TET2* allele was more variable, suggesting that mutations are acquired at different time points during the progression of disease and that multiple clones may exist in the same individual [53]. This suggests that the *TET2* gene is very prone to mutations and that cells carrying a *TET2* mutation are easily accumulated.

10. *TET2* GENE MUTATIONS AND SOLID TUMORS

TET2 mutations are not limited to hematological malignancies. A review of the COSMIC database (Catalog of Somatic Mutations in Cancer, http://www.sanger.ac.uk/cosmic/June 2014) reveals that *TET2* mutation has not only an overall mutation rate of 16.3% in hematopoietic and lymphoid tumors, but also a significant frequency of mutation in cancers from other tissues producing solid tumors such as endometrial (7.38%), urinary track (3.88%), large intestine (3.76%), lung (1.7%), and liver (1.42%). More importantly, copy number variation of the *TET2* gene appears to be very prominent in solid tumors. In fact, loss of the *TET2* gene was found in 67.5% of ovarian cancers, 45.6% of lung cancers, 29.6% of breast cancers, 12.2% of large intestine cancers, 10.6% of endometrial cancers, and 10.1% of kidney cancers. Since haplo–insufficiency causes hematological malignancies, *TET2* mutations may have a broad implication for the formation of many solid tumors. Indeed, a study by Yang et al. demonstrated a good association of tumor development with the decrease of *TET* gene expression and 5-methylcytosine hydroxylation [66]. Furthermore, *TET2* mutations appeared to define a subset of metastatic tumors in castration-resistant prostate cancers [67].

11. COOPERATION OF *TET2* GENE MUTATION WITH OTHER ONCOGENIC MUTATIONS

The presence of *TET2* mutations in various hematological malignancies suggests that *TET2* alterations may be an early event in tumorigenesis. The finding of recurrent somatic *TET2* mutations in normal elderly individuals with clonal hematopoiesis further supports such a notion [68]. In this regard, *TET2* mutations may provide a background mutation for various malignancies, which are dictated by the accumulation of additional mutations. This appears to be the case in KITD816V-positive mastocytosis. *TET2* mutations are found in ~20% of mastocytosis patients, mostly in aggressive forms of the disease, with a demonstrated oncogenic cooperation with KITD816V in mast cells [69,70]. Furthermore, upon transfection with KITD816V, primary mouse bone marrow-derived mast cells from *TET2*$^{-/-}$ mice, but not those from wild-type control mice, gained growth capability in the absence of added cytokines [70]. A similar oncogenic cooperation has also been observed with TET2 and JAK2V617F. By using JAK2V617F mouse models and *TET2* knockout mice, two studies demonstrated that TET2 deficiency enhanced both the severity of Jak2V617F-induced MPN phenotypes, and the self-renewal properties of the Jak2V617F hematopoietic stem cells enhanced disease initiating potential of JAK2VF-positive stem cells, though did not lead to leukemic transformation [71,72]. By crossing our JAK2V617F transgenic mice with *TET2* knockout mice [45,73], we also found that the loss of TET2 significantly accelerated the progression of JAK2V617F-induced phenotypes (Chen et al. unpublished). Clinical implication of these findings remains to be demonstrated in MPN patients. Interestingly, a recent study demonstrated that the order in which JAK2 and TET2 mutations are acquired in MPN patients affects clinical features, responses to targeted therapy, the biology of stem and progenitor cells, and clonal evolution [74].

12. *TET2* AND *IDH1/2* GENE MUTATIONS

Another important implication of TET2 in hematological malignancies lies in involvement in the tumorigenic effects caused by isocitrate dehydrogenase IDH1 and IDH2. IDH1 and IDH2 are enzymes involved in citrate metabolism, a critical step in the Krebs cycle. Mutations in the *IDH1/2* genes were initially described in the majority of lower grade gliomas and secondary glioblastomas [75,76]. Subsequently, *IDH1* and *IDH2* mutations were observed in myeloid malignancies including de novo and secondary AML and preleukemic clonal malignancies, including MDS and MPNs [77,78]. The normal function of IDH1 and IDH2 is to catalyze the oxidative decarboxylation of isocitrate, producing 2-oxoglutarate in an NADP-dependent manner. The most common *IDH1/2* mutations in brain tumors and AML affect R132 of IDH1 or R140 and R172 of IDH2, which reside in the active site of the enzyme and participate in isocitrate binding. Importantly, the mutant enzymes acquire a neomorphic activity by catalyzing the NADPH-dependent reduction of 2-oxoglutarate to R(−)-2-hydroxyglutarate, a so-called oncometabolite [30,31]. A cell-permeable (R) enantiomer of 2-hydroxyglutarate was shown to be sufficient to promote leukemogenesis [79]. As described above, 2-oxoglutarate is a cofactor for catalytic activity of TET enzymes, and 2-hydroxyglutarate acts as an inhibitor of TET2 by competing with 2-oxoglutarate for binding to the enzyme. This potentially leads to attenuation of TET2 activity, causing an effect equivalent to a loss-of-function mutation of *TET2*. In fact, *TET2* mutations and *IDH1/2* mutations appear to be mutually exclusive in AML, and *TET2* loss-of-function mutations and *IDH1/2* gain-of-function mutations are associated with similar epigenetic defects [80]. This suggests an overlapping biologic effect of *TET2* and *IDH1/2* mutations.

13. PROGNOSTIC IMPACT OF *TET2* GENE MUTATIONS

Mutated in various hematological malignancies, the *TET2* gene may not only serve as a marker for diagnosis, but also has prognostic value. Indeed, studies have demonstrated that *TET2* gene mutations were associated with poor prognosis in AML [81–83]. The prognostic effect of *TET2* mutations in AML patients is most notable in patients with normal cytogenetics without the *FLT3-ITD* mutations [81]. In cytogenetically normal AML patients who received cytarabine/daunorubicin-based first-line therapy, *TET2* mutations adversely affected overall survival [82]. A more recent large scale meta-analysis covering 2552 patients with AML provides further support for the adverse impact of *TET2* mutations on prognosis [83]. These data indicate that *TET2* mutations are important in AML prognosis and may help to guide risk-adapted therapeutic strategies for treatment of AML.

Although *TET2* mutations have the highest frequency in CMML, analyses of large cohorts failed to demonstrate any prognostic impact of *TET2* mutations in this disease [84]. Likewise, studies failed to identify a strong association of *TET2* alteration with clinical phenotype and overall survival in MDS patients [62], although the *TET2* status may predict a better response to the demethylating agent azacitidine in higher risk MDS patients [65]. *TET2* mutations do not cluster in any particular subtype of classical MPNs, which includes polycythemia vera, essential thrombocytosis, and primary myelofibrosis, and have no clear prognostic impact on the diseases [85,86].

Finally, loss-of-function mutations in *TET2* and gain-of-function mutations in *IDH1/2* are mutually exclusive of one another [80]. This is thought to be due to a convergent mechanism of transformation related to the fact that TET2 function is impaired by 2-hydroxyglutarate produced in the presence of *IDH1/2* mutations. Despite this, the prognostic impact of *TET2* and *IDH1/2* mutations in AML appear to be opposite. In contrast to *TET2* mutations, mutations in *IDH1/2* appear to confer improved outcome in AML patients [87,88]. Studies have also established a firm association between *IDH1/2* mutations and serum 2-hydroxyglutarate concentrations in AML, and as expected, higher serum 2-hydroxyglutarate concentrations conferred better prognosis [89,90]. Whether 2-hydroxyglutarate targets something else in addition to TET2, is not known.

14. CONCLUSION

Recent studies on the TET family proteins have provided new insight into the mechanism by which DNA modification at the epigenetic level controls gene expression and cell behaviors. These enzymes oxidize methylcytosines and mediate DNA demethylation. The *TET2* gene apparently has a major role in hematopoiesis. How TET proteins contribute to regulation of specific genes in different cell types remains to be elucidated. This process is presumably fine-tuned by various factors and proteins that regulate gene expression and protein activity. More comparative study of the three TET family members and thorough examination of cells lacking one or more of the TET proteins will provide better clues. Genome-wide mapping of oxidized methylcytosines with improved technology should help to advance the field.

The presence of *TET2* mutations in patients with various hematological malignancies and the phenotypes of *TET2* knockout mice provide further evidence that deregulation at the epigenetic level can cause cancers. The successful introduction of DNA methylation and histone deacetylase inhibitors for the treatment of several subtypes of cancer has paved the way for development of more anticancer drugs targeting epigenetic regulators. Whether TET2 can be targeted remains to be demonstrated. On the one hand, as a loss-of-function mutant, *TET2* may not be directly targeted. Instead, effort should be focused on targeting cellular components that are unregulated due to loss of TET2 function. On the other hand, TET2 does not act like a typical tumor suppressor because its haplo-insufficiency causes disease phenotypes. In this regard, drugs that can stimulate *TET2* gene expression or activate TET2 enzymatic activity may be used to compensate for the haplo-insufficiency.

High-throughput DNA sequencing has led to discovery of multiple oncogenic mutations associated with specific types of cancers, and single-cell analysis has helped to define the order of various genetic alteration events. Mutations of *TET2* have been identified in a broad range of hematological malignancies, and loss of the *TET2* gene is found in many solid tumors. Therefore, TET2 should have general implications for development of various types of cancers. TET2 alteration appears to be a very early event during the course of tumorigenesis. Loss of TET2 function alone may not be sufficient to cause cell transformation, but rather require cooperation with other oncogenic partners to cause full-blown cell transformation. Studying the interplay between loss of TET2 function and gain-of-function of other oncogenic mutations will help to understand the pathogenesis of cancers and to find more effective targets for therapeutic drug development.

Much effort has been put toward correlating *TET2* mutational status with diagnosis and prognosis in clinical studies. While *TET2* mutations clearly predict poor outcome for AML patients, no prognostic association has been found with other hematological malignancies, including CMML, MDS, and MPNs. This may imply that loss of TET2 function has a different impact on preleukemic hematological malignancies than on AML. Alternatively, *TET2* mutation analyses at the DNA level may not be sufficient to completely capture functional deregulation of *TET2*. In this regard, analysis of TET2 in the protein and enzymatic activity level may be more informative. In any case, the pathological impact of *TET2* mutations requires further investigation. In addition, the involvement of TET2 in cancers related to *IDH1/2* mutation is also intriguing. The tumorigenic function of *IDH1/2* mutants is thought to be mediated by TET2 because the lack of 2-oxoglutarate, and the presence of the competitive inhibitor 2-hydroxyglutarate in cells expressing mutant IDH1/2 attenuate TET2 activity, and consequently TET2-dependent demethylation of genomic DNA. Evidently, *TET2* mutations and *IDH1/2* mutations have been found to be mutually exclusive in cancer patients, suggesting an overlapping biologic effect of these two genetic events. However, the prognostic impacts of *TET2* and *IDH1/2* mutations on AML are opposite. Whether *IDH1/2* mutations have biological effects other than causing inhibition of TET2 remains to be investigated.

REFERENCES

[1] Ono R, Taki T, Taketani T, Taniwaki M, Kobayashi H, Hayashi Y. Leukemia-associated protein with a CXXC domain, is fused to MLL in acute myeloid leukemia with trilineage dysplasia having t(10;11) (q22;q23). Cancer Res 2002;62:4075–80.

[2] Ito S, D'Alessio AC, Taranova OV, Hong K, Sowers LC, Zhang Y. Role of Tet proteins in 5mC to 5hmC conversion, ES-cell self-renewal and inner cell mass specification. Nature 2010;466:1129–33.

[3] Tahiliani KP, Koh Y, Shen WA, et al. Conversion of 5-methylcytosine to 5-hydroxymethylcytosine in mammalian DNA by MLL partner TET1. Science 2009;324:930–5.

[4] He YF, Li BZ, Li Z, et al. Tet-mediated formation of 5-carboxylcytosine and its excision by TDG in mammalian DNA. Science 2011;333:1303–7.

[5] Ito S, Shen L, Dai Q, et al. Tet proteins can convert 5-methylcytosine to 5-formylcytosine and 5-carboxylcytosine. Science 2011;333:1300–3.

[6] Solary E, Bernard OA, Tefferi A, Fuks F, Vainchenker W. The Ten-Eleven Translocation-2 (TET2) gene in hematopoiesis and hematopoietic diseases. Leukemia 2014;28:485–96.

[7] Ko M, An J, Bandukwala HS, et al. Modulation of TET2 expression and 5-methylcytosine oxidation by the CXXC domain protein IDAX. Nature 2013;497:122–6.

[8] Allen MD, Grummitt CG, Hilcenko C, et al. Solution structure of the nonmethyl-CpG-binding CXXC domain of the leukaemia-associated MLL histone methyltransferase. EMBO J 2006;25:4503–12.

[9] Hu L, Li Z, Cheng J, et al. Crystal structure of TET2-DNA complex: insight into TET-mediated 5mC oxidation. Cell 2013;155:1545–55.

[10] Iyer LM, Tahiliani M, Rao A, Aravind L. Prediction of novel families of enzymes involved in oxidative and other complex modifications of bases in nucleic acids. Cell Cycle 2009;8:1698–710.

[11] Kriaucionis S, Heintz N. The nuclear DNA base 5-hydroxymethylcytosine is present in Purkinje neurons and the brain. Science 2009;324:929–30.

[12] Schiesser S, Hackner B, Pfaffeneder T, et al. Mechanism and stem-cell activity of 5-carboxycytosine decarboxylation determined by isotope tracing. Angew Chem Int Ed Engl 2012;51:6516–20.

[13] Liutkeviciute Z, Lukinavicius G, Masevicius V, Daujotyte D, Klimasauskas S. Cytosine- 5-methyltransferases add aldehydes to DNA. Nat Chem Biol 2009;5:400–2.

[14] Chen CC, Wang KY, Shen CK. The mammalian de novo DNA.methyltransferases, DNMT3A and DNMT3B are also DNA 5-hydroxymethylcytosine dehydroxymethylases. J Biol Chem 2012;287: 33116–21.

[15] Spruijt CG, Gnerlich F, Smits AH, et al. Dynamic readers for 5-(hydroxy)methylcytosine and its oxidized derivatives. Cell 2013;152:1146–59.

[16] Zhou T, Xiong J, Wang M, et al. Structural basis for hydroxymethylcytosine recognition by the SRA domain of UHRF2. Mol Cell 2014;54:879–86.

[17] Xu Y, Wu F, Tan L, et al. Genome-wide regulation of 5hmC, 5mC, and gene expression by Tet1 hydroxylase in mouse embryonic stem cells. Mol Cell 2011;42:451–64.

[18] Xu Y, Xu C, Kato A, et al. Tet3 CXXC domain and dioxygenase activity cooperatively regulate key genes for Xenopus eye and neural development. Cell 2012;151:1200–13.

[19] Koh KP, Yabuuchi A, Rao S, et al. Tet1 and Tet2 regulate 5-hydroxymethylcytosine production and cell lineage specification in mouse embryonic stem cells. Cell Stem Cell 2011;8:200–13.

[20] Huang Y, Chavez L, Chang X, et al. Distinct roles of the methylcytosine oxidases Tet1 and Tet2 in mouse embryonic stem cells. Proc Natl Acad Sci USA 2014;111:1361–6.

[21] Wu H, D'Alessio AC, Ito S, et al. Genome-wide analysis of 5-hydroxymethylcytosine distribution reveals its dual function in transcriptional regulation in mouse embryonic stem cells. Genes Dev 2011;25:679–84.

[22] Deplus R, Delatte B, Schwinn MK, et al. TET2 and TET3 regulate GlcNAcylation and H3K4 methylation through OGT and SET1/COMPASS. EMBO J 2013;32:645–55.

[23] Chen Q, Chen Y, Bian C, Fujiki R, Yu X. TET2 promotes histone O-GlcNAcylation during gene transcription. Nature 2013;493:561–4.

[24] Vella P, Scelfo A, Jammula S, et al. Tet proteins connect the O-linked N-acetylglucosamine transferase Ogt to chromatin in embryonic stem cells. Mol Cell 2013;49:645–56.

[25] Hart GW, Slawson C, Ramirez-Correa G, Lagerlof O. Cross talk between O-GlcNAcylation and phosphorylation: roles in signaling, transcription, and chronic disease. Annu Rev Biochem 2011;80:825–58.

[26] Fujiki R, Hashiba W, Sekine H, et al. GlcNAcylation of histone H2B facilitates its monoubiquitination. Nature 2011;480:557–60.

[27] Bauer C, Göbel K, Nagaraj N, et al. Phosphorylation of TET proteins is regulated via O-GlcNAcylation by the O-linked N-acetylglucosamine transferase (OGT). J Biol Chem 2015;290:4801–12.

[28] Nakagawa T, Lv L, Nakagawa M, Yu Y, et al. CRL4^VprBP E3 ligase promotes monoubiquitylation and chromatin binding of TET dioxygenases. Mol Cell 2015;57:247–60.

[29] Ko M, Huang Y, Jankowska AM, et al. Impaired hydroxylation of 5-methylcytosine in myeloid cancers with mutant TET2. Nature 2010;468:839–43.

[30] Dang L, White DW, Gross S, et al. Cancer-associated IDH1 mutations produce 2-hydroxyglutarate. Nature 2009;462:739–44.

[31] Ward PS, Patel J, Wise DR, et al. The common feature of leukemia-associated IDH1 and IDH2 mutations is a neomorphic enzyme activity converting alpha-ketoglutarate to 2-hydroxyglutarate. Cancer Cell 2010;17:225–34.

[32] Yin R, Mao SQ, Zhao B, et al. Ascorbic acid enhances tet-mediated 5-methylcytosine oxidation and promotes DNA demethylation in mammals. J Am Chem Soc 2013;135:10396–403.

[33] Blaschke K, Ebata KT, Karimi MM, et al. Vitamin C induces Tet-dependent DNA demethylation and a blastocyst-like state in ES cells. Nature 2013;500:222–6.

[34] Minor EA, Court BL, Young JI, Wang G. Ascorbate induces ten-eleven translocation (Tet) methylcytosine dioxygenase-mediated generation of 5-hydroxymethylcytosine. J Biol Chem 2013;288:13669–74.

[35] Shih AH, Abdel-Wahab O, Patel JP, Levine RL. The role of mutations in epigenetic regulators in myeloid malignancies. Nat Rev Cancer 2012;12:599–612.

[36] Pastor WA, Aravind L, Rao A. TETonic shift: biological roles of TET proteins in DNA demethylation and transcription. Nat Rev Mol Cell Biol 2013;14:341–56.

[37] Song SJ, Ito K, Ala U, et al. The oncogenic microRNA miR-22 targets the TET2 tumor suppressor to promote hematopoietic stem cell self-renewal and transformation. Cell Stem Cell 2013;13:87–101.

[38] Song SJ, Poliseno L, Song MS, et al. MicroRNA-antagonism regulates breast cancer stemness and metastasis via TET-family-dependent chromatin remodeling. Cell 2013;154:311–24.

[39] Cheng J, Guo S, Chen S, et al. An extensive network of TET2-targeting MicroRNAs regulates malignant hematopoiesis. Cell Rep 2013;5:471–81.

[40] Wang Y, Zhang Y. Regulation of TET protein stability by calpains. Cell Rep 2014;6:278–84.

[41] Zhang Q, Liu X, Gao W, et al. Differential regulation of the ten-eleven translocation (TET) family of dioxygenases by O-linked β-N-acetylglucosamine transferase (OGT). J Biol Chem 2014;289:5986–96.

[42] Quivoron C, Couronné L, Della Valle V, et al. TET2 inactivation results in pleiotropic hematopoietic abnormalities in mouse and is a recurrent event during human lymphomagenesis. Cancer Cell 2011; 20:25–38.

[43] Moran-Crusio K, Reavie L, Shih A, et al. Tet2 loss leads to increased hematopoietic stem cell self-renewal and myeloid transformation. Cancer Cell 2011;20:11–24.

[44] Ko M, Bandukwala HS, An J, et al. Ten-Eleven-Translocation 2 (TET2) negatively regulates homeostasis and differentiation of hematopoietic stem cells in mice. Proc Natl Acad Sci USA 2011;108:14566–71.

[45] Li Z, Cai X, Cai CL, et al. Deletion of Tet2 in mice leads to dysregulated hematopoietic stem cells and subsequent development of myeloid malignancies. Blood 2011;118:4509–18.

[46] Shide K, Kameda T, Shimoda H, et al. TET2 is essential for survival and hematopoietic stem cell homeostasis. Leukemia 2012;26:2216–23.

[47] Dawlaty MM, Ganz K, Powell BE, et al. Tet1 is dispensable for maintaining pluripotency and its loss is compatible with embryonic and postnatal development. Cell Stem Cell 2011;9:166–75.

[48] Yamaguchi S, Hong K, Liu R, et al. Tet1 controls meiosis by regulating meiotic gene expression. Nature 2012;492:443–7.

[49] Gu TP, Guo F, Yang H, et al. The role of Tet3 DNA dioxygenase in epigenetic reprogramming by oocytes. Nature 2011;477:606–10.

[50] Hu X, Zhang L, Mao SQ, et al. Tet and TDG mediate DNA demethylation essential for mesenchymal-to-epithelial transition in somatic cell reprogramming. Cell Stem Cell 2014;14:512–22.

[51] Langlois T, da Costa Reis Monte Mor B, Lenglet G, et al. TET2 deficiency inhibits mesoderm and hematopoietic differentiation in human embryonic stem cells. Stem Cells 2014;32:2084–97.

[52] Delhommeau F, Dupont S, Della Valle V, et al. Mutation in TET2 in myeloid cancers. N Engl J Med 2009;360:2289–301.

[53] Langemeijer SM, Kuiper RP, Berends M, et al. Acquired mutations in TET2 are common in myelodysplastic syndromes. Nat Genet 2009;41:838–42.

[54] Abdel-Wahab O, Mullally A, Hedvat C, et al. Genetic characterization of TET1, TET2, and TET3 alterations in myeloid malignancies. Blood 2009;114:144–7.

[55] Abdel-Wahab O, Patel J, Levine RL. Clinical implications of novel mutations in epigenetic modifiers in AML. Hematol Oncol Clin N Am 2011;25:1119–33.

[56] Lemonnier F, Couronné L, Parrens M, et al. Recurrent TET2 mutations in peripheral T-cell lymphomas correlate with TFH-like features and adverse clinical parameters. Blood 2012;120:1466–9.

[57] Couronné L, Bastard C, Bernard OA. TET2 and DNMT3A mutations in human T-cell lymphoma. N Engl J Med 2012;366:95–6.

[58] Quivoron C, Couronné L, Della Valle V, et al. TET2 inactivation results in pleiotropic hematopoietic abnormalities in mouse and is a recurrent event during human lymphomagenesis. Cancer Cell 2011; 20:25–38.

[59] Asmar F, Punj V, Christensen J, et al. Genome-wide profiling identifies a DNA methylation signature that associates with TET2 mutations in diffuse large B-cell lymphoma. Haematologica 2013;98:1912–20.

[60] Delhommeau F, LeCoué dic JP, Dawson MA, Chen E, Bareford D, et al. Two routes to leukemic transformation after a JAK2 mutation-positive myeloproliferative neoplasm. Blood 2010;115:2891–900.

[61] Schaub FX, Looser R, Li S, et al. Clonal analysis of TET2 and JAK2 mutations suggests that TET2 can be a late event in the progression of myeloproliferative neoplasms. Blood 2010;115:2003–7.

[62] Bejar R, Stevenson K, Abdel-Wahab O, et al. Clinical effect of point mutations in myelodysplastic syndromes. N Engl J Med 2011;364:2496–506.

[63] Tefferi A, Lim KH, Abdel-Wahab O, et al. Detection of mutant TET2 in myeloid malignancies other than myeloproliferative neoplasms: CMML, MDS, MDS/MPN and AML. Leukemia 2009;23:1343–5.

[64] Kosmider O, Gelsi-Boyer V, Cheok M, et al. TET2 mutation is an independent favorable prognostic factor in myelodysplastic syndromes (MDSs). Blood 2009;114:3285–91.

[65] Itzykson R, Kosmider O, Cluzeau T, et al. Impact of TET2 mutations on response rate to azacitidine in myelodysplastic syndromes and low blast count acute myeloid leukemias. Leukemia 2011;25:1147–52.

[66] Yang H, Liu Y, Bai F, et al. Tumor development is associated with decrease of TET gene expression and 5-methylcytosine hydroxylation. Oncogene 2013;32:663–9.

[67] Nickerson ML, Im KM, Misner KJ, et al. Somatic alterations contributing to metastasis of a castration-resistant prostate cancer. Hum Mutat 2013;34:1231–41.

[68] Busque L, Patel JP, Figueroa ME, et al. Recurrent somatic TET2 mutations in normal elderly individuals with clonal hematopoiesis. Nat Genet 2012;44:1179–81.

[69] Tefferi A, Levine RL, Lim KH, et al. Frequent TET2 mutations in systemic mastocytosis: clinical, KITD816V and FIP1L1-PDGFRA correlates. Leukemia 2009;23:900–4.

[70] Soucie E, Hanssens K, Mercher T, et al. In aggressive forms of mastocytosis, TET2 loss cooperates with c-KITD816V to transform mast cells. Blood 2012;120:4846–9.

[71] Chen E, Schneider RK, Breyfogle LJ, et al. Distinct effects of concomitant Jak2V617F expression and Tet2 loss in mice promote disease progression in myeloproliferative neoplasms. Blood 2015;125:327–35.

[72] Kameda T, Shide K, Yamaji T, et al. Loss of TET2 has dual roles in murine myeloproliferative neoplasms: disease sustainer and disease accelerator. Blood 2015;125:304–15.

[73] Xing S, Wanting TH, Zhao W, et al. Transgenic expression of JAK2V617F causes myeloproliferative disorders in mice. Blood 2008;111:5109–17.

[74] Ortmann CA, Kent DG, Nangalia J, et al. Effect of mutation order on myeloproliferative neoplasms. N Engl J Med 2015;372:601–12.

[75] Parsons DW, Jones S, Zhang X, et al. An integrated genomic analysis of human glioblastoma multiforme. Science 2008;321:1807–12.

[76] Yan H, Parsons DW, Jin G, et al. IDH1 and IDH2 mutations in gliomas. N Engl J Med 2009;360:765–73.

[77] Mardis ER, Ding L, Dooling DJ, et al. Recurring mutations found by sequencing an acute myeloid leukemia genome. N Engl J Med 2009;361:1058–66.

[78] Tefferi A, Lasho TL, Abdel-Wahab O, et al. IDH1 and IDH2 mutation studies in 1473 patients with chronic-, fibrotic- or blast-phase essential thrombocythemia, polycythemia vera or myelofibrosis. Leukemia 2010;24:1302–9.

[79] Losman JA, Looper RE, Koivunen P, Lee S, Schneider RK, McMahon C, et al. (R)-2-hydroxyglutarate is sufficient to promote leukemogenesis and its effects are reversible. Science 2013;339:1621–5.

[80] Figueroa ME, Abdel-Wahab O, Lu C, et al. Leukemic IDH1 and IDH2 mutations result in a hypermethylation phenotype, disrupt TET2 function, and impair hematopoietic differentiation. Cancer cell 2010;18:553–67.

[81] Chou WC, Chou SC, Liu CY, et al. TET2 mutation is an unfavorable prognostic factor in acute myeloid leukemia patients with intermediate-risk cytogenetics. Blood 2011;118:3803–10.

[82] Metzeler KH, Maharry K, Radmacher MD, et al. TET2 mutations improve the new European leukemianet risk classification of acute myeloid leukemia: a cancer and leukemia group B study. J Clin Oncol 2011;29:1373–81.

[83] Liu WJ, Tan XH, Luo XP, et al. Prognostic significance of Tet methylcytosine dioxygenase 2 (TET2) gene mutations in adult patients with acute myeloid leukemia: a meta-analysis. Leuk Lymphoma 2014;55:2691–8.

[84] Itzykson R, Kosmider O, Renneville A, et al. Prognostic score including gene mutations in chronic myelomonocytic leukemia. J Clin Oncol 2013;31:2428–36.

[85] Tefferi A, Pardanani A, Lim KH, et al. TET2 mutations and their clinical correlates in polycythemia vera, essential thrombocythemia and myelofibrosis. Leukemia 2009;23:905–11.

[86] Vannucchi AM, Lasho TL, Guglielmelli P, et al. Mutations and prognosis in primary myelofibrosis. Leukemia 2013;27:1861–9.

[87] Green CL, Evans CM, Zhao L, et al. The prognostic significance of IDH2 mutations in AML depends on the location of the mutation. Blood 2011;118:409–12.

[88] Patel JP, Gönen M, Figueroa ME, et al. Prognostic relevance of integrated genetic profiling in acute myeloid leukemia. N Engl J Med 2012;366:1079–89.

[89] DiNardo CD, Propert KJ, Loren AW, et al. Serum 2-hydroxyglutarate levels predict isocitrate dehydrogenase mutations and clinical outcome in acute myeloid leukemia. Blood 2013;121:4917–24.

[90] Wang JH, Chen WL, Li JM, et al. Prognostic significance of 2-hydroxyglutarate levels in acute myeloid leukemia in China. Proc Natl Acad Sci USA 2013;110:17017–22.

CHAPTER 18

Epigenetic drugs for cancer therapy

Bowen Yan, Xuehui Li, Alta Johnson, Yurong Yang, Wei Jian, Yi Qiu
Department of Anatomy and Cell Biology, College of Medicine, University of Florida, Gainesville, Florida, USA

Contents

1. EPIGENETIC PATHWAYS TARGETED FOR CANCER THERAPY

1.1 DNA methylation

1.1.1 DNA methylation and DNA methyltransferases

DNA methylation is a covalent chemical modification of DNA occurring at cytosine residues in CpG dinucleotides. Approximately 70–80% of cytosine in CpG dyads is methylated on both strands in human somatic cells. DNA methylation is a stable epigenetic mark that is linked to the maintenance of chromatin in a silent state. Therefore, it regulates chromatin structure and gene expression, which subsequently involves processes such as X-chromosome inactivation, genomic imprinting, embryogenesis, gametogenesis, and silencing of repetitive DNA elements. The mammalian DNA methyltransferases (DNMTs) are enzymes that catalyze the transfer of a methyl group from S-adenosyl-ʟ-methionine to cytosine. Among the three enzymatically active DNMTs, DNMT1 is thought to

Epigenetic Gene Expression and Regulation
http://dx.doi.org/10.1016/B978-0-12-799958-6.00018-4

function as the major maintenance methyltransferase. This enzyme maintains DNA methylation at hemimethylated DNA after DNA replication [1], and it is responsible for copying the preexisting methylation patterns to the newly synthesized strand [2]. DNMT3A and DNMT3B are de novo methyltransferases active on unmethylated DNA. Both of them are responsible for establishing new DNA methylation patterns during early development [3], as well as maintaining these patterns during mitosis [4]. DNMT3L is homologous to DNMT3A and DNMT3B within the N-terminal regulatory region and is highly expressed in germ cells. Although catalytically inactive, DNMT3L regulates DNMT3A and DNMT3B by stimulating their catalytic activity [5].

1.1.2 Aberrant DNA methylation and DNMTs recruitment in cancer

Deregulation of DNA methylation directly affects mammalian development, and in particular the development of cancer [6]. Altered methylation patterns are found in the majority of tumor types. Methylation changes generally occur early in cancer development, which supports the hypothesis that epigenetic changes precede cancer development [7–10]. A hallmark of many cancers is global hypomethylation, but regional hypermethylation of CpG islands. This local hypermethylation in cancer is usually found on CpG islands associated with promoters of tumor suppressors or other genes involved in cell cycle regulation, leading to inactivation of these genes (Figure 1). Promoter hypermethylation also facilitates mutations in these regions, as methylated cytosine residues are spontaneously deaminated to thymine residues, causing mutational silencing of the genes [11–13]. This hypothesis explains the high incidence of CpG to TpG transition mutations observed in the promoters of tumor suppressors, for example, the p53 tumor suppressor gene [13].

The effects of global hypomethylation are more varied and not well understood. In mouse models, hypomethylation has been shown to induce genomic instability and tumorigenesis [14]. It has been suggested that global hypomethylation can induce reexpression of normally silenced genes, some of which may be oncogenic. Genes reactivated by global hypomethylation can include silenced oncogenes, imprinted genes, genes on the inactivated X chromosome [15], endogenous retroviruses and transposons [16], as well as silenced drug resistance genes [17]. For example, a cell–cell adhesion glycoprotein P-cadherin is often overexpressed in breast cancer, but not in normal breast tissue. The aberrant expression of P-cadherin in breast cancer is regulated by gene promoter hypomethylation [18]. A similar mechanism regulates overexpression of cyclin D2 in gastric and ovarian cancer [19,20], and melanoma associated antigen (MAGE) in melanomas [21]. Global hypomethylation also induces chromosomal instability by a mechanism that is not well understood; one possible cause is a large number of derepressed transposons and retroviruses created by this hypomethylated state [22,23].

Tumor heterogeneity is a major barrier to effective cancer diagnosis and treatment. Recent studies suggest that methylation patterns vary in different cancer types and

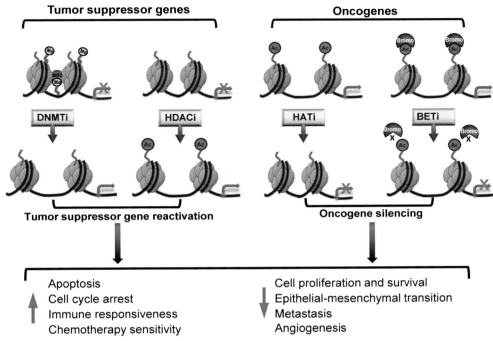

Figure 1 Transcriptional regulation of tumor suppressor genes and oncogenes by epigenetic inhibitors. Tumor suppressor genes are reactivated in response to the DNA methyltransferase inhibitors (DNMTi) and histone deacetylase inhibitors (HDACi). In reverse, oncogenes are silenced by histone acetyltransferase inhibitors (HATi) or bromodomain/extra-terminal inhibitors (BETi). Ac, acetylation; Me, DNA methylation; MBD, methyl-CpG binding protein.

tumor stages [24]. Epigenetic analysis of large gene panels and genome-wide screening of DNA methylation levels discovered that overall methylation patterns can be used as biomarkers for cancer risk and/or tumor type [25–28]. Cancer-specific differentially DNA-methylated regions (cDMRs) were identified, and stochastic methylation variation of the same cDMRs distinguishes various cancers from normal tissue, with intermediate variation in adenomas [29]. Whole-genome bisulfite sequencing shows these variable cDMRs are related to loss of sharply delimited methylation boundaries at CpG islands. It is suggested that the loss of epigenetic stability of well-defined genomic domains underlies increased methylation variability in cancer and may contribute to tumor heterogeneity. The distinct methylation patterns can be used not only to differentiate carcinoma from other tumor types, but also to predict tumor progression stage, with potential clinical applications in diagnosis and prognosis [30].

Exact nature of the defect in the cellular methylation machinery in tumor cells remains unknown. It is proposed that inappropriate DNA methyltransferase (DNMT) expression pattern or timing during the cell cycle could disrupt the regulation of DNA methylation patterns as DNMT1, 3A, and 3B are expressed differentially during the cell cycle [31]. Global hypomethylation in cancer cells may also be due to upregulation of the DNA demethylase system [32]. Increased expression of DNMTs can result in hyper-methylation of CpG islands in cancer cells and may play important roles in malignant progression of cancer, leading to aberrant methylation in many important tumor sup-pressor genes. In fact, it has been shown that DNMT overexpression is an early and significant event in urothelial, hepatic, gastric, pancreatic, lung, breast, and uterine cervix carcinogenesis [33].

Although the DNMT1 and the DNMT3 family of proteins have been considered either maintenance or de novo methyltransferases, respectively, it is likely that all three DNMTs possess both functions in vivo, particularly during carcinogenesis [31]. DNMT1 has been shown to be essential for the survival and proliferation of human cancer cells [34]. Increased DNMT1 protein expression correlates significantly with frequent DNA hypermethylation of multiple CpG islands, poor tumor differentiation, and malignant progression [35–37]. In bladder cancer, progressive increase of DNMT1 protein expression occurs during the precancerous stages [38]. Depletion of DNMT1 resulted in lower cel-lular maintenance methyltransferase activity, global and gene-specific demethylation and reexpression of tumor suppressor genes in human cancer cells. Specific depletion of DNMT1, but not DNMT3A or DNMT3B, markedly potentiated the ability of 5-aza-2′-deoxycytidine (decitabine), a DMNT inhibitor, to reactivate silenced tumor suppres-sor genes, indicating that inhibition of DNMT1 function is the principal means by which decitabine reactivates genes. These results indicate that DNMT1 is necessary and sufficient to maintain global methylation and aberrant CpG island methylation in human cancer cells [39].

DNMT3B depletion reactivated methylation-silenced gene expression but did not induce global or juxta-centromeric satellite demethylation, as specific depletion of DNMT1 did. This indicates that DNMT3B has significant site selectivity, which is dis-tinct from DNMT1 [40]. It has been shown that DNMT3B, but not DNMT3A, effi-ciently methylates the same set of genes in tumors and in nontumor tissues, thus demonstrating de novo methyltransferases can initiate methylation and silencing of spe-cific genes in phenotypically normal cells. This suggests that DNA methylation patterns in cancer are the result of specific targeting of at least some tumor suppressor genes, such as Sfrp2 (secreted frizzled-related protein 2), Sfrp4, and Sfrp5, rather than of random, stochastic methylation followed by clonal selection, due to a proliferative advantage caused by tumor suppressor gene silencing [7].

1.2 Histone deacetylation

The reversible acetylation of histones and nonhistone proteins by histone acetyltransfer-ases (HATs) and deacetylases (HDACs) plays a critical role in transcriptional regulation and many other cellular processes in eukaryotic cells. Acetylation of histones is commonly associated with the transcriptional activation of genes, and is thought to be responsible for the formation of a local open chromatin structure required for the binding of multiple transcription factors [41]. In contrast, the removal of acetyl groups by histone deacetylases frequently accompanies the suppression of gene activity [42]. However, nonhistone protein lysine acetylation plays a diverse role in the regulation of all aspects of cellular processes [43]. HDACs catalyze the removal of acetyl groups from the ε-amino groups of lysine residues. Mammalian HDACs are classified into four classes (I, II, III, and IV) based on the sequence homology of the yeast histone deacetylases Rpd3 (reduced potassium dependency), Hda1 (histone deacetylase1), and Sir2 (silent information regulator 2), respectively. Class I HDACs include HDAC1, 2, 3, and 8. Class II HDACs contain HDACs 4, 5, 6, 7, 9, and 10. Class III enzymes, however, require the coenzyme NAD+ as a cofactor. HDAC11 belongs to the class IV family (reviewed in Ref [44]). Although the precise cellular functions of the different HDAC enzymes are still poorly understood, evidence suggests that different members of the HDAC family have distinct functions [45,46].

1.2.1 Aberrant HDACs recruitment in cancer

It has become increasingly clear that class I HDAC enzymes are clinically relevant to cancer therapy [47,48]. Increased HDAC1 expression levels have been reported in a variety of cancers, such as gastric [49], prostate [50], colon [51], and breast [52] carcinomas. Overexpression of HDAC2 has been found in cervical [53], gastric [54], and colorectal carcinoma [55]. Other studies have reported high levels of HDAC3 [51] expression in colon cancer specimens. HDAC8 has been found to be associated with various types of leukemia [56]. HDAC1, 6, and 8 could also be important for breast cancer invasion [57]. These observations suggest that transcriptional repression of tumor suppressor genes by overexpression or aberrant recruitment of HDACs to their promoter regions could be a common phenomenon in tumor onset and progression. It is now known that HDACs have been associated with the deregulation of a number of well-characterized cellular oncogenes and tumor suppressor genes. For example, class I HDACs promote cell proliferation by inhibiting p21 and p27 promoter activity [51,58]. In some tumors, is epigenetically inactivated by hypoacetylation of the promoter, and treatment with HDAC inhibitors leads to inhibition of tumor cell growth and an increase in both acetylation of the promoter and gene expression [59]. The transcription factor snail recruits HDAC1, HDAC2, and the corepressor complex mSin3A to the E-cadherin promoter to repress its expression [60]. Downregulation or loss of function of E-cadherin has been implicated in the acquisition of invasive potential by carcinomas [61], and so aberrant recruitment of HDACs to this promoter may have a crucial role in tumor invasion and metastasis. The role of HDACs in cancer is not restricted to their contribution to histone deacetylation, but also includes their role in deacetylation of nonhistone proteins. For example, HDAC1 interacts with the tumor suppressor p53 and deacetylates it in vivo and in vitro [62,63]. p53 is phosphorylated and acetylated under stress conditions. Since lysine residues acetylated in p53 overlap with those that are ubiquitinated, p53 acetylation serves to promote protein stability and activation, inducing checkpoints in the cell-division cycle, permanent cell-division arrest, and cell death. Aberrant recruitment of HDACs to specific promoters through the interaction with fusion proteins that result from chromosomal translocations in hematological malignancies has also been intensively studied. In acute promyelocytic leukemia, leukemic fusion between the PML (promyelocytic leukemia) gene and the retinoic acid receptor (RAR) gene suppresses transcription through recruitment of HDACs. Thus, cancer cells are unable to undergo differentiation, leading to excessive proliferation [64,65]. Similar phenomena have been described for the RARalpha–PLZF (promyelocytic leukemia zinc finger protein) fusion, the AML1 (acute myelocytic leukemia protein 1)-ETO (8; 21) fusion, and for the myc/Mad/Max (Mad, human dimerization protein 1; Max, human Myc-associated factor X) signaling pathway involved in solid malignancies [66–69].

Class II HDACs have also been shown to associate with various cancer types. Inhibition of class II HDACs induces p21 expression in breast cancer cell lines, suggesting that class II HDAC subfamily may exert specific roles in breast cancer progression [70]. HDAC4 inhibits p21 gene expression through interaction with Sp1(specificity protein 1) at p21 proximal promoter. Induction of p21, mediated by silencing of HDAC4 arrested cancer cell growth in vitro and inhibited tumor growth in an in vivo human glioblastoma model [71]. HDAC4 also interacts with PLZF and represses RARalpha–PLZF fusion protein activity [72,73]. In the prostate cancer model, HDAC4 is recruited to the nuclei of cancer cells, where it may exert an inhibitory effect on differentiation and contribute to the development of the aggressive phenotype during late stage of prostate cancer [74]. HDAC5 and HDAC9 are significantly upregulated in high-risk medulloblastoma in comparison with low-risk medulloblastoma and their expressions are associated with poor survival [75]. Higher expression of HDAC7 and HDAC9 is associated with pancreatic adenocarcinomas and poor prognosis in childhood acute lymphoblastic leukemia (ALL) [76,77].

Class IIb deacetylase HDAC6 is linked to breast cancer. HDAC6 is expressed at a significantly higher level in breast cancer patients with small tumors and low histologic grade, and in estrogen receptor alpha- and progesterone receptor-positive tumors. Furthermore, patients with high levels of HDAC6 mRNA tended to be more responsive to endocrine treatment than those with low levels, indicating that HDAC6 may be an early prognosis marker [78,79]. Overexpression of HDAC6 in MCF-7 breast cancer cells increased cell motility, suggesting a role for HDAC6 in metastases [79]. HDAC6 has additional functions in integrating signaling and cytoskeleton remodeling. It is shown that cortactin, a genuine substrate of HDAC6, is overexpressed in several carcinomas [80,81]. Therefore, HDAC6 could be a viable target for cancer therapy. There is emerging evidence that inhibiting the HDAC6-mediated aggresome pathway leads to the accumulation of misfolded proteins and apoptosis in tumor cells through autophagy [82].

Class III deacetylases, the NAD+ dependent SIRT (Sirtuin; silent mating type information regulation 2 homolog) proteins are also connected to cancer. For instance, SIRT1 is upregulated in human lung cancer, prostate cancer, and leukemia, and has been found to be downregulated in colon tumors (reviewed in Ref [48]). SIRT1 is also responsible for the loss of the acetylation levels of K16-H4 and K9-H3, which is common in human cancer at early cancer development [83]. Upregulation of SIRT1 expression in human cancer can also induce deregulation of key proteins, such as p53 and E2F [84,85].

1.3 Histone acetylation

The lysine acetylation is a major histone modification involved in transcription activation, chromatin structure, and DNA repair. Histone acetylation is often associated with a more open chromatin conformation. Therefore, histone acetylation is often present at promoters and enhancers and, in some cases, throughout the transcribed region of active genes [86,87]. Importantly, lysine acetylation also occurs on a wide range of nonhistone proteins, which adds to the diversity of HATs' function [88]. Lysine acetylation can serve as the binding motif for bromodomain (BrD) containing proteins, which often refer as readers of the histone code [89]. The BrD-containing protein family includes various transcriptional coregulators, chromatin-modifying enzymes and nuclear scaffold proteins that are able to specifically recognize acetylated lysine residues on histone tails. BrD containing proteins can also bind acetylated lysine residues on nonhistone proteins [90].

There are two major classes of HATs. Type-A is primarily nuclear and can be further divided into three families depending on their structural homology: Gcn5-related N-acetyltransferases (GNAT), MYST (acronym for the founding members MOZ (monocytic leukemia zinc finger protein), Ybf2 (indentical with Sas2), Sas2 (something about silencing 2), TIP60 (Tat-interactive protein 60 kD)), and p300/CBP (CREB-binding protein). Type-B is mainly cytoplasmic localized and includes HAT1 and HAT2 [91].

1.3.1 HATs and cancer

Considering the cellular functions of HATs and their target proteins, it emerges that these enzymes can act as either tumor suppressors or oncogenes, depending on the cellular or molecular context and cancer type. To date, altered HATs have been reported in several types of cancers. Several primary mutations such as amplifications, deletions, point mutations, or translocations of HATs have been described [92]; however, in many cancers, an altered expression profile of HATs without mutation has been reported [93–96]. Therefore, HATs can play a central role in the physiopathology of cancers. Overexpression of p300 is observed in breast cancer [95], hepatocellular carcinoma [94], small and non-small lung cancer [97], prostate cancer [98], and colon cancer [93]. All of which are associated with poor prognosis and survival. Other HATs, such as p/CAF (p300/CBP associated factor) and GCN5, are also associated with various types of cancer [99–101].

HATs' contribution to malignant transformation may be through the contribution of altered patterns in acetylation on histone and nonhistone proteins. It has been shown that global histone acetylation patterns are perturbed in cancers [83,102]; it is also well established that acetylation patterns of nonhistone proteins, including many important oncogenes and tumor suppressors such as MYC, p53, and steroid hormone receptors, can be dynamically modulated in cancer [103]. p300/CBP acetylates steroid hormone receptors, such as androgen receptor (AR) and estrogen receptor (ER), whose aberrant stimulation is known to induce proliferation in breast and prostate cancers [95,98,104]. The acetylation mimic mutants of the AR enhance prostate cancer cell growth, and that acetylation of the AR may contribute to the development of androgen-independent prostate cancer [104]. In addition, p300 is required at an early stage of chromatin remodeling and transcription complex assembly after binding of androgen receptor to the gene but before many critical histone modifications occur [105]. It has been shown that the ability of BRCA1 (breast cancer 1, early onset) to repress ER-alpha activity correlates with its ability to induce downregulation of the cellular levels of the transcriptional coactivator p300 in breast and prostate cancer cells. When the tumor suppressor gene BRCA1 is mutated, p300/CBP acetylation activates ER in human breast and prostate cancer cell lines, whereas reintroduction of BRCA1 downregulates p300 [106]. p300/CBP also plays a role in the development of drug resistance. p300 mediates cellular resistance to doxorubicin in bladder cancer [107], and the interaction between CBP and β-catenin is associated with drug resistance in breast cancer and ALL [103].

2. EPIGENETIC DRUGS IN CLINIC AND CLINICAL TRIALS

2.1 DNA methylation inhibitors

The most common DMNT inhibitors in clinical use, 5-azacytidine (5-aza, azacitidine) and 5-aza-2′-deoxycytidine (decitabine), were synthesized almost 50 years ago [108]. 5-aza is a cytidine analog. It was originally developed as a nucleoside antimetabolite that could be incorporated into nucleic acids to induce chromosome breakage and mutations, and inhibit protein synthesis by interfering with tRNA and rRNA function [109–111]. Subsequently, it was shown that 5-aza incorporates into nucleic acids and covalently binds to DNMTs, leading to a rapid loss of methylation as a result of DNMT depletion (reviewed in Ref [112]). Decitabine is a very structurally similar compound that was suggested as a less toxic and more specific alternative to 5-aza, as it is not integrated into RNA [112–114]. 5-aza and decitabine have both been approved by the FDA for treatment of myelodysplastic syndromes (MDS) and acute myeloid leukemia (AML), as well as in clinical trials for other type of cancers (Table 1). However, both 5-aza and decitabine are toxic and highly unstable in aqueous solutions (reviewed in Ref [115]). This makes

Table 1 DNA methyltransferase (DNMT) inhibitors in clinical trials

Drugs	Disease	Clinical trials (.gov) identifier
Azacitidine (5–aza)	Leukemia or AML or MDS, myelodysplastic syndrome, chronic myelomonocytic leukemia, acute myeloid leukemia or myelodysplastic syndrome, recurrent non-small cell lung cancer or stage IV non-small cell lung cancer, pancreatic cancer, resected pancreatic adenocarcinoma	NCT00887068, NCT00387647, NCT01566695, NCT01908387, NCT00721214, NCT00660400, NCT01757535, NCT01995578, NCT01835587, NCT01305460, NCT01599325, NCT00761722, NCT01462578, NCT01652781, NCT01290302, NCT01747499, NCT01338337, NCT01350947, NCT01541280, NCT00897130, NCT01152346, NCT01201811, NCT01281124, NCT01167816, NCT01845805
Decitabine	Ovarian cancer, myelodysplastic syndrome, acute myeloid leukemia, chronic myelomonocytic leukemia, primary myelofibrosis	NCT00477386, NCT01400633, NCT01633099, NCT01809392, NCT01758367, NCT00492401, NCT01133886, NCT02109744, NCT01277484, NCT00349596, NCT01149408, NCT01786343, NCT01687400, NCT00941109, NCT00986804, NCT01251627, NCT01333449, NCT00095784
Decitabine and azacitidine	Leukemia	NCT01720225
SGI-110	Hepatocellular carcinoma, MDS or CMML or AML	NCT01752933, NCT01261312

them difficult to use in clinical settings, especially in solid tumors, so there is a need for other DNMT inhibitors with more favorable properties. Several new DNMT inhibitors have been designed in the last two decades, such as zebularine and SGI-110 [116,117]. Zebularine is also a cytidine analog, but it is more stable and less toxic than 5-aza and decitabine. Zebularine was originally developed as a cytidine deaminase inhibitor, but was later shown to potently inhibit DNMT activity [118] and cancer cell growth [119–121]. SGI-110 (previously called S110) is a dinucleotide that includes a deoxyguanosine (5′-DACpG-3′) and is largely resistant to cytidine deaminase [117]. SGI-110 has shown activity in cancer xenograft models and exhibits DNA hypomethylation activity in primates [122], and is currently being studied in a phase I clinical trial focused in patients with MDS and AML (Table 1).

2.2 Histone deacetylase inhibitors

Histone deacetylase (HDAC) inhibitors were discovered in the 1970s, when it was shown that treatment of cells with sodium butyrate led to hyperacetylation of histones [123]. Sodium butyrate was found to be difficult to use clinically due to its poor pharmacological properties [124]. In the following two decades, several more promising antitumor agents that inhibit HDACs were discovered. These include valproic acid (VPA), suberoylanilide hydroxamic acid (SAHA, trade name Vorinostat), Trichostatin A (TSA), and depsipeptide or FK228 (FR901228, trade name Romidepsin). All of them inhibit HDACs by binding to the active site (reviewed in [125,126]), resulting in release of epigenetic repression. HDAC inhibitors also synergize with DNA-damaging treatments such as radiotherapy or chemotherapy with nucleoside analogs [127,128], probably because HDACs are also involved in DNA replication and DNA repair [129]. The three most popular HDAC inhibitors in widespread clinical use today are VPA, SAHA, and depsipeptide, while Trichostatin A is not used clinically due to toxic side effects. VPA has the longest clinical history as it has been used for treatment of epilepsy since the 1960s. It inhibits proliferation of cultured cancer cells at millimolar concentrations and shows synergistic effects in combination with decitabine or hydralazine, and it is currently in clinical trials for multiple types of cancer (Table 4). SAHA is more potent, as it was shown to induce growth inhibition, differentiation, or apoptosis in cultured cancer cells [130,131] and inhibit cancer cell growth synergistically with decitabine or zebularine at micromolar concentrations; however, it has a short biological half-life of about 2h [132]. Depsipeptide, discovered in 1994 [133], is a cyclic tetrapeptide that preferentially targets class I HDACs [134]. It shows synergistic effects in combination with decitabine, zebularine, Trichostatin A, and 5-aza, and inhibits cancer cell proliferation at submicromolar concentrations (Table 4). Currently, SAHA and depsipeptide have been approved by the FDA for treatment against refractory cutaneous and peripheral T-cell lymphoma [135,136].

The newer compounds such as panobinostat (LBH589), mocetinostat (MGCD01030), belinostat (PXD101), pracinostat (SB939), and entinostat (MS275) have been extensively studied in the clinic [137,138]. The second generation of orally available HDAC inhibitors (HDACi) have been developed including AR-42, hydroxamides quisinostat (JNJ-26481585), abexinostat (PCI-24781), SHP141, and recently developed CXD101 with the hope that these agents have higher potency with less toxicity than the parental compounds [138]. These inhibitors have entered the clinic for the treatment of hematological malignancy and advanced cancer (Table 2).

2.3 HAT inhibitors

Curcumin and anacardic acid are two natural compounds with potent HAT inhibitory effect on p300 and CBP [139,140]. Curcumin is a natural product from the spice turmeric and is a specific inhibitor of p300/CBP HAT activity, and it represses the p300/CBP HAT-dependent transcriptional activation from chromatin [140]. Curcumin inhibits tumor growth and reduces vasculogenic mimicry (VM) in a murine choroidal melanoma model [5]. It is showed to sensitize other anticancer agents via inhibition of NF-kB (nuclear factor kappa light chain enhancer of activated B cells) pathway in cancer cell lines and is therefore used in clinical studies (Table 3) [91,92]. However, curcumin appears to have multiple unrelated protein targets [141]. Anacardic acid is a simple salicylic acid derivative (nonadecyl salicylic acid) that is isolated and purified from cashew nut shell liquid. Anacardic acid is a potent inhibitor of p300 and p/CAF (p300/CBP associated chromatin assembly factor; lysine acetyltransferase 2B). Although it does not affect DNA transcription, HAT-dependent transcription from a chromatin template was strongly inhibited by anacardic acid. However, since the cellular permeability of anacardic acid is very low, in vivo effects could not be evaluated [91,139]. Garcinol, a polyisoprenylated benzophenone, is also a natural product [142]. Garcinol is a potent inhibitor of HAT activity with an IC50 of 7 μM. Although it permeated cells easily, it was found to be highly toxic. Garcinol is a potent apoptosis-inducing agent as it upregulates many apoptosis pathway-related genes [142]. Garcinol has been used as a scaffold for designing derivatives that may be more useful as histone acetyltransferase inhibitors (HATi) in biological systems. One of its derivatives, LTK-14, has been selected to be highly specific against p300 with low toxicity [143]. LTK-14 is a p300-selective noncompetitive inhibitor for both acetyl-CoA and a histone that inhibits the HAT activity of p300 but not p/CAF and shows similar effects as curcumin [144]. LTK-14 represses the p300-mediated acetylation of p53 in vivo, inhibits histone acetylation of HIV-infected cells, and consequently inhibits the multiplication of HIV [143]. The recently developed compound C646 targets the p300/CBP HAT domain, and induces apoptosis in prostate cancer cells, suggesting that p300/CBP HAT inhibitors could serve as new anti-tumor therapeutics [145] (Table 3).

Table 2 Histone deacetylase (HDAC) inhibitors in clinical trials

Drugs	Disease	Clinical trials (gov) identifier
Valproic acid (VPA, valproate)	Breast cancer, hematuria bladder cancer, head and neck cancer, oral cavity cancer, or oropharyngeal cancer, high grade sarcoma	NCT01007695, NCT01738815, NCT01695122, NCT01010958
Vorinostat (SAHA, MK-0683)	Children with relapsed solid tumor or lymphoma or leukemia, advanced non-small cell lung cancer, non-small cell lung cancer, breast cancer, advanced breast cancer, recurrent and/or metastatic breast cancer, women who are undergoing surgery for newly diagnosed stage I, stage II, or stage III breast cancer, sickle cell disease or sickle cell anemia, cutaneous T-cell lymphoma, primary cutaneous T-cell lymphoma, advanced melanoma, brain metastases, low-grade non-Hodgkin's lymphoma, metastatic intraocular melanoma or recurrent intraocular melanoma, advanced, recurrent, or metastatic adenoid cystic carcinoma, soft tissue sarcoma, recurrent or stage IV renal cell cancer	NCT01422499, NCT01059552, NCT00735826, NCT00788112, NCT00719875, NCT00416130, NCT00262834, NCT01000155, NCT01728805, NCT00121225, NCT00958074, NCT00838929, NCT00253630, NCT01587352, NCT01175980, NCT00918489, NCT00278395
Romidepsin (FK228)	Cutaneous T-cell lymphoma or peripheral T-cell lymphoma	NCT00007345
Panobinostat (LBH589)	Sickle cell disease, relapsed or refractory non-Hodgkin's lymphoma, recurrent glioma high-grade meningioma or brain metastasis, myelofibrosis, myelodysplastic syndromes, chronic myelomonocytic leukemia, or acute myeloid leukemia)	NCT01245179, NCT01261247, NCT01324635, NCT01298934, NCT01613976
Belinostat (PXD101)	Thymoma/thymic carcinoma, neoplasms or lymphomas, peripheral T-cell lymphoma	NCT00589290, NCT01273155, NCT00865969
Pracinostat (SB939)	Recurrent or metastatic prostate cancer	NCT01075308
Entinostat (SNDX-275, MS275)	Hodgkin's lymphoma, clear cell renal cell carcinoma, or stage IV renal cell cancer	NCT00866333, NCT01038778
AR-42 (OSU-HDAC42)	Advanced or relapsed multiple myeloma, chronic lymphocytic leukemia, or lymphoma	NCT01129193
Hydroxamides quisinostat (JNJ 26481585)	Cutaneous T-cell lymphoma	NCT01486277
SHP141	Stage IA, IB, or IIA cutaneous T-cell lymphoma	NCT01433731
CXD101	Advanced cancer	NCT01977638

Table 3 HAT and BET inhibitors in clinical trials

Drugs	Drug type	Disease	Clinical trials (.gov) identifier
CPI-0610	BET inhibitor*	Lymphoma	NCT01949883
GSK525762	BET inhibitor	Carcinoma, relapsed or refractory hematologic malignancies	NCT01587703, NCT01943851
TEN-010	BET inhibitor	Malignancies	NCT01987362
OTX015	Bromodomain (BrD) inhibitor	Acute leukemia, other hematological malignacies	NCT01713582
Curcumin	p300 inhibitor	Colon cancer, prostate cancer, breast cancer, Pancreatic neoplasms or adenocarcinoma	NCT01294072, NCT01917890, NCT02064673, NCT01740323, NCT00094445
Dietary supplement: microgranular curcumin C3 complex®	p300 inhibitor	Head and neck cancer	NCT01160302

*Inhibitor of bromodomain and extra-terminal (BET) proteins

2.4 Bromodomain inhibitors

The first small-molecule inhibitor of BrDs was developed targeting the p/CAF BrD [146]. This work was followed by the discovery of CBP BrD inhibitor ischemin [147]. Ischemin alters posttranslational modifications on p53 and histones, inhibits p53 interaction with CBP and transcriptional activity in cells, and prevents apoptosis in ischemic cardiomyocytes [148]. Recently, two independent groups developed nanomolar, cell permeable inhibitors of BrDs in the BET (bromodomain and extra-terminal) family, the JQ1 and I-BET series, showing for the first time that BrD inhibitors may be used as a cancer drug [149,150]. JQ1 binds to selective BrD proteins, such as BRD2, 3 and 4 (bromodomain-containing 2, 3, and 4), with the strongest affinity for BRD4. BRD4 translocation has been seen in an incurable subtype of human squamous carcinoma. JQ1 displaces the BRD4 fusion oncoprotein from chromatin, prompting squamous differentiation and specific antiproliferative effects in BRD4-dependent cell lines and patient-derived xenograft models [149]. In the case of I-BET series, it was shown that I-BET762 provides protection against lipopolysaccharide (LPS)-induced endotoxic shock and bacteria-induced sepsis [150]. It has been shown that I-BET is able to displace the BETs and disrupt the association of MLL fusion and super elongation complexes from the target oncogenes, therefore inducing cell cycle arrest and apoptosis [151]. BET inhibition reduced the expression of MYC in a variety of malignant hematopoietic cell lines, including MLL-translocated acute myeloid leukemia [151,152], multiple myeloma [153], and Burkitt's lymphoma [154]. Other newly developed BET inhibitors, such as OTX-015, CPI-0610, and TEN-010, are in clinical trials for treatment of AML and other hematological malignancies (Table 3).

3. COMBINED THERAPY

3.1 DNA methylation and HDAC inhibitors in combination in cancer therapy

Although DNA methylation and histone deacetylation are carried out by different chemical reactions and require different sets of enzymes, it seems that there is a biological relationship between the two systems in modulating gene repression programming. Accumulating evidence also suggests that this epigenetic cross-talk may be involved in gene transcription and aberrant gene silencing in tumors. Thus, combined therapy with both DNMT and HDAC inhibitors is favored in clinical trials (Table 4). DNMT inhibitors and HDAC inhibitors synergistically affect chromatin states and lead to a more pronounced reexpression of epigenetically silenced tumor suppressor genes and cell cycle regulators [155]. For example, decitabine plus VPA synergistically induce NY-ESO-1 (New York esophageal squamous cell carcinoma 1) antigen in glioma cells [156], making them a target for immunotherapy. The combined treatment is more effective when key tumor suppressor genes, such as p16 (cyclin-dependent kinase (CDK) inhibitors 2A, CDKN2A) and p18, are epigenetically silenced [157], suggesting that the induction of silenced tumor suppressor genes may be critical for therapeutic effect. It was shown that non-small cell lung cancer cells with methylated CDKN2A are significantly more sensitive to the treatment than cell lines with deleted CDKN2A, and the combination of zebularine/depsipeptide results in a synergistic effect on cell growth inhibition that is also linked with the presence of epigenetically silenced CDKN2A [158].

DNMT inhibitors and HDAC inhibitors do not necessarily act only on DNA methylation and histone acetylation, respectively, but have other functions as well. It was shown that HDAC inhibitors can affect DNA methylation [159], and DNMT inhibitors can affect histone methylation [25]. When these agents are used in combination, one inhibitor can affect epigenetic changes brought about by the other [156,160]. DNMT inhibitors can also inhibit proliferation of cancer cells by causing DNA damage and chromosomal instability. Most DNMT inhibitors, such as decitabine, 5-aza, and zebularine, can integrate into DNA and introduce perturbations in its normal structure, leading to strand breaks [111], or derepress endogenous transposons and retroviral sequences silenced by methylation [161], which can lead to DNA damage and chromosomal instability as well. These effects are likely augmented by HDAC inhibitors, which can introduce further changes into DNA structure, contribute to the activation of silenced transposons and retroviral sequences, or inhibit DNA damage

Table 4 Combined DNMT inhibitors and HDAC inhibitors in clinical trials

Drugs	Disease	Clinical trials (.gov) identifier
Azacitidine and vorinostat	Recurrent or metastatic nasopharyngeal cancer or nasal natural killer T-cell lymphoma, myelodysplastic syndromes or acute myeloid leukemia, locally recurrent or metastatic nasopharyngeal cancer or nasal natural killer T-cell lymphoma, advanced solid tumors, relapsed or refractory non–Hodgkin's lymphoma, or acute myeloid leukemia, diffuse large B-cell lymphoma (DLBCL)	NCT00336063, NCT01748240, NCT00392353, NCT01617226, NCT01748240, NCT00336063, NCT00275080, NCT00948064, NCT01120834
Azacitidine and entinostat	Myelodysplastic syndromes, chronic myelomonocytic leukemia or acute myeloid leukemia, advanced breast cancer, newly diagnosed stage IA–IIIA non-small lung cancer undergoing surgery, advanced non-small cell lung cancer	NCT00101179, NCT00313586, NCT01305499, NCT01349959, NCT01886573, NCT00387465
Azacitidine and romidepsin	Solid tumors or non–small cell lung cancer, lymphoid malignancies or lymphoma or Hodgkin's lymphoma or non–Hodgkin's lymphoma	NCT01537744, NCT01998035
Azacitidine and belinostat	Chronic myelogenous leukemia, acute myeloid leukemia, acute promyelocytic leukemia, or primary myelofibrosis, myelodysplastic syndromes	NCT00351975
Azacitidine and mocetinostat	Myelodysplastic syndrome	NCT02018926
Azacitidine and pracinostat (LBH589)	Myelodysplastic syndromes, chronic myelomonocytic leukemia, acute myeloid leukemia	NCT01873703, NCT00946647, NCT01912274
Azacitidine, decitabine and pracinostat	Myelodysplastic syndrome	NCT01993641
Decitabine and vorinostat	Advanced solid tumors or relapsed or refractory non–Hodgkin's lymphoma, acute myeloid leukemia, acute lymphocytic leukemia, or chronic myelogenous leukemia	NCT00275080
Decitabine and valproic acid	Myelodysplastic syndrome or acute myelogenous leukemia	NCT00414310
Decitabine and AR–42	Acute myeloid leukemia	NCT01798901
Decitabine and LBH589	Leukemia, myeloid, acute myelodysplastic syndromes in patients age older than 60 years	NCT00691938

repair. For example, it was shown that depsipeptide can slow down removal of decitabine from the DNA, suggesting an explanation for synergy between these compounds as decitabine staying on the DNA longer would lead to greater DNA demethylation and DNA damage [162].

3.2 Epigenetic drugs with other cancer treatment agents in clinical trials

HDAC inhibitors and DNMT inhibitors can also synergize with radiotherapy [127,128], JAK (Janus kinase) inhibitors [163], mTOR (mammalian target of rapamycin) inhibitors [164], CDK inhibitors [165], TRAIL (TNF-related apoptosis-inducing ligand) cytokine [166], and conventional chemotherapy agents such as cisplatin [167], paclitaxel, doxorubicin, and 5-fluorouracil [168]. Furthermore, they can induce a response in tumors, which are resistant to chemotherapeutic agents [169]. It is suggested that the open chromatin structure induced by epigenetic drugs causes the DNA to be more accessible for chemotherapy drugs and radiation [170]. These results show clinical potential for using combinations of epigenetic drugs and conventional agents, especially to overcome or prevent chemotherapy resistance [171,172] (Table 5).

4. CLOSING REMARKS AND FUTURE PERSPECTIVES

It has been widely accepted that the alteration of the epigenome in cancer is a major driver of tumor onset and progression. Therefore, epigenetic modifiers have become attractive cancer therapeutic targets. The landmark of this progressive approach has been the FDA approval of DNA methyltransferase inhibitors and HDAC inhibitors for treatment of hematological malignancies. The increased number of inhibitors targeting DNMTs or HDACs with improved cytotoxicity and specificity have been developed and tested in clinical trials. The clinical trials with inhibitors that specifically target the BrDs of BET proteins or HATs are further additions in our ability to manipulate the epigenome for therapeutic goals. In addition, the combinations of epigenetic-based therapy with other anticancer therapies have been found to be more effective than monotherapeutic approaches. Beside the epigenetic targets discussed in this chapter, accumulating evidence has indicated the importance of histone methyltransferase and histone demethylase in some types of cancers [173–175]. The recently developed LSD1 (lysine-specific demethylase 1A) inhibitor in combination with other epigenetic drugs has been shown to be directly cytotoxic to cancer cells by reactivating multiple epigenetically silenced genes [176,177]. Although no clinical study directly targeting histone methylation has been conducted thus far, the histone demethylating agents will be worth further investigation as an anticancer therapy.

Table 5 Combined therapy of epigenetic inhibitors and other drugs in clinical trials

Drugs	Drug type	Diseases	Clinical trials (.gov) identifier
Azacitidine (AZA) and ruxolitinib (RUX)	DNMT inhibitor and JAK inhibitor	Myelofibrosis and myelodysplastic syndrome/myeloproliferative neoplasm	NCT01787487
Decitabine and ruxolitinib	DNMT inhibitor and JAK inhibitor	Myeloproliferative neoplasms	NCT02076191
Panobinostat and ruxolitinib	HDAC inhibitor and JAK inhibitor	Myelofibrosis, primary myelofibrosis, post-polycythemia vera myelofibrosis or post-essential thrombocythemia myelofibrosis	NCT01693601NCT01433445
Entinostat, lapatinib ditosylate and trastuzumab	HDAC inhibitor, tyrosine kinase inhibitor, and monoclonal antibody	HER2-positive breast cancer, inflammatory breast cancer, male breast cancer, recurrent breast cancer, stage IV breast cancer	NCT01434303
Vorinostat and lenalidomide	HDAC inhibitor and immunomodulatory drug	Multiple myeloma, plasma cell neoplasm	NCT00729118
Vorinostat and bevacizumab	HDAC inhibitor and Monoclonal antibodies, angiogenesis inhibitor	Clear cell renal cell carcinoma, recurrent renal cell cancer, stage III renal cell cancer, stage IV renal cell cancer, recurrent glioblastoma multiforme, malignant glioma, adult brain tumor	NCT00324870NCT01738646
Belinostat (PDX101), cisplatin and etoposide PCI-24781 and PZP115891	HDAC inhibitor and topoisomerase inhibitor HDAC inhibitor and tyrosine kinase inhibitor	Small cell lung carcinoma, malignant epithelial neoplasms Metastatic solid tumors	NCT00926640 NCT01543763
Azacytidine and PKC412	DNMT inhibitor and multi-target protein kinase inhibitor	Leukemia	NCT01202877
Azacitidine and sirolimus	DNMT inhibitor and mTOR inhibitor	Adult acute myeloid leukemia or myelodysplastic syndromes	NCT01869114

Continued

Table 5 Combined therapy of epigenetic inhibitors and other drugs in clinical trials—cont'd

Drugs	Drug type	Diseases	Clinical trials (.gov) identifier
Azacitidine (Vidaza) and nab-paclitaxel (Abraxane)	DNMT inhibitor and mitotic inhibitor	Advanced or metastatic solid tumors, advanced or metastatic breast cancer	NCT00748553
Azacitidine and bevacizumab	DNMT inhibitor and humanized monoclonal antibody, angiogenesis inhibitor	Renal cell carcinoma	NCT00934440
Azacitidine and volasertib	DNMT inhibitor and inhibitor of the PLK1 (polo-like kinase 1)	Myelodysplastic syndromes, leukemia, myelomonocytic, chronic	NCT01957644
Azacitidine and bortezomib	DNMT inhibitor and proteasome inhibitor	Leukemia; myelodysplastic syndromes myelodysplastic/myeloproliferative neoplasms	NCT00062936
Decitabine and bortezomib	DNMT inhibitor and proteasome inhibitor	Acute myeloid leukemia	NCT01420926
Decitabine and dasatinib	DNMT inhibitor and tyrosine kinase inhibitor	Leukemia	NCT01498445
Decitabine and iodine I 131	DNMT inhibitor and radiation	Metastatic papillary thyroid cancer or follicular thyroid cancer	NCT00085293
Decitabine, LHB589, and tamoxifen	DNMT inhibitor, HDAC inhibitor and antagonist of the estrogen receptor	Breast cancer, breast tumors, breast neoplasms	NCT01194908
Curcumin and irinotecan	P300 inhibitor and topoisomerase 1 inhibitor	Advanced colorectal cancer	NCT01859858
Curcumin* and cholecalciferol	P300 inhibitor and vitamin D	Stage 0–II chronic lymphocytic leukemia or small lymphocytic lymphoma	NCT02100423
CUDC-907	Phosphoinositide 3-kinase (PI3K) and histone deacetylase (HDAC) inhibitor	Multiple myeloma or lymphoma	NCT01742988

*Dietary Supplement of curcumin

REFERENCES

[1] Pradhan S, Bacolla A, Wells RD, Roberts RJ. Recombinant human DNA (cytosine-5) methyltransferase. I. Expression, purification, and comparison of de novo and maintenance methylation. J Biol Chem November 12, 1999;274(46):33002–10.

[2] Chen T, Li E. Structure and function of eukaryotic DNA methyltransferases. Curr Top Dev Biol 2004;60:55–89.

[3] Okano M, Bell DW, Haber DA, Li E. DNA methyltransferases Dnmt3a and Dnmt3b are essential for de novo methylation and mammalian development. Cell October 29, 1999;99(3):247–57.

[4] Chen T, Ueda Y, Dodge JE, Wang Z, Li E. Establishment and maintenance of genomic methylation patterns in mouse embryonic stem cells by Dnmt3a and Dnmt3b. Mol Cell Biol August 2003;23(16):5594–605.

[5] Chen LX, He YJ, Zhao SZ, Wu JG, Wang JT, Zhu LM, et al. Inhibition of tumor growth and vasculogenic mimicry by curcumin through down-regulation of the EphA2/PI3K/MMP pathway in a murine choroidal melanoma model. Cancer Biol Ther January 15, 2011;11(2):229–35.

[6] Gopalakrishnan S, Van Emburgh BO, Robertson KD. DNA methylation in development and human disease. Mutat Res December 1, 2008;647(1–2):30–8.

[7] Linhart HG, Lin H, Yamada Y, Moran E, Steine EJ, Gokhale S, et al. Dnmt3b promotes tumorigenesis in vivo by gene-specific de novo methylation and transcriptional silencing. Genes Dev December 1, 2007;21(23):3110–22.

[8] Belinsky SA, Nikula KJ, Palmisano WA, Michels R, Saccomanno G, Gabrielson E, et al. Aberrant methylation of p16(INK4a) is an early event in lung cancer and a potential biomarker for early diagnosis. Proc Natl Acad Sci USA September 29, 1998;95(20):11891–6.

[9] Derks S, Postma C, Moerkerk PT, van den Bosch SM, Carvalho B, Hermsen MA, et al. Promoter methylation precedes chromosomal alterations in colorectal cancer development. Cell Oncol 2006;28(5–6):247–57.

[10] Palmisano WA, Divine KK, Saccomanno G, Gilliland FD, Baylin SB, Herman JG, et al. Predicting lung cancer by detecting aberrant promoter methylation in sputum. Cancer Res November 1, 2000;60(21):5954–8.

[11] Sulewska A, Niklinska W, Kozlowski M, Minarowski L, Naumnik W, Niklinski J, et al. DNA methylation in states of cell physiology and pathology. Folia Histochem Cytobiol 2007;45(3):149–58.

[12] Fryxell KJ, Zuckerkandl E. Cytosine deamination plays a primary role in the evolution of mammalian isochores. Mol Biol Evol September 2000;17(9):1371–83.

[13] Rideout 3rd WM, Coetzee GA, Olumi AF, Jones PA. 5-Methylcytosine as an endogenous mutagen in the human LDL receptor and p53 genes. Science September 14, 1990;249(4974):1288–90.

[14] Eden A, Gaudet F, Waghmare A, Jaenisch R. Chromosomal instability and tumors promoted by DNA hypomethylation. Science April 18, 2003;300(5618):455.

[15] Sharp AJ, Stathaki E, Migliavacca E, Brahmachary M, Montgomery SB, Dupre Y, et al. DNA methylation profiles of human active and inactive X chromosomes. Genome Res October 2011;21(10):1592–600.

[16] Yoder JA, Walsh CP, Bestor TH. Cytosine methylation and the ecology of intragenomic parasites. Trends Genet August 1997;13(8):335–40.

[17] Chekhun VF, Lukyanova NY, Kovalchuk O, Tryndyak VP, Pogribny IP. Epigenetic profiling of multidrug-resistant human MCF-7 breast adenocarcinoma cells reveals novel hyper- and hypomethylated targets. Mol Cancer Ther March 2007;6(3):1089–98.

[18] Paredes J, Albergaria A, Oliveira JT, Jeronimo C, Milanezi F, Schmitt FC. P-cadherin overexpression is an indicator of clinical outcome in invasive breast carcinomas and is associated with CDH3 promoter hypomethylation. Clin Cancer Res August 15, 2005;11(16):5869–77.

[19] Sakuma M, Akahira J, Ito K, Niikura H, Moriya T, Okamura K, et al. Promoter methylation status of the Cyclin D2 gene is associated with poor prognosis in human epithelial ovarian cancer. Cancer Sci March 2007;98(3):380–6.

[20] Oshimo Y, Nakayama H, Ito R, Kitadai Y, Yoshida K, Chayama K, et al. Promoter methylation of cyclin D2 gene in gastric carcinoma. Int J Oncol December 2003;23(6):1663–70.

[21] De Smet C, Loriot A, Boon T. Promoter-dependent mechanism leading to selective hypomethylation within the 5′ region of gene MAGE-A1 in tumor cells. Mol Cell Biol June 2004;24(11):4781–90.

[22] Florl AR, Lower R, Schmitz-Drager BJ, Schulz WA. DNA methylation and expression of LINE-1 and HERV-K provirus sequences in urothelial and renal cell carcinomas. Br J Cancer July 1999;80(9):1312–21.

[23] Howard G, Eiges R, Gaudet F, Jaenisch R, Eden A. Activation and transposition of endogenous retroviral elements in hypomethylation induced tumors in mice. Oncogene January 10, 2008;27(3):404–8.

[24] Wermann H, Stoop H, Gillis AJ, Honecker F, van Gurp RJ, Ammerpohl O, et al. Global DNA methylation in fetal human germ cells and germ cell tumours: association with differentiation and cisplatin resistance. J Pathol August 2010;221(4):433–42.

[25] Kondo Y, Issa JP. DNA methylation profiling in cancer. Expert Rev Mol Med 2010;12:e23.

[26] Figueroa ME, Lugthart S, Li Y, Erpelinck-Verschueren C, Deng X, Christos PJ, et al. DNA methylation signatures identify biologically distinct subtypes in acute myeloid leukemia. Cancer Cell January 19, 2010;17(1):13–27.

[27] Hawes SE, Stern JE, Feng Q, Wiens LW, Rasey JS, Lu H, et al. DNA hypermethylation of tumors from non-small cell lung cancer (NSCLC) patients is associated with gender and histologic type. Lung Cancer August 2010;69(2):172–9.

[28] Worthley DL, Whitehall VL, Buttenshaw RL, Irahara N, Greco SA, Ramsnes I, et al. DNA methylation within the normal colorectal mucosa is associated with pathway-specific predisposition to cancer. Oncogene March 18, 2010;29(11):1653–62.

[29] Hansen KD, Timp W, Bravo HC, Sabunciyan S, Langmead B, McDonald OG, et al. Increased methylation variation in epigenetic domains across cancer types. Nat Genet 2011;43(8):768–75.

[30] Hernandez-Vargas H, Lambert MP, Le Calvez-Kelm F, Gouysse G, McKay-Chopin S, Tavtigian SV, et al. Hepatocellular carcinoma displays distinct DNA methylation signatures with potential as clinical predictors. PLoS One 2010;5(3):e9749.

[31] Robertson KD. DNA methylation, methyltransferases, and cancer. Oncogene May 28, 2001; 20(24):3139–55.

[32] Rai K, Sarkar S, Broadbent TJ, Voas M, Grossmann KF, Nadauld LD, et al. DNA demethylase activity maintains intestinal cells in an undifferentiated state following loss of APC. Cell September 17, 2010;142(6):930–42.

[33] Daniel FI, Cherubini K, Yurgel LS, de Figueiredo MA, Salum FG. The role of epigenetic transcription repression and DNA methyltransferases in cancer. Cancer February 15, 2011;117(4):677–87.

[34] Chen T, Hevi S, Gay F, Tsujimoto N, He T, Zhang B, et al. Complete inactivation of DNMT1 leads to mitotic catastrophe in human cancer cells. Nat Genet March 2007;39(3):391–6.

[35] Etoh T, Kanai Y, Ushijima S, Nakagawa T, Nakanishi Y, Sasako M, et al. Increased DNA methyltransferase 1 (DNMT1) protein expression correlates significantly with poorer tumor differentiation and frequent DNA hypermethylation of multiple CpG islands in gastric cancers. Am J Pathol February 2004;164(2):689–99.

[36] Nakagawa T, Kanai Y, Ushijima S, Kitamura T, Kakizoe T, Hirohashi S. DNA hypermethylation on multiple CpG islands associated with increased DNA methyltransferase DNMT1 protein expression during multistage urothelial carcinogenesis. J Urol May 2005;173(5):1767–71.

[37] Saito Y, Kanai Y, Nakagawa T, Sakamoto M, Saito H, Ishii H, et al. Increased protein expression of DNA methyltransferase (DNMT) 1 is significantly correlated with the malignant potential and poor prognosis of human hepatocellular carcinomas. Int J Cancer July 1, 2003;105(4):527–32.

[38] Nakagawa T, Kanai Y, Saito Y, Kitamura T, Kakizoe T, Hirohashi S. Increased DNA methyltransferase 1 protein expression in human transitional cell carcinoma of the bladder. J Urol December 2003;170 (6 Pt 1):2463–6.

[39] Robert MF, Morin S, Beaulieu N, Gauthier F, Chute IC, Barsalou A, et al. DNMT1 is required to maintain CpG methylation and aberrant gene silencing in human cancer cells. Nat Genet January 2003;33(1):61–5.

[40] Beaulieu N, Morin S, Chute IC, Robert MF, Nguyen H, MacLeod AR. An essential role for DNA methyltransferase DNMT3B in cancer cell survival. J Biol Chem August 2, 2002;277(31):28176–81.

[41] Sterner DE, Berger SL. Acetylation of histones and transcription-related factors. Microbiol Mol Biol Rev June 2000;64(2):435–59.

[42] Ng HH, Bird A. Histone deacetylases: silencers for hire. Trends Biochem Sci March 2000;25(3):121–6.

[43] Glozak MA, Sengupta N, Zhang X, Seto E. Acetylation and deacetylation of non-histone proteins. Gene December 19, 2005;363:15–23.

[44] Yang XJ, Seto E. The Rpd3/Hda1 family of lysine deacetylases: from bacteria and yeast to mice and men. Nat Rev Mol Cell Biol March 2008;9(3):206–18.

[45] Cho Y, Griswold A, Campbell C, Min KT. Individual histone deacetylases in Drosophila modulate transcription of distinct genes. Genomics November 2005;86(5):606–17.

[46] Foglietti C, Filocamo G, Cundari E, De Rinaldis E, Lahm A, Cortese R, et al. Dissecting the biological functions of Drosophila histone deacetylases by RNA interference and transcriptional profiling. J Biol Chem June 30, 2006;281(26):17968–76.

[47] Haberland M, Johnson A, Mokalled MH, Montgomery RL, Olson EN. Genetic dissection of histone deacetylase requirement in tumor cells. Proc Natl Acad Sci USA May 12, 2009;106(19):7751–5.

[48] Ropero S, Esteller M. The role of histone deacetylases (HDACs) in human cancer. Mol Oncol June 2007;1(1):19–25.

[49] Choi JH, Kwon HJ, Yoon BI, Kim JH, Han SU, Joo HJ, et al. Expression profile of histone deacetylase 1 in gastric cancer tissues. Jpn J Cancer Res December 2001;92(12):1300–4.

[50] Halkidou K, Gaughan L, Cook S, Leung HY, Neal DE, Robson CN. Upregulation and nuclear recruitment of HDAC1 in hormone refractory prostate cancer. Prostate May 1, 2004;59(2):177–89.

[51] Wilson AJ, Byun DS, Popova N, Murray LB, L'Italien K, Sowa Y, et al. Histone deacetylase 3 (HDAC3) and other class I HDACs regulate colon cell maturation and p21 expression and are deregulated in human colon cancer. J Biol Chem May 12, 2006;281(19):13548–58.

[52] Zhang Z, Yamashita H, Toyama T, Sugiura H, Ando Y, Mita K, et al. Quantitation of HDAC1 mRNA expression in invasive carcinoma of the breast*. Breast Cancer Res Treat November 2005;94(1):11–6.

[53] Huang BH, Laban M, Leung CH, Lee L, Lee CK, Salto-Tellez M, et al. Inhibition of histone deacetylase 2 increases apoptosis and p21Cip1/WAF1 expression, independent of histone deacetylase 1. Cell Death Differ April 2005;12(4):395–404.

[54] Song J, Noh JH, Lee JH, Eun JW, Ahn YM, Kim SY, et al. Increased expression of histone deacetylase 2 is found in human gastric cancer. APMIS April 2005;113(4):264–8.

[55] Ashktorab H, Belgrave K, Hosseinkhah F, Brim H, Nouraie M, Takkikto M, et al. Global histone H4 acetylation and HDAC2 expression in colon adenoma and carcinoma. Dig Dis Sci October 2009; 54(10):2109–17.

[56] Balasubramanian S, Ramos J, Luo W, Sirisawad M, Verner E, Buggy JJ. A novel histone deacetylase 8 (HDAC8)-specific inhibitor PCI-34051 induces apoptosis in T-cell lymphomas. Leukemia May 2008;22(5):1026–34.

[57] Park SY, Jun JA, Jeong KJ, Heo HJ, Sohn JS, Lee HY, et al. Histone deacetylases 1, 6 and 8 are critical for invasion in breast cancer. Oncol Rep June 2011;25(6):1677–81.

[58] Hauser C, Schuettengruber B, Bartl S, Lagger G, Seiser C. Activation of the mouse histone deacetylase 1 gene by cooperative histone phosphorylation and acetylation. Mol Cell Biol November 2002;22(22):7820–30.

[59] Gui CY, Ngo L, Xu WS, Richon VM, Marks PA. Histone deacetylase (HDAC) inhibitor activation of p21WAF1 involves changes in promoter-associated proteins, including HDAC1. Proc Natl Acad Sci USA February 3, 2004;101(5):1241–6.

[60] Peinado H, Ballestar E, Esteller M, Cano A. Snail mediates E-cadherin repression by the recruitment of the Sin3A/histone deacetylase 1 (HDAC1)/HDAC2 complex. Mol Cell Biol January 2004; 24(1):306–19.

[61] Hajra KM, Fearon ER. Cadherin and catenin alterations in human cancer. Genes Chromosomes Cancer July 2002;34(3):255–68.

[62] Juan LJ, Shia WJ, Chen MH, Yang WM, Seto E, Lin YS, et al. Histone deacetylases specifically down-regulate p53-dependent gene activation. J Biol Chem July 7, 2000;275(27):20436–43.

[63] Luo J, Su F, Chen D, Shiloh A, Gu W. Deacetylation of p53 modulates its effect on cell growth and apoptosis. Nature November 16, 2000;408(6810):377–81.

[64] Lin RJ, Sternsdorf T, Tini M, Evans RM. Transcriptional regulation in acute promyelocytic leukemia. Oncogene October 29, 2001;20(49):7204–15.

[65] He LZ, Tolentino T, Grayson P, Zhong S, Warrell Jr RP, Rifkind RA, et al. Histone deacetylase inhibitors induce remission in transgenic models of therapy-resistant acute promyelocytic leukemia. J Clin Invest November 2001;108(9):1321–30.

[66] Minucci S, Nervi C, Lo Coco F, Pelicci PG. Histone deacetylases: a common molecular target for differentiation treatment of acute myeloid leukemias? Oncogene May 28, 2001;20(24):3110–5.

[67] Ferrara FF, Fazi F, Bianchini A, Padula F, Gelmetti V, Minucci S, et al. Histone deacetylase-targeted treatment restores retinoic acid signaling and differentiation in acute myeloid leukemia. Cancer Res January 1, 2001;61(1):2–7.

[68] Kitamura K, Hoshi S, Koike M, Kiyoi H, Saito H, Naoe T. Histone deacetylase inhibitor but not arsenic trioxide differentiates acute promyelocytic leukaemia cells with t(11;17) in combination with all-trans retinoic acid. Br J Haematol March 2000;108(4):696–702.

[69] David G, Alland L, Hong SH, Wong CW, DePinho RA, Dejean A. Histone deacetylase associated with mSin3A mediates repression by the acute promyelocytic leukemia-associated PLZF protein. Oncogene May 14, 1998;16(19):2549–56.

[70] Duong V, Bret C, Altucci L, Mai A, Duraffourd C, Loubersac J, et al. Specific activity of class II histone deacetylases in human breast cancer cells. Mol Cancer Res December 2008;6(12):1908–19.

[71] Mottet D, Pirotte S, Lamour V, Hagedorn M, Javerzat S, Bikfalvi A, et al. HDAC4 represses p21(WAF1/Cip1) expression in human cancer cells through a Sp1-dependent, p53-independent mechanism. Oncogene January 15, 2009;28(2):243–56.

[72] Chauchereau A, Mathieu M, de Saintignon J, Ferreira R, Pritchard LL, Mishal Z, et al. HDAC4 mediates transcriptional repression by the acute promyelocytic leukaemia-associated protein PLZF. Oncogene November 18, 2004;23(54):8777–84.

[73] Yuki Y, Imoto I, Imaizumi M, Hibi S, Kaneko Y, Amagasa T, et al. Identification of a novel fusion gene in a pre-B acute lymphoblastic leukemia with t(1;19)(q23;p13). Cancer Sci June 2004;95(6):503–7.

[74] Halkidou K, Cook S, Leung HY, Neal DE, Robson CN. Nuclear accumulation of histone deacetylase 4 (HDAC4) coincides with the loss of androgen sensitivity in hormone refractory cancer of the prostate. Eur Urol March 2004;45(3):382–9.

[75] Milde T, Oehme I, Korshunov A, Kopp-Schneider A, Remke M, Northcott P, et al. HDAC5 and HDAC9 in medulloblastoma: novel markers for risk stratification and role in tumor cell growth. Clin Cancer Res June 15, 2010;16(12):3240–52.

[76] Ouaissi M, Sielezneff I, Silvestre R, Sastre B, Bernard JP, Lafontaine JS, et al. High histone deacetylase 7 (HDAC7) expression is significantly associated with adenocarcinomas of the pancreas. Ann Surg Oncol August 2008;15(8):2318–28.

[77] Moreno DA, Scrideli CA, Cortez MA, de Paula Queiroz R, Valera ET, da Silva Silveira V, et al. Differential expression of HDAC3, HDAC7 and HDAC9 is associated with prognosis and survival in childhood acute lymphoblastic leukaemia. Br J Haematol September 2010;150(6):665–73.

[78] Zhang Z, Yamashita H, Toyama T, Sugiura H, Omoto Y, Ando Y, et al. HDAC6 expression is correlated with better survival in breast cancer. Clin Cancer Res October 15, 2004;10(20):6962–8.

[79] Saji S, Kawakami M, Hayashi S, Yoshida N, Hirose M, Horiguchi S, et al. Significance of HDAC6 regulation via estrogen signaling for cell motility and prognosis in estrogen receptor-positive breast cancer. Oncogene June 30, 2005;24(28):4531–9.

[80] Zhang X, Yuan Z, Zhang Y, Yong S, Salas-Burgos A, Koomen J, et al. HDAC6 modulates cell motility by altering the acetylation level of cortactin. Mol Cell July 20, 2007;27(2):197–213.

[81] Luxton GW, Gundersen GG. HDAC6-pack: cortactin acetylation joins the brew. Dev Cell August 2007;13(2):161–2.

[82] Rodriguez-Gonzalez A, Lin T, Ikeda AK, Simms-Waldrip T, Fu C, Sakamoto KM. Role of the aggresome pathway in cancer: targeting histone deacetylase 6-dependent protein degradation. Cancer Res April 15, 2008;68(8):2557–60.

[83] Fraga MF, Ballestar E, Villar-Garea A, Boix-Chornet M, Espada J, Schotta G, et al. Loss of acetylation at Lys16 and trimethylation at Lys20 of histone H4 is a common hallmark of human cancer. Nat Genet April 2005;37(4):391–400.

[84] Chen WY, Wang DH, Yen RC, Luo J, Gu W, Baylin SB. Tumor suppressor HIC1 directly regulates SIRT1 to modulate p53-dependent DNA-damage responses. Cell November 4, 2005;123(3):437–48.

[85] Wang C, Chen L, Hou X, Li Z, Kabra N, Ma Y, et al. Interactions between E2F1 and SirT1 regulate apoptotic response to DNA damage. Nat Cell Biol September 2006;8(9):1025–31.

[86] Heintzman ND, Stuart RK, Hon G, Fu Y, Ching CW, Hawkins RD, et al. Distinct and predictive chromatin signatures of transcriptional promoters and enhancers in the human genome. Nat Genet March 2007;39(3):311–8.

[87] Wang Z, Zang C, Cui K, Schones DE, Barski A, Peng W, et al. Genome-wide mapping of HATs and HDACs reveals distinct functions in active and inactive genes. Cell September 4, 2009;138(5):1019–31.

[88] Choudhary C, Kumar C, Gnad F, Nielsen ML, Rehman M, Walther TC, et al. Lysine acetylation targets protein complexes and co-regulates major cellular functions. Science August 14, 2009; 325(5942):834–40.

[89] Rice JC, Allis CD. Histone methylation versus histone acetylation: new insights into epigenetic regulation. Curr Opin Cell Biol June 2001;13(3):263–73.

[90] Barbieri I, Cannizzaro E, Dawson MA. Bromodomains as therapeutic targets in cancer. Brief Funct Genomics May 2013;12(3):219–30.

[91] Selvi RB, Kundu TK. Reversible acetylation of chromatin: implication in regulation of gene expression, disease and therapeutics. Biotechnol J March 2009;4(3):375–90.

[92] Di Cerbo V, Schneider R. Cancers with wrong HATs: the impact of acetylation. Brief Funct Genomics May 2013;12(3):231–43.

[93] Ishihama K, Yamakawa M, Semba S, Takeda H, Kawata S, Kimura S, et al. Expression of HDAC1 and CBP/p300 in human colorectal carcinomas. J Clin Pathol November 2007;60(11):1205–10.

[94] Li M, Luo RZ, Chen JW, Cao Y, Lu JB, He JH, et al. High expression of transcriptional coactivator p300 correlates with aggressive features and poor prognosis of hepatocellular carcinoma. J Transl Med 2011;9:5.

[95] Xiao XS, Cai MY, Chen JW, Guan XY, Kung HF, Zeng YX, et al. High expression of p300 in human breast cancer correlates with tumor recurrence and predicts adverse prognosis. Chin J Cancer Res September 2011;23(3):201–7.

[96] Hou X, Li Y, Luo RZ, Fu JH, He JH, Zhang LJ, et al. High expression of the transcriptional coactivator p300 predicts poor survival in resectable non-small cell lung cancers. Eur J Surg Oncol June 2012;38(6):523–30.

[97] Gao Y, Geng J, Hong X, Qi J, Teng Y, Yang Y, et al. Expression of p300 and CBP is associated with poor prognosis in small cell lung cancer. Int J Clin Exp Pathol 2014;7(2):760–7.

[98] Debes JD, Sebo TJ, Lohse CM, Murphy LM, Haugen DA, Tindall DJ. p300 in prostate cancer proliferation and progression. Cancer Res November 15, 2003;63(22):7638–40.

[99] Ozdag H, Teschendorff AE, Ahmed AA, Hyland SJ, Blenkiron C, Bobrow L, et al. Differential expression of selected histone modifier genes in human solid cancers. BMC Genomics 2006;7:90.

[100] Yu C, Gong AY, Chen D, Solelo Leon D, Young CY, Chen XM. Phenethyl isothiocyanate inhibits androgen receptor-regulated transcriptional activity in prostate cancer cells through suppressing PCAF. Mol Nutr Food Res October 2013;57(10):1825–33.

[101] Chen L, Wei T, Si X, Wang Q, Li Y, Leng Y, et al. Lysine acetyltransferase GCN5 potentiates the growth of non-small cell lung cancer via promotion of E2F1, cyclin D1, and cyclin E1 expression. J Biol Chem May 17, 2013;288(20):14510–21.

[102] Seligson DB, Horvath S, Shi T, Yu H, Tze S, Grunstein M, et al. Global histone modification patterns predict risk of prostate cancer recurrence. Nature June 30, 2005;435(7046):1262–6.

[103] Wang F, Marshall CB, Ikura M. Transcriptional/epigenetic regulator CBP/p300 in tumorigenesis: structural and functional versatility in target recognition. Cell Mol Life Sci November 2013; 70(21):3989–4008.

[104] Fu M, Wang C, Zhang X, Pestell RG. Acetylation of nuclear receptors in cellular growth and apoptosis. Biochem Pharmacol September 15, 2004;68(6):1199–208.

[105] Ianculescu I, Wu DY, Siegmund KD, Stallcup MR. Selective roles for cAMP response element-binding protein binding protein and p300 protein as coregulators for androgen-regulated gene expression in advanced prostate cancer cells. J Biol Chem February 3, 2012;287(6):4000–13.

[106] Fan S, Ma YX, Wang C, Yuan RQ, Meng Q, Wang JA, et al. p300 modulates the BRCA1 inhibition of estrogen receptor activity. Cancer Res January 1, 2002;62(1):141–51.

[107] Takeuchi A, Shiota M, Tatsugami K, Yokomizo A, Tanaka S, Kuroiwa K, et al. p300 mediates cellular resistance to doxorubicin in bladder cancer. Mol Med Rep January 2012;5(1):173–6.

[108] Sorm F, Piskala A, Cihak A, Vesely J. 5-Azacytidine, a new, highly effective cancerostatic. Experientia April 15, 1964;20(4):202–3.

[109] Viegas-Pequignot E, Dutrillaux B. Segmentation of human chromosomes induced by 5-ACR (5-azacytidine). Hum Genet December 15, 1976;34(3):247–54.

[110] Cihak A. Biological effects of 5-azacytidine in eukaryotes. Oncology 1974;30(5):405–22.

[111] Karon M, Benedict WF. Chromatid breakage: differential effect of inhibitors of DNA synthesis during G 2 phase. Science October 6, 1972;178(4056):62.

[112] Christman JK. 5-Azacytidine and 5-aza-2′-deoxycytidine as inhibitors of DNA methylation: mechanistic studies and their implications for cancer therapy. Oncogene August 12, 2002;21(35):5483–95.

[113] Vesely J, Cihak A. Incorporation of a potent antileukemic agent, 5-aza-2′-deoxycytidine, into DNA of cells from leukemic mice. Cancer Res October 1977;37(10):3684–9.

[114] Bouchard J, Momparler RL. Incorporation of 5-Aza-2′-deoxycytidine-5′-triphosphate into DNA. Interactions with mammalian DNA polymerase alpha and DNA methylase. Mol Pharmacol July 1983;24(1):109–14.

[115] Stresemann C, Lyko F. Modes of action of the DNA methyltransferase inhibitors azacytidine and decitabine. Int J Cancer July 1, 2008;123(1):8–13.

[116] Brueckner B, Kuck D, Lyko F. DNA methyltransferase inhibitors for cancer therapy. Cancer J January–February 2007;13(1):17–22.

[117] Chuang JC, Warner SL, Vollmer D, Vankayalapati H, Redkar S, Bearss DJ, et al. S110, a 5-Aza-2′-deoxycytidine-containing dinucleotide, is an effective DNA methylation inhibitor in vivo and can reduce tumor growth. Mol Cancer Ther May 2010;9(5):1443–50.

[118] Zhou L, Cheng X, Connolly BA, Dickman MJ, Hurd PJ, Hornby DP. Zebularine: a novel DNA methylation inhibitor that forms a covalent complex with DNA methyltransferases. J Mol Biol August 23, 2002;321(4):591–9.

[119] Cheng JC, Weisenberger DJ, Gonzales FA, Liang G, Xu GL, Hu YG, et al. Continuous zebularine treatment effectively sustains demethylation in human bladder cancer cells. Mol Cell Biol February 2004;24(3):1270–8.

[120] Balch C, Yan P, Craft T, Young S, Skalnik DG, Huang TH, et al. Antimitogenic and chemosensitizing effects of the methylation inhibitor zebularine in ovarian cancer. Mol Cancer Ther October 2005;4(10):1505–14.

[121] Yang PM, Lin YT, Shun CT, Lin SH, Wei TT, Chuang SH, et al. Zebularine inhibits tumorigenesis and stemness of colorectal cancer via p53-dependent endoplasmic reticulum stress. Sci Rep 2013;3:3219.

[122] Lavelle D, Saunthararajah Y, Vaitkus K, Singh M, Banzon V, Phiasivongsva P, et al. S110, a novel decitabine dinucleotide, increases fetal hemoglobin levels in baboons (P. anubis). J Transl Med 2010;8:92.

[123] Candido EP, Reeves R, Davie JR. Sodium butyrate inhibits histone deacetylation in cultured cells. Cell May 1978;14(1):105–13.

[124] Miller AA, Kurschel E, Osieka R, Schmidt CG. Clinical pharmacology of sodium butyrate in patients with acute leukemia. Eur J Cancer Clin Oncol September 1987;23(9):1283–7.

[125] Marks PA, Dokmanovic M. Histone deacetylase inhibitors: discovery and development as anticancer agents. Expert Opin Investig Drugs December 2005;14(12):1497–511.

[126] Martinet N, Bertrand P. Interpreting clinical assays for histone deacetylase inhibitors. Cancer Manag Res 2011;3:117–41.

[127] Chinnaiyan P, Vallabhaneni G, Armstrong E, Huang SM, Harari PM. Modulation of radiation response by histone deacetylase inhibition. Int J Radiat Oncol Biol Phys May 1, 2005;62(1):223–9.

[128] Munshi A, Kurland JF, Nishikawa T, Tanaka T, Hobbs ML, Tucker SL, et al. Histone deacetylase inhibitors radiosensitize human melanoma cells by suppressing DNA repair activity. Clin Cancer Res July 1, 2005;11(13):4912–22.

[129] Spange S, Wagner T, Heinzel T, Kramer OH. Acetylation of non-histone proteins modulates cellular signalling at multiple levels. Int J Biochem Cell Biol January 2009;41(1):185–98.

[130] Richon VM. Cancer biology: mechanism of antitumour action of vorinostat (suberoylanilide hydroxamic acid), a novel histone deacetylase inhibitor. Br J Cancer 2006;95(S1):2.

[131] Butler LM, Agus DB, Scher HI, Higgins B, Rose A, Cordon-Cardo C, et al. Suberoylanilide hydroxamic acid, an inhibitor of histone deacetylase, suppresses the growth of prostate cancer cells in vitro and in vivo. Cancer Res September 15, 2000;60(18):5165–70.

[132] Kelly WK, O'Connor OA, Krug LM, Chiao JH, Heaney M, Curley T, et al. Phase I study of an oral histone deacetylase inhibitor, suberoylanilide hydroxamic acid, in patients with advanced cancer. J Clin Oncol June 10, 2005;23(17):3923–31.

[133] Ueda H, Nakajima H, Hori Y, Goto T, Okuhara M. Action of FR901228, a novel antitumor bicyclic depsipeptide produced by *Chromobacterium violaceum* no. 968, on Ha-ras transformed NIH3T3 cells. Biosci Biotechnol Biochem September 1994;58(9):1579–83.

[134] Furumai R, Matsuyama A, Kobashi N, Lee KH, Nishiyama M, Nakajima H, et al. FK228 (depsipeptide) as a natural prodrug that inhibits class I histone deacetylases. Cancer Res September 1, 2002;62(17):4916–21.

[135] Piekarz RL, Frye R, Turner M, Wright JJ, Allen SL, Kirschbaum MH, et al. Phase II multi-institutional trial of the histone deacetylase inhibitor romidepsin as monotherapy for patients with cutaneous T-cell lymphoma. J Clin Oncol November 10, 2009;27(32):5410–7.

[136] Duvic M, Talpur R, Ni X, Zhang C, Hazarika P, Kelly C, et al. Phase 2 trial of oral vorinostat (suberoylanilide hydroxamic acid, SAHA) for refractory cutaneous T-cell lymphoma (CTCL). Blood January 1, 2007;109(1):31–9.

[137] Plumb JA, Finn PW, Williams RJ, Bandara MJ, Romero MR, Watkins CJ, et al. Pharmacodynamic response and inhibition of growth of human tumor xenografts by the novel histone deacetylase inhibitor PXD101. Mol Cancer Ther August 2003;2(8):721–8.

[138] West AC, Johnstone RW. New and emerging HDAC inhibitors for cancer treatment. J Clin Invest January 2, 2014;124(1):30–9.

[139] Balasubramanyam K, Swaminathan V, Ranganathan A, Kundu TK. Small molecule modulators of histone acetyltransferase p300. J Biol Chem May 23, 2003;278(21):19134–40.

[140] Balasubramanyam K, Varier RA, Altaf M, Swaminathan V, Siddappa NB, Ranga U, et al. Curcumin, a novel p300/CREB-binding protein-specific inhibitor of acetyltransferase, represses the acetylation of histone/nonhistone proteins and histone acetyltransferase-dependent chromatin transcription. J Biol Chem December 3, 2004;279(49):51163–71.

[141] Zhou H, Beevers CS, Huang S. The targets of curcumin. Curr Drug Targets March 1, 2011;12(3):332–47.

[142] Balasubramanyam K, Altaf M, Varier RA, Swaminathan V, Ravindran A, Sadhale PP, et al. Polyisoprenylated benzophenone, garcinol, a natural histone acetyltransferase inhibitor, represses chromatin transcription and alters global gene expression. J Biol Chem August 6, 2004;279(32):33716–26.

[143] Mantelingu K, Reddy BA, Swaminathan V, Kishore AH, Siddappa NB, Kumar GV, et al. Specific inhibition of p300-HAT alters global gene expression and represses HIV replication. Chem Biol June 2007;14(6):645–57.

[144] Arif M, Pradhan SK, Thanuja GR, Vedamurthy BM, Agrawal S, Dasgupta D, et al. Mechanism of p300 specific histone acetyltransferase inhibition by small molecules. J Med Chem January 22, 2009;52(2):267–77.

[145] Santer FR, Hoschele PP, Oh SJ, Erb HH, Bouchal J, Cavarretta IT, et al. Inhibition of the acetyltransferases p300 and CBP reveals a targetable function for p300 in the survival and invasion pathways of prostate cancer cell lines. Mol Cancer Ther September 2011;10(9):1644–55.

[146] Zeng L, Li J, Muller M, Yan S, Mujtaba S, Pan C, et al. Selective small molecules blocking HIV-1 Tat and coactivator PCAF association. J Am Chem Soc March 2, 2005;127(8):2376–7.

[147] Sachchidanand, Resnick-Silverman L, Yan S, Mutjaba S, Liu WJ, Zeng L, et al. Target structure-based discovery of small molecules that block human p53 and CREB binding protein association. Chem Biol January 2006;13(1):81–90.

[148] Borah JC, Mujtaba S, Karakikes I, Zeng L, Muller M, Patel J, et al. A small molecule binding to the coactivator CREB-binding protein blocks apoptosis in cardiomyocytes. Chem Biol April 22, 2011;18(4):531–41.

[149] Filippakopoulos P, Qi J, Picaud S, Shen Y, Smith WB, Fedorov O, et al. Selective inhibition of BET bromodomains. Nature December 23, 2010;468(7327):1067–73.

[150] Nicodeme E, Jeffrey KL, Schaefer U, Beinke S, Dewell S, Chung CW, et al. Suppression of inflammation by a synthetic histone mimic. Nature December 23, 2010;468(7327):1119–23.

[151] Dawson MA, Prinjha RK, Dittmann A, Giotopoulos G, Bantscheff M, Chan WI, et al. Inhibition of BET recruitment to chromatin as an effective treatment for MLL-fusion leukaemia. Nature October 27, 2011;478(7370):529–33.

[152] Zuber J, Shi J, Wang E, Rappaport AR, Herrmann H, Sison EA, et al. RNAi screen identifies Brd4 as a therapeutic target in acute myeloid leukaemia. Nature October 27, 2011;478(7370):524–8.

[153] Delmore JE, Issa GC, Lemieux ME, Rahl PB, Shi J, Jacobs HM, et al. BET bromodomain inhibition as a therapeutic strategy to target c-Myc. Cell September 16, 2011;146(6):904–17.

[154] Mertz JA, Conery AR, Bryant BM, Sandy P, Balasubramanian S, Mele DA, et al. Targeting MYC dependence in cancer by inhibiting BET bromodomains. Proc Natl Acad Sci USA October 4, 2011;108(40):16669–74.

[155] Cameron EE, Bachman KE, Myohanen S, Herman JG, Baylin SB. Synergy of demethylation and histone deacetylase inhibition in the re-expression of genes silenced in cancer. Nat Genet January 1999;21(1):103–7.

[156] Oi S, Natsume A, Ito M, Kondo Y, Shimato S, Maeda Y, et al. Synergistic induction of NY-ESO-1 antigen expression by a novel histone deacetylase inhibitor, valproic acid, with 5-aza-2′-deoxycytidine in glioma cells. J Neurooncol March 2009;92(1):15–22.

[157] Gore SD, Baylin S, Sugar E, Carraway H, Miller CB, Carducci M, et al. Combined DNA methyltransferase and histone deacetylase inhibition in the treatment of myeloid neoplasms. Cancer Res June 15, 2006;66(12):6361–9.

[158] Chen M, Voeller D, Marquez VE, Kaye FJ, Steeg PS, Giaccone G, et al. Enhanced growth inhibition by combined DNA methylation/HDAC inhibitors in lung tumor cells with silenced CDKN2A. Int J Oncol October 2010;37(4):963–71.

[159] Sarkar S, Abujamra AL, Loew JE, Forman LW, Perrine SP, Faller DV. Histone deacetylase inhibitors reverse CpG methylation by regulating DNMT1 through ERK signaling. Anticancer Res September 2011;31(9):2723–32.

[160] Dobosy JR, Selker EU. Emerging connections between DNA methylation and histone acetylation. Cell Mol Life Sci May 2001;58(5–6):721–7.

[161] Groudine M, Eisenman R, Weintraub H. Chromatin structure of endogenous retroviral genes and activation by an inhibitor of DNA methylation. Nature July 23, 1981;292(5821):311–7.

[162] Chai G, Li L, Zhou W, Wu L, Zhao Y, Wang D, et al. HDAC inhibitors act with 5-aza-2′-deoxycytidine to inhibit cell proliferation by suppressing removal of incorporated abases in lung cancer cells. PLoS One 2008;3(6):e2445.

[163] Novotny-Diermayr V, Hart S, Goh KC, Cheong A, Ong LC, Hentze H, et al. The oral HDAC inhibitor pracinostat (SB939) is efficacious and synergistic with the JAK2 inhibitor pacritinib (SB1518) in preclinical models of AML. Blood Cancer J May 2012;2(5):e69.

[164] Ellis L, Ku SY, Ramakrishnan S, Lasorsa E, Azabdaftari G, Godoy A, et al. Combinatorial antitumor effect of HDAC and the PI3K-Akt-mTOR pathway inhibition in a Pten deficient model of prostate cancer. Oncotarget December 2013;4(12):2225–36.

[165] Almenara J, Rosato R, Grant S. Synergistic induction of mitochondrial damage and apoptosis in human leukemia cells by flavopiridol and the histone deacetylase inhibitor suberoylanilide hydroxamic acid (SAHA). Leukemia July 2002;16(7):1331–43.

[166] Kaminskyy VO, Surova OV, Vaculova A, Zhivotovsky B. Combined inhibition of DNA methyltransferase and histone deacetylase restores caspase-8 expression and sensitizes SCLC cells to TRAIL. Carcinogenesis October 2011;32(10):1450–8.

[167] Shang D, Liu Y, Matsui Y, Ito N, Nishiyama H, Kamoto T, et al. Demethylating agent 5-aza-2′-deoxycytidine enhances susceptibility of bladder transitional cell carcinoma to Cisplatin. Urology June 2008;71(6):1220–5.

[168] Mirza S, Sharma G, Pandya P, Ralhan R. Demethylating agent 5-aza-2-deoxycytidine enhances susceptibility of breast cancer cells to anticancer agents. Mol Cell Biochem September 2010;342(1–2):101–9.

[169] Plumb JA, Strathdee G, Sludden J, Kaye SB, Brown R. Reversal of drug resistance in human tumor xenografts by 2′-deoxy-5-azacytidine-induced demethylation of the hMLH1 gene promoter. Cancer Res November 1, 2000;60(21):6039–44.

[170] Falk M, Lukasova E, Kozubek S. Chromatin structure influences the sensitivity of DNA to gamma-radiation. Biochim Biophys Acta December 2008;1783(12):2398–414.

[171] Gravina GL, Festuccia C, Marampon F, Popov VM, Pestell RG, Zani BM, et al. Biological rationale for the use of DNA methyltransferase inhibitors as new strategy for modulation of tumor response to chemotherapy and radiation. Mol Cancer 2010;9:305.

[172] Thurn KT, Thomas S, Moore A, Munster PN. Rational therapeutic combinations with histone deacetylase inhibitors for the treatment of cancer. Future Oncol February 2011;7(2):263–83.

[173] Ding J, Zhang ZM, Xia Y, Liao GQ, Pan Y, Liu S, et al. LSD1-mediated epigenetic modification contributes to proliferation and metastasis of colon cancer. Br J Cancer August 20, 2013;109(4):994–1003.

[174] Yu Y, Wang B, Zhang K, Lei Z, Guo Y, Xiao H, et al. High expression of lysine-specific demethylase 1 correlates with poor prognosis of patients with esophageal squamous cell carcinoma. Biochem Biophys Res Commun July 26, 2013;437(2):192–8.

[175] Lynch JT, Harris WJ, Somervaille TC. LSD1 inhibition: a therapeutic strategy in cancer? Expert Opin Ther Targets December 2012;16(12):1239–49.

[176] Fiskus W, Sharma S, Shah B, Portier BP, Devaraj SG, Liu K, et al. Highly effective combination of LSD1 (KDM1A) antagonist and pan-histone deacetylase inhibitor against human AML cells. Leukemia November 2014;28(11):2155–64.

[177] Han H, Yang X, Pandiyan K, Liang G. Synergistic re-activation of epigenetically silenced genes by combinatorial inhibition of DNMTs and LSD1 in cancer cells. PLoS One 2013;8(9):e75136.

LIST OF ACRONYMS AND ABBREVIATIONS

22G RNAs A class of small interfering RNAs (siRNAs)

26G RNAs A class of small interfering RNAs (siRNAs)

2i Small-molecule kinase inhibitors

3'HS1 3'-DNase I-Hypersensitive Site 1

3'RR 3'-Regulatory Region

3'UTR 3'-Untranslated Region of mRNA

3C Chromosome Conformation Capture

3D Three-dimensional

3P-seq Polyadenylated ends sequencing

4C Circular Chromosome Conformation Capture

5–azaC 5-Azacytidine, a cytidine analog

5'HS 5'-DNase I-Hypersensitive Site

5'HS4 5'-DNase I-Hypersensitive Site 4; DNA insulator element; bound by CTCF and located at the 5'-end of the chicken *β-globin* locus

5'UTR 5'-Untranslated Region of mRNA

5C Chromosome Conformation Capture Carbon Copy

5caC 5-Carboxylcytosine

5fC 5-Formylcytosine

5hmC 5-Hydroxymethylcytosine

5mC 5-Methylcytosine; 5meC

5meC 5-Methylcytosine; 5mC

α-KG Alpha-Ketoglutaric acid

Abs Antibodies

ACF ATP-dependent Chromatin Assembly Factor, large subunit

ACH Active Chromatin Hub

AD Alzheimer's Disease

ADIPOQ Adiponectin, C1Q and collagen domain containing (*Homo sapiens*)

ADP Adenosine Diphosphate

AEBP2 AE Binding Protein 2

AGO3 Argonaute 3

AHSP Alpha-Hemoglobin Stabilizing Protein, DPE-deficient

AIRN Antisense IGF2R (insulin-like growth factor 2 receptor) RNA

AKT1 v-akt murine Thymoma Viral Oncogene Homolog 1

ALA δ-Aminolevulinic Acid

ALAS ALA Synthase; iosforms ALAS1 and ALAS2

ALAS1 ALA Synthase, ubiquitously-expressed

ALAS2 ALA synthase, erythroid-specific

alncRNA-EC7 Activation-lncRNA-EC7; specifically required for neighboring BAND3 gene, transcribed from an enhancer

AML Acute Myeloid Leukemia

AML1 Acute Myelocytic Leukemia Protein 1

AML1-ETO AML1 (Acute Myelocytic Leukemia Protein 1)-ETO (8;21) fusion

ANRASSF1 Antisense Noncoding RASSF1 (Ras Association (RalGDS/AF-6) domain family member 1); intronic lncRNA

ANRIL Antisense noncoding RNA in the INK4b-ARF-INK4a Locus; lncRNA; p15AS, unspliced isoform able to silence the tumor suppressor gene *p15*

antagoNAT Antagonistic Natural Antisense Transcripts

AR Androgen Receptor

ARF Alternate Reading Frame; p19^{ARF}

ARID AT-Rich Interactive Domain

ARID1A AT-Rich Interactive Domain 1A (SWI-like) (*Homo sapiens*); also called BAF250, SMARCF1; osa (*Drosophila*)

ARID2 AT-Rich Interactive Domain 2 (ARID, RFX-like) (*Homo sapiens*); PBAF complex protein (*Drosophila*); also called BAF200, p200

ARID5B AT-Rich Interactive Domain-Containing Protein 5B

as-lncRN Antisense long noncoding RNA

ASC Apoptosis-Associated Speck-Like Protein Containing a CARD gene; also called PYCARD

Ash1 Absent, Small, or Homeotic Discs 1

ASH1L Ash1(Absent, Small, or Homeotic Discs 1)-Like

ASH2L Ash2(Absent, Small, or Homeotic Discs 2)-Like

asRNA Antisense RNA

ASXL1 Additional Sex Combs-Like 1; transcriptional regulator

ASXL2 Additional Sex Combs-Like 2; Transcriptional Regulator

ATF3 Activating Transcription Factor 3 (*Homo sapiens*)

ATM Ataxia Telangiectasia Mutated

ATP Adenosine Triphosphate

Atpif1 ATPase Inhibitory Factor 1

AUB Aubergine

B-cells B-lymphocytes

B6 Pyridoxine

BACH BTB and CNC Homology, Basic Leucine Zipper transcription factor (*Homo sapiens*)

BACH1 BTB and CNC Homology 1, Basic Leucine Zipper transcription factor 1 (*Homo sapiens*)

BACH2 BTB and CNC Homology 2, Basic Leucine Zipper transcription factor 2 (*Homo sapiens*)

Bad BCL2-associated agonist of cell death gene

BAF47 SWI/SNF-related, matrix-associated, actin-dependent regulator of chromatin, subfamily B, member 1 (*Homo* sapiens); also called INI1, SMARCB1, SNF5

BAF180 Polybromo 1 (*Homo sapiens*); PBAF complex protein (*Drosophila*); also called PB1, PBRM1

BAF190 Brahma Related Gene 1; also called SWI/SNF-related, matrix-associated, actin-dependent regulator of chromatin, subfamily A, member 4 (*Homo sapiens*); also called BG1, SMARCA2, SMARCA4, SNF2, SWI2

BAF200 AT-Rich Interactive Domain 2 (ARID, RFX-Like) (*Homo sapiens*); PBAF complex protein (*Drosophila*); also called ARID2, p200

BAF250 AT-Rich Interactive Domain 1A (SWI-Like) (*Homo sapiens*); also called ARID1A, SMARCF1; *osa* (*Drosophila*)

BAP1 BRCA1 Associated Protein-1

BAX BCL2-Associated X Protein

BC1 Brain Cytoplasmic RNA 1; found in dendrites that regulates the post-synaptic signaling

BCL B-cell Lymphoma

BCL-XL BCL2-like isoform 1

BCL11A B-cell CLL/Lymphoma 11A (zinc finger protein) (*Homo sapiens*)

BCL2 B-cell CLL/Lymphoma 2 (*Homo sapiens*)

BDNF Brain-Derived Neurotrophic Factor

BDNF-AS Brain-Derived Neurotrophic Factor Antisense lncRNA

BER Base Excision Repair Mechanism

BET Bromodomain and Extra-Terminal; family of proteins BRD2, BRD3, BRD4, and BRDt

BET(i) Bromodomain and Extra-Terminal Inhibitors

BFU-E Blast Forming Unit Erythroid

bHLH Basic Region-Helix Loop Helix proteins

BHMT Betaine Homocysteine Methyltransferase

Bid BH3 interacting domain death agonist gene

bim Bcl2-interacting mediator of cell death gene

Bmi1 B lymphoma Mo-MLV insertion region 1 homolog

BPA Bisphenol A

BPTF Bromodomain PHD Finger Transcription Factor (*Homo sapiens*); also called NURF301

BRCA1 Breast Cancer 1, early onset

BRCA2 Breast Cancer 2, early onset

BRD Bromodomain

BRD2/3/4 Bromodomain 2, 3, 4

BRDT Bromodomain, Testis-specific (*Homo Sapiens*); also called BRDt

BRG1 Brahma Related Gene 1; also called SWI/SNF-related, matrix-associated, actin-dependent regulator of chromatin, subfamily A, Member 4 (*Homo sapiens*); also called BAF190, SMARCA4, SNF2, SWI2

brm Brahma gene (*Drosophila*)

BRM ATP-dependent Helicase BRAHMA (*Arabidopsis*)

BTB Broad Complex Tramtrack Bric a brac

BURs Base Unpairing Regions

Bvht Braveheart lncRNA (*Mus*); lncRNA

BWS Beckwith–Wiedemann Syndrome

bZip Basic Region-Leucine Zipper Motif

c-Maf Cellular v-maf avian Musculoaponeurotic Fibrosarcoma Oncogene Homolog (*Homo sapiens*)

c-Myb Cellular Myeloblastosis oncogene; v-myb avian myeloblastosis viral oncogene homolog (*Homo sapiens*); belongs to the myeloblastosis family of transcription factors; also called Myb or MYB

c-Myc Cellular Myelocytomatosis, avian myelocytomatosis viral oncogene homolog (*Homo sapiens*); transcription factor that is a helix-loop-helix (HLH) protein; also called myc or MYC

CAF1 Chromatin Assembly Factor 1, Subunit B (*Homo sapiens*, gene name CHAF1B); also called Nurf55 (*Drosophila*)

CAGE-seq Capped RNA sequencing

CARD C-terminal Caspase-Recruitment Domain

CBP CREB binding protein (*Homo sapiens*); also called CEBP, CREB, CREBBP, KAT3A; Lysine (K)-specific Histone Acetyltransferase (HAT)

CBS Cystathionine beta-Synthase

cbx Crossbronx gene (*Drosophila*); also called Cbx1, Chromobox 1 (*Mus*); also called CBX1, Chromobox Homolog 1 (*Homo sapiens*)

CBX1 Chromobox Homolog 1 (*Homo sapiens*)

CCC Clear Cell Carcinoma

CCCH-type A type of zinc finger proteins, capable of binding nucleotides and are involved in the RNA pathways

CD4+ Cluster of Differentiation 4; glycoprotein

CD4+ T-cells Cluster of Differentiation 4 T helper cells

CDC25A Cell Division Cycle 25A

CDH1 Cadherin 1, Type 1, E-Cadherin (epithelial) (*Homo sapiens*)

CDH13 Cadherin-13

CDK Cyclin-Dependent Kinase

CDKN1A Cyclin-Dependent Kinase Inhibitor 1A

CDKN2A Cyclin-Dependent Kinase (CDK) Inhibitors 2A, also known as p16

CDKN2A/2B Cyclin-Dependent Kinase (CDK) Inhibitors p14^{ARF}; also called p19^{ARF} (*Mus*); also called p15^{INK4B}, and p16^{INK4A}; function as tumor suppressors and prevent CDK4-dependent phosphorylation of RB (Retinoblastoma Protein)

CEBP CREB binding protein; CCAAT enhancer binding protein; also called CBP, CREB, CREBBP, KAT3A

Ced-3 Cell Death Protein 3

Ced-4 Cell Death Protein 4

Ced-9 Cell Death Protein 9

CENPA Centromere Protein A

CENPE Centromere Protein E

CENPF Centromere Protein F

ceRNA Competing Endogenous RNA, or sponge noncoding RNA; competes for the binding of key microRNAs

CFU-E Colony Forming Unit Erythroid

CHD1 Chromodomain Helicase DNA Binding Protein 1

CHD4 Chromodomain Helicase DNA Binding Protein 4; also called Mi-2b, Mi-2β, Mi2-BETA (*Homo sapiens*)

CHD8 Chromodomain Helicase DNA Binding Protein 8 (*Homo sapiens*)

ChiA-PET Chromatin Interaction Analysis by Paired-End Tag Sequencing

ChIP Chromatin Immunoprecipitation

ChIP-seq Combines ChIP with massively parallel sequencing to identify interactions between DNA and RNA and proteins

ChIRP Chromatin Isolation by RNA Purification

Cip1 CDK-Interacting Protein 1

circRNAs Circular RNAs

CJD Creutzfeld–Jakob Disease

CKD Chronic Kidney Disease

CLL Chronic Lymphocytic Leukemia

CLP Common Lymphoid Progenitor

CMML Chronic Myeloid Monocytic Leukemia

CMP Common Myeloid Progenitor

CNC Cap'n'Collar, a type of leucine zipper proteins or domains

CNV Copy Number Variation

COMPASS Complex SET1 (WRD5, RBBP5, ASH2L, DPY30); H3K4 Histone Methyltransferase (*Saccharomyces*)

CoREST Corepressor for Element-1-Silencing Transcription Factor

CP Coproporphy

CP Coproporphyrinogen

CpG Cytosine-Phosphate-Guanine Dinucleotide

CpNpG Cytosine-Phosphate-[any nucleotide]-Phosphate-Guanine

CR Caloric Restriction

CREB CREB-binding protein (*Homo sapiens*); CCAAT enhancer-binding Protein; also called CBP, CEBP, CREBBP, KAT3A; Lysine (K)-specific Histone Acetyltransferase (HAT)

CREBBP CREB-binding protein (*Homo sapiens*); CCAAT enhancer-binding protein; also called KAT3A, CBP, CEBP; Lysine (K)-specific Histone Acetyltransferase (HAT)

CRL4^{VprBP} VprBP-DDB1-CUL4-ROC1 E3 ubiquitin ligase

CRPC Calcitonin Gene-Related Peptide-Receptor Component Protein
CTBP1 C-Terminal-Binding Protein 1; involved in the stimulation of cell proliferation
CTBP1-AS C-Terminal-Binding Protein 1; antisense lncRNA
CTCF CCCTC-DNA Binding Factor Protein, containing 11 zinc fingers
CTD RNA Pol II C-Terminal Domain
CTH Cystathionase; Cystathionine Gamma-Lyase
CTNNB1 Catenin Beta 1
CTs Chromosome Territories
CVD Cardiovascular Disease
CXXC1 CXXC Finger Protein 1 (*Homo sapiens*); a component of mammalian SETD1A/B complexes
CXXC4 CXXC Finger Protein 4 (*Homo sapiens*)
Cys Cysteine
DAF-12 Dafachronic Acids Nuclear Receptor; transcription factor, activated by Dafachronic acids
DAPK Death-Associated Protein Kinase 1
DAs Dafachronic Acids
DEAD-box RNA Helicase p68 [DEAD (Asp-Glu-Ala-Asp) box helicase 5]; RNA splicing Protein p68 (*Homo sapiens*); also called SRA
DPE Downstream Promoter Element
$D_H J_H$ Ig Heavy (H) chain variable (D)–Ig Heavy (H) chain variable (J), where D = Diversity and J = Joining
Dicer Type of double-stranded RNA-specific endoribonuclease type III
DLBCL Diffuse Large B-cell Lymphoma
DMD Duchenne Muscular Dystrophy
DMRs Differentially Methylated Regions
dMyc *Drosophila* avian myelocytomatosis viral oncogene homolog
DNA Deoxyribonucleic Acid
DNaseI Deoxyribonuclease I
DNMT DNA Methyltransferase
DNMT1 DNA (Cytosine-5)-Methyltransferase 1 (*Homo sapiens*); recognize hemi-methylated DNA postreplication and copies pattern to new strand, a maintenance methyltransferase
DNMT3A/B DNA (Cytosine-5)-Methyltransferase 3 alpha (*Homo sapiens*); capable of methylation of unmethylated DNA
DNMT3B DNA (Cytosine-5)-Methyltransferase 3 beta (*Homo sapiens*); also called ICF1
DNMTi DNMT Inhibitor
DOT1 Histone methyltransferase
DOT1L DOT1-Like Histone H3K79 methyltransferase (*Homo sapiens*)
dpc Days Post-Coitum
DPE Downstream Promoter Element
DPY30 Component of WRAD module, required to increase enzymatic activity of the WRAD complex (*Homo sapiens*); also known as the WRAD module
DRED Direct Repeat Erythroid Definitive protein
dRING Sex Combs Extra, Sce (*Drosophila*); also called RING1 (*Drosophila*); E3 ubiquitin ligase
DRM2 Domains Rearranged Methyltransferase 2
Drosha A type of double-stranded RNA-specific endoribonuclease type III
DSBH Double-Stranded β Helix; jelly-roll motif
dSET1 *Drosophila* SET1
dSfmbt Member of one of main PcG Complexes (*Drosophila*)
DsRed *Discoma* sp. expression Red (Fluorescent Protein Reporter System)
DUB Deubiquintinas
Duffy Duffy blood group factor

E(z) Enhancer of Zeste (*Drosophila*)

E2-2 Transcription factor 4 (*Homo sapiens*); member of E family of transcription factors

E2A Transcription factor 3 (*Homo sapiens*); member of E family of transcription factors, critical regulators in lymphocyte development, dimerize to bind Ebox motif

E2F1 E2F Transcription Factor 1

Eμ IGH Intronic Enhancer

Eμ-*myc* IGH intronic enhancer-myelocytomatosis

Eμ-PAIR *Eμ* and antisense transcripts from PAIR ncRNAs; interactions between these components facilitate locus compaction

EB3 Erythrocyte Band 3; also called SLC4A1

Ebox Sequence CANNTG

EED Embryonic Ectoderm Development Protein; homolog of esc, extra sex combs (*Drosophila*)

EGCG Epigallocatechin-3-Gallate

egl-1 EGg Laying Defective gene; also called *BH3-only* gene

eIF2α Eukaryotic Translation Initiation Factor 2, subunit 1 alpha, 35 kDa (*Homo sapiens*)

EHD Eyelid Homolog Domain

EHMT2 Euchromatic Histone-Lysine N-Methyltransferase 2 (*Homo sapiens*); also called G9a, Histone methyltransferase; homolog and heterodimer to GLP, CG2995 gene product from transcript CG2995-RB (*Drosophila melanogaster*)

endo-siRNAs Endogenous Small Interfering RNAs

EP300 E1A binding protein p300 (*Homo sapiens*); also called KAT3B (Lysine (K)-Specific Histone Acetyltransferase [HAT] 3B), p300, RSTS2

epiRILs Epigenetic Recombinant Inbred Lines

EpiSC Epiblast Stem Cell

EPO Erythropoietin

ER-α Estrogen Receptor-α

eRNA Enhancer RNA

esc Extra Sex Combs (*Drosophila*); homolog of EED

ESCs Embryonic Stem Cells

ESR1 Estrogen Receptor 1

ETO 8; 21

ETO2 Core-binding factor, runt domain, alpha subunit 2; translocated to chr3 (*Homo sapiens*); gene name *CBFA2T3* (*Homo sapiens*), gene name *Cbfa2t3* (*Mus*)

EZH Enhancer of Zeste protein (*Homo sapiens*); also called E(z) (*Drosophila*); catalyzes mono-, di- and trimethylation of H3K27, H3K27me1/2/3

EZH1 Enhancer of Zeste Protein 1 (*Homo sapiens*)

EZH1/2 Enhancer of Zeste Homolog 1/2 (*Homo sapiens*)

EZH2 Enhancer of Zeste Protein 2 (*Homo sapiens*); subunit of the PRC2 complex; also called KMT6A, Lysine (K)-Specific Methyltransferase 6A

FAD Flavin Adenine Dinucleotide, fully oxidized quinone form

FADH+ Flavin Adenine Dinucleotide, redox semiquinone form

FADH2 Flavin Adenine Dinucleotide, reduced hydroquinone form

fas Fas (faint sausage) cell surface death receptor (*Mus*)

FECH Ferrochelatase

Fendrr FOXF1 Adjacent Noncoding Developmental Regulatory RNA; lncRNA

FGF2 Fibroblast Growth Factor 2

FISH Fluorescence in situ Hybridization

FK228 Depsipeptide, FR901228, trade name Romidep

FOG-1 Friend of GATA1, Protein FOG-1 (*C. elegans*)

FOXO Forkhead Box, sub-group O (*Drosophila*)

FOXO1 Forkhead Box O1 (*Homo sapiens*)

FOXO3 Forkhead Box O3

FOXO3A Forkhead Box O3, same as FOXO3

G-patch Glycine-rich signature domain associated with eukaryotic post-translational processing

G1E GATA-1- Erythroid cells, generated from in vitro differentiated GATA1 mESCs, and complete maturation upon GATA1 function restoration

G9a Histone methyltransferase, homolog and heterodimer to GLP; CG2995 gene product from transcript CG2995-RB (*Drosophila*); also called EHMT2, Euchromatic Histone-Lysine N-Methyltransferase 2 (*Homo sapiens*)

GATA1 GATA Binding Protein 1 (Globin Transcription Factor 1) (*Homo sapiens*)

GATA1/TAL1 complex GATA, Globin Transcription Factor 1/T-cell Acute Lymphocytic Leukemia 1

GATA2 GATA Binding Protein 2 (Globin Transcription Factor 2) (*Homo sapiens*)

GATA3 GATA Binding Protein 3 (Globin Transcription Factor 3) (*Homo sapiens*)

GATA4 GATA Binding Protein 4 (Globin Transcription Factor 4) (*Homo sapiens*)

GATAD2A GATA Zinc Finger Domain-Containing Protein 2A

GCLC and GCLM Catalytic and modulatory subunits of γ-glutamylcysteine synthetase

GCN5 SAGA complex histone acetyltransferase catalytic subunit Gnc5 (*S. pombe*); also called KAT2A (*Homo sapiens*)

GGT γ-Glutamyltranspeptidase

GLP Histone methyltransferase, homolog and heterodimer to G9a; CG2995 gene product from transcript CG2995-RB (*Drosophila melanogaster*); also called EHMT2, Euchromatic Histone-lysine N-Methyltransferase 2 (*Homo sapiens*)

GLS Glutaminase

GMP Granulocyte/Monocyte Precursor

GNAT Gcn5-Related N-Acetyltransferases

GNMT Glycine N-Methyltransferase

grim CG4345 gene product from transcript CG4345-RA (*Drosophila*), proapoptotic gene

GSH Glutathione Synthesis

GSK Glucokinase

GSS Glutathione Synthetase

GTPs Green Tea Polyphenols; an inhibitor of DNA methyltransferases (DNMTs)

GWAS Genome-Wide Association Studies

H element Transacting enhancer element

H19 H19, (noncoding) imprinted maternally-expressed transcript (*Homo sapiens*, locus 11p15.5 near *Igf2*); (*Mus*); lncRNA

H2A.Z Variant of histone H2A

H3.3 Variant of histone H3

H3K27 Histone 3 Lysine-27

H3K27me3 Histone 3 Lysine-27 Trimethylation

H3K9 Histone 3 Lysine-9

H3K9me2 Histone 3 Lysine-9 Dimethylation

H3K9me3 Histone 3 Lysine-9 Trimethylation

HAT Histone Acetyltransferase

HATi Histone Acetyltransferase Inhibitors

HBS1L HBS1-Like Translational GTPase gene (*Homo sapiens*); also found in (*Mus*)

HCF1 Host Cell Factor 1 (*C. elegans*); subunit of MLL complexes

HCPs Promoters with a high CpG content

HCys Homocysteine, L-Homocysteine, L-HCys

Hda1 Histone Deacetylase1, Yeast Histone Deacetylase

HDAC Histone deacetylase

HDAC class III Class 3 mammalian histone deacetylase, requires coenzyme NAD+

HDAC-Sin3A Histone deacetylase–SIN3 transcription regulator family member A chromatin repressive complex

HDAC1 Histone deacetylase 1 (*Homo sapiens*); class I mammalian histone deacetylase

HDAC2 Histone deacetylase 2 (*Homo sapiens*); class I mammalian histone deacetylase

HDAC3 Histone deacetylase 3 (*Homo sapiens*); class I mammalian histone deacetylase

HDAC4 Histone deacetylase 4 (*Homo sapiens*); class II mammalian histone deacetylase

HDAC5 Histone deacetylase 5 (*Homo sapiens*); class II mammalian histone deacetylase

HDAC6 Histone deacetylase 6 (*Homo sapiens*); class II mammalian histone deacetylase

HDAC7 Histone deacetylase 7 (*Homo sapiens*); class II mammalian histone deacetylase

HDAC8 Histone deacetylase 8 (*Homo sapiens*); class I mammalian histone deacetylase

HDAC9 Histone deacetylase 9 (*Homo sapiens*); class II mammalian histone deacetylase

HDAC10 Histone deacetylase 10 (*Homo sapiens*); class II mammalian histone deacetylase

HDAC11 Histone deacetylase 11 (*Homo sapiens*); class IV mammalian histone deacetylase

HDACi Histone deacetylase Inhibitors

HDM Histone demethylase

HEB Transcription factor 12, member of E family of transcription factors (*Homo sapiens*)

HEN1 miRNA methyltransferase protein

HGPS Hutchinson–Gilford Progeria

Hi-C Based on 3C technology, an extension, that allows for the unbiased identification long range interactions using a genome-wide approach

hid Head Involution Defective gene (*Drosophila*)

HIRA Histone Cell Cycle Regulation-Defective Homolog A; histone chaperone

HLH Helix-Loop-Helix

HMB Hydroxymethylbilane

HMBS HMB Synthase

HMG High Mobility Group

HMT Histone Metyltransferase

hnRNP-K Heterogeneous Nuclear Ribonucleoprotein K

HOTAIR HOX Transcript Antisense RNA; lncRNA; epigenetic silencing is mediated by the physical attraction of both the polycomb repressive PRC2 complex and LSD1-CoREST to its 5′ and 3′ regions, respectively

HOTTIP HOXA Transcript at the Distal Tip; lncRNA; regulates the human HOXA locus; binds with WDR5 to recruit H3K4me3-associated MLL1 complex

HOX Homeobox genes

HOXA2 Homeobox A2

HOXB4 Homeobox B4 (*Homo sapiens*); also called Hoxb4 (*Mus*)

HP1 Heterochromatin Protein 1

HPFH Hereditary Persistence of Fetal Hemoglobin

HPL-1 HP1-Like-1 (*C. elegans*)

HPL-2 HP1-Like-2 (*C. elegans*)

HRI Heme Regulated eIF2α Kinase

HSs DNase I-Hypersensitive Sites

HS2 DNase I-Hypersensitive Site 2; erythroid-specific gene enhancers, contains a palindromic GATA1 binding element with a high affinity for its binding site and is capable of positively mediating the autoregulation of *GATA1* expression; *GATA1*-HS2 located 200-bp from the gene *GATA1* gene

HSC Hematopoietic Stem Cell

HSP70 Heat Shock Protein 70

HSPC Hematopoietic Stem/Progenitors Cell

HuR Human Antigen R

ICD-10 code ICD-10 is the 10th International Classification of Diseases, established by World Health Organisation (*Homo sapiens*)

ICM Inner Cell Mass

IDH Isocitrate Dehydrogenase

IFITM1 Interferon Induced Transmembrane Protein 1

IGF Insulin-Like Growth Factor

IGF2 Insulin-Like Growth Factor 2 Receptor; Locus 11p15.5; only expressed from the paternally-inherited chromosome

IGFBP3 Insulin-Like Growth Factor-Binding Protein 3

IGH Immunoglobulin Heavy Locus; also called IgH

IKAROS Transcription factor, tumor suppressor, IKAROS family zinc finger 1 (*Homo sapiens*); (*Mus*)

IKAROS Family Transcription factors Aiolos, Eos, Helios, Ikaros, and Pegasus

INI1 SWI/SNF-related, matrix-associated, actin-dependent regulator of chromatin, subfamily B, member 1 (*Homo sapiens*); also called SNF5, BAF47, SMARCB1

INK4a Cyclin-Dependent Kinase Inhibitor 2A (inhibitor of CDK4)

INO80 INO80 complex subunit (*Homo sapiens*); ATPase-dependent chromatin remodeling enzymes subfamily member; YY1 is supposedly involved in recruitment for binding of INO80 to DNA and activation of specific genes

iPSC Induced Pluripotent Stem Cell

IRER Irradiation Responsive Enhancer Region

iRNAs Interfering RNAs

ISWI Imitation SWI; Imitation SWItching

JAK Janus Kinase

JAK2 Janus Kinase 2

JARID1A Jumonji, AT-Rich Interactive Domain 1A

JARID1C Lysine (K)-Specific Demethylase 5C (*Homo sapiens*); Smcy (Y-linked structural maintenance of chromosomes) homolog, X-linked (*Mus*); also called SMCX, KDM5C

JARID1D Smcy (Y-linked structural maintenance of chromosomes) Homolog, X-linked (*Mus*); also called KDM5D

JARID2 Jumonji, AT-Rich Interactive Domain 2

JBP1 J-Binding Protein 1

JHDM1E Lysine (K)-Specific Demethylase 7C; also called PHF2, KDM7C

JHDM2A Lysine (K)-Specific Demethylase 3A (*Homo sapiens*) also called KDM3A

JmjC Jumonji C; histone demethylases

JMJD3 Jumonji domain containing 3 lysine-specific demethylase 3 (*Xenopus*); the H3K9 histone demethylase

JQ1 BET (Bromodomain and Extra-Terminal Inhibitors) Bromodomain Inhibitor

JunB Jun B Proto-Oncogene

K-Ras Kirsten Rat Sarcoma; also called Kras

KAT Lysine (K)-Specific Histone Acetyltransferase (HAT);

KAT2A SAGA complex histone acetyltransferase catalytic subunit Gnc5 (*S. pombe*); also called GCN5; called Lysine (K)-Specific Histone Acetyltransferase (HAT);

KAT3A CREB-binding protein (*Homo sapiens*); also called CREB, CBP; Lysine (K)-Specific Histone Acetyltransferase (HAT)

KAT3B Lysine (K)-Specific Histone Acetyltransferase (HAT) 3B; also called EP300, E1A binding Protein p300 (*Homo sapiens*); p300, RSTS2

kb Kilo base Pairs, 1000 base pairs

Kcnqot1 Potassium Voltage-Gated Channel, Subfamily Q, Member 1 Overlapping Transcript 1; lncRNA

Kcnqt1 Potassium Voltage-Gated Channel, Subfamily Q, Member 1

kD kilo Dalton, also kDa, 1000 × unit of atomic mass of a molecule (usually large molecules)

KDM Lysine Demethylase

KDM1A Lysine (K)-Specific Demethylase 1A; also called LSD1

KDM2-7 Jumonji C Histone Lysine Demethylases

KDM3A Lysine (K)-Specific Demethylase 3A (*Homo sapiens*); also called JHDM2A

KDM4C Lysine (K)-Specific Demethylase 4C

KDM5A Lysine (K)-Specific Demethylase 5A

KDM5C Lysine (K)-Specific Demethylase 5C (*Homo sapiens*); also called SMCX, JARID1C

KDM5D Smcy (Y-linked structural maintenance of chromosomes) homolog, X-linked (*Mus*); also called JARID1D

KDM6A Lysine (K)-Specific Demethylase 6A; also called UTX

KDM7C Lysine (K)-Specific Demethylase 7C; also called PHF2, JHDM1E

KLF Krüpple-Like Factor

KLF1 Krüpple-Like Factor 1, Erythroid (*Homo sapiens*); EKLF

KLF3 Krüpple-Like Factor 3, Basic (*Homo sapiens*)

KLF4 Krüpple-Like Factor 4, Gut (*Homo sapiens*)

KMT Lysine Methyltransferase

KMT2A Lysine (K)-Specific Methyltransferase 2A; member of Trithorax complex; also called MLL1

KMT2B Lysine (K)-Specific Methyltransferase 2B; also called MLL4 (*Homo sapiens*)

KMT2C Lysine (K)-Specific Methyltransferase 2C; also called MLL3 (*Homo sapiens*)

KMT2D Lysine (K)-Specific Methyltransferase 2D; also called MLL2 (*Homo sapiens*)

KMT3A Lysine (K)-Specific Methyltransferase 3A; also called SETD2, SET Domain Containing 2

KMT3B Lysine (K)-Specific Methyltransferase 3B; also called NSD1 Nuclear Receptor Binding SET Domain Protein 1

KMT6A Lysine (K)-Specific Methyltransferase 6A; also called EZH2, Enhancer of Zeste Protein 2 (*Homo sapiens*); subunit of the PRC2 complex

Kras Kirsten Rat Sarcoma; also called K-Ras

L3MBTL2 l(3)mbt (Lethal (3) Malignant Brain Tumor (*Drosophila*))-Like 2 (*Homo sapiens*)

LADs Lamina-Associated Domains

LCR Locus Control Region

LCR HS Locus Control Region DNaseI Hypersensitive Sites

Ldb1 *lim* Domain Binding 1

let-7 Lethal-7

LIM PDZ and LIM Domain 5 (*Homo sapiens*); also called lim and PDLIM5; transcription factor

linc-MD1 Muscle Differentiation 1, linked to myogenesis

lincRNA Long Intergenic Noncoding RNA

lincRNA-EPS lincRNA Erythroid Prosurvival

lincRNA-p21 Tumor protein p53 pathway corepressor 1 (nonprotein coding); also called TP53COR1 (*Homo sapiens*)

lincRNA-RoR Long Intergenic Noncoding RNA Regulator of Reprogramming

LINE1 Long Interspersed Nuclear Element

LMNA Lamin A/C

LMO2 Lim Only 2

lncRNA Long Noncoding RNA

lncRNA-ES1/2/3 lncRNAs associated with the CNS; involved in neural stem cell self-renewal capacity and neural differentiation; associated with pluripotent stem cells, since within their regulatory elements they have binding sites for the plutipotency transcription factors OCT4 and NANOG

lncRNA-N1/2/3 lncRNAs associated with neuronal differentiation

LOCKS Large Organized Chromatin K9-modifications; domains which are G9a-dependent and enriched for H3K9me2

LOX Lysyl Oxidase

LPS Lipopolysaccharide

LSD1 Lysine (K)-Specific Demethylase 1A; also known as KDM1A

LSD1-CoREST Lysine (K)-Specific Demethylase 1A–Corepressor for Element-1-Silencing Transcription Factor Complex

LT-HSC Long-Term Repopulating Activity HSC

LTK-14 Garcinol (polyisoprenylated benzophenone) Derivative; p300-selective noncompetitive inhibitor for acetyl-CoA and histone, represses p300-mediated acetylation of p53, inhibits acetylation of HIV cells

Mad Human dimerization protein 1

Mad/Max Active transcription repressor dimer, transcription factors, dimer blocks c-myc access to Max

macroH2A Histone H2A variant, found on the X-Chromosome

Maf v-maf avian musculoaponeurotic fibrosarcoma oncogene homolog (*Homo sapiens*); subunit of NF-E2 heterodimer, contains basic region-leucine zipper (bZip) motif, nuclear factor, erythroid derived 2 (*Homo sapiens*); (*Mus*); small Maf proteins (p18, MafF, MafG, and MafK)

MAGE Melanoma-Associated Antigen

MALAT1 Metastasis Associated Lung Adenocarcinoma Transcript 1; lncRNA; also called NEAT2

MAML1 Mastermind-Like Protein 1

MARE Maf Recognition Element

MARs Matrix-Associated/Attachment Regions

mascRNA MALAT1-associated small cytoplasmic RNA

MAT Methionine Adenosyltransferase

Max Human Myc-Associated Factor X

Mb Mega base pairs

MBD Methyl-CpG Binding Protein

MBD2 Methyl CpG Binding Domain Protein 2

MCF-7 Breast cancer cell line

MDS Myelodysplasia Syndrome

MeCP2 Methyl CpG Binding Protein 2; causes more activation than repression

MED1 *Mediator* Complex Subunit 1

MED12 *Mediator* Complex Subunit 12

MEF Mouse Embryonic Fibroblasts

MEF2C Myocyte-Specific Enhancer Factor 2C

MEF2D Myocyte-Specific Enhancer Factor 2D

MEP Myeloid Erythroid Precursor

MET1 DNA Methyltransferase 1

MetS Metabolic Syndrome

MGMT O-6-MethyGuanine-DNA Methyltransferase

MHCs Myosin Heavy Chain Genes

mHRM Heme Regulatory Motif

Mi-2/NuRD Nucleosome Remodeling and Deacetylase Complex

Mi2 Chromodomain helicase DNA binding protein

Mi-2β Chromodomain helicase DNA binding protein 4; also called Mi-2b, Mi2-BETA (*Homo sapiens*), CHD4

MIM# Mendielian Inheritance in Man; accession number from *OMIM*

miR, miRNA, or microRNA A class of very short abundant RNAs that tend to negatively regulate transcription, approximately 22 nt in length

miR-133/135 microRNA-133 and microRNA-135, post-transcriptional regulators of the mastermind-like protein 1 (MAML1) and myocyte-specific enhancer factor 2C (MEF2C), respectively

miR-675-3p/5p Skeletal muscle differentiation promoting Mirnas

MIWI Murine Piwi

MIWI2 Murine Piwi 2

MLL Mixed Lineage Leukemia, involved in the maintenance of transcriptionally active state of specific genes, will bind to promoters of HOX genes

MLL complex Complex with H3K4 Histone Methyltransferase activity and capable of positively regulating gene transcription

MLL1 Lysine (K)-specific Methyltransferase 2A (*Homo sapiens*); member of Trithorax complex; also called KMT2A

MLL2 Lysine (K)-specific Methyltransferase 2D; also called KMT2D (*Homo sapiens*)

MLL3 Lysine (K)-specific Methyltransferase 2C; also called KMT2C (*Homo sapiens*)

MLL4 Lysine (K)-specific Methyltransferase 2B; also called KMT2B (*Homo sapiens*)

MMTV Mouse Mammary Tumor Virus

MOF Males-Absent on the First Protein

Mop1 Mediator of Paramutation 1

MOZ Monocytic Leukemia Zinc Finger Protein

MPN Myeloproliferative Neoplasia

mRNA Messenger RNA

MS Methionine Synthase

MSC Multipotent Stem Cell

mSin3A Transcriptional Regulator, SIN3A (yeast), (*Mus*)

mtDNA Mitochondrial DNA

MTF2 Polycomb-Like Protein PCL2

MTHF 5-Methyltetrahydrofolate

MTHFR Methylenetetrahydrofolate reductase gene; reduces 5,10-methylenetetrahydrofolate to 5-methyl-tetrahydrofolate

mTOR Mammlian Target of Rapamycin

Mxi c-Myc antagonist

Myb Myeloblastosis oncogene; also called c-Myb

Myc Myelocytomatosis; avian myelocytomatosis viral oncogene homolog; bHLH-leucine zipper proteins

myc/Mad/Max Signaling pathway, transcription Factors

MYCN v-myc myelocytomatosis viral related oncogene, neuroblastoma-derived (Avian)

MyoD Myogenic Differentiation Protein

MYOD1 Myogenic Differentiation 1 (*Homo sapiens*)

MYST Acronym for the complex (MOZ, Ybf2, Sas2, TIP60)

NAB Sodium Butyrate

NAD Nicotinamide Adenine Dinuceotide

NAD+ Nicotinamide Adenine Dinuceotide, oxidized form

NADH Nicotinamide Adenine Dinuceotide, reduced form

NAM Nicotinamide Adenine Mononucleotide

nano-ChIP seq A type of chromatin immunoprecipitation which utilizes fewer cells

NANOG Nanog Homeobox

NATS Natural Antisense Transcripts

ncRNA Noncoding RNA

ncRNA-a ncRNA-activating; class of lncRNAs

ncRNA-a3 ncRNA-activating3; lncRNA that regulates TAL1

ND Neurodegenerative Disease

NF-E2 motif DNA sequence motif (GC TGA G/C TCA T); related NF-E2 proteins (NRF1 and NRF2)

NFE2 Heterodimer, consists of p45 and small Maf subunits, Nuclear Factor, Erythroid 2 (*Homo sapiens*); also called p45, NF-E2

NFκB Nuclear factor kappa-light-chain-enhancer of activated B cells

NOTCH1 Notch1

noxa Phorbol-12-myristate-13-acetate-induced protein 1 gene

NPC Nuclear Pore Complex

NPCs Neural Progenitor Cells

NRF Nucleosome Free Region

NSD1 Nuclear Receptor Binding SET Domain Protein 1; also called KMT3B, Lysine (K)-specific methyltransferase 3B;

NSD3 Wolf–Hirschhorn Syndrome (WHS) candidate 1-like 1 (*Homo sapiens*); histone methyltransferase protein

NTDs Neural Tube Defects

NUPs Nucleoporins

NuRD Nucleosome Remodeling and Deacetylase

NURF55 Chromatin assembly factor 1, subunit B (*Drosophila*); also called CAF1 (*Drosophila*); also called CHAF1B (*Homo sapiens*); also called RBBP4, RBAP48, core component of PRC2

NUT Nuclear protein in testis (*Homo sapiens*); locus at 15q14

NUT-NMC Nuclear protein in testis (NUT)-Midline Carcinoma (*Homo sapiens*); 75% caused by single translocation producing a chimeric gene, NUT-BRD4 fusion, t(15q14; 19p13.1); remaining cases involve NUT fused with BRD3 or other partner genes, such as NSD3 (Wolf–Hirschhorn syndrome (WHS) candidate 1-like 1, a histone methyltransferase)

NY-ESO-1 New York Esophageal Squamous Cell Carcinoma 1

OCT4 Octamer-Binding Transcription Factor 4

OGT O-linked β-d-N-Acetylglucosamine Transferase

OMIM Online Mendielian Inheritance in Man; online database of mendelian phenotypes and genes initially created by Dr Victor McKusick, John Hopkins University

OR Olfactory Receptors

ORF Open Reading Frame

ORPHA# *Orpha number* refers to the *Orphanet* classification of diseases; www.orpha.net portal for Rare Diseases and Orphan Drugs (*Homo sapiens*)

osa Protein of the Trithorax Group with ARID domain and Eyelid (EHD) domain homology; also called BAF250, similar to ARID1A and SMARCF1

OSKM OCT4, SOX2, KLF4 and c-MYC; pluripotency factors

oxBs-seq Oxidative Bisulfate Sequencing; both of the following methods discern between 5mC and 5hmC, (1) TrueMethyl Seq, CEGX, Cambridge Epigenetix; and (2) oxBS-450K, modified oxBs-seq for Infinium HumanMethylation450K BeadChip (Illumnia)

p/CAF p300/CBP Associated Chromatin Assembly Factor; Lysine acetyltransferase 2B (*Homo sapiens*)

p21 Murine Cyclin-Dependent Kinase Inhibitor 1A

p27 An Inhibitor of Cyclin-Dependent Kinases

p45 Subunit of NF-E2 heterodimer, contains Basic Region-Leucine Zipper (bZip) Motif, Nuclear Factor, Erythroid-Derived 2 (*Homo sapiens*); (*Mus*); also called NF-E2

p53 Transformation-Related Protein 53 (Trp53, p53, *Mus*); also called TP53, TRP53, P53 (*Homo sapiens*)

p68 DEAD (Asp-Glu-Ala-Asp)-Box Helicase 5 (*Homo sapiens*); DEAD-box RNA helicase p68; RNA splicing protein p68

p200 AT-Rich interactive domain 2 (ARID, RFX-like) (*Homo sapiens*); also called ARID2, BAF200

p300 E1A Binding Protein p300 (*Homo sapiens*); also called EP300, RSTS2, KAT3B (Lysine [K]-specific Histone Acetyltransferase [HAT] 3B)

PA Physical Activity

PAIR PAX (Paired Box)–Activated Intergenic Repeat Elements

PAIR4 PAX4 (Paired Box 4)–Activated Intergenic Repeat Element

PAIR5 PAX5 (Paired Box 5)–Activated Intergenic Repeat Element

PAIR6 PAX6 [Paired Box 6]–Activated Intergenic Repeat Element

PAR–CLIP Photoactivatable Ribonucleoside-Enhanced Crosslinking and Immunoprecipitation

PAX3 Paired Box 3; miRNA

PAX4 Paired Box 4; miRNA

PAX5 Paired Box 5; miRNA

PAX6 Paired Box 6; miRNA

PAX7 Paired Box 7; miRNA

PB1 Polybromo 1 (*Homo sapiens*); also called BAF180, PBRM1

PBA 4-Phenylbutyrate

PBAF Polybromo-Associated BAF

PBG Porphobilinogen

PBGS PBG Synthase

PBRM1 Polybromo 1 (*Homo sapiens*); also called BAF180, PB1

Pc Polycomb Protein, member of the core complex of PRC1 in *Drosophila*

Pc bodies Polycomb bodies

PCAT1 Prostate Cancer-Associated ncRNA Transcript 1; lncRNA; 8q24 (*Homo sapiens*); also called PCAT-1

PcG Polycomb Group Proteins (EZH1 or EZH2, ESC, SUZ12, RBBP7/4 or RBAP46/48); also incudes Pho, Pc, PCGF1/3/5/6

PCGEM1 Prostate-Specific Transcript 1

PCGF1/3/5/6 Polycomb Group Ring Finger 1/3/5/6

PCL Polycomb-Like Proteins

PCL1 Polycomb-Like Protein; also called PHF1

PCL2 Polycomb-Like Protein; also called MTF2

PCL3 Polycomb-Like Protein; also called PHF19

PDK4 Pyruvate Dehydrogenase Lipoamide Kinase Isozyme 4

PGC–1α Peroxisome Proliferator-Activated Receptor Gamma Coactivator-1 Alpha

PGC7 Maternal Factor that protects the maternal pronucleus from demethylation of 5mC to 5hmC

PGCs Primordial Germ Cells

PHC1-3 Polyhomeotic Homolog (*Homo sapiens*)

PHD Polyhomeotic Distal Protein (*Drosophila*)

PHF1 PHD Finger Protein 1 (*Homo sapiens*); also called PCL1

PHF2 PHD Finger Protein 2 (*Homo sapiens*); Lysine (K)-Specific Demethylase 7C; also called JHDM1E, KDM7C

PHF19 PHD Finger Protein 2 (*Homo sapiens*); also called PCL3

Pho Pleiohometic protein; acts upstream of PRC2; member of one of main PcG complexes (*Drosophila*); also called YY1, Yin Yang 1; transcription factor

PIK3CA Phosphatidylinositol-4,5-Bisphosphate 3-Kinase, catalytic subunit alpha

piRNA Piwi-Interacting RNA

PIWI P-Element Induced Wimpy Testis (*Drosophila*)

PLP Pyridoxal 5′-Phosphate

PML Promyelocytic Leukemia protein

PPAR–δ Peroxisome Proliferator-Activated Receptor Delta

PPOX Protoporphyrinogen Oxidase

PRC Polycomb Repressor Complex

PRC1 Polycomb Repressive Complex 1

PRC2 Polycomb Repressive Complex 2 (includes EZH2 or EZH1, Esc, SUZ12, RBBP7/4 or RBAP46/48)
Pre-B Cells Precursor B-Cells
PRG-1 Piwi-Related gene 1 (*C. elegans*)
PRMT4 Coactivator-Associated Arginine Methyltransferase 1; also called CARM1
Pro-B Cells Progenitor B-Cells
Psc Posterior Sex Combs (*Drosophila*)
PSCs Pluripotent Stem Cells
PSF Phosphotyrosine Binding-Associated Splicing Factor
pTEFB Transcription Elongation Factor
PTEN Phosphatase and TENsin homolog
PTMs Post-Translational Modifications
PU.1 Spleen Focus Forming Virus (SFFV) Proviral Integration Oncogene (*Mus*); also called SPI1 (Spi-1 proto-oncogene (*Homo sapiens*)
puma p53 Upregulated Modulator of Apoptosis gene
PYCARD Apoptosis-Associated Speck-Like Protein Containing a CARD gene; also called ASC
PYD PYRIN-PAAD-DAPIN Domain
PYR-complex Complex that binds specifically to the Ikaros-like DNA sites and includes the long polypyrimidine-rich region upstream of the human δ-globin gene; a chromatin remodeling complex with IKAROS (Transcription Factors Aiolos, Eos, Helios, Ikaros, and Pegasus) as a part of the DNA Binding subunit
RAD21 Mitotic cohesin complex, non-SMC subunit Rad21 [*S. pombe*] homolog (*Homo sapiens*)
RAD23A RAD23 Homolog A [*S. cerevisiae*] (*Homo sapiens*); RAD23 = repairosome nucleotide excision repair protein (ubiquitin-like protein)
RAG2 Recombination Activating Gene 2 (*Homo sapiens*); a subunit of the V(D)J recombinase
RAR Retinoic Acid Receptor
RARalpha-PLZF Promyelocytic Leukemia Zinc Finger Protein Fusion
Ras Rat Sarcoma
RASSF1 Ras Association (Ralgds/AF-6) Domain Family Member 1; also called RASSF1A
Rb Retinoblastoma Protein
RbAp46 Retinoblastoma Binding Protein 46; (*Homo sapiens*); Histone Chaperones (*Homo sapiens*); core component of PRC2; also called RBBP7
RbAp48 Retinoblastoma Binding Protein 48; (*Homo sapiens*); Histone Chaperones (*Homo sapiens*); core component of PRC2; also called RBBP4, NURF55
RbAp46/48 Retinoblastoma Binding Protein 46/48; (*Homo sapiens*); Histone Chaperones (*Homo sapiens*); core component of PRC2; also called RBBP7/4
RBBP4 Retinoblastoma Binding Protein 7/4, Histone Chaperones (*Homo sapiens*); core component of PRC2; also called RBAP48, NURF55
RBBP5 Retinoblastoma Binding Protein 5, Histone Chaperones (*Homo sapiens*)
RBBP7 Retinoblastoma Binding Protein 7/4, Histone Chaperones (*Homo sapiens*); core component of PRC2; also called RBAP46
RBBP7/4 Retinoblastoma Binding Protein 7/4, Histone Chaperones (*Homo sapiens*); core component of PRC2; also called RBAP46/48
RBC Red Blood Cell
RBP2 Retinol-Binding Protein 2 (*Homo sapiens*); RNA-Binding Protein 2 (*Drosophila*)
RdDM RNA-Directed DNA Methylation
rDNA Ribosomal DNA
reaper Reaper Gene (*Drosophila*), proapoptotic gene
RelA v-rel Reticuloendotheliosis Viral Oncogene Homolog A
RelB v-rel Reticuloendotheliosis Viral Oncogene Homolog B

REST RE1-Silencing Transcription Factor (*Homo sapiens*)

RFX Regulatory Factor X, complex of proteins associated with the regulation of major histocompatibility class II genes within the mammalian immune system

RFX-like Regulatory Factor X-like

Rh Rhesus Protein

RhAG Rhesus-Associated Glycoprotein

RING Ring Finger Protein; also called Sce (*Drosophila*); also called dRING

RING1 Ring Finger Protein 1 (*Homo sapiens*); a putative E3 ubiquitin-protein ligase

RING1A/B Ring Finger Protein 1 (*Arabidopsis*); a putative E3 ubiquitin-protein ligase

Riok3 RIO Kinase 3; protein involved in nuclear condensation; (*Homo sapiens*); (*Mus*)

RISC RNA-Induced Silencing Complex

RMST Rhabdomyosarcoma 2 Associated Transcript; lncRNAs associated with neuronal differentiation

RNA-seq RNA Sequencing

RNAe RNA-Induced Epigenetic Silencing

RNAi RNA Interference

RNAPII RNA Polymerase II

Rpd3 Reduced Potassium Dependency 3; yeast histone deacetylase

RSTS2 E1A Binding Protein p300 (*Homo sapiens*); also called EP300, p300, KAT3B (Lysine (K)-Specific Histone Acetyltransferase (HAT) 3B)

RUNX1 Runt-Related Transcription Factor 1; miRNA

RUNX3 Runt-Related Transcription Factor 3; miRNA

S value A unit of measure (force) based on sedimentation rate in a gravitational field or centrifuge called the Svedberg constant, and is dependent on the mass and geometrical shape of the molecule, e.g., such as in 28S RNA

S/MARs Scaffold/Matrix Associated/Attachment Regions; also called SMARs

SA1 Stromal Antigen 1; suggested name–cohesin subunit SA-1

SA2 Stromal Antigen 2, cohesin subunit SA-2

SAGA Spt-Ada-Gcn5 Acetyltransferase

SAH S-Adenosylhomocysteine

SAHA Suberoylanilide Hydroxamic Acid, trade name Vorinostat, MK-0683

SAHF Senescence-Associated Heterochromatin Foci

SAHH SAH Hydrolase

SAL-RNA1 Senescence Associated Long Noncoding RNA 1

SAM S-Adenosylmethionine

SARs Scaffold-Associated/Attachment Regions

Sas2 Something about silencing 2

SATB1 Special AT-rich Sequence-Binding Protein-1

SATB2 Special AT-rich Sequence-Binding Protein-2

SBS Sequencing by Synthesis Reaction

SC35 domains Serine/Arginine-Rich Splicing Factor 2

SCC Squamous Cell Carcinoma

SCC3 Sister-Chromatid Cohesin Protein 3

Sce Sex Combs Extra (*Drosophila*); also called RING or dRING; E3 ubiquitin ligase

SCL Stem Cell Leukemia

SCNT Somatic Cell Nuclear Transfer

SEC Super Elongation Complex

SET acronym for a domain containing Su(var)3–9, Enhancer-of-Zeste (E[z]), and Trithorax (Trx)

SET1 Complex of proteins, core members include WDR5, RBBP5, ASH2L, DPY30; related to SET1 COMPASS Complex (*Saccharomyces*); H3K4 histone methyltransferase

SET1A SET Domain Containing 1A, gene name *SETD1A* (*Homo sapiens*)

SETD2 SET Domain Containing 2; also called KMT3A, Lysine (K)-Specific Methyltransferase 3A

SFFV Spleen Focus Forming Virus

SFN Sulforaphane; an inhibitor of histone deacetylases (HDACs)

Sfrp Secreted Frizzled-Related Protein 2, tumor suppressor genes

Sfrp4 Secreted Frizzled-Related Protein 4, tumor suppressor genes

Sfrp5 Secreted Frizzled-Related Protein 5, tumor suppressor genes

SFSF3 Serine/Arginine-Rich Splicing Factor 3

SFSR3 Serine/Arginine-Rich Splicing Factor 3

SGI-110 Dinucleotide which includes deoxyguanosine (5′-DACpG-3′), largely resistant to cytidine deaminase, (previously called S110)

SHU Shutdown

sickle Sickle Gene (*Drosophila*), proapoptotic gene

Sin3A SIN3 Transcription Regulator Family Member A (*Homo sapiens*)

SINE Short Interspersed Nuclear Element

Sir2 Silent Mating Type Information Regulator 2; silent information regulator 2, yeast histone deacetylase

siRISC Small Interfering RNA-induced Silencing Complex

siRNA Small Interfering RNA

SIRT Sirtuin Proteins (Silent mating type information regulation 2 homolog); SIRT1

SKY Spectral Karyotyping; a derivative of FISH, allowing for the visualization of interphase chromosomes

SMAD2 SMAD Family Member 2

SMARCA2 SWI/SNF-related, matrix-associated, actin-dependent regulator of chromatin, subfamily A, member 2, also called BAF190, SWI2, SNF2

SMARCA4 Brahma related gene 1; also called SWI/SNF-related, matrix-associated, actin-dependent regulator of chromatin, subfamily A, member 4 (*Homo sapiens*); also called BRG1, BAF190, SNF2, SWI2

SMARCB1 SWI/SNF-related, matrix-associated, actin-dependent regulator of chromatin, subfamily B, member 1 (*Homo sapiens*); also called SNF5, BAF47, INI1

SMARCC1 SWI/SNF-related, matrix-associated, actin-dependent regulator of chromatin, subfamily C, member 1 (*Homo sapiens*); also called SWI3

SMARCF1 AT-Rich Interactive Domain 1A (SWI-like) (*Homo sapiens*); also called ARID1A, BAF250

SMARs Scaffold/Matrix Associated/Attachment Regions; also called S/MARs

SMC1 Structural Maintenance of Chromosome Protein 1

SMC1A Structural Maintenance of Chromosome Protein 1A

SMC3 Structural Maintenance of Chromosome Protein 3

SmcHD1 Structural Maintenance of Chromosomes Flexible Hinge Domain Containing sRNA

SMCX Lysine (K)-Specific Demethylase 5C (*Homo sapiens*); also called KDM5C, JARID1C

Smcy (Y-linked structural maintenance of chromosomes) homolog, X-linked (*Mus*); Lysine (K)-Specific Demethylase 5D (*Homo sapiens*); also called KDM5D, JARID1D

SNAIL Snail family zinc finger 1 (*Homo sapiens*)

SNF2 Brahma related gene 1; also called SWI/SNF-related, matrix-associated, actin-dependent regulator of chromatin, subfamily A, member 4 (*Homo sapiens*); also called BRG1, BAF190, SMARCA2, SMARCA4, SWI2

SNF5 Sucrose Nonfermenting 5 (*Saccharomyces*); SWI/SNF subunit of the chromatin remodeling complex; interacts with SNF2 and SNF6 in transcriptional activation and is involved in transcriptional regulation; under hypoxic conditions, will relocate to the cytosol; also called SMARCB1, BAF47, INI1

SNF6 SWI/SNF subunit of the chromatin remodeling complex; interacts with SNF2 and SNF5 in transcriptional activation and is involved in transcriptional regulation; under hypoxic conditions, will relocate to the cytosol

snoRNAs Small Nucleolar RNAs

SNPs Single Nucleotide Polymorphisms

snRNAs Small nuclear RNAs

SOCS1 Suppressor of cytokine signaling 1

SOCS3 Suppressor of cytokine signaling 3

SOX2 SRY (Sex Determining Region Y)-Box 2

SOX6 SRY (Sex Determining Region Y)-Box 6 (*Homo sapiens*)

SOX9 SRY (Sex Determining Region Y)-Box 9; a type of miRNA

Sp1 Specificity Protein 1

SRA Steroid Receptor RNA Activator 1

SRC-1 Steroid Receptor Coactivator-1

sRNA Small RNA

SRS Silver–Russell Syndrome

SRY Sex-Determining Region Y

ST-HSC Short-Term Repopulating Activity HSC

STAG1 Cohesin subunit SA-1

STE24 Zinc metallopeptidase STE24 (*Homo sapiens*); also called ZMPSTE24

STIL SCL/TAL1 Interrupting Locus; (Stem Cell Leukemia/T-Cell Acute Lymphocytic Leukemia 1)-Interrupting Locus

Su(var)3–9 Suppressor of Variegation 3–9

SUMO Small Ubiquitin-Like Modifier Protein

SUV39H1 Suppressor of Variegation 3–9 Homolog 1

SUZ12 Suppressor of Zeste 12 protein (*Homo sapiens*); also called Su(z)12 (*Drosophila*); member of polycomb PRC2 complex

SVM Support Vector Machine, algorithm used in computational software tools for epigenetics analysis

SWI SWI/SNF-related, matrix-associated, actin-dependent regulator of chromatin, subfamily A, member 1 (*Homo sapiens*, SMARC family)

SWI/SNF SWItching/Sucrose Nonfermenting; protein complex composed of 10 subunits including SWI1, SWI2/SNF2, SWI3, SNF5, and SNF6 (*Saccharomycetes*); SWI/SNF-related, matrix-associated, actin-dependent regulator of chromatin, subfamily A, member 1

SWI1 Switching Deficient (*Saccharomyces*); subunit of the SWI/SNF chromatin remodeling complex; regulates transcription by remodeling chromatin; required for transcription of numerous key housekeeping genes

SWI2 Brahma related gene 1; also called SWI/SNF-related, matrix-associated, actin-dependent regulator of chromatin, subfamily A, Member 4 (*Homo sapiens*); also called BRG1, BAF190, SMARCA2, SMARCA4, SNF2

SWI3 SWI/SNF-related, matrix-associated, actin-dependent regulator of chromatin, subfamily C, member 1 (SMARCC1, *Homo sapiens*)

T-ALL T-Cell Acute Lymphoblastic Leukemia

T-cells T-Lymphocytes

Tab-seq Tet-Assisted Bisulfite Sequencing, Epigenie

TADs Topologically Associated Domains

TAF TBP-Associated Factor

TAIR The Arabidopsis Information Resource

TAL1 T-cell Acute Lymphocytic Leukemia 1 (*Homo sapiens*, locus 1p32); also called SCL/TAL-1 Factor; helix-loop-helix protein

Tas2r110 Taste-Bud Specific Chemoreceptor gene (*Mus*)

TBP TFII-D Subunit TATA-Binding-Protein

TCA Tricarboxylic Acid Cycle; also called the Kreb's cycle

TDG Thymine-DNA Glycosylase

TEs Transposable Elements

TET Ten-Eleven Translocation

TFs Transcription Factors

TFII-D Transcription Factor TFII-D

TFII-H Transcription Factor TFII-H

TFII-I Transcription Factor TFII-I, capable of binding Ebox elements and interact with upstream regulatory factors (e.g., USF1 and c-myc)

TFAM Mitochondrial Transcription Factor A

THF Tetrahydrofolate

TIG1 Tazarotene-Induced Gene-1

Tip60 Tat-Interactive Protein 60 kD

TOR Target of Rapamycin

TR2 Nuclear receptor subfamily 2, group C, member 1

TR4 Nuclear receptor subfamily 2, group C, member 2

TRAIL TNF-Related Apoptosis-Inducing Ligand

Trr Trithorax-Related protein (*Drosophila*)

Trx Trithorax protein (*Drosophila*)

TrxG Trithorax Group proteins

TSA Trichostatin A

TSSs Transcription Start Sites

Ubx Ultrabithorax (*Drosophila*)

URO Uroporphyrinogen

UROD URO Decarboxylase

UROS URO Synthase

USF Upstream Stimulatory Factor; heterodimer composed of USF1 and USF2, interacts with NF-E2

USF1 Upstream Transcription Factor 1

USF2 Upstream Transcription Factor 2, c-fos interacting

UTR Untranslated Region

UTX Lysine-specific Demethylase 6A; also called KDM6A

VEGFA Vascular Endothelial Growth Factor A

V_H Ig Heavy (H) chain variable (V) gene; where V =Variable, C = Constant, D = Diversity, and J = Joining

VHL von Hippel–Lindau

VM Vasculogenic Mimicry

VPA Valproic Acid, Valproate

WAF1 Wild-type P53 Activated Protein 1

WAGO9 Argonaute Protein (*C. elegans*)

WD40 repeat domain Domain 40–60 repeats of the dipeptide Tryptophan-Aspartate (WD)

WDR5 WD Repeat-containing Protein 5; binds to MLL complex; binds lncRNA HOTTIP; essential for the histone mark H3K4me3 in ESCs; >100 lncRNAs bind to WRD5

WDR82 WD40 Repeat Domain 82 (*Homo sapiens*)

WGATAR Consensus DNA sequence (A/T)GATA(A/G)

WHS Wolf–Hirschhorn Syndrome

Wnt Wingless-type MMTV (Mouse Mammary Tumor Virus) integration site family (*Mus*)

WRAD module Conserved components of SET1 family (WDR5, RBBP5, ASH2L, and DPY30) related to the yeast SET1 COMPASS Complex

WRAP53 WD40 Repeat-Containing Protein Encoding RNA Antisense to P53

WRN Werner Syndrome gene; Werner syndrome, RecQ helicase-like

XCI X-Chromosome Inactivation

XICs Xist-encoding Inactivation Centers

Xist Noncoding X-Inactivation Specific Transcript, derived from the inactive X-chromosome

XK X-linked Kx constituent of the two-component Kell blood group linked to McLeod syndrome

Ybf2 Something about Silencing 2; also called Sas2

YY1 Yin Yang 1; transcription factor; also called Pho, Pleiohometic Protein; acts upstream of PRC2; member of one of main PcG complexes (*Drosophila*)

z Zeste (*Drosophila*), gene

ZF Zinc Finger

ZFHX4 Zinc Finger Homeobox 4

ZGPAT Zinc Finger, CCCH-type with G-Patch (G = Glycine-rich) domain containing transcription repressor (*Homo sapiens*); recruits Mi-2/NuRD; also called ZIP

ZIP Zinc Finger, CCCH-type with G-Patch (G = Glycine-rich) domain containing transcription repressor (*Homo sapiens*); also called ZGPAT

ZMPSTE24 Zinc Metallopeptidase STE24

Zuc Zucchini

GLOSSARY

Acetylation The addition of an acetyl ($O=C-CH_3$) moiety to a chemical entity.

Acute Myeloid Leukemia (AML) A fast-growing cancer that affects the bone marrow resulting in abnormal myeloblasts, red blood cells, and platelets. It is the most common acute leukemia in adults. Myleloid leukemia primarily affects granulocytes and monocytes.

ADP-ribosylation The addition of one or more ADP-ribose units to a protein.

Aggresome Misfolded proteins generated by a highly regulated process that serves to sequester proteins during protein degradation.

Alopecia Loss of hair.

Alzheimer's Disease (AD) A neurodegenerative disease of the brain associated with progressive loss of memory, judgment, and the ability to function.

Androgenetic Having a duplicated male genome.

Antagonistic Natural Antisense Transcripts antagoNATs, oligonucleotides designed against Natural Antisense Transcript selective targets via strand-specific oligonucleotides.

Apoptosis Programmed cell death.

Arabisopsis Information Resource (TAIR) www.arabidopsis.org

Astrocytes Collectively two types of glial cells of the CNS, fibrous and protoplasmic.

B-cell Lymphoma (BCL) A lymphoma of the B cells.

Beckwith-Wiedemann Syndrome (BWS) An overgrowth syndrome with a complex genetic cause that affects several parts of the body. Cancerous and noncancerous tumors occur more frequently in children.

Biomarker A biological marker that can be used to indicate a biological state.

Bisphenol A (BPA) A compound that is known to be endocrine-active and causes adverse effects on the development of animals, including reproduction.

Bowtie 2 Backward search alignment tool for short reads, allows alignment to a reference genome; used in conjunction with BWA; software available at ftp://ftp.cs.ucl.ac.uk//genetic/gp-code/bowtie2gp/

Braveheart (Bvht, *Mus*) lncRNA required for the progression of the development of mesodermal cells to make the commitment to develop as cardiac structures.

Breast Cancer 1/2 (BRCA1/2, early onset) Two types of early onset breast cancer (BRCA1 and BRCA2) associated with the loss of either the BRCA1 gene or the BRCA2 genes.

Bromodomains and Extra-Terminal family of proteins (BET: BRD2, BRD3, BRD4, and BRDt) Regulate key oncogene transcription and thereby play a critical role in the processes of transcriptional elongation and cell-cycle progression.

Butyrylation The addition of the acyl moiety butyryl. For histones it is usually targeted to a lysine residue in the N-terminal tail.

BWA Burrows-Wheeler Aligner; used in conjunction with the Bowtie search tool, allows for rapid/accurate short read alignments; bio-bwa.sourceforge.net/

Cajal bodies Structural compartments containing an enrichment for the components of the splicing machinery.

Cardiovascular Disease (CVD) A broad term encompassing disease of the heart and blood vessels.

Centenarian A person aged 100 or above.

Centromere The central region of chromosome linking the sister chromatids.

ChIP-seq Combines ChIP with massively parallel sequencing to identify interactions between DNA and RNA and proteins.

ChromHMM Chromatin–Hidden Markov model system; unsupervised automated system for computational analysis to learn chromatin states; compbio.mit.edu/ChromHMM

ChroModule Six-module method used to annotate epigenetic states in eight different ENCODE cell types using a supervised machine-learning approach.

Chromosome Territories or CTs Minimally overlapping domains occupied discretely by chromosomes.

Chronic Kidney Disease (CKD) A broad term describing the progressive failure of proper renal function over time.

Chronic Myelomonocytic Leukemia (CMML) A slow-growing cancer that affects the white blood cells, resulting in abnormal granulocytes and monocytes.

Circular RNAs (circRNA) From exons or introns generated by back splicing and are associated with miRNAs related to cancer pathways.

Cis-acting With activity on the same chromosome, sometimes on a nearby gene; as opposed to *trans*-acting.

Citrullination The conversion of the amino acid arginine to citrulline. In the case of histone modifications, the reaction usually involves the N-terminal tails of the histone and results in demethylation of the residue.

Clear Cell Carcinoma (CCC) A broad term describing a variety of cancers that have a clear cytoplasm. They are typically rare, aggressive, and malignant.

CNCI Coding-Non-Coding Index; SVM (Support Vector Machine algorithm) classification tool based on sequence intrinsic composition to differentiate protein-coding sequences from noncoding sequences; www.bioinfo.org/software/cnci/

COMPASS complex SET1; H3K4 Histone Methyltransferase (*Saccharomycetes*).

Competing endogenous RNA: (ceRNA) or sponge noncoding RNA Competes for the binding of key microRNAs.

COSMIC database Catalog of somatic mutations in cancer, http://cancer.sanger.ac.uk/cosmic/

CPAT Coding Potential Assessment Tool; A logistic regression model built using four sequence-based linguistic features; lilab.research.bcm.edu/cpat/

CPC Coding Potential Calculator; an SVM (Support Vector Machine algorithm) classification tool based on six biological sequence features; cpc.cbi.pku.edu.cn/

Creutzfeldt-Jakob Disease (CJD) A rare neurodegenerative disorder of the brain that usually occurs late in life. It caused by the prion PrP, which leads to rapid progressive dementia, eventually resulting in death.

Crotonylation The addition of the acyl moiety crotonyl. For histones, it is usually targeted to a lysine residue in the N-terminal tail.

CSF Codon Substitution Frequencies sequence alignment tool.

CSF & PhyloCSF CSF (Codon Substitution Frequencies) & PhyloCSF (Phylogenetic Codon Substitution Frequencies); A multispecies alignment metric based on the difference in codon substitution frequencies between coding and noncoding sequences without regard to known protein sequence homologies; compbio.mit.edu/PhyloCSF

Cufflinks v7 Bipartitie graph-based method used in reference-based transcriptome assemblies to both assemble and estimate abundance of RNA sequences in samples; www.broadinstitute.org/cancer/software/genepattern/modules/docs/Cufflinks/7

Cystathionase; Cystathionine Gamma-Lyase An enzyme which catalyzes the breakdown of cystathionine to cysteine and alpha-ketobutyrate.

Cytosine One of the four principal nucleotide bases of DNA and RNA.

Dafachronic acids A steroid acid derived from the oxidation of cytochrome P450.

Databases on Medicine and Molecular Biology Resource for over 900 databases on the Internet, www.meddb.info/index.php.en?cat=6

Deacetylases A class of enzymes responsible for removal of acetyl groups from proteins.

Decitabine 5-aza-2'-deoxycytidine, DNMT inhibitor.

Dementia Impairment of a person's cognitive and daily activities.

Diffuse Large B-cell Lymphoma (DLBCL) A lymphoma of the B cells characterized by the diffuse distribution of large malignant lymphocytes. It is the most common form of non-Hodgkin's lymphoma.

DiseaseMeth The human disease methylation database; 202.97.205.78/diseasemeth/

DNA methylation Methylation of Cytosine (5mC), primarily in CpG dinucleotides, which promotes long-term gene or region silencing by affecting histone modifications and preventing the binding of transcriptional activation factors. DNA methylation functions as a stable epigenetic mark that is associated with maintaining chromatin in a silenced state. It is involved in processes including genomic imprinting, embryogenesis, and X-inactivation.

Drosha A type of double-stranded RNA-specific endoribonuclease type III.

Duchenne Muscular Dystrophy (DMD) An X-linked recessive decease caused by an abnormal dystrophin gene that results in muscle degeneration and early death.

ENCODE Encyclopedia of DNA Elements Consortium; genome.ucsc.edu/ENCODE/

Enhancer RNA (eRNA) A special class of short noncoding RNAs transcribed from enhancer regions of DNA that function as transcriptional enhancers, 50–2000 nt in length.

Epigenetics The study of mechanisms that stably regulate DNA function and the mitotically and/or meiotically heritable changes in gene activity and subsequent phenotypes that occur in the absence of changes in DNA sequence, regardless of transmittable status.

Epigenie Epigenetic databases, tools and resources; www.epigenie.com/epigenetic-tools-and-databases/

Euchromatin The form of chromatin that undergoes active transcription.

Ensembl Genome database for vertebrates and other eukaryotes; public gene annotation database; www.ensembl.org/index.html

FANTOM Functional Annotation of the Mammalian Genome Consortium; fantom.gsc.riken.jp/, FANTOM5, fantom.gsc.riken.jp/5/

Formylation The addition a formyl group. For histones, it is usually occurs on a lysine residue in the N-terminal tail.

FlyBase A database for *Drosophila* genes and genomes; www.flybase.org

Gametogenesis *The process* by which diploid precursor cells undergo cell division and differentiation to form mature haploid gametes.

GENCODE Encyclopaedia of Genes and Gene Variants Consortium; www.gencodegenes.org/

Genistein Soybean phytoestrogen, a bioactive botanical isoflavone that is known for its roles an inhibition mechanism associated with chronic noncommunicable diseases and cancers.

Genome The entire genetic makeup of an organism.

Genomic imprinting A mechanism by which specific alleles of genes are expressed based on the parental chromosomal origin of the allele.

Genotoxic An agent posing damage to the DNA.

Glycosylation The addition of N-acetylglucosamine. For histones, this modification is usually targeted to serine or threonine in the N-terminal tails of histones.

GRACOMICS Graphical comparison software package for visualizing multiple graphical results; bibs.snu.ac.kr/software/GRACOMICS/gracomics.php

GSNAP & STAR GSNAP (Genomic Short-Read Nucleotide Alignment Program) & STAR (Spliced Transcripts Alignments to a Reference); One-step method used for splice junction discovery involving read alignment and simultaneous splice junction calls; research-pub.gene.com/gmap/, STAR code.google.com/p/rna-star/

GWAS Genome-Wide Association Studies; studies that use several different statistical approaches to the analysis of genome-wide population-based data analysis.

Gynogenetic Having a duplicated female genome.

Haematopoiesis (Hematopoiesis) Process of formation of the cellular components of blood.

Hereditary Persistence of Fetal Hemoglobin (HPFH) A condition in which the gene for fetal hemoglobin (HbF) fails to be suppressed after birth, and is expressed into adulthood rather than being replaced by expression beta or delta hemoglobin subunits.

Heterochromatin The complex of proteins and DNA that carries the genetic information.

Histome The *H*istone *I*nfobase, a database of human histones, their post-translational modifications and modifying enzymes; www.actrec.gov.in/histonme/index.php

Histone Database NHGRI Histone Database; genome.nhgri.nih.gov/histones/

HistoneHits Histone Systematic Mutation Database: 54.235.254.95/histonehits/#home

histoneHMM A bivariate Hidden Markov Model algorithm for the differential analysis of histone modifications on a genome-wide level of distribution; histonehmm.molgen.mpg.de/

Histones A group of eukaryotic proteins, highly alkaline in characteristic, which package the DNA into nucleosomes.

Homeobox genes (*HOX* genes) A multigene family that has been highly conserved; its members are organized into gene clusters, where each gene cluster located on different chromosomes has sets of transcriptional regulators; the clusters are involved in different developmental programs and associated with the regulation of cell specification processes and head-to-tail body axis.

HOTAIR HOX Transcript Antisense RNA; lncRNA; epigenetic silencing is mediated by the physical attraction of both the Polycomb repressive PRC2 complex and LSD1-CoREST to its 5′ and 3′ regions, respectively.

HOTTIP HOXA Transcript at the Distal Tip; lncRNA; regulates the human HOXA locus; binds with WDR5 to recruit H3K4me3-associated MLL1 complex.

Human Histone Modification Database HHMD; 202.97.205.78/hhmd/index.jsp

Hutchinson–Gilford Progeria (HGPS) A rare genetic condition caused by mutations in the LMNA gene that cause symptoms associated with premature aging in children.

Hydroxylation A modification that involves the addition of a hydroxyl group to tyrosine ring.

Hypermethylation An enhancement in the methylation of cytosine and adenine residues of DNA.

Hypoacetylation A decrease in the acetylation level of proteins.

Hypomethylation A reduction in the methylation of cytosine and adenine residues of DNA.

ICD-10 (code#) ICD-10 is the 10th International Classification of Diseases (International Statistical Classification of Diseases and Related Health Problems), maintained by the World Health Organization; code numbers are associated with each accession or record description of a disease classification; ICD-10 version: 2015, apps.who.int/classifications/icd10/browse/2015/en

Implantation The stage in which the embryo adheres to the wall of the uterus.

Insulator A regulatory element that exhibits a phenotype of interference between other regulatory sequences.

IsoLasso RNA-seq assembler based on the LASSO algorithm; alumni.cs.ucr.edu/~liw/isolasso.html

Kell blood group A two antigen group containing Kell and XK, where XK is linked to McLeod syndrome (neuroacanthocytosis).

Kirsten Rat Sarcoma (K-Ras) One of the two first types of virus-induced forms of cancer sarcomas observed in rats caused by the Kirsten sarcoma virus.

Lamina-Associated Domains or LADs Silent heterochromatin domains of 0.1–10 Mb in size, have been identified that associate with the nuclear lamina.

Laminopathy A rare genetic disorder arising from mutations in genes expressing the nuclear lamina proteins.

Lasso Least Absolute Shrinkage and Selection Operator, an algorithm for linear regression that involves a reduction and selection approach.

Leukemia Cancer of the blood and bone marrow, typically involving white blood cells.

Long noncoding RNA (lncRNA) Noncoding transcript usually greater than 200 nt; this class includes activating long noncoding RNA (alncRNA), long intergenic noncoding RNA (lincRNAs), and antisense long noncoding RNA (as-lncRNA).

MACS Model-based Analysis of ChIP-seq; computational algorithm used for the identification and localization of histone modifications, and transcription and/or chromatin binding factors; liulab.dfci.harvard.edu/MACS

Malonylation The addition of the acyl moiety malonyl. For histones, it is usually targeted to a lysine residue in the N-terminal tail.

MapSplice & Tophat Two-step method used for splice junction discovery involving read alignments followed by splice inference; MapSplice www.netlab.uky.edu/p/bioinfo/MapSplice; TopHat, tophat.cbcb.umd.edu

McLeod syndrome (Neuroacanthocytosis) A neurological condition that occurs primarily in males, involving movement and abnormal erythrocytes cells (in this case, star-shaped), and is linked to the XK antigen of the Kell blood group.

Mediator **complex** A 26-subunit, 1.2 MDa complex, required for protein-coding gene expression.

Mesenchyme The tissue constituting loosely associated cells, which do not contain polarity and are encompassed by the extracellular matrix.

Metabolic Syndrome (MetS) A constellation of conditions that include hypertriglyceridemia, low high-density lipoprotein levels, abdominal obesity, elevated blood pressure, elevated fasting blood serum glucose and is associated with a higher risk of diabetes, and cardiovascular disease.

MethDB DNA Methylation and Environmental Epigenetic Effects Database; www.methdb.de/

MethHC DNA Methylation and Gene Expression in Human Cancer Database; MethHC.mbc.nctu.edu.tw

Methylase An enzyme catalyzing methylation.

Methylation The addition of a methyl (CH_3) group to a chemical entity.

MethylomeDB Brain Methylome Database; www.neuroepigenomics.org/methylomedb/

MGI Mouse Genome Informatics; www.informatics.jax.org/

micro RNA (miRNA or miR) A class of very short abundant RNAs that tend to negatively regulate transcription, approximately 22 nt in length.

Mira Also called *Mistral*; lncRNA.

Mistral lncRNA, intergenic transcript from region between *Hoxa6* and *Hoxa7* genes, forms RNA-DNA hybrid at site of synthesis, which then recruits MLL1 complex proteins; also called *Mira*.

MITIE Mixes Integer Transcript IdEntification; biocomputational tool for quantification and transcript reconstruction; bioweb.me/mitie

Muscular atrophy Decrease in the muscular mass of an individual.

Myelodysplasia Syndrome (MDS) A broad term describing the underproduction of all blood cells.

Myeloproliferative Neoplasia (MPN) A broad term describing the overproduction of one or more types of blood cells nano-ChIP seq: a type of chromatin immunoprecipitation which utilizes fewer cells.

NCBI epigenomics Genome maps of DNA and histone modifications; www.ncbi.nlm.nih.gov/epigenomics

NCBI gene Gene information across a wide range of species; www.ncbi.nlm.nih.gov/gene/

Neural Tube Defects (NTDs) A condition affecting the spinal cord or brain where the neural tube fails to close completely during development.

Neurodegeneration The gradual loss of neuronal structure and/or function.

Noncoding RNA All RNAs that do not encode proteins.

Nuclear lamina A meshwork of intermediate filaments (e.g., lamins) lining the inside of the nuclear envelope at the nuclear periphery.

Nuclear pore complexes Complexes of proteins involved in the export of transcripts into the cytoplasm.

O-linked N-acetylglucosamine (O-GlcNAc) O-linkage of N-acetylglucosamine. In the case of histones, this involves O-linkage of N-acetylglucosamine to serine or threonine, usually in the N-terminal tails of histones.

Oases De novo RNA-seq assembler; www.ebi.ac.uk/~zerbino/oases/

Online Mendelian Inheritance in Man *OMIM* is the online database of mendelian phenotypes and genes initially created by *Dr McKusick* (*Homo sapiens*); each accession or record is assigned a unique number (MIM#); www.omim.org

ORMAN Optimal Resolution of Multimapping Ambiguity of RNA-seq Reads; algorithm for the computation of the minimal number of potential transcription products for a gene, and multimapping of the transcripts based on distribution in the region covered; orman.sf.net

Orphanet Rare Diseases and Orphan Drugs (*Homo sapiens*); each accession or record is assigned a unique *Orpha number* (ORPHA#), which refers to the *Orphanet* classification of diseases; www.orpha.net

Oxidative stress An imbalance between the creation and detoxification of reactive oxygen species and subsequent repair, which leads to DNA damage.

Pericentromere The chromosomal region situated near the centromere.

Phosphorylation The addition of a phosphate $\left(PO_4^{3-}\right)$ group to a protein or an organic entity.

PIWI System including a class of genes and associated noncoding RNAs with functions in stem cells and in the germ line.

PlantGDB (AtGDB) *Arabidopsis* Genome Database; www.plantgdb.org/AtGDB/

Pluripotency A cellular state in which the potential to generate different cell types exists.

Polycomb (Pc) Bodies Corresponds to discrete foci of silent genes bound by the polycomb group (PcG) proteins.

Polyester RNA-seq simulation software; bioconductor.org/

Polymerase An enzyme catalyzing the synthesis of nucleic acid polymers.

PomBase The scientific resource for fission yeast; www.pombase.org

Prion Protein capable of distinct structures due to different folding patterns. One of the folded structures is self propagating and transmissible.

Progeria *See* Hutchinson–Gilford Progeria: (HGPS).

Proline isomerization The *cis-trans* isomerization of proline. This leads to structural changes in the protein and has been shown to have regulatory functions in the histone tails.

Promyelocytic Leukemia (PML) Phosophoprotein, which acts as both a tumor suppressor and transcription factor and contains BOX, RING, and zinc-binding domains. It is associated with acute promyocytic leukemia (APL).

Pronucleus Designates the nucleus of the sperm or oocyte after fertilization and before they fuse to generate the diploid embryonic nucleus.

Propionylation The addition of the acyl moiety propionyl. For histones it is usually targeted to a lysine residue in the N-terminal tail.

Rat Sarcoma (Ras) Usually a malignant tumor affecting connective and supportive tissues in rats.

RefSeq NCBI Reference Sequence Database; public gene annotation database; www.ncbi.nlm.nih.gov/refseq/

Replicative senescence The restriction in a cell's ability to divide, which results in senescence.

Reprogramming The epigenetic reorganization that accompanies the return of differentiated cells to a more pluripotent state.

RGD Rat Genome Database; rgd.mcw.edu/

RNAs A term referring to a class of noncoding RNAs with sizes ranging from ~15 kb to more than 100 kb.

Saccharomyces **Genome Database** SGD; www.yeastgenome.org

Scripture A comprehensive ab initio statistical transcriptome reconstruction algorithm that uses a segmentation method involving spliced/unspliced reads within a connectivity graph to infer transcripts used in reference-based transcriptome assemblies; www.broadinstitute.org/software/Scripture/

Senescence The process of biological aging.

SeqMINER Integrative qualitative and quantitative ChIP-seq data interpretation platform; bips.u-strasbg.fr/seqminer/

SICER Spatial clustering approach for the identification of ChIP-enriched regions for use with histone modification datasets; transcript inference tool that utilizes read enrichments from neighbor bins to call up genomic regions enriched with NGS (next-generation sequence) short reads; home.gwu.edu/~wpeng/Software.htm

Silver-Russell Syndrome (SRS) A type of primordial dwarfism. Thought to be caused by an imprinting error involving hypomethylation of H19 and IGF2.

Sirtuins Sir2 proteins; Class III histone deacetylases.

Spectacle A spectacle-learning approach to the analysis of chromatin state; github.com/jiminsong/Spectacle

Splicing speckles SC35 Domains or nuclear speckles, or interchromatin granule clusters; irregular shaped structures that are enriched with pre-mRNA splicing factors.

Squamous Cell Carcinoma (SCC) A type of skin cancer involving abnormal growth of squamous cells. It is the second most common skin cancer.

STAR code.google.com/p/rna-star/

Stem Cell Leukemia (SCL) A type of leukemia characterized by abnormal precursor cells of lymphoblasts, myeloblasts, or monoblasts.

StringTie Transcript assembler utilizing an optimization of network flow algorithm for assembly of complex datasets; ccb.jhu.edu/software/stringtie/

Succinylation The addition of the acyl moiety succinyl. For histones it is usually targeted to a lysine residue in the N-terminal tail.

Sulforaphane An organosulfur compound with the isothiocyanate group that has antioxidant properties.

Sumoylation The covalent attachment of SUMO proteins to other proteins, (similar to ubiquintination).

Telomere The repetitive nucleotide sequences found at the ends of chromatid.

TopHat Alignment software, tophat.cbcb.umd.edu

Transcription factories Specialized units for specific RNA Polymerase-related transcription of genes and coregulated genes in unique regions or territories of the nucleus.

Transcription factors A sequence-specific DNA-binding protein that controls the production of RNA.

Transposons A class of parasitic DNA sequences that can relocate to other positions in the genome.

Traph Transcripts in gRAPHs; a biocomputational tool for estimating RNA-seq transcript expression; www.cs.helsinki.fi/en/gsa/traph/

Trinity & Velvet Transcript inference tool used to partition the genome into de Bruijn graphs (Velvet) from raw reads, then process those graphs to infer transcripts; used in nonreference genome transcriptome assembly (Trinity); Trinity trinityrnaseq.sf.net; Velvet, www.ebi.ac.uk/~zerbino/velvet

Tudor domain Mediates binding to the histone marks H3K36me2 and H3K36me3, typically enriched in coding regions of active genes.

Tumorigenesis The formation of a tumor.

Ubiquitination Addition of a ubiquitin moiety to the lysine residue of proteins or peptides.

Velvet De Bruijn graph software; www.ebi.ac.uk/~zerbino/velvet

Werner Syndrome (WRN) A genetic disorder caused by mutations in the WRN gene, which causes the rapid develop of symptoms associated with aging in early adulthood.

Wolf–Hirschhorn Syndrome (WHS) A genetic syndrome usually caused by a cytogenetic deletion with the short arm of chromosome 4. Individuals with the disorder have distinctive facial features, delayed growth and development, seizures, and mental retardation.

WormBase Nematode Information Resource; www.wormbase.org

Wormpath Software package for the discovery of molecular interaction networks utilizing WormBase (*C. elegans*); bifacility.uni-koeln.de/wormpath/

X-inactivation The dosage compensation mechanism by which mammalian females inactivate one copy of their X chromosomes to equalize the single copy of the X borne by the males.

Xenbase *Xenopus* Genome Database; www.xenbase.org/entry/

Xenograft Transplantation of a graft of tissues from one species into another.

Zebularine Cytidine analog.

ZFIN Zebrafish Model Organism Database; zfin.org/

INDEX

Note: 'Page numbers followed by "f" indicate figures and "t" indicate tables.'

Printed in the United States
By Bookmasters